The section where the symbol is first used is shown in parentheses.

$\begin{vmatrix} a_1 & b_1 \\ a_2 & b_2 \end{vmatrix}$ second-order determinant (9.3)

$\begin{vmatrix} a_1 & b_1 & c_1 \\ a_2 & b_2 & c_2 \\ a_3 & b_3 & c_3 \end{vmatrix}$ third-order determinant (9.3)

$\begin{bmatrix} a_1 & b_1 \\ a_2 & b_2 \end{bmatrix}$ second-order matrix (9.4)

$\begin{bmatrix} a_1 & b_1 & c_1 \\ a_2 & b_2 & c_2 \\ a_3 & b_3 & c_3 \end{bmatrix}$ third-order matrix (9.4)

$\cap$ the intersection of (10.1)

$\varnothing$ empty set, or null set (10.1)

$\cup$ the union of (10.1)

$|a - b|$ distance between a and b (10.2)

s_n nth term of a sequence (12.1)

S_n sum of n terms of a sequence (12.2)

$\sum$ the sum (12.2)

S_∞ infinite sum (12.2)

$n!$ n factorial:
$$n! = n(n - 1)(n - 2) \cdots 3 \cdot 2 \cdot 1 \quad (12.5)$$

$\binom{n}{k}$ binomial coefficient:
$$\binom{n}{k} = \frac{n!}{k!(n - k)!} \quad (12.5)$$

Modeling, Functions, and Graphs
ALGEBRA FOR COLLEGE STUDENTS

Books in This Series

College Algebra, Seventh Edition, by Michael D. Grady, Irving Drooyan, and Edwin F. Beckenbach

Intermediate Algebra, Seventh Edition, by Irving Drooyan and Katherine Franklin

Modeling, Functions, and Graphs: Algebra for College Students, by Katherine Franklin and Irving Drooyan

Modeling, Functions, and Graphs
ALGEBRA FOR COLLEGE STUDENTS

KATHERINE FRANKLIN
Los Angeles Pierce College

IRVING DROOYAN
Los Angeles Pierce College, Emeritus

Wadsworth Publishing Company
Belmont, California ■ A Division of Wadsworth, Inc.

Mathematics Editor: Anne Scanlan-Rohrer
Editorial Assistant: Leslie With
Assistant Editor: Tamiko Verkler
Production: Greg Hubit Bookworks
Print Buyer: Martha Branch
Copy Editor: Patricia Harris
Technical Illustrator: Lotus Art
Compositor: Beacon Graphics
Cover Painting: Dina Herrmann, *The Song She Never Sang* (oil on canvas, $22\frac{3}{4}'' \times 26\frac{1}{2}''$)
Signing Representative: Cynthia Berg

© 1991, 1988, 1984, 1980 by Wadsworth, Inc.

© 1976, 1972, 1968, 1965, 1962 by Wadsworth Publishing Company, Inc. All rights reserved. No part of this book may be reproduced, stored in a retrieval system, or transcribed, in any form or by any means, electronic, mechanical, photocopying, recording, or otherwise, without the prior written permission of the publisher, Wadsworth Publishing Company, Belmont, California 94002, a division of Wadsworth, Inc.

Printed in the United States of America

1 2 3 4 5 6 7 8 9 10—95 94 93 92 91

Library of Congress Cataloging in Publication Data

Franklin, Katherine.
    Modeling, functions, and graphs: algebra for college students / Katherine Franklin, Irving Drooyan.
      p. cm.
    Includes index.
    ISBN 0-534-13284-7
    1. Algebra.   I. Franklin, Katherine.   II. Title.
QA154.2.D77   1991
512.9—dc20   90-42689
                                                                                 CIP

ISBN 0-534-13284-7

Contents

Preface, ix
Review of Elementary Topics, xiii

PART 1 FUNDAMENTALS 1

CHAPTER 1 POLYNOMIALS 3

1.1 Preliminary Concepts, 3
1.2 Sums and Differences, 11
1.3 Products, 21
1.4 Factoring, 29
1.5 Special Products and Factors, 37
   Chapter Review, 47

CHAPTER 2 RATIONAL EXPRESSIONS 51

2.1 Preliminary Concepts, 51
2.2 Products and Quotients, 61
2.3 Sums and Differences, 70
2.4 Complex Fractions, 78
2.5 Integer Exponents, 84
   Chapter Review, 96

CHAPTER 3 EXPONENTS, ROOTS, AND RADICALS 100

3.1 nth Roots and Irrational Numbers, 100
3.2 Rational Exponents, 110
3.3 Simplifying Radicals, 117
3.4 Operations on Radical Expressions, 125
3.5 Complex Numbers, 132
Chapter Review, 137

CHAPTER 4 EQUATIONS IN ONE VARIABLE 141

4.1 Linear Equations, 141
4.2 Equations Containing Fractions, 153
4.3 Quadratic Equations, 164
4.4 Completing the Square; Quadratic Formula, 176
4.5 Other Nonlinear Equations, 189
Chapter Review, 199

Summary for Part I, 204
Cumulative Review Exercises for Part I, 206

PART II FUNCTIONS AND GRAPHS 211

CHAPTER 5 EQUATIONS IN TWO VARIABLES 213

5.1 Graphs, 213
5.2 Graphs of Linear Equations, 225
5.3 Point-Slope Formula for Lines, 234
5.4 Further Properties of Lines, 247
Chapter Review, 258

CHAPTER 6 FUNCTIONS 263

6.1 Definitions and Notation, 263
6.2 Graphs of Functions, 275
6.3 Some Basic Graphs, 286
6.4 Functions as Mathematical Models, 293
6.5 Quadratic Functions, 305
Chapter Review, 318

CHAPTER 7 MORE ABOUT FUNCTIONS AND GRAPHS 323

7.1 Graphing Techniques, 323
7.2 Polynomial Functions, 330
7.3 Graphing Polynomial Functions, 338
7.4 Rational Functions, 344
7.5 Inverse Functions, 353
Chapter Review, 362

CHAPTER 8 EXPONENTIAL AND LOGARITHMIC FUNCTIONS 364

8.1 Exponential Growth and Decay, 364
8.2 Exponential Functions, 378
8.3 Logarithms, 386
8.4 Logarithmic Functions, 394
8.5 Properties of Logarithms, 404
8.6 The Natural Base, 412
Chapter Review, 421

Summary for Part II, 424
Cumulative Review Exercises for Part II, 426

PART III ADDITIONAL ALGEBRAIC TOPICS 429

CHAPTER 9 SYSTEMS OF LINEAR EQUATIONS 431

9.1 Systems in Two Variables, 431
9.2 Systems in Three Variables, 445
9.3 Determinants and Cramer's Rule, 459
9.4 Solution of Systems Using Matrices, 470
Chapter Review, 478

CHAPTER 10 INEQUALITIES 482

10.1 Linear Inequalities, 482
10.2 Absolute-Value Inequalities, 492
10.3 Nonlinear Inequalities, 500
10.4 Linear Inequalities in Two Variables; Systems, 509
Chapter Review, 515

CHAPTER 11 CONIC SECTIONS 518

    11.1 Circles and Ellipses, 519
    11.2 Parabolas, 532
    11.3 Hyperbolas, 537
    11.4 Systems Involving Quadratic Equations, 548
           Chapter Review, 556

CHAPTER 12 SEQUENCES AND SERIES 560

    12.1 Sequences, 560
    12.2 Series, 565
    12.3 Arithmetic Progressions, 570
    12.4 Geometric Progressions, 580
    12.5 The Binomial Expansion, 592
           Chapter Review, 602

Summary for Part III, 606
Cumulative Review Exercises for Part III, 608

APPENDIX USING EXPONENTIAL AND LOGARITHMIC TABLES 612

ANSWERS A-1

INDEX AN-1

Preface

Modeling, Functions, and Graphs is intended for students who have demonstrated competence with basic algebraic skills, either by *recent* completion of a first course in algebra or by adequate performance on a placement examination. The material treated corresponds to the second algebra course at a community college or the introductory math course at a four-year institution.

In preparing the manuscript we have tried to address four main issues:

1. *Many algebra courses have become burdened with a lengthy review of elementary topics.*

    We have condensed and strengthened the coverage of basic algebraic skills so that more time is available for new material.

2. *Traditional word problems are not an adequate introduction to the notion of mathematical modeling.*

    We have concentrated on writing and interpreting functional relationships, starting with simple variable expressions in Part I and progressing to basic linear, quadratic, and exponential models in Part II.

3. *Many trigonometry students have difficulty mastering the subject because of a poor grasp of functions.*

    We have introduced a variety of simple functions and their graphs to familiarize students with these ideas before they begin their study of the more abstract circular functions.

4. *Students who complete a second algebra course as a prerequisite to a business calculus course often do not have sufficient experience with functions and graphing to succeed in calculus.*

    We have emphasized the understanding of functions through numerous examples and exercises involving modeling and graphing with basic functions.

The text is divided into three parts. (See the table of contents for the location of specific topics.)

Part I, *Fundamentals,* reviews and extends students' algebraic skills. The material is organized into four chapters so that students can cover this material in less time than is normally given to review. In each chapter the construction of mathematical "models" for a variety of relationships is emphasized. Traditional "word problems" (coin, mixture, and age problems) are for the most part avoided.

Part II, *Functions and Graphs,* provides an introduction to these fundamental notions and is designed to prepare students for further study of functions in trigonometry, precalculus, or "soft" calculus courses. The concept of function is introduced in Chapter 6, after students have had some experience working with graphs and the coordinate system in Chapter 5. Functions are used to model applied problems, and the graphs of common functions are explored. More graphing practice, including graphing of polynomial and rational functions, is provided in Chapter 7. This chapter will be particularly helpful to students who proceed directly to a soft calculus course without any further study of functions. Chapter 7 can be omitted in courses that lead to precalculus or college algebra. Exponential functions and logarithms are treated in Chapter 8.

Part III, *Additional Algebraic Topics,* includes topics often covered in courses called "College Algebra." The chapters in this section are independent of each other, so that any chapter can be omitted without loss of continuity, in order to meet course objectives or time limitations.

Pedagogical Features

- The "Review of Elementary Topics" provides a reference for useful facts from elementary algebra and geometry.
- Common student errors are highlighted in the exposition.
- Each exercise set includes two types of problems: "A" exercises are skill oriented and referenced to examples in the text. "B" exercises are more challenging and are not specifically referenced to examples.
- Each chapter includes a set of review exercises.
- A summary and cumulative review exercises follow each of the three parts.

Supplemental Materials

The following ancillaries are provided for adopters of the text:

Instructor's Manual, prepared by Steven Blasberg, includes worked-out solutions to the even-numbered problems in the text as well as one multiple-choice and one fill-in test per chapter.

EXP-Test, a computerized test bank for IBM PCs and compatible hardware, contains all of the test questions from the Instructor's Manual.

Software from *The Math Lab* by Avery, Barker, and Soler is provided by the publisher. This software follows an exploratory model that allows students to go beyond algebraic concepts. The student focuses on the correct approach to a problem, and the underlying principles of a problem take precedence over algebraic procedure.

Acknowledgments

We benefited greatly from the comments and suggestions of the many instructors who reviewed drafts of the manuscript. They include:

Wayne Andrepont, University of Southwestern Louisiana
Bill Attebery, Emeritus, Louisiana Tech University
Kathleen Bavelas, Manchester Community College
Sandra Beken, Horry Georgetown Technical College
Doyle Bostic, Southeastern Oklahoma State University
Charles Carico, Emeritus, Los Angeles Pierce College
Barbara Chudilowsky, De Anza College
Harold Farmer, Wallace State College
Dorothy Gotway, University of Missouri–St. Louis
P. A. Kendrick, Brevard Community College
Robert McMillan, Oklahoma Christian University of Science and Arts
John Martin, Santa Rosa Junior College
Glenn Oubre, University of Southwestern Louisiana
Thomas Ribley, Valencia Community College
Gloria Rivkin, Lawrence Institute of Technology
Chantal Shafroth, North Carolina Central University
Henry Smith, University of New Orleans
Elias Toubassi, University of Arizona
Deborah Vrooman, Coastal Carolina College of the University of South Carolina
Judith Willoughby, Minneapolis Community College
C. T. Wolf, Millersville University
Paige Yuhn, Santa Barbara City College

We would especially like to thank Kathleen Bavelas, Steven Blasberg (West Valley Community College), and John Martin, who checked the text and exercises for accuracy.

Review of Elementary Topics

We include here for reference some fundamental ideas from elementary algebra and geometry.

Notation

$=$ is equal to or equals
$\neq$ is not equal to
$<$ is less than
$\leq$ is less than or equal to
$>$ is greater than
$\geq$ is greater than or equal to
$|a|$ absolute value of a

$$|a| = \begin{cases} a & \text{if } a \geq 0 \\ -a & \text{if } a < 0 \end{cases}$$

$a + b$ the sum of a and b
$a - b$ the difference of b subtracted from a
$a \cdot b$ the product of a and b (also written ab)
$\dfrac{a}{b}$ the quotient of a divided by b
$-a$ the negative (or opposite) of a; $-a = (-1)a$
$\sqrt{a}, \; a \geq 0$ the principal square root of a; $\sqrt{a} \cdot \sqrt{a} = a$
$\angle A$ (or A)   angle A

Properties of the Real Numbers

The four arithmetic operations (addition, subtraction, multiplication, and division) and the notions of equality and order in the set of real numbers are governed by a number of properties. Several important properties are given below.

$a + b = b + a$ Commutative properties
$ab = ba$

$(a + b) + c = a + (b + c)$ Associative properties
$(ab)c = a(bc)$

$a(b + c) = ab + ac$ Distributive property

$a + 0 = a$ Identity properties
$a \cdot 1 = a$

If $a = b$ and $b = c$, then $a = c$.   Transitive property
If $a < b$ and $b < c$, then $a < c$.

If $a = b$, then b may be replaced by a or a by b in any statement without altering the truthfulness of the statement.   Substitution property

Subtraction and division are defined in terms of addition and multiplication, respectively, as follows:

$$a - b = a + (-b) \quad \text{and} \quad \frac{a}{b} = a\left(\frac{1}{b}\right) \quad (b \neq 0).$$

Division by zero is not defined.

Because subtraction is the inverse operation for addition and division is the inverse operation for multiplication, we have the following inverse properties:

$a + (-a) = 0$ Negative (or additive-inverse) property

$a \cdot \dfrac{1}{a} = 1 \quad (a \neq 0)$ Reciprocal (or multiplicative-inverse) property

In view of the definitions and properties above we can state the rules of operation for signed numbers as follows.

1. To add two numbers with like signs, add their absolute values and use the same sign for the sum.

2. To add two numbers with opposite signs, subtract their absolute values and for the sum use the sign of the number with the larger absolute value.
3. To subtract one number from another, change the subtraction sign to addition and add the opposite of the second number.
4. The product or quotient of two numbers with like signs is positive; the product or quotient of two numbers with opposite signs is negative.

Elements of Geometry

Triangles

In any triangle the sum of the angles is 180°.

Equilateral triangle 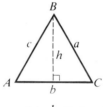 Three sides of equal length; three angles of equal measure:

$$a = b = c; \quad \angle A = \angle B = \angle C.$$

Isosceles triangle 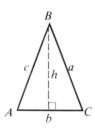 At least two sides of equal length; angles opposite these sides have equal measure:

$$a = c; \quad \angle A = \angle C.$$

Right triangle

One angle is a right angle; the side opposite the right angle is called the **hypotenuse**; the other two sides are called the **legs**.

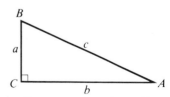

Pythagorean theorem: If a and b denote the legs of a right triangle and c denotes the hypotenuse, then

$$a^2 + b^2 = c^2.$$

Similar Triangles

Two triangles are said to be **similar** if their corresponding angles are equal. In similar triangles the ratios of corresponding sides are equal. For example, the two triangles below are similar:

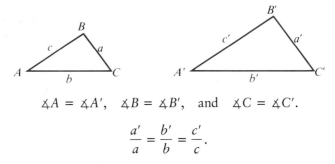

$\angle A = \angle A'$, $\angle B = \angle B'$, and $\angle C = \angle C'$.

$$\frac{a'}{a} = \frac{b'}{b} = \frac{c'}{c}.$$

Perimeter and Area of Plane Figures

1. Triangle

   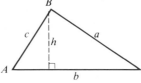

   Perimeter: $P = a + b + c$
   Area: $A = \frac{1}{2}bh$

2. Rectangle

   Perimeter: $P = 2l + 2w$
   Area: $A = lw$

3. Parallelogram

   Perimeter: $P = 2a + 2b$
   Area: $A = bh$

4. Trapezoid

   Perimeter: $P = a + b + c + d$
   Area: $A = \frac{1}{2}h(a + b)$

5. Circle

   Circumference: $C = 2\pi r$
   Area: $A = \pi r^2$

Volume and Surface Area of Solid Figures

1. Rectangular prism

   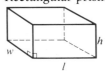

   Volume: $V = lwh$
   Surface area: $S = 2lw + 2lh + 2wh$

2. Right circular cylinder

Volume: $V = \pi r^2 h$
Surface area: $S = 2\pi r^2 + 2\pi rh$

3. Sphere

Volume: $V = \dfrac{4}{3}\pi r^3$
Surface area: $S = 4\pi r^2$

4. Right circular cone

Volume: $V = \dfrac{1}{3}\pi r^2 h$
Surface area: $S = \pi r^2 + \pi rs$

Metric Conversions

Length

1 centimeter	0.394 inch
1 kilometer	0.621 mile
1 inch	2.54 centimeters
1 mile	1.609 kilometers

Volume

1 liter	1.057 quarts
1 milliliter	1 cubic centimeter
1 quart	0.946 liter

Mass (Weight)

1 gram	0.035 ounce
1 kilogram	2.205 pounds
1 ounce	28.350 grams
1 pound	0.454 kilogram

PART I

◀ *Fundamentals*

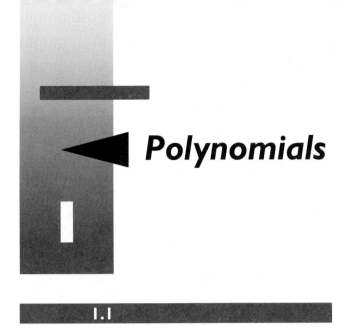

1 Polynomials

1.1

PRELIMINARY CONCEPTS

In this section we give a brief review of concepts and skills from elementary algebra that you will find immediately useful as you begin this course.

Subsets of the Real Numbers

In algebra we work primarily with collections, or **sets**, of numbers. Our study of algebra will deal almost exclusively with the **real numbers**, denoted by **R**. The real numbers are classified according to their properties as follows.

- The set **N** of **natural, or counting, numbers**, as its name suggests, consists of the numbers $1, 2, 3, 4, \ldots$, and so on, where "..." indicates that the list continues without end.
- The set **W** of **whole numbers** consists of the natural numbers and zero: $0, 1, 2, 3, \ldots$.
- The set **J** of **integers** consists of the natural numbers, their negatives, and zero: $\ldots, -3, -2, -1, 0, 1, 2, 3, \ldots$.
- The set **Q** of **rational numbers** includes all numbers that can be represented in the form a/b, where a and b are integers and b does not equal zero. Examples are $-3/4$, $18/27$, 3, and -6. (The integers 3 and -6 are rational because they can be written in the form $3/1$ and $-6/1$.) All rational numbers, when represented in decimal form, either terminate or exhibit a repeating pattern. For example, $-3/4$ is equivalent to the terminating decimal -0.75, and $2/3$ is equivalent to the repeating decimal $0.666\ldots$. (A repeating decimal is denoted by a bar over the repeating digit or block of digits; thus, $0.666\ldots = 0.\overline{6}$.)
- The set **H** of **irrational numbers** includes all numbers whose decimal representations are nonterminating and nonrepeating. An irrational

number *cannot* be written in the form a/b, where a and b are integers. Examples of irrational numbers are $\sqrt{15}$, π, and $-\sqrt[3]{7}$. (Note that $\sqrt{9}$ is *not* irrational, since $\sqrt{9} = 3$, a rational number.)

Each item in a set is called an **element** or **member** of the set. For example, -4 is an element of the set of integers. We use the symbol $\in$ to mean "is an element of"; thus, we might write $-4 \in J$ for the statement above.

If all the elements of a set are also members of a larger set, that set is a **subset** of the larger set. The different subsets of the real numbers are related as shown in Figure 1.1. Notice that the natural numbers are a subset of the whole numbers, the whole numbers are a subset of the integers, and the integers are a subset of the rational numbers. Also observe that every real number is either rational or irrational.

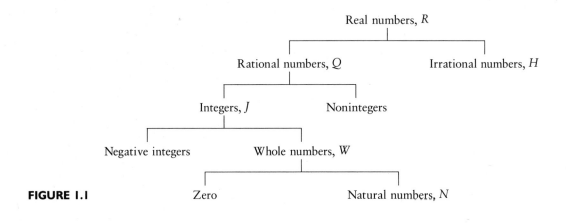

FIGURE 1.1

EXAMPLE 1
a. 2 is an element of N, W, J, Q, and R.
b. $\sqrt{15}$ is an element of H and R.
c. -5.73 is an element of Q and R.
d. The number π, whose decimal representation begins $3.14159\ldots$, is an element of H and R.

Exponential Notation

Recall that repeated multiplication can be indicated by an **exponent**. For example, the expression 3^4 means $3 \cdot 3 \cdot 3 \cdot 3$. The expression 3^4 is called a **power**, the number 3 is called the **base** of the power, and the exponent is 4. In general, a power is defined as follows.

> If n is a natural number,
>
> $$a^n = a \cdot a \cdot a \cdots \cdot a \qquad (n \text{ factors}).$$

EXAMPLE 2

a. $\left(\dfrac{2}{3}\right)^3 = \left(\dfrac{2}{3}\right)\left(\dfrac{2}{3}\right)\left(\dfrac{2}{3}\right) = \dfrac{8}{27}$

b. $(-3)^4 = (-3)(-3)(-3)(-3) = 81$

c. $(-4)^3 = (-4)(-4)(-4) = -64$

Order of Operations

We simplify expressions involving the four arithmetic operations and powers by performing the indicated operations in the following order.

> **Order of Operations**
>
> 1. First, simplify any expression within a grouping symbol (parentheses, brackets, fraction bars, etc.). Start with the innermost grouping symbol and work outward.
> 2. Next, evaluate all powers.
> 3. Next, perform multiplications and divisions as encountered in order from left to right.
> 4. Last, perform additions and subtractions in order from left to right.

EXAMPLE 3

$\dfrac{4 + 3 \cdot 6}{2} - \dfrac{9 - 2 \cdot 2}{5}$ Simplify expressions grouped by the fraction bars. Multiply first.

$= \dfrac{4 + 18}{2} - \dfrac{9 - 4}{5}$ Add and subtract.

$= \dfrac{22}{2} - \dfrac{5}{5}$ Divide.

$= 11 - 1$ Subtract.

$= 10$

Common Error

Note that $4 + 3 \cdot 6 \neq 7 \cdot 6$ and that $9 - 2 \cdot 2 \neq 7 \cdot 2$.

An exponent applies only to its base. For example, $-3^2 = -9$, but $(-3)^2 = (-3)(-3) = 9$. The following examples show that the base may consist of a single number or an expression enclosed in parentheses.

EXAMPLE 4

a. $5 - 3 \cdot 4^2$
$= 5 - 3 \cdot 16$
$= 5 - 48 = -43$

c. $(5 - 3 \cdot 4)^2$
$= (5 - 12)^2$
$= (-7)^2 = 49$

b. $5 - (3 \cdot 4)^2$
$= 5 - 12^2$
$= 5 - 144 = -139$

d. $(5 - 3) \cdot 4^2$
$= 2 \cdot 16$
$= 32$

Common Error

Note in Example 4b that in the expression $5 - 12^2$, the exponent 2 applies only to 12, not to -12. Thus, $5 - 12^2 \neq 5 + 144$.

More complicated calculations can be simplified by carefully following the order of operations.

EXAMPLE 5

$$\left[\frac{8 - 3^2(5 - 7)^3}{4^3 - 3 \cdot 2^3}\right]\left[\frac{11 - 3^2(-3)^2}{-2^2 - (6 - 9)^3 - 3^2}\right]$$

$$= \left[\frac{8 - 9(-2)^3}{64 - 3(8)}\right]\left[\frac{11 - 9(9)}{-4 - (-3)^3 - 9}\right]$$

$$= \left[\frac{8 - 9(-8)}{64 - 24}\right]\left[\frac{11 - 81}{-4 - (-27) - 9}\right]$$

$$= \left(\frac{80}{40}\right)\left(\frac{-70}{14}\right)$$

$$= (2)(-5) = -10$$

Algebraic Expressions

Recall that we use **variables** to indicate an unspecified or unknown element of a set. Variables are usually lowercase letters like a, b, c, x, y, or z. Any meaningful collection of numbers, variables, and symbols of operation is called an **algebraic expression**. We evaluate an algebraic expression for given values of the variable or variables by substituting the values into the expression and then simplifying according to the order of operations. The use of a calculator will often facilitate computations, particularly if the expression involves decimal fractions.

EXAMPLE 6   a. For $F = 144$,

$$\frac{5(F - 32)}{9} = \frac{5(144 - 32)}{9}$$
$$= \frac{5(112)}{9} = 62.222\ldots.$$

b. For $g = 32$ and $t = \tfrac{3}{2}$,

$$\tfrac{1}{2}gt^2 + 8t = \tfrac{1}{2}(32)\left(\tfrac{3}{2}\right)^2 + 8\left(\tfrac{3}{2}\right)$$
$$= \tfrac{1}{2}(32)\left(\tfrac{9}{4}\right) + 8\left(\tfrac{3}{2}\right)$$
$$= 36 + 12 = 48.$$

c. For $r = 2.8$ and $h = 6.1$, and using the approximation 3.14 for π,

$$\pi r^2 + 2\pi r h = (3.14)(2.8)^2 + 2(3.14)(2.8)(6.1)$$
$$= (3.14)(7.84) + 2(3.14)(2.8)(6.1)$$
$$= 24.6176 + 107.2624 = 131.88.$$

Mathematical Modeling

Mathematics provides us with a symbolic language for solving applied problems of many kinds. The process of describing a physical situation in mathematical terms is called **mathematical modeling**. The exercise set following this section includes some examples of this process in which the mathematical models are simple algebraic expressions.

EXAMPLE 7 The area of a circle is given by π times the square of its radius.

a. Choose a variable to represent the radius of a circle and write an algebraic expression for the area.
b. Using the approximation 3.14 for π, find the area of a circle of radius 5 centimeters.

Solutions a. Let r stand for the radius of the circle. Then the area is given by πr^2.
b. If $r = 5$ centimeters, the area of the circle is approximately

$$\pi r^2 = (3.14)(5)^2$$
$$= (3.14)(25) = 78.5 \text{ square centimeters.}$$

EXERCISE 1.1

A

■ *Name the subsets of the real numbers to which each of the following numbers belongs. See Example 1.*

1. $-\dfrac{5}{8}$
2. 137
3. $\sqrt{8}$
4. $2.71828\ldots$
5. -36
6. $\sqrt{49}$
7. 0
8. $0.\overline{357}$
9. $13.\overline{289}$
10. $\sqrt{\dfrac{4}{9}}$
11. $6.468725\ldots$
12. $\dfrac{13}{7}$

■ *Simplify. See Example 2.*

13. -5^2
14. $(-5)^2$
15. $(-3)^4$
16. -3^4
17. -4^3
18. $(-4)^3$

■ *Simplify. See Example 3.*

19. $\dfrac{3(6-8)}{-2} - \dfrac{6}{-2}$
20. $\dfrac{5(3-5)}{2} - \dfrac{18}{-3}$
21. $6[3 - 2(4 + 1)]$
22. $5[3 + 4(6 - 4)]$
23. $(4 - 3)[2 + 3(2 - 1)]$
24. $(8 - 6)[5 + 7(2 - 3)]$
25. $64 \div (8[4 - 2(3 + 1)])$
26. $27 \div (3[9 - 3(4 - 2)])$
27. $5[3 + (8 - 1)] \div (-25)$
28. $-3[-2 + (6 - 1)] \div (-9)$
29. $[-3(8 - 2) + 3] \cdot [24 \div 6]$
30. $[-2 + 3(5 - 8)] \cdot [-15 \div 3]$

31. $\left[\dfrac{7-(-3)}{5-3}\right]\left[\dfrac{4+(-8)}{3-5}\right]$

32. $\left[\dfrac{12+(-2)}{3+(-8)}\right]\left[\dfrac{6+(-15)}{8-5}\right]$

33. $\left(3-2\left[\dfrac{5-(-4)}{2+1}-\dfrac{6}{3}\right]\right)$

34. $\left(7+3\left[\dfrac{6+(-18)}{4+2}\right]-5\right)+3$

■ See Examples 4 and 5.

35. $\dfrac{4\cdot 2^3}{16}+3\cdot 4^2$

36. $\dfrac{4\cdot 3^2}{6}+(3\cdot 4)^2$

37. $\dfrac{3^2-5}{6-2^2}-\dfrac{6^2}{3^2}$

38. $\dfrac{3^2\cdot 2^2}{4-1}+\dfrac{(-3)(2)^3}{6}$

39. $\dfrac{(-5)^2-3^2}{4-6}+\dfrac{(-3)^2}{2+1}$

40. $\dfrac{7^2-6^2}{10+3}-\dfrac{8^2\cdot(-2)}{(-4)^2}$

41. $\dfrac{8^2+6\left(\dfrac{5^2+3}{4-2^3}\right)-3}{-2^3+4\left(\dfrac{3-3^3}{1-4}\right)+6}$

42. $\dfrac{12+3\left(\dfrac{12-20}{3^2-1}\right)^2-1}{-8+6\left(\dfrac{12-30}{2^4-5^2}\right)^2+1}$

43. $\dfrac{3(3+2)^2-3^2\cdot 3+2}{3\cdot 2^3+2(2-1)-1}$

44. $\dfrac{6^2-2\left(\dfrac{4+6}{5}\right)^3+8}{3^2-3\cdot 2+2^2}$

■ Evaluate each expression for the given values of the variables. See Example 6.

45. $\dfrac{5(F-32)}{9}$; $F=212$

46. $\dfrac{R+r}{r}$; $R=12$ and $r=2$

47. $\dfrac{E-e}{R}$; $E=18$, $e=2$, and $R=4$

48. $\dfrac{a-4s}{1-r}$; $r=2$, $s=12$, and $a=4$

49. $P+Prt$; $P=1000$, $r=0.04$, and $t=2$

50. $R(1+at)$; $R=2.5$, $a=0.05$, and $t=20$

51. $\dfrac{1}{2}gt^2$; $g=32$ and $t=2$

52. $\dfrac{1}{2}gt^2-12t$; $g=32$ and $t=3$

53. $\dfrac{1}{2}gt^2-12t$; $g=32$ and $t=\dfrac{3}{4}$

54. $\dfrac{Mv^2}{g}$; $M=64$, $v=2$, and $g=32$

55. $\dfrac{32(V-v)^2}{g}$; $V=12.78$, $v=4.26$, and $g=32$

56. $\dfrac{32(V-v)^2}{g}$; $V = 38.3$, $v = -6.7$, and $g = 9.8$

57. ar^{n-1}; $a = 2.14$, $r = 3.7$, and $n = 4$

58. ar^{n-1}; $a = -8.0$, $r = 0.35$, and $n = 6$

59. $\dfrac{a - ar^n}{1 - r}$; $a = 42.98$, $r = 0.26$, and $n = 3$

60. $\dfrac{a - ar^n}{1 - r}$; $a = 6.3$, $r = -0.85$, and $n = 6$

■ Evaluate the following expressions for the volumes and surface areas of some common solids. Use the approximation 3.14 for π.

61. a. Volume of a sphere: $\frac{4}{3}\pi r^3$, for $r = 1.2$ meters

   b. Surface area of a sphere: $4\pi r^2$, for $r = 0.7$ centimeters

62. a. Volume of a rectangular prism (or box): lwh, for $l = 12.3$ inches, $w = 4$ inches, and $h = 7.3$ inches

   b. Surface area of a box: $2lw + 2lh + 2wh$, for $l = 6.2$ feet, $w = 5.8$ feet, and $h = 2.6$ feet

63. a. Volume of a right circular cylinder: $\pi r^2 h$, for $r = 6$ meters and $h = 23.2$ meters

   b. Surface area of a right circular cylinder: $2\pi r^2 + 2\pi rh$, for $r = 15.3$ inches and $h = 4.5$ inches

64. a. Volume of a right circular cone: $\frac{1}{3}\pi r^2 h$, for $r = 4.6$ feet and $h = 8.1$ feet

   b. Surface area of a right circular cone: $\pi r^2 + \pi rs$, for $r = 16$ centimeters and $s = 42$ centimeters

■ For Problems 65–70 see Example 7.

65. The perimeter of a rectangle is given by twice its length plus twice its width.

   a. Choose variables to represent the length and width of a rectangle. Write an algebraic expression for its perimeter.

   b. Determine the perimeter of a rectangle of length 16 centimeters and width 12 centimeters.

66. The area of a trapezoid is given by the product of half its height times the sum of its bases.

   a. Choose variables to represent the height of a trapezoid and each of its bases. Write an algebraic expression for the area of the trapezoid.

   b. Determine the area of a trapezoid with bases of 10 and 12 centimeters and height of 14 centimeters.

67. The pressure exerted by a gas is given by a constant k times the temperature of the gas, divided by the volume it occupies.

   a. Choose variables to represent the temperature and volume of a gas. Write an algebraic expression for the pressure it exerts.

   b. Determine the pressure in pounds per square inch exerted by 200 cubic inches of a gas at 400° Kelvin if the value of the gas constant is 20.

68. An electrical circuit contains two resistors. The net resistance in the circuit is given by the product of the two individual resistances divided by their sum.
   a. Choose variables to represent the two given resistances. Write an algebraic expression for the net resistance in the circuit.
   b. Determine the net resistance if the two individual resistances are 10 and 20 ohms.

69. When a sum of money is invested in an interest-bearing account, the simple interest accumulated is given by the product of the original investment (called the principal), the interest rate, and the time in years that the money is invested. The sum of the principal and the interest is called the amount.
   a. Choose variables to represent the principal, the interest rate, and the time the money is invested. Write algebraic expressions for the interest and the amount.
   b. Determine the amount accumulated if $800 is invested at 9% for 3 years.

70. The expansion in length that a section of highway will experience on a very hot day is given by a constant k times the length of the section of highway times the difference between the present temperature and the temperature when the highway was built.
   a. Choose variables to represent the length of a highway, the present temperature, and the temperature when the highway was built. Write an algebraic expression for the amount the highway will expand.
   b. Determine the amount that a 1000-foot section of highway will expand at 105°F if the highway was built at 65°F and the value of the constant is 0.000012.

1.2

SUMS AND DIFFERENCES

Factors and Terms

Algebraic expressions that are multiplied together are known as **factors**, while expressions that are added together are called **terms**. For example,

$3x + 4yz$ contains two terms; the first term, $3x$, contains two factors, and the second term, $4yz$, contains three factors;

$3(x - 4y^2)$ consists of a single term; however, the factor $x - 4y^2$ contains two terms, x and $-4y^2$.

The **numerical coefficient**, or simply **coefficient**, of a term is its numerical factor. For example,

the coefficient of x in the term $6x$ is 6;

the coefficient of a^2b in the term $-2a^2b$ is -2;

the coefficient of xy in the term xy is 1.

Since any difference $a - b$ can be rewritten as a sum $a + (-b)$, we can regard any algebraic expression as a *sum* of its terms. For example,

$$2ab^2 - 4a^2b = 2ab^2 + (-4a^2b)$$

and

$$3st^3 - 5s^2t^2 - st = 3st^3 + (-5s^2t^2) + (-st).$$

With this point of view the signs in the expression indicate positive or negative coefficients, and all operations are understood to be addition.

EXAMPLE 1 a. The expression $2x^2 - 3y$, or $2x^2 + (-3y)$, consists of two terms, $2x^2$ and $-3y$. The coefficient of x^2 is 2 and the coefficient of y is -3.

b. The expression $x^2 - xy + 6$, or $x^2 + (-xy) + 6$, consists of three terms, x^2, $-xy$, and 6. The coefficient of x^2 is 1, the coefficient of $-xy$ is -1, and 6 is the constant term.

Polynomials

An algebraic expression consisting of one term of the form cx^n, where c is a constant and n is a whole number, or a product of such expressions, is called a **monomial**. For example,

$$y^3, \quad -3x^4, \quad \text{and} \quad 2x^2y^3$$

are monomials. A **polynomial** is a sum of one or more monomials. Thus,

$$x^2, \quad \frac{1}{5}x - 2, \quad 3x^2 - 2x + 1, \quad \text{and} \quad x^3 - 2x^2 + 1$$

are polynomials.

If a polynomial contains two or three terms, we refer to it as a **binomial** or a **trinomial**, respectively. Polynomials with four or more terms do not have special names.

The **degree** of a monomial in one variable is given by the exponent on the variable. Thus, $3x^4$ is of fourth degree and $7x^5$ is of fifth degree. The degree of a polynomial in one variable is given by the largest exponent in the polynomial.

EXAMPLE 2 a. $2z - 3$ is a first-degree binomial.

b. $3x^2 - 2x + 1$ is a second-degree trinomial.

c. $y^5 - y - 4$ is a fifth-degree trinomial.
d. $8x - 5x^3 + 9$ is a third-degree trinomial.

Since each term of a polynomial in one variable must be of the form cx^n, where n is a whole number, expressions containing terms in which the variable occurs in the denominator or under a radical, or in which the variable has a fractional or negative exponent, are *not* polynomials.

EXAMPLE 3 a. The expressions $x^2 + 7x - 2/x^2$ and $3x + \sqrt{x}$ are *not* polynomials.
b. The expressions $2x^{-1} + x$ and $4x^2 - x^{2/3}$ are *not* polynomials.

Combining Like Terms

Terms that differ at most in their numerical coefficients are called **like terms**. For example,

$$3x^2y \quad \text{and} \quad -x^2y \quad \text{are like terms}$$

and

$$-xy^3 \quad \text{and} \quad 4xy^3 \quad \text{are like terms,}$$

while

$$3x^2y \quad \text{and} \quad 3x^2y^2 \quad \text{are not like terms}$$

because their variable factors are not identical.

We can use the distributive law in the form

$$ba + ca = (b + c)a$$

to rewrite a sum of like terms as a single term. For example,

$$2x + 3x = (2 + 3)x = 5x.$$

When we simplify expressions in this way, we are *combining like terms*.

EXAMPLE 4 a. $3x + 2y + x - 3y = 3x + x + 2y - 3y$
$$= 4x - y$$
 b. $x^2y - 4y^2 - 3x^2y = x^2y - 3x^2y - 4y^2$
$$= -2x^2y - 4y^2$$

Sums of Polynomials

We find the sum of two polynomials by removing the parentheses and adding any like terms.

EXAMPLE 5 a. $(y^2 - 2y) + (4y^2 + 3y)$ b. $(3x^3 + 2x^2 - x + 1) + (3x^2 + 2x)$
$\quad\quad = y^2 - 2y + 4y^2 + 3y \quad\quad\quad\quad = 3x^3 + 2x^2 - x + 1 + 3x^2 + 2x$
$\quad\quad = y^2 + 4y^2 - 2y + 3y \quad\quad\quad\quad = 3x^3 + 2x^2 + 3x^2 - x + 2x + 1$
$\quad\quad = 5y^2 + y \quad\quad\quad\quad\quad\quad\quad\quad = 3x^3 + 5x^2 + x + 1$

Differences of Polynomials

Since the difference $a - b$ can be regarded as the sum $a + (-b)$, we can write the difference of two polynomials as a sum by changing the sign of *each term* of the second polynomial and removing parentheses. We then add all like terms.

EXAMPLE 6 a. $(x^2 + 2x) - (2x^2 - 3x) = x^2 + 2x - 2x^2 + 3x$
$$= -x^2 + 5x$$
 b. $(5y^2 - 2) - (y^2 - 2y) + (2y^2 - 4y + 3)$
$$= 5y^2 - 2 - y^2 + 2y + 2y^2 - 4y + 3$$
$$= 6y^2 - 2y + 1$$

Common Error

Note that in Example 6a $-(2x^2 - 3x) \neq -2x^2 - 3x$.

Sums and differences can also be obtained by using a vertical format, with like terms vertically aligned. For example, the sum

$$(2x^2 - 3x + 1) + (-x + 3) + (x^3 + 5x^2 - 2)$$

can be written as

$$2x^2 - 3x + 1$$
$$- x + 3$$
$$\underline{x^3 + 5x^2 - 2}$$
$$x^3 + 7x^2 - 4x + 2.$$

If a difference is written in vertical form, we replace each term in the subtracted polynomial by its opposite (or negative) and then add.

EXAMPLE 7

$$3t^2 - 4t + 3$$
$$\underline{(-) \; -5t^2 + 2t - 2}$$

can be written as the sum

$$3t^2 - 4t + 3$$
$$\underline{(+) \; 5t^2 - 2t + 2}$$
$$8t^2 - 6t + 5$$

Simplifying Expressions

When simplifying expressions that involve "nested" parentheses or other grouping symbols it is usually best to start with the innermost parentheses and work outward. It is helpful to simplify an expression inside grouping symbols before removing the parentheses or brackets.

EXAMPLE 8

a. $3x - [2 - (3x + 1)]$
$= 3x - [2 - 3x - 1]$
$= 3x - [1 - 3x]$
$= 3x - 1 + 3x$
$= 6x - 1$

b. $x^2 - [3x - (x^2 - 2)]$
$= x^2 - [3x - x^2 + 2]$
$= x^2 - 3x + x^2 - 2$
$= 2x^2 - 3x - 2$

Evaluation of Polynomials

Polynomials can be evaluated for given values of their variable or variables just as any other algebraic expression is evaluated, by substituting the values into the polynomial and simplifying the resulting expression.

EXAMPLE 9 a. For $x = 2$,

$$3x^2 - 2x + 1 = 3(2)^2 - 2(2) + 1$$
$$= 12 - 4 + 1 = 9.$$

b. For $y = -1$,

$$y^5 - y - 4 = (-1)^5 - (-1) - 4$$
$$= -1 + 1 - 4 = -4.$$

c. For $z = \tfrac{3}{4}$,

$$2z - 3 = 2\left(\frac{3}{4}\right) - 3$$
$$= \frac{3}{2} - 3 = -\frac{3}{2}.$$

Modeling with Polynomials   Many applied problems can be modeled with polynomial expressions in one or more variables.

EXAMPLE 10 A mountain climber throws a rope down from the top of an 80-foot cliff with an initial, or starting, velocity of 8 feet per second. (The velocity of the rope increases as it falls, due to the acceleration of gravity.) Using physics, it can be shown that the height of the rope t seconds after it is thrown is given by

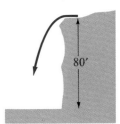

$$80 - 8t - 16t^2.$$

Thus, after 1 second the rope is

$$80 - 8(1) - 16(1)^2$$

or 56 feet above the ground, and after 2 seconds it is

$$80 - 8(2) - 16(2)^2$$

or 0 feet above the ground; that is, the rope reaches the ground in 2 seconds.

EXAMPLE 11 The cubicles in a large office have 8-foot ceilings. To calculate the amount of paint needed to paint a cubicle, we add the areas of the ceiling and of each wall and subtract the area of the door. If the cubicle is x feet long and y feet wide, then the ceiling has an area of xy square feet, two of the walls have areas of $8x$ square feet each, and the remaining two walls have areas of $8y$ square feet each. The door is 7 feet tall by 3 feet wide, so its area is 21 square feet. Thus, we need enough paint to cover

$$xy + 2 \cdot 8x + 2 \cdot 8y - 21, \quad \text{or} \quad xy + 16x + 16y - 21,$$

square feet.

EXAMPLE 12 The profit earned by a company through the sale of a product or service is calculated by subtracting its costs from its revenue (the amount of money obtained from sales). Thus,

$$\text{profit} = \text{revenue} - \text{cost}.$$

Usually these quantities depend on the number of items sold. A furniture company finds that it costs $30x + 2000$ dollars to produce x coffee tables, and the sale of the tables brings in $120x - 0.02x^2$ dollars in revenue. The profit earned from the coffee tables is then

$$(120x - 0.02x^2) - (30x + 2000)$$

or $-0.02x^2 + 90x - 2000$ dollars. If the company produces and sells 100 tables, its profit is

$$-0.02(100)^2 + 90(100) - 2000 = \$6800.$$

EXERCISE 1.2

A

■ *Identify each polynomial as a monomial, a binomial, or a trinomial. Give the degree of each polynomial and the coefficient of each term. See Examples 1 and 2.*

1. $2x^3 - x^2$
2. $x^2 - 2x + 1$
3. $5n^4$
4. $3n + 1$
5. $3r^2 - r + 2$
6. r^3
7. $y^3 - 2y^2 - y$
8. $3y^2 + 1$

■ *Which of the following are not polynomials? See Example 3.*

9. a. $1 - 4x^2y$ b. $3x^2 - 4x + \dfrac{2}{x}$

    c. $2\sqrt{x} - 7x^3 + 2$ d. $\sqrt{2}\,x^3 + \dfrac{3}{4}x^2 - x$

10. a. $\sqrt{3}\,x^2 - 7x + 2$ b. $2x^{4/3} + 6x^{1/3} - 2$

    c. $\dfrac{2}{x^2 - 6x + 5}$ d. $\dfrac{1}{4}x^{-2} + 3x^{-1} + 4$

■ *Simplify each expression by combining like terms. See Example 4.*

11. $3x^2 + 4x^2$ 12. $7x^3 - 3x^3$ 13. $-6y^3 + 3y^3$
14. $-5y^2 - 6y^2$ 15. $8z^2 - 8z^2 + z^2$ 16. $-6z^3 + 6z^2 - z^2$
17. $3x^2y + 4x^2y - 2x$ 18. $6xy^2 - 4xy^2 + 3y$ 19. $3r^2 + (3r^2 + 4r)$
20. $r^2 + (2r^2 + r)$ 21. $(s^2 - s) - 3s$ 22. $(s^2 + s) - 3s^2$

■ *Simplify each expression. See Examples 5 and 6.*

23. $(2t^2 + 3t - 1) - (t^2 + 1)$ 24. $(t^2 - 4t + 1) - (2t^2 - 2)$
25. $(u^2 - 3u - 2) - (3u^2 - 2u + 1)$ 26. $(2u^2 + 4u + 2) - (u^2 - 4u - 1)$
27. $(2x^2 - 3x + 5) - (3x^3 + x - 2)$ 28. $(4y^4 - 3y^2 - 7) - (6y^2 - y + 2)$
29. $(4t - 3t^2 + 2t^4 - 1) - (5t^3 - 2 + t + t^2)$ 30. $(4s^3 - 3s + 2s^2 - 1) - (2s + s^3 - s^2 - 1)$

■ *Use a vertical format to perform the indicated operations. See Example 7.*

31. $(4a^2 + 6a - 7) - (a^2 + 5a + 2)$
32. $(7c^2 - 10c + 8) - (-6c^2 - 3c - 2)$
33. $(4x^2y - 3xy + xy^2) - (-5x^2y + xy - 2xy^2) - (x^2y - 3xy)$
34. $(m^2n^2 - 2mn + 7) - (-2m^2n^2 + mn - 3) - (3m^2n^2 - 4mn + 2)$

■ *Perform the indicated operations. See Examples 5, 6, and 7.*

35. Subtract $4x^2 - 3x + 2$ from the sum of $x^2 - 2x + 3$ and $x^2 - 4$.
36. Subtract $2t^2 + 3t - 1$ from the sum of $2t^2 - 3t + 5$ and $t^2 + t + 2$.
37. Subtract the sum of $2b^2 - 3b + 2$ and $b^2 + b - 5$ from $4b^2 + b - 2$.
38. Subtract the sum of $7c^2 + 3c - 2$ and $3 - c - 5c^2$ from $2c^2 + 3c + 1$.

■ *Simplify. See Example 8.*

39. $y - [2y + (y + 1)]$ 40. $3x + [2x - (x + 4)]$
41. $3 - [2x - (x + 1) + 2]$ 42. $5 - [3y + (y - 4) - 1]$
43. $(3x + 2) - [x + (2 + x) + 1]$ 44. $-(x - 3) + [2x - (3 + x) - 2]$
45. $[x^2 - (2x + 3)] - [2x^2 + (x - 2)]$ 46. $[2y^2 - (4 - y)] + [y^2 - (2 + y)]$
47. $[2x^2 - 2x - (x^2 - 3x + 7)] - [4x^2 - (x^2 + 3)]$
48. $-(x^2 + 3x) - [3x^2 - (6x - x^2 + 2)]$
49. $3y - (2x - y) - (y - [2x - (y - 2x)] + 3y)$

50. $[x - (y + x)] - (2x - [3x - (x - y)] + y)$
51. $[x - (3x + 2)] - (2x - [x - (4 + x)] - 1)$
52. $-(2y - [2y - 4y + (y - 2)] + 1) + [2y - (4 - y) + 1]$

■ *Evaluate each polynomial for the given values of the variable. See Example 9.*

53. $x^3 - 3x^2 + x + 1$;   a. $x = 2$,   b. $x = -2$
54. $2x^3 + x^2 - 3x + 4$;   a. $x = 3$,   b. $x = -3$
55. $t^2 + 3t + 1$;   a. $t = \dfrac{1}{2}$,   b. $t = -\dfrac{1}{3}$
56. $2t^2 - t + 1$;   a. $t = \dfrac{1}{4}$,   b. $t = -\dfrac{1}{2}$
57. $3z^3 - 2z^2 + 3$;   a. $z = 1.8$,   b. $z = -2.6$
58. $z^3 + 4z - 2$;   a. $z = 2.1$,   b. $z = -3.1$
59. $a^6 - a^5$;   a. $a = -1$,   b. $a = -2$
60. $a^5 - a^4$;   a. $a = -1$,   b. $a = -2$

■ *For Problems 61 and 62 use the following fact. If an object is thrown into the air from a height s_0 above the ground with an initial velocity v_0, then its height t seconds later is given by the polynomial $-\frac{1}{2}gt^2 + v_0 t + s_0$. See Example 10.*

61. a. Write a polynomial that gives the height of a tennis ball thrown into the air with an initial velocity of 16 feet per second from a height of 8 feet. The value of g is 32.
    b. Find the height of the tennis ball at $t = 1$ second and at $t = \frac{1}{2}$ second.
62. a. Write a polynomial that gives the height of a satellite launched with an initial velocity of 10,000 meters per second from the top of a booster rocket 46 meters high. The value of g is 9.8.
    b. Find the height of the satellite at $t = 4$ seconds and at $t = 10$ seconds.

■ *For Problems 63–66 see Example 11.*

63. a. Write a polynomial that gives the surface area of a box with a square base and top.
    b. What is the surface area in square inches of a box of length and width 18 inches and height 8 inches? What is the surface area in square feet?

64. a. Write a polynomial that gives the surface area of an empty swimming pool that is of uniform depth and twice as long as it is wide.
    b. What is the surface area in square feet of the pool if its width is 12 feet and its depth is 6 feet? What is its surface area in square yards?

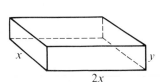

65. a. The sides and top of the box in Problem 63a are constructed from cardboard that costs 2 cents per square inch, and its base is constructed from reinforced cardboard that costs 10 cents per square inch. Write a polynomial that gives the cost of the box.
    b. How much does the box in Problem 63b cost?

66. a. The swimming pool in Problem 64a is painted with sealers that cost $0.80 per square foot for the sides and $1.20 per square foot for the bottom of the pool. Write a polynomial that gives the cost of sealing the pool.
    b. How much would it cost to seal the pool in Problem 64b?

■ *For Problems 67–70 see Example 12.*

67. Writewell, Inc. makes fountain pens. It costs Writewell $5x + 400$ dollars to manufacture x pens, and the company receives $13x - 0.005x^2$ dollars from the sale of the pens.
    a. Write an expression for the profit Writewell earns from the sale of x pens.
    b. What is the profit (or loss) on the sale of 600 pens?

68. It costs The Sweetshop $4x + 200$ dollars to produce x pounds of chocolate creams. The company brings in $8.8x - 0.004x^2$ dollars on the sale of the chocolates.
    a. Write an expression for The Sweetshop's profit from the sale of x pounds of chocolate creams.
    b. What is the profit (or loss) on the sale of 200 pounds?

69. It costs an appliance manufacturer $300x + 200$ dollars per week to produce x top-loading washing machines, which will then bring in weekly revenues of $500x - 0.08x^2$ dollars.
    a. Write an expression for the weekly profit from producing x washing machines.
    b. What is the profit (or loss) from 60 washing machines?

70. A company can produce x lawn mowers per week for $120x + 80$ dollars. The sale of the lawn mowers will generate $200x - 0.3x^2$ dollars in weekly revenue.
    a. Write an expression for the weekly profit from producing x lawn mowers.
    b. What is the profit (or loss) from 200 lawn mowers?

■ *For Problems 71–74 refer to the geometric formulas in the Review of Elementary Topics.**

71. a. Write a polynomial that gives the area of the speaker frame (the shaded region) pictured.
    b. If $x = 8$ inches, find the area of the frame. Use $\pi = 3.14$.

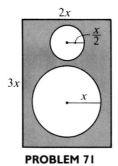

PROBLEM 71

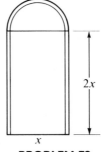

PROBLEM 72

*The Review of Elementary Topics, which follows the preface, provides a reference for basic notions from elementary algebra and geometry.

72. a. A Norman window is shaped like a rectangle whose length is twice its width, surmounted by a semicircle. Write a polynomial that gives its area.
   b. If $x = 3$ feet, find the area of the window.
73. a. A grain silo is built in the shape of a cylinder with a hemisphere on top. Write an expression for the volume of the silo in terms of the radius and height of the cylindrical portion.
   b. If the total height of the silo is five times its radius, write a polynomial in one variable for its volume.
74. a. A cold medication capsule is made in the form of a cylinder with a hemispherical cap on each end. Write an expression for the volume of the capsule in terms of the radius and length of the cylindrical portion.
   b. If the radius of the capsule is one-fourth of its overall length, write a polynomial in one variable for its volume.

1.3

PRODUCTS

Products of Monomials

Consider the following examples of products:

1. $x^2 x^3 = (xx)(xxx) = xxxxx = x^5$;
2. $(x^2)^3 = (x^2)(x^2)(x^2) = (xx)(xx)(xx) = x^6$;
3. $(xy)^3 = (xy)(xy)(xy) = (xxx)(yyy) = x^3 y^3$.

These calculations suggest the three **laws of exponents**, which are used to simplify products of powers.

For all natural numbers m and n,

$$a^m a^n = a^{m+n}, \qquad (1)$$

$$(a^m)^n = a^{mn}, \qquad (2)$$

and

$$(ab)^n = a^n b^n. \qquad (3)$$

EXAMPLE 1 a. From (1),

$$x^2 x^3 = x^{2+3} = x^5 \quad \text{and} \quad y^3 y^4 y^2 = y^{3+4+2} = y^9.$$

b. From (2),

$$(x^2)^3 = x^6 \quad \text{and} \quad (x^5)^2 = x^{10}.$$

c. From (3),

$$(xy)^3 = x^3 y^3 \quad \text{and} \quad (3x)^4 = 3^4 x^4 = 81 x^4.$$

d. From (2) and (3),

$$(x^2 y^3)^3 = x^6 y^9 \quad \text{and} \quad (2x^3 y^2 z)^2 = 2^2 x^6 y^4 z^2 = 4 x^6 y^4 z^2.$$

Common Error Note that although we can simplify the product $x^2 x^3$ as x^5, we cannot simplify the *sum* $x^2 + x^3$, because x^2 and x^3 are not like terms.

We can use the associative and commutative properties of multiplication with the first law of exponents to multiply two or more monomials.

EXAMPLE 2 a. $(3x^2 y)(2xy^2)$
$\quad = 3 \cdot 2 x^2 x y y^2$
$\quad = 6 x^3 y^3$

b. $(-2xy^4)(x^2 y)(4y^2)$
$\quad = -2 \cdot 4 x x^2 y^4 y y^2$
$\quad = -8 x^3 y^7$

Products of Polynomials To multiply polynomials containing more than one term we use a generalized form of the distributive property:

$$a(b + c + d + \cdots) = ab + ac + ad + \cdots.$$

Thus, to find the product of a monomial and a polynomial we multiply each term of the polynomial by the monomial.

EXAMPLE 3 a. $3x(x + y + z) = 3x(x) + 3x(y) + 3x(z)$
$$= 3x^2 + 3xy + 3xz$$

b. $-2ab^2(3a^2b - ab + 2ab^2)$
$$= -2ab^2(3a^2b) - 2ab^2(-ab) - 2ab^2(2ab^2)$$
$$= -6a^3b^3 + 2a^2b^3 - 4a^2b^4$$

We can also use the distributive property to simplify more general polynomial products by multiplying each term of the second polynomial by each term of the first.

EXAMPLE 4 a. $(x + 2)(x^2 - x + 1)$
$$= x(x^2 - x + 1) + 2(x^2 - x + 1)$$
$$= x^3 - x^2 + x + 2x^2 - 2x + 2$$
$$= x^3 + x^2 - x + 2$$

b. $(2x - 3)(x^2 + 2x - 1)$
$$= 2x(x^2 + 2x - 1) - 3(x^2 + 2x - 1)$$
$$= 2x^3 + 4x^2 - 2x - 3x^2 - 6x + 3$$
$$= 2x^3 + x^2 - 8x + 3$$

Products of Binomials

Products of binomials occur so frequently that it is worthwhile to learn a "shortcut" for this type of multiplication. We can use the following scheme to perform the multiplication mentally:

$$(3x - 2y)(x + y) = 3x^2 + 3xy - 2xy - 2y^2$$
$$= 3x^2 + xy - 2y^2.$$

This process is sometimes called the "FOIL" method, where "FOIL" represents

the product of the First terms;
the product of the Outer terms;
the product of the Inner terms;
the product of the Last terms.

EXAMPLE 5 a. $(2x - 1)(x + 3)$
$= 2x^2 + 6x - x - 3$
$= 2x^2 + 5x - 3$

b. $(3x + 1)(2x - 1)$
$= 6x^2 - 3x + 2x - 1$
$= 6x^2 - x - 1$

Simplifying Expressions

The distributive property is often useful in simplifying expressions involving grouping symbols. For example, to simplify

$$2[x - 3y + 3(y - x)] - 2(2x + y),$$

we begin by applying the distributive property to $3(y - x)$, that is, to the *inner* set of parentheses, and then we combine like terms.

EXAMPLE 6 $2[x - 3y + 3(y - x)] - 2(2x + y)$
$= 2[x - 3y + 3y - 3x] - 2(2x + y)$
$= 2[-2x] - 2(2x + y)$
$= -4x - 4x - 2y$
$= -8x - 2y$

The laws of exponents and the distributive property can also be used to find products of polynomials with variable exponents.

EXAMPLE 7 a. $x^n x^{n+3}$
$= x^{n+n+3}$
$= x^{2n+3}$

b. $(y^{2n+1})^3$
$= y^{3(2n+1)}$
$= y^{6n+3}$

c. $(x^{n+1} y^{n-1})^2$
$= x^{2(n+1)} y^{2(n-1)}$
$= x^{2n+2} y^{2n-2}$

EXAMPLE 8 a. $2a^{2n}(3a^n - 2) = 2a^{2n}(3a^n) - 2a^{2n}(2)$
$= 6a^{3n} - 4a^{2n}$
b. $(2a^n + 1)(a^n - 2) = 2a^{2n} - 4a^n + a^n - 2$
$= 2a^{2n} - 3a^n - 2$

Modeling

Expressions involving products of polynomials often occur in mathematical models.

EXAMPLE 9 The total revenue obtained from selling a product is found by multiplying the price of one item by the number of items sold. Thus

revenue = (price of one item) (number of items sold).

A clothing company sells $600 - 15x$ pairs of jeans per week if it charges x dollars per pair.

a. Write an expression for the weekly revenue obtained from charging x dollars for a pair of jeans.
b. Find the weekly revenues if the company charges $15 per pair, $20 per pair, and $25 per pair, respectively.

Solutions a. The total revenue is given by

revenue = (price of one pair) (number of pairs sold)
$= x(600 - 15x)$
$= 600x - 15x^2$.

b. The revenue obtained by charging $15 per pair is

$$600(15) - 15(15)^2 = \$5625.$$

If the company charges $20 per pair, the revenue is

$$600(20) - 15(20)^2 = \$6000.$$

If the company charges $25 per pair, the revenue is

$$600(25) - 15(25)^2 = \$5625.$$

EXAMPLE 10 The local thespian society sold tickets to its opening night performance for $5 and drew an audience of 100 people. The next night it reduced the price by $0.25, and 10 more people attended; that is, 110 people bought tickets at $4.75 apiece. In fact, for each $0.25 reduction in price, 10 additional tickets were sold. Write an expression for the society's revenue from ticket sales on any particular night.

Solution Let x stand for the *number* of $0.25 reductions in price. Then the price of a ticket can be expressed as

$$5.00 - 0.25x.$$

At this price, the society can expect to sell

$$100 + 10x$$

tickets—10 additional tickets for each $0.25 price reduction. The total revenue is then given by

$$\text{revenue} = (\text{price per ticket})(\text{number of tickets sold})$$
$$= (5.00 - 0.25x)(100 + 10x),$$

or $500 + 25x - 2.5x^2$ dollars when the ticket price is reduced by $0.25x$.

EXERCISE 1.3

A

■ *Simplify each product. See Example 1.*

1. $(y^2z)^3$
2. $(yz^4)^2$
3. $(2xy^2z)^2$
4. $(3x^2yz^2)^3$
5. $(-2ab^3c)^3$
6. $(-3a^2bc^3)^3$

■ *See Example 2.*

7. $(7t)(-2t^2)$
8. $(4c^3)(2c)$
9. $(4a^2b)(-10ab^2c)$
10. $(-6r^2s^2)(5rs^3)$
11. $2(3x^2y)(x^3y^4)$
12. $-5(ab^3)(-3a^2bc)$
13. $(-r^3)(-r^2s^4)(-2rt^2)$
14. $(-5mn)(2m^2n)(-n^3)$
15. $(y^2z)(-3x^2z^2)(-y^4z)$
16. $(-3xy)(2xz^4)(3x^3y^2z)$
17. $(2rt)(-3r^2t)(-t^2)$
18. $-a^2(ab^2)(2a)(-3b^2)$

■ *Simplify each expression.*

19. $(x^2y)^2 + (xy)^3$
20. $(2xy^3)^4 + (3xy)^2$
21. $(2rs^2t)^2 - (rst^2)^2$
22. $(3r^2st)^3 - (4r^2s^2t^2)^2$
23. $(-mn)^2 - mn^2(mn)^2$
24. $(m^2n)(mn) + (mn^2)^2$
25. $(2x^2y)^2(xy) + (xy^2)$
26. $(xy)^2 + (-x^2y)^2(-xy^2)$
27. $(2xy)^2 C\ 3x(x^2y)^2 + 4x(xy^2)$
28. $3(x^2y)^2 + x(x^2y) - x^2(x^2y^2)$
29. $(u^2v)^2 - 2u^2(u^2v^2) + 3uv(u^3v)$
30. $2u^2(v^3) + 4v(uv)^2 - uv(uv^2)$

■ *Write each product as a polynomial and simplify. See Example 3.*

31. $4y(x - 2y)$
32. $3x(2x + y)$
33. $-6x(2x^2 - x + 1)$
34. $-2y(y^2 - 3y + 2)$
35. $a^2b(3a^2 - 2ab - b)$
36. $ab^3(-a^2b^2 + 4ab - 3)$
37. $2x^2y^3(4xy^4 - 2x^2y - 3x^3y^2)$
38. $5x^2y^2(3x^4y^2 + 3x^2y - xy^6)$

■ *See Example 4.*

39. $(y + 2)(y^2 - 2y + 3)$
40. $(t + 4)(t^2 - t - 1)$
41. $(x - 3)(x^2 + 5x - 6)$
42. $(x - 7)(x^2 - 3x + 1)$
43. $(x - 2)(x - 1)(x + 3)$
44. $(y - 2)(y + 2)(y + 4)$
45. $(z - 3)(z + 2)(z + 1)$
46. $(z - 5)(z + 6)(z - 1)$
47. $(2x + 3)(3x^2 - 4x + 2)$
48. $(3x - 2)(4x^2 + x - 2)$
49. $(2a^2 - 3a + 1)(3a^2 + 2a - 1)$
50. $(b^2 - 3b + 5)(2b^2 - b + 1)$

■ *See Example 5.*

51. $(n + 2)(n + 8)$
52. $(r - 1)(r - 6)$
53. $(r + 5)(r - 2)$
54. $(z - 3)(z + 5)$
55. $(2z + 1)(z - 3)$
56. $(3t - 1)(2t + 1)$
57. $(4r + 3s)(2r - s)$
58. $(2z - w)(3z + 5w)$
59. $(2x - 3y)(3x - 2y)$
60. $(3a + 5b)(3a + 4b)$
61. $(3t - 4s)(3t + 4s)$
62. $(2x - 3z)(2x + 3z)$
63. $(2a^2 + b^2)(a^2 - 3b^2)$
64. $(s^2 - 5t^2)(3s^2 + 2t^2)$

■ *Simplify each expression. See Example 6.*

65. $2[a - (a - 1) + 2]$
66. $3[2a - (a + 1) + 3]$
67. $a[a - (2a + 3) - (a - 1)]$
68. $-2a[3a + (a - 3) - (2a + 1)]$
69. $2(x - [x - 2(x + 1) + 1] + 1)$
70. $-4(4 - [3 - 2(x - 1) + x] + x)$
71. $-x(x - 3[2x - 3(x + 1)] + 2)$
72. $x(4 - 2[3 - 4(x + 1)] - x)$
73. $-4[2x^2 - 2(x + 1)(x - 2) - 4x]$
74. $-3[2x^2 - 3(x - 2)(x + 3) + 3x]$
75. $x(x[x(x - 2) + 1] - 3) + 4$
76. $x(x[x(x + 3) - 2] - 3) - 5$

■ *Solve. See Examples 9 and* 10.

77. A football conference is made up of *n* teams. The expression $(\frac{1}{2})n(n - 1)$ gives the number of games that must be played in order for each team to play each other team once.

    a. Write the expression as a binomial.
    b. How many games must be played if the conference has 10 teams?

78. The expression $(\frac{1}{6})n(n-1)(n-2)$ gives the number of different three-item pizzas that can be created from a list of n toppings.
   a. Write the expression as a trinomial.
   b. If Mario's Pizza offers 12 different toppings, how many different three-item pizzas can be made?
79. A publishing house finds that it sells $1200 - 10x$ copies per week of a new best-seller if it charges x dollars per copy.
   a. Write an expression for the revenue obtained by charging x dollars per copy.
   b. What is the revenue when the publisher charges $20 per copy?
80. Compuquik sells $800 - 4x$ programmable calculators per month when it charges x dollars per calculator.
   a. Write an expression for the revenue obtained by charging x dollars per calculator.
   b. What is Compuquik's revenue when it charges $45 for a calculator?
81. A farmer inherits an apple orchard on which 60 trees are planted per acre. Each tree yields 12 bushels of apples. Experimentation has shown that for each additional tree planted per acre, the yield per tree decreases by ½ bushel.
   a. Write expressions for the number of trees per acre and for the yield per tree if x additional trees are planted per acre.
   b. Write a polynomial for the total yield per acre if x additional trees are planted per acre.
   c. What is the yield per acre if 10 additional trees are planted per acre? What yield per acre would you expect if 10 trees per acre are removed?
82. An entrepreneur buys a motel with 40 units. The previous owner charged $24 per night for a room and on the average filled 32 rooms per night at that price. The entrepreneur discovers that for every $2 he raises the price, another room stands vacant.
   a. Write expressions for the price of a room and the number of rooms that will be occupied per night if the new owner makes x price increases of $2 each.
   b. Write a polynomial for the total income received from the motel per night if the owner raises the price of a room by $2x$ dollars.
   c. What income will the motel generate if the owner charges $30 for a room? If he charges $20 for a room, what income would you expect?
83. A small company manufactures radios. When it charges $20 for a radio, it sells 500 radios per month. For each dollar the price is increased, 15 fewer radios are sold per month.
   a. Write expressions for the price per radio and the number of radios sold per month if the company raises the price of a radio x dollars.
   b. Write a polynomial for the company's total monthly income if it raises the price of a radio by x dollars.
   c. What income can the company expect if it charges $25 for a radio? If it charges $30?

84. A travel agency offers a group rate of $600 per person for a weekend in Lake Tahoe if 20 people sign up. For each additional person who signs up, the price per person is reduced by $10.
   a. Write expressions for the size of the group and the price per person if x additional people sign up.
   b. Write a polynomial for the travel agency's total income if x additional people sign up for the trip.
   c. If 25 members of a ski club sign up for the weekend, what is the travel agency's income? If 30 members sign up?

B

■ *Simplify each expression. Assume that all exponents denote natural numbers. See Examples 7 and 8.*

85. $a^{2n}a^{n-3}$
86. $b^n b^{2n+1}$
87. $y^{2n+6} y^{4-n}$
88. $a^{2n-2} a^{n+3}$
89. $(x^{2n} y)^3$
90. $(xy^{3n})^2$
91. $(x^{2n+1} y^{n-1})^3$
92. $(x^{n-2} y^{2n+1})^2$
93. $x^n(2x^n - 1)$
94. $3t^n(2t^n + 3)$
95. $a^{2n+1}(a^n + a)$
96. $b^{2n+2}(b^{n-1} + b^n)$
97. $(1 + a^n)(2 - a^n)$
98. $(a^n - 3)(a^n + 2)$
99. $(2a^n - b^n)(a^n + 2b^n)$
100. $(a^{2n} - 2b^n)(a^{3n} + b^{2n})$

■ *Verify each product.*

101. $(x + a)(x - a) = x^2 - a^2$
102. $(x - a)^2 = x^2 - 2ax + a^2$
103. $(x + a)^2 = x^2 + 2ax + a^2$
104. $(x + a)(x + b) = x^2 + (a + b)x + ab$
105. $(x + a)(x^2 - ax + a^2) = x^3 + a^3$
106. $(x - a)(x^2 + ax + a^2) = x^3 - a^3$

1.4

FACTORING

It is sometimes useful to write a polynomial as a single *term* composed of two or more *factors*. This process is the reverse of multiplication and is called *factoring*. For example, observe that

$$3x^2 + 6x = 3x(x + 2).$$

Of course, we can also write

$$3x^2 + 6x = 6\left(\frac{1}{2}x^2 + x\right),$$

or

$$3x^2 + 6x = 3x^2\left(1 + \frac{2}{x}\right) \quad (x \neq 0)$$

or any of an infinite number of such expressions. However, we are primarily interested in factoring a polynomial into a unique form (except for signs and order of factors) referred to as the **completely factored form**. A polynomial with integer coefficients is in completely factored form if

1. it is written as a product of polynomials with integer coefficients;
2. no polynomial factor can be further factored.

Common Factors

We can factor a common factor from a polynomial by using the distributive property in the form

$$ab + ac = a(b + c).$$

We first identify the common factor. For example, observe that the polynomial

$$6x^3 + 9x^2 - 3x$$

contains the monomial $3x$ as a factor of each term. We therefore write

$$6x^3 + 9x^2 - 3x = 3x(\quad\quad)$$

and then insert the appropriate polynomial factor within the parentheses. This factor can be determined by inspection. We ask ourselves for monomials that multiply $3x$ to yield $6x^3$, $9x^2$, and $-3x$, respectively, to obtain

$$6x^3 + 9x^2 - 3x = 3x(2x^2 + 3x - 1).$$

We can always check the result of factoring an expression by multiplying the factors. In the example above,

$$3x(2x^2 + 3x - 1) = 6x^3 + 9x^2 - 3x.$$

EXAMPLE 1

a. $18x^2y - 24xy^2$
$= 6xy(? - ?)$
$= 6xy(3x - 4y)$
because
$6xy(3x - 4y)$
$= 18x^2y - 24xy^2.$

b. $y(x - 2) + z(x - 2)$
$= (x - 2)(? + ?)$
$= (x - 2)(y + z)$
because
$(x - 2)(y + z)$
$= y(x - 2) + z(x - 2).$

It is often useful to factor -1 from the terms of a binomial.

$$a - b = (-1)(-a + b)$$
$$= (-1)(b - a)$$
$$= -(b - a)$$

Hence, we have the following important relationship:

$$a - b = -(b - a).$$

That is, $a - b$ and $b - a$ are negatives of each other.

EXAMPLE 2 a. $3x - y = -(y - 3x)$ b. $a - 2b = -(2b - a)$

Factoring Quadratic Polynomials

A useful type of factoring involves quadratic (second-degree) binomials or trinomials. For example, consider the trinomial

$$x^2 + 6x - 16. \quad (1)$$

We desire, if possible, to find two binomial factors,

$$(x + a)(x + b),$$

whose product is the given trinomial. Now,

$$(x + a)(x + b) = x^2 + (a + b)x + ab. \quad (2)$$

Comparing (1) and (2) we seek two integers a and b such that $a + b = 6$ and $ab = -16$; that is, their sum must be the coefficient of the linear term, $6x$, and their product must be the constant term -16. By inspection, or by trial and error, we determine that the two numbers are 8 and -2, so that

$$x^2 + 6x - 16 = (x + 8)(x - 2).$$

Checking, we note that $(x + 8)(x - 2) = x^2 + 6x - 16$.

EXAMPLE 3 Factor.

a. $x^2 - 7x + 12$ b. $x^2 - x - 12$

Solutions   a. Find two numbers whose product is 12 and whose sum is -7. Since the product is positive and the sum is negative, the two numbers must both be negative. By inspection, or by trial and error, we find that the two numbers are -4 and -3. Hence,

$$x^2 - 7x + 12 = (x - 4)(x - 3).$$

b. Find two numbers whose product is -12 and whose sum is -1. Since the product is negative, the two numbers must be of opposite sign and their sum must be -1. By inspection, or by trial and error, we find that the two numbers are -4 and 3. Hence,

$$x^2 - x - 12 = (x - 4)(x + 3).$$

Although we have not specifically noted the check in the examples above, the check should be done mentally for each factorization.

Often we are confronted with a quadratic trinomial in which the coefficient of the second-degree term is other than 1.

EXAMPLE 4 Factor $8x^2 - 9 - 21x$.

Solution   1. Write the trinomial in decreasing powers of x.

$$8x^2 - 21x - 9$$

2. Consider possible combinations of first-degree factors of the first term:

$$(8x \quad)(x \quad)$$
$$(4x \quad)(2x \quad).$$

3. Consider possible factorizations of the last term: 9 may be factored as $9 \cdot 1$ or as $3 \cdot 3$. Form all possible pairs of binomial factors using these factorizations:

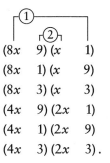

$(8x \quad 9)(x \quad 1)$
$(8x \quad 1)(x \quad 9)$
$(8x \quad 3)(x \quad 3)$
$(4x \quad 9)(2x \quad 1)$
$(4x \quad 1)(2x \quad 9)$
$(4x \quad 3)(2x \quad 3).$

4. Select the combinations of products ① and ② whose sum or difference could be the second term, $-21x$:

$(8x \quad 3)(x \quad 3).$

5. Insert the proper signs:

$(8x + 3)(x - 3).$

With practice, factoring trinomials of the form $Ax^2 + Bx + C$ can usually be done mentally. The following observations may help.

1. If both B and C are positive, both signs in the factored form are positive. For example, as a first step in factoring $6x^2 + 11x + 4$ we could write

$(\quad + \quad)(\quad + \quad).$

2. If B is negative and C is positive, both signs in the factored form are negative. Thus, as the first step in factoring $6x^2 - 11x + 4$ we could write

$(\quad - \quad)(\quad - \quad).$

3. If C is negative, the signs in the factored form are opposite. Thus, as a first step in factoring $6x^2 - 5x - 4$ we could write

$(\quad + \quad)(\quad - \quad) \quad \text{or} \quad (\quad - \quad)(\quad + \quad).$

EXAMPLE 5   a. $6x^2 + 5x + 1$
$= (\ +\ )(\ +\ )$
$= (3x + 1)(2x + 1)$
   b. $6x^2 - 5x + 1$
$= (\ -\ )(\ -\ )$
$= (3x - 1)(2x - 1)$
   c. $6x^2 - x - 1$
$= (\ +\ )(\ -\ )$
$= (3x + 1)(2x - 1)$
   d. $6x^2 - xy - y^2$
$= (\ +\ )(\ -\ )$
$= (3x + y)(2x - y)$

Factoring by Grouping

Sometimes a polynomial can be factored by grouping. For example, we can factor a from the first two terms of

$$ax + ay + bx + by$$

and b from the last two terms to obtain

$$a(x + y) + b(x + y).$$

Note that this last expression is not yet in factored form, since it contains two terms. Now, since $(x + y)$ is a factor of both terms, we can write the expression as

$$(x + y)(a + b).$$

Thus,

$$ax + ay + bx + by = (a + b)(x + y).$$

EXAMPLE 6   a. $x^2 - xb - ax + ab = x(x - b) - a(x - b)$
$= (x - b)(x - a)$
   b. $2x^3 + 3x^2 - 6x - 9 = x^2(2x + 3) - 3(2x + 3)$
$= (2x + 3)(x^2 - 3)$

Sometimes it is necessary to rearrange the terms of an expression so that each group of terms contains a common factor.

1.4 ■ FACTORING

EXAMPLE 7 To factor $3x^2y + 2y + 3xy^2 + 2x$ first rewrite the expression in the form

$$3x^2y + 2x + 3xy^2 + 2y.$$

Then factor the common monomial x from the first pair of terms and the common monomial y from the second pair of terms to obtain

$$x(3xy + 2) + y(3xy + 2).$$

Now factor the common binomial, $(3xy + 2)$, from each term, to obtain

$$(3xy + 2)(x + y).$$

Often more than one factoring technique can be applied to a given polynomial. It is best to remove any common monomial factors first before looking for other factors.

EXAMPLE 8 a. $32x^2 - 84x - 36 = 4(8x^2 - 21x - 9)$
$ = 4(8x + 3)(x - 3)$
b. $x^2y + xy - xy^2 - y^2 = y(x^2 + x - xy - y)$
$ = y[x(x + 1) - y(x + 1)]$
$ = y(x + 1)(x - y)$

The following examples illustrate factorization of polynomials in which the exponents include variables.

EXAMPLE 9 a. $x^{n+2} - 2x^{2n}$
$ = x^n x^2 - 2x^n x^n$
$ = x^n(? - ?)$
$ = x^n(x^2 - 2x^n)$
  b. $x^{2n} + x^{n+1} - x^n$
$ = x^n x^n + x^n x - x^n$
$ = x^n(? + ? - ?)$
$ = x^n(x^n + x - 1)$

EXERCISE 1.4

A

■ *Factor completely. Check by multiplying factors. See Example 1.*

1. $4x^2z + 8xz$
2. $3x^2y + 6xy$
3. $3n^4 - 6n^3 + 12n^2$
4. $2x^4 - 4x^2 + 8x$
5. $15r^2s + 18rs^2 - 3r$
6. $2x^2y^2 - 3xy + 5x^2$
7. $3m^2n^4 - 6m^3n^3 + 14m^3n^2$
8. $6x^3y - 6xy^3 + 12x^2y^2$
9. $15a^4b^3c^4 - 12a^2b^2c^5 + 6a^2b^3c^4$
10. $14xy^4z^3 + 21x^2y^3z^2 - 28x^3y^2z^5$
11. $a(a + 3) + b(a + 3)$
12. $b(a - 2) + a(a - 2)$
13. $y(y - 2) - 3x(y - 2)$
14. $2x(x + 3) - y(x + 3)$
15. $4(x - 2)^2 - 8x(x - 2)^3$
16. $6(x + 1) - 3x(x + 1)^2$
17. $x(x - 5)^2 - x^2(x - 5)^3$
18. $x^2(x + 3)^3 - x(x + 3)^2$
19. $(x - 1)^2 - (x - 1)^2(x + 3)$
20. $(x + 2)^2(x - 1) - (x + 2)^2$
21. $4(x - 1)(x + 3)^2 + 2(x + 1)^2(x + 3)$
22. $3(x + 2)^2(x - 4) + 6(x + 2)(x + 1)^2$

■ *Supply the missing factors or terms. See Example 2.*

23. $3m - 2n = -(?)$
24. $2a - b = -(?)$
25. $-2x + 2 = -2(?)$
26. $-6x - 9 = -3(?)$
27. $-ab - ac = ?(b + c)$
28. $-a^2 + ab = ?(a - b)$
29. $2x - y + 3z = -(?)$
30. $3x + 3y - 2z = -(?)$

■ *Factor completely. See Examples 3, 4, and 5.*

31. $x^2 + 5x + 6$
32. $x^2 + 5x + 4$
33. $y^2 - 7y + 12$
34. $y^2 - 7y + 10$
35. $x^2 - 6 - x$
36. $x^2 - 15 - 2x$
37. $2x^2 + 3x - 2$
38. $3x^2 - 7x + 2$
39. $7x + 4x^2 - 2$
40. $1 - 5x + 6x^2$
41. $9y^2 - 21y - 8$
42. $10y^2 - 3y - 18$
43. $10u^2 - 3 - u$
44. $8u^2 - 3 + 5u$
45. $21x^2 - 43x - 14$
46. $24x^2 - 29x + 5$
47. $5a + 72a^2 - 12$
48. $-30a + 72a^2 - 25$
49. $12 - 53x + 30x^2$
50. $39x + 80x^2 - 20$
51. $-30t - 44 + 54t^2$
52. $48t^2 - 122t + 39$
53. $3x^2 - 7ax + 2a^2$
54. $9x^2 + 9ax - 10a^2$
55. $15x^2 - 4xy - 4y^2$
56. $12x^2 + 7xy - 12y^2$
57. $18u^2 + 20v^2 - 39uv$
58. $24u^2 - 20v^2 + 17uv$
59. $12a^2 - 14b^2 - 13ab$
60. $24a^2 - 15b^2 - 2ab$
61. $10a^2b^2 - 19ab + 6$
62. $12a^2b^2 - ab - 20$
63. $56x^2y^2 - 2xy - 4$
64. $54x^2y^2 + 3xy - 2$
65. $22a^2z^2 - 21 - 19az$
66. $26a^2z^2 - 24 + 23az$

■ *Factor completely. See Examples 6 and 7.*

67. $ax + a + b + bx$
68. $ax^2 + x + a^2x + a$
69. $x^2 - ax + xy - ay$
70. $x^3 - x^2y + xy - y^2$
71. $3x - 2xy - 6x^2 + y$
72. $5xz + y - 5yz - x$
73. $a^3 + 2ab^2 - 2a^2b - 4b^3$
74. $2a^2 + 3a - 2ab - 3b$
75. $6x^2y - 4xy^2 + 3x - 2y$

76. $3x^2 - 3x + 2xy - 2y$
77. $x^3y^2 + x^3 - 3y^2 - 3$
78. $12 - 4y^3 + 3x^2 - x^2y^3$
79. $x^3 + 8 + 2x^2 + 4x$
80. $x^3 + 9 + 3x^2 + 3x$
81. $2x^3 - 3 - 3x^2 + 2x$
82. $2x^3 + 14 + 7x^2 + 4x$

■ Factor completely. See Example 8.

83. $3x^2y + 12xy + 12y$
84. $2x^2y + 6xy - 20y$
85. $2a^3 + 15a^2 + 7a$
86. $2a^3 - 8a^2 - 10a$
87. $40a^2 - 80ab + 40b^2$
88. $20a^2 + 60ab + 45b^2$
89. $6x^3y - 11x^2y^2 + 3xy^3$
90. $9x^3y + 18x^2y^2 + 8xy^3$
91. $6u^3v^2 - 15u^2v + 6u$
92. $9u^2v^3 + 12uv^2 - 12v$
93. $12s^4t^4 - 10s^3t^3 + 2s^2t^2$
94. $16s^3t^3 - 16s^2t^2 - 12st$

B

■ Factor completely. Assume that all variables in exponents denote natural numbers. See Example 9.

95. $x^{2n} - x^n$
96. $x^{4n} + x^{2n}$
97. $x^{3n} - x^{2n} - 2x^n$
98. $4y^{4n} + 3y^{3n} + 2y^{2n}$
99. $2x^{n+2} + 4x^n - 2x^2$
100. $6x^{n+2} - 3x^{n+1} - 3x^n$

1.5

SPECIAL PRODUCTS AND FACTORS

The products below are special cases of the multiplication of binomials. They occur so often that you should learn to recognize them on sight:

> I. $(x + a)^2 = (x + a)(x + a) = x^2 + 2ax + a^2$;
> II. $(x - a)^2 = (x - a)(x - a) = x^2 - 2ax + a^2$;
> III. $(x + a)(x - a) = x^2 - a^2$.

Common Errors   Note that in (I) $(x + a)^2 \neq x^2 + a^2$ and in (II) $(x - a)^2 \neq x^2 - a^2$.

EXAMPLE 1 a. $3(x + 4)^2$
$= 3(x^2 + 2 \cdot 4x + 4^2)$
$= 3x^2 + 24x + 48$

b. $(y + 5)(y - 5)$
$= y^2 - 5^2$
$= y^2 - 25$

c. $(3x - 2y)^2 = (3x)^2 - 2(3x)(2y) + (2y)^2$
$= 9x^2 - 12xy + 4y^2$

d. $3x - x[3 - (x + 2)^2] = 3x - x[3 - (x^2 + 4x + 4)]$
$= 3x - x[3 - x^2 - 4x - 4]$
$= 3x - x[-x^2 - 4x - 1]$
$= 3x + x^3 + 4x^2 + x$
$= x^3 + 4x^2 + 4x$

Of course, each of the formulas above, when viewed from right to left, also represents a special case of factoring quadratic polynomials:

I. $x^2 + 2ax + a^2 = (x + a)^2$;
II. $x^2 - 2ax + a^2 = (x - a)^2$;
III. $x^2 - a^2 = (x + a)(x - a)$.

The trinomials in (I) and (II) are sometimes called **perfect-square trinomials** because they are squares of binomials. Note that the expression $x^2 + a^2$ *cannot* be factored.

EXAMPLE 2 Factor.

a. $x^2 + 8x + 16$ b. $x^2 - 10x + 25$
c. $4a^2 - 12ab + 9b^2$ d. $25a^2b^2 + 20ab + 4$

Solutions   a. Observe that 16 is equal to 4^2 and 8 is equal to $2 \cdot 4$. Hence,

$$x^2 + 8x + 16 = x^2 + 2 \cdot 4x + 4^2$$
$$= (x + 4)^2.$$

b. Observe that 25 is equal to 5^2 and 10 is equal to $2 \cdot 5$. Hence,

$$x^2 - 10x + 25 = x^2 - 2 \cdot 5x + 5^2$$
$$= (x - 5)^2.$$

c. Observe that $4a^2 = (2a)^2$, $9b^2 = (3b)^2$, and $12ab = 2(2a)(3b)$. Hence,

$$4a^2 - 12ab + 9b^2 = (2a)^2 - 2(2a)(3b) + (3b)^2$$
$$= (2a - 3b)^2.$$

d. Observe that $25a^2b^2 = (5ab)^2$, $4 = 2^2$, and $20ab = 2(5ab)(2)$. Hence,

$$25a^2b^2 + 20ab + 4 = (5ab)^2 + 2(5ab)(2) + 2^2$$
$$= (5ab + 2)^2.$$

Binomials of the form $x^2 - a^2$ are often called the **difference of two squares**.

EXAMPLE 3 Factor if possible.

   a. $x^2 - 81$ b. $4x^2 - 9y^2$ c. $x^2 + 81$

Solutions   a. $x^2 - 81$ can be written as the difference of two squares, $x^2 - 9^2$, and thus can be factored according to (III) above:

$$x^2 - 81 = x^2 - 9^2$$
$$= (x + 9)(x - 9).$$

   b. $4x^2 - 9y^2$ can be written as the difference of two squares, $(2x)^2 - (3y)^2$, and thus can be factored as

$$4x^2 - 9y^2 = (2x)^2 - (3y)^2$$
$$= (2x + 3y)(2x - 3y).$$

   c. $x^2 + 81$, equivalent to $x^2 + 0x + 81$, is *not* of form (III). It is *not* factorable, because no two real numbers have a product of 81 and a sum of 0.

The factors $x + 9$ and $x - 9$ in Example 3a are called **conjugates** of each other. In general, any binomials of the form $a - b$ and $a + b$ are called a **conjugate pair**.

Common Error

$x^2 + 81 \neq (x + 9)(x + 9)$, which you can verify by performing the indicated multiplication.

Factoring by Grouping

Sometimes polynomials of four terms can be factored by grouping the terms into a trinomial and a monomial and then using the formulas above.

EXAMPLE 4 Factor.

    a. $x^2 - 2x + 1 - y^2$ b. $x^2 - y^2 + 4yz - 4z^2$

Solutions a. Group the first three terms of the expression to obtain the form of a perfect-square trinomial:

$$(x^2 - 2x + 1) - y^2 = (x - 1)^2 - y^2.$$

This last expression is the difference of two squares, so

$$\begin{aligned} x^2 - 2x + 1 - y^2 &= (x - 1)^2 - y^2 \\ &= (x - 1 + y)(x - 1 - y). \end{aligned}$$

b. A similar factorization can be performed here. Group the last three terms as shown to get

$$\begin{aligned} x^2 - y^2 + 4yz - 4z^2 &= x^2 - (y^2 - 4yz + 4z^2) \\ &= x^2 - (y - 2z)^2 \\ &= (x + y - 2z)(x - y + 2z). \end{aligned}$$

Sum or Difference of Cubes

The two products shown below are useful because the first product results in the sum of two cubes, $x^3 + y^3$, and the second results in the difference of two cubes, $x^3 - y^3$:

$$\begin{aligned} (x + y)(x^2 - xy + y^2) &= x^3 - x^2y + xy^2 + x^2y - xy^2 + y^3 \\ &= x^3 + y^3 \end{aligned} \tag{1}$$

and

$$\begin{aligned} (x - y)(x^2 + xy + y^2) &= x^3 + x^2y + xy^2 - x^2y - xy^2 - y^3 \\ &= x^3 - y^3. \end{aligned} \tag{2}$$

EXAMPLE 5 a. $(x + 3)(x^2 - 3x + 9)$
$= (x + 3)(x^2 - 3x + 3^2)$
$= x^3 + 3^3 = x^3 + 27$

b. $(2y - 1)(4y^2 + 2y + 1)$
$= (2y - 1)[(2y)^2 + 2y + 1]$
$= (2y)^3 - 1^3 = 8y^3 - 1$

Viewing Equations (1) and (2) from right to left we have the following special factorizations for the sum and difference of two cubes:

I. $x^3 + y^3 = (x + y)(x^2 - xy + y^2);$
II. $x^3 - y^3 = (x - y)(x^2 + xy + y^2).$

EXAMPLE 6 a. $8a^3 + b^3 = (2a)^3 + b^3$
$= (2a + b)[(2a)^2 - 2ab + b^2]$
$= (2a + b)(4a^2 - 2ab + b^2)$

b. $x^3 - 27y^3 = x^3 - (3y)^3$
$= (x - 3y)[x^2 + 3xy + (3y)^2]$
$= (x - 3y)(x^2 + 3xy + 9y^2)$

We now summarize the techniques of factoring introduced in this and the previous section.

Suggestions for Factoring Polynomials

1. Write the polynomial in descending powers of one of its variables.
2. Factor out any factors common to each term of the polynomial.
3. A binomial may be one of the following factorable forms: difference of squares, sum of cubes, or difference of cubes.
4. A quadratic trinomial can often be factored as a product of two binomials.
5. Polynomials of four terms can sometimes be factored by grouping.
6. Check the result of the factorization by multiplying the factors.

CHAPTER 1 ■ POLYNOMIALS

Most of our work to this point has involved second- and third-degree polynomials. However, the methods of factoring that we have studied also apply to certain polynomials of higher degree.

EXAMPLE 7

a. $x^6 + 2x^3 + 1 = (x^3 + 1)(x^3 + 1)$
$= (x^3 + 1)^2$
$= (x + 1)^2(x^2 - x + 1)^2$

b. $x^4 - 3x^2 - 4 = (x^2 - 4)(x^2 + 1)$
$= (x + 2)(x - 2)(x^2 + 1)$

c. $4x^6y^4 - x^4y^6 = x^4y^4(4x^2 - y^2)$
$= x^4y^4(2x - y)(2x + y)$

Modeling

Many savings institutions offer accounts on which the interest is *compounded annually*. This means that at the end of each year the interest accrued is added to the principal, and the interest for the next year is computed on this larger amount. An expression for the amount of money in the account after n years is

$$P(1 + r)^n,$$

where P is the original principal and r is the interest rate.

EXAMPLE 8 Carmella invests $2000 in an account that offers an interest rate r compounded annually.

a. Write expressions for the amount of money in Carmella's account after 2 years and after 3 years.
b. Write the expressions found in (a) as polynomials.

Solutions

a. Using the expression $P(1 + r)^n$, we find that after 2 years Carmella's account contains

$$2000(1 + r)^2 \text{ dollars},$$

and after 3 years it contains

$$2000(1 + r)^3 \text{ dollars}.$$

b. Expand the expressions found in (a):

$$2000(1 + r)^2 = 2000(1 + 2r + r^2)$$
$$= 2000 + 4000r + 2000r^2$$

and

$$2000(1 + r)^3 = 2000(1 + r)(1 + 2r + r^2)$$
$$= 2000(1 + 3r + 3r^2 + r^3)$$
$$= 2000 + 6000r + 6000r^2 + 2000r^3.$$

EXERCISE 1.5

A

■ *Write each expression as a polynomial and simplify. See Example 1.*

1. $(x + 3)^2$
2. $(y - 4)^2$
3. $(2y - 5)^2$
4. $(3x + 2)^2$
5. $(x + 3)(x - 3)$
6. $(x - 7)(x + 7)$
7. $(3t - 4s)(3t + 4s)$
8. $(2x + a)(2x - a)$
9. $(5a - 2b)^2$
10. $(4u + 5v)^2$
11. $(8xz + 3)^2$
12. $(7yz - 2)^2$
13. $2[4x + (x + 1)^2]$
14. $3[2x + (x + 2)^2]$
15. $-x + 2x[4 - (x - 3)^2]$
16. $-2x + x[3 - (x + 4)^2]$
17. $-2x[x + (2x - 1)^2 - 4]$
18. $-x[2x - (2x + 1)^2 + 3]$

■ *Factor completely. See Examples 2 and 3.*

19. $x^2 - 25$
20. $x^2 - 36$
21. $x^2 - 24x + 144$
22. $x^2 + 26x + 169$
23. $x^2 - 4y^2$
24. $9x^2 - y^2$
25. $4x^2 + 12x + 9$
26. $4y^2 + 4y + 1$
27. $9u^2 - 30uv + 25v^2$
28. $16s^2 - 56st + 49t^2$
29. $4a^2 - 25b^2$
30. $16a^2 - 9b^2$
31. $x^2y^2 - 81$
32. $x^2y^2 - 64$
33. $9x^2y^2 + 6xy + 1$
34. $4x^2y^2 + 12xy + 9$
35. $16x^2y^2 - 1$
36. $64x^2y^2 - 1$
37. $(x + 2)^2 - y^2$
38. $x^2 - (y - 3)^2$

■ *See Example 4.*

39. $x^2 + 2x + 1 - y^2$
40. $x^2 - 6x + 9 - y^2$
41. $y^2 - x^2 + 2x - 1$
42. $y^2 - x^2 + 4x - 4$
43. $4x^2 + 4x + 1 - 4y^2$
44. $9x^2 - 6x + 1 - 9y^2$

■ *Write each expression as a polynomial and simplify. See Example 5.*

45. $(x - 1)(x^2 + x + 1)$
46. $(x + 2)(x^2 - 2x + 4)$
47. $(2x + 1)(4x^2 - 2x + 1)$
48. $(3x - 1)(9x^2 + 3x + 1)$
49. $(3a - 2b)(9a^2 + 6ab + 4b^2)$
50. $(2a + 3b)(4a^2 - 6ab + 9b^2)$

■ *Factor completely. See Example 6.*

51. $x^3 + 27$
52. $y^3 - 1$
53. $(2x)^3 - y^3$
54. $y^3 + (3x)^3$
55. $a^3 - 8b^3$
56. $27a^3 + b^3$
57. $x^3y^3 - 1$
58. $8 + x^3y^3$
59. $27a^3 + 64b^3$
60. $8a^3 - 125b^3$
61. $125a^3b^3 - 1$
62. $64a^3b^3 + 1$
63. $x^3 + (x - y)^3$
64. $(x + y)^3 - z^3$
65. $(x + 2y)^3 - 8x^3$
66. $x^3 + (x - 2y)^3$
67. $(x + 1)^3 - (x - 1)^3$
68. $(2y - 1)^3 + (y - 1)^3$

■ *See Example 7.*

69. $y^4 - 9$
70. $y^4 - 49$
71. $a^4 + 3a^2 + 2$
72. $a^4 - 5a^2 + 6$
73. $3x^4 + 7x^2 + 2$
74. $4x^4 - 11x^2 - 3$
75. $x^4 - 16$
76. $x^4 - 81$
77. $x^6 + 3x^3 - 4$
78. $x^6 - 6x^3 - 27$
79. $u^8 - 5u^4 + 4$
80. $u^8 - 13u^4 + 36$
81. $4x^2y^3 - 36y^3$
82. $x^3 - 4x^3y^2$
83. $x^4y^2 - x^2y^2$
84. $x^3y - xy^3$
85. $24a^6b^2 - 3b^2$
86. $2a^3 - 54a^3b^9$
87. $12a^6b^2 - 6a^4b^2 - 6a^2b^2$
88. $9a^3b^6 + 3a^3b^4 - 6a^3b^2$
89. $9a^2x^8 + 9a^4x^6 - 18a^6x^4$
90. $6ax^5 + 9a^3x^3 - 6a^5x$
91. $6x^9 - 22x^6 - 8x^3$
92. $4x^8 - 30x^5 - 54x^2$
93. $x^6 - y^6$
94. $x^9 - y^9$

■ *Solve. See Example 8.*

95. Jack invests $500 in an account bearing interest rate r, compounded annually.
    a. Write expressions for the amount of money in Jack's account after 2 years, after 3 years, and after 4 years.
    b. Write the expressions found in (a) as polynomials.
    c. Using either the expressions found in (a) or the polynomials found in (b), find the amount in Jack's account after 2 years, after 3 years, and after 4 years at an interest rate of 8%.

96. A small company borrows $800 for start-up costs and agrees to repay the loan at interest rate r, compounded annually.
    a. Write expressions for the amount the company will owe if it repays the loan after 2 years, after 3 years, or after 4 years.
    b. Write the expressions found in (a) as polynomials.
    c. Using either the expressions found in (a) or the polynomials found in (b), find the amount the company owes after 2 years, after 3 years, or after 4 years at an interest rate of 12%.

1.5 ■ SPECIAL PRODUCTS AND FACTORS   45

■ *Solve. See the list of geometric formulas in the Review of Elementary Topics.*

97. A camper shell is made from a square sheet of metal measuring 9 feet on a side as follows. First, four equal squares are cut from the corners. The sides are then bent down and riveted together as shown. The cut-out squares are x feet on a side.

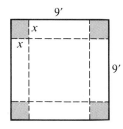
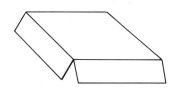

   a. Write expressions for the length, width, and height of the camper shell.
   b. Write an expression for the volume enclosed by the camper shell. Express the volume as a polynomial.
   c. Write an expression for the surface area of the camper shell. Express the surface area as a polynomial.

98. A square metal cake pan is made from a sheet of metal measuring 20 inches by 20 inches as follows. First, four equal squares are cut from the corners. The sides are then bent up and welded together as shown. The cut-out squares are x inches on a side.

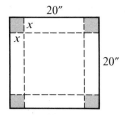
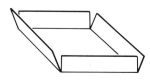

   a. Write expressions for the length, width, and height of the cake pan.
   b. Write an expression for the volume of the cake pan. Express the volume as a polynomial.
   c. Write an expression for the surface area of the cake pan. Express the surface area as a polynomial.

99. Three cylindrical kitchen canisters are each 18 centimeters tall but are of different radii. The radius of the second canister is 2 centimeters greater than the

radius of the first, and the radius of the third canister is 2 centimeters greater than the radius of the second.

    a. Write algebraic expressions for the radii of the three canisters.

    b. Write an expression for the volume of each canister.

    c. Write an expression for the total volume of the set of canisters. Express the volume as a polynomial.

100. A company makes cylindrical storage drums in three different sizes: the smallest drum is 2 feet tall, the middle-sized drum is 3 feet tall and has a radius 6 inches greater than that of the smallest, and the largest drum is 4 feet tall and has a radius 1 foot greater than that of the smallest.

    a. Write algebraic expressions for the radii of the three drums.

    b. Write an expression for the volume of each drum.

    c. What is the total volume of a set of storage drums, one of each size? Express the volume as a polynomial.

101. a. Are the expressions $x^2 - y^2$ and $(x - y)^2$ equivalent?

    b. Factor $x^2 - y^2$, if possible.

    c. Expand $(x - y)^2$.

102. a. Are the expressions $x^2 + y^2$ and $(x + y)^2$ equivalent?

    b. Factor $x^2 + y^2$, if possible.

    c. Expand $(x + y)^2$.

103. a. Are the expressions $x^3 - y^3$ and $(x - y)^3$ equivalent?

    b. Factor $x^3 - y^3$, if possible.

    c. Expand $(x - y)^3$.

104. a. Are the expressions $x^3 + y^3$ and $(x + y)^3$ equivalent?

    b. Factor $x^3 + y^3$, if possible.

    c. Expand $(x + y)^3$.

B

105. a. Write an expression for the area of the square.

    b. Express the area as a polynomial.

    c. Divide the square into four pieces whose areas are given by the terms of your answer to (b).

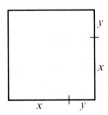

106. a. Write an expression for the area of the shaded region.
    b. Express the area in factored form.
    c. By making one cut in the shaded region, rearrange the pieces into a rectangle whose area is given by your answer to (b).

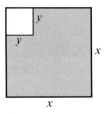

107. The sail pictured is a right triangle of base and height x. It has a colored stripe along the hypotenuse and a white triangle of base and height y in the lower corner.
    a. Write an expression for the area of the colored stripe.
    b. Express the area of the stripe in factored form.
    c. If the sail is 7½ feet high and the white triangle is 4½ feet high, use your answer to (b) to calculate mentally the area of the stripe.

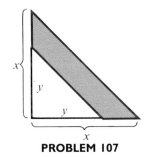
PROBLEM 107

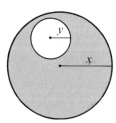
PROBLEM 108

108. An hors d'oeuvres tray has radius x, and the dip container has radius y.
    a. Write an expression for the area available for the chips (shaded region).
    b. Express the area in factored form.
    c. If the tray has radius 8½ inches and the space for the dip has radius 2½ inches, use your answer to (b) to calculate mentally the area for chips. (Express your answer as a multiple of π.)

CHAPTER REVIEW

A

[1.1]

■ *Simplify.*

1. -6^2
2. $(-6)^2$
3. $(-2)^3$
4. -2^3
5. $(4-2)[3-2(3-4)]$
6. $2[1+(6-2)] \div (-4)$
7. $\left[\dfrac{6-(-2)}{4-2}\right]\left[\dfrac{3+(-6)}{2-5}\right]$
8. $\left(6+2\left[\dfrac{4-(-6)}{6-4}\right]+2\right)-3$

9. $\dfrac{2 \cdot 3^2}{6} - 3 \cdot 2^2$
10. $-2^3 + 3\left[\dfrac{5^2 + 3}{4 - (-3)}\right] + (-3)^2$

■ *Evaluate each expression.*

11. $\dfrac{1}{2}gt^2 - 6t$, for $g = 32$ and $t = 2$

12. $\dfrac{a - ar^n}{1 - r}$, for $a = 2.1$, $r = 0.5$, and $n = 3$

[1.2]
■ *Simplify each expression.*

13. $(x^2 - 2x + 3) - (2x^2 + x - 4)$
14. $(2y^3 - y^2 + y - 1) - (y^3 + 2y^2 + 3y - 4)$
15. $y - [3y - (y + 4)]$
16. $4 - [2y + (y - 3) - 5]$
17. $[x - (x - y)] - (2x - [x - (x + y)] - y)$
18. $-(y - [2y + y - (y - 2)] + 1) + [3y - (3 - 2y) - 2]$

■ *Evaluate each polynomial.*

19. $x^3 - 2x^2 - x + 1$, for $x = -2$
20. $2t^3 - t^2 + 2t - 3$, for $t = 1.2$
21. $y^4 - y^3$, for $y = -3$
22. $x^5 - x^2$, for $x = -2$

[1.3]
■ *Multiply.*

23. $(-2a^3)(-a^2b)(3ab^2)$
24. $-b^2(ab)(3a^2)a^2$
25. $(3xy^2)^2(xy) - x(x^2y)$
26. $(-2y)^2(x^3y) + (x^2y)^2 - y^3$
27. $ab^2(2a^2 - 3ab + b)$
28. $-a^2b(3a^2 - 2ab^2 - b^2)$
29. $(3x - 2y)(2x + 4y)$
30. $(x + 4)(x^2 - 3x + 2)$
31. $(2x - 3)^3$
32. $(3x^3 - 1)(3x^3 + 1)$
33. $a[a - 2(a + 1) - (a - 3)]$
34. $-b[2b^2 + b(b - 1) + b]$

[1.4]
■ *Factor completely.*

35. $a(a - 3)^3 - a^2(a - 3)^2$
36. $2(b + 1)^2(b - 3) + 4(b + 1)^3$
37. $14x^2 + 19x - 3$
38. $6x^2y^2 - 13xy - 5$
39. $6xy + 4x - 3y - 2$
40. $2x^3 - 4x^2 + 6x - 12$
41. $3x^4 - 4x^3 - 4x^2$
42. $x^3y - x^2y^2 - 2xy^3$
43. $y^{4n} - y^n$
44. $x^{n+1} + x^n + x$

[1.5]
■ *Write each expression as a polynomial and simplify.*

45. $2[x + (x - 1)^2]$
46. $-y[y - (y - 2)^2 + 3]$

■ *Factor completely.*

47. $4x^2 - 49y^2$
48. $y^2 - (y - 1)^2$
49. $(a + b)^3 - 8$
50. $a^3 + (a^2 - b)^3$

51. The volume of a pyramid with a square base is given by $\frac{1}{3}s^2h$, where s is the side of the base and h is the height. Find the volume of the Great Pyramid of Cheops in Egypt if its height is 250 yards and the side of its base is 160 yards.

52. The harmonic mean of two numbers is given by twice their product divided by their sum.
    a. Choose variables and write an algebraic expression for the harmonic mean.
    b. Find the harmonic mean of 6 and 12.

53. a. Write an algebraic expression for the surface area of a rectangular tank whose length is 4 feet greater than its width.
    b. Write an algebraic expression for the cost of the tank if the bottom and sides of the tank cost $2 per square foot and the top costs $0.80 per square foot.
    c. How much will the tank described above cost if it is 8 feet wide and 4 feet tall?

54. Senator Fogbank sells silk-screened T-shirts to support his campaign. It costs $2x + 700$ dollars to produce x T-shirts, and their sale will bring in $8x - 0.01x^2$ dollars.
    a. Write an expression for the profit expected from the sale of x T-shirts.
    b. What profit will be earned from the sale of 1000 T-shirts?

55. An ice cream vendor sells $200 - 2x$ Fudgesicles per day if he charges x cents per Fudgesicle.
    a. Write an expression for the vendor's income from the sale of Fudgesicles when he charges x cents.
    b. Make a table showing his income at different selling prices x for 10-cent increments in x.

56. In the spring the Neighborhood Nursery takes in $400x - 80x^2$ dollars per week from the sale of blue iris bulbs. Use factoring to write an expression for the number of bulbs sold per week at a price of x dollars per bulb.

57. An airline offers a charter flight to Hawaii for $500 per person if 25 people sign up. For each additional person who signs up, the fare is reduced by $5 per person.
    a. Write a polynomial for the airline's total income from the flight if x additional people sign up.
    b. If 37 people sign up, what is the airline's income from the flight?

58. If the interest on an account is computed every 6 months and added to the principal, the interest is said to be compounded semiannually. At the end of n years the amount of money in the account is given by

$$P\left(1 + \frac{r}{2}\right)^{2n},$$

where P is the original principal and r is the interest rate.
    a. Write polynomials for the amount of money in the account after 1 year and the amount after 18 months.
    b. If $500 is invested at 6% interest, find the amount after 1 year and the amount after 18 months.

59. A greeting card company makes boxes for its cards by cutting a square out of each corner of a 10-inch by 12-inch sheet of cardboard, then folding up the sides and gluing them.
  a. Write a polynomial for the volume of a box if the side of the cut-out square is x inches.
  b. Write a polynomial for the surface area of the box (the top is not included).

60. a. Write a polynomial for the area between two concentric circles if the radius of the inner circle is 8 feet less than the radius of the outer circle. What is the area of a circular driveway as described above if its outer radius is 24 feet?
  b. Write a polynomial for the volume between two concentric spheres if the radius of the outer sphere is h units greater than the radius of the inner sphere. What is the volume of the atmosphere of the earth if the earth's radius is 4000 miles and the atmosphere is 250 miles thick?

B

■ *Simplify.*

61. $a^n(2a^n - a)$ **62.** $(3a^n + 2)(a^n - 3)$

63. $2b^{3n-2} \cdot 3b^{2n-3}$ **64.** $(b^{2n-1})^3$

■ *Verify each product.*

65. $(x + a)^3 = x^3 + 3ax^2 + 3a^2x + a^3$ **66.** $(x - a)^3 = x^3 - 3ax^2 + 3a^2x - a^3$

■ *Factor completely.*

67. $2x^n - 4x^{2n}$ **68.** $6x^{n+2} + x^{n+1} - x^n$

■ *Show that the shaded areas are equal.*

69.

70.

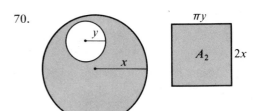

2 Rational Expressions

Polynomials and quotients of polynomials are called **algebraic fractions** or **rational expressions**. For example,

$$x^2 + 2x, \quad 3y, \quad \frac{y}{y+1}, \quad \text{and} \quad \frac{x^2 - 2x + 1}{2x - 1}$$

are rational expressions. Since division by zero does not represent a real number, a rational expression is said to be *undefined* at any value of the variable(s) for which the denominator vanishes (is equal to zero). Thus, the fraction

$$\frac{y}{2y+1}$$

is undefined for $y = -\frac{1}{2}$, and the fraction

$$\frac{x^2 - 2x + 1}{x^2 - 4}$$

is undefined for $x = 2$ and $x = -2$.

2.1

PRELIMINARY CONCEPTS

Every fraction represents either a positive number or a negative number or zero. By the rules for division of signed numbers:

$$\frac{a}{b} = \frac{-a}{-b} \qquad (1)$$

and

$$-\frac{a}{b} = \frac{-a}{b} = \frac{a}{-b} \quad (b \neq 0). \qquad (2)$$

EXAMPLE 1 a. $\dfrac{2}{3} = \dfrac{-2}{-3} = -\dfrac{2}{-3} = -\dfrac{-2}{3}$ b. $-\dfrac{2}{3} = \dfrac{-2}{3} = \dfrac{2}{-3}$

Thus, if any two signs of a fraction (the sign of the numerator, the sign of the denominator, or the sign that precedes the fraction) are changed, an equivalent fraction results.

The forms a/b and $-a/b$ are generally the simplest and most convenient representations for fractions. Particular care should be taken in simplifying a fraction when the numerator or denominator contains more than one term. For example,

$$-\frac{a-b}{c} = \frac{-(a-b)}{c},$$

where the minus sign on the right side precedes *the entire numerator $a - b$*. This fraction can now be written as

$$\frac{-a+b}{c} \quad \text{or} \quad \frac{b-a}{c}.$$

Common Error In particular, note that $-\dfrac{a-b}{c} \neq \dfrac{-a-b}{c}.$

EXAMPLE 2 Write each fraction on the left as an equal fraction with the denominator shown on the right.

a. $\dfrac{x-1}{-3};\ \dfrac{}{3}$ b. $-\dfrac{a}{b-a};\ \dfrac{}{a-b}$

2.1 ■ PRELIMINARY CONCEPTS

c. $\dfrac{3}{3x-2y}; \quad \dfrac{}{2y-3x}$

d. $\dfrac{x-y}{y-x}; \quad \dfrac{}{x-y}$

Solutions

a. $\dfrac{x-1}{-3}$

$= \dfrac{-(x-1)}{3}$

$= \dfrac{1-x}{3}$

b. $-\dfrac{a}{b-a}$

$= \dfrac{a}{-(b-a)}$

$= \dfrac{a}{a-b} \quad (a \neq b)$

c. $\dfrac{3}{3x-2y}$

$= \dfrac{-3}{-(3x-2y)}$

$= \dfrac{-3}{2y-3x} \quad (2y \neq 3x)$

d. $\dfrac{x-y}{y-x}$

$= \dfrac{x-y}{-(x-y)}$

$= -1 \quad (x \neq y)$

Fundamental Principle of Fractions

There are infinitely many fractions equivalent to a given quotient. Thus, for example,

$$\dfrac{1}{2} = \dfrac{2}{4} = \dfrac{3}{6} = \dfrac{4}{8} = \cdots \quad \text{and} \quad \dfrac{3}{5} = \dfrac{6}{10} = \dfrac{9}{15} = \dfrac{12}{20} = \cdots.$$

The following property, called the **fundamental principle of fractions**, enables us to write equivalent forms of fractions.

> **Fundamental Principle of Fractions**
>
> *An equivalent fraction is obtained if the numerator and the denominator of a fraction are each multiplied or divided by the same nonzero number.*

This property can be expressed in symbols as follows:

$$\dfrac{a}{b} = \dfrac{ac}{bc} \quad (b, c \neq 0).$$

Reducing Fractions

A fraction is said to be in lowest terms if the numerator and denominator do not contain common factors. To express a given fraction in lowest terms (to **reduce** the fraction), we look for common factors in the numerator and denominator and then apply the fundamental principle of fractions.

EXAMPLE 3   a. $\dfrac{yz^2}{y^3z} = \dfrac{z \cdot yz}{y^2 \cdot yz}$

$= \dfrac{z}{y^2}$ $(y, z \neq 0)$

b. $\dfrac{8x^3y}{6x^2y^3} = \dfrac{4x \cdot 2x^2y}{3y^2 \cdot 2x^2y}$

$= \dfrac{4x}{3y^2}$ $(x, y \neq 0)$

We use slash lines to indicate the division of numerator and denominator by a common factor. For example, instead of writing

$$\dfrac{y}{y^2} = \dfrac{1 \cdot y}{y \cdot y} = \dfrac{1}{y} \quad (y \neq 0),$$

we can write

$$\dfrac{y}{y^2} = \dfrac{\cancel{y}}{\underset{y}{\cancel{y^2}}} = \dfrac{1}{y} \quad (y \neq 0).$$

If the numerator of the fraction has more than one term, it is especially important to *factor* the numerator before attempting to divide out common factors.

EXAMPLE 4   a. $\dfrac{6y - 3}{3} = \dfrac{\cancel{3}(2y - 1)}{\cancel{3}}$

$= 2y - 1$

b. $\dfrac{9x^3 - 6x^2 + 3x}{3x}$

$= \dfrac{\cancel{3x}(3x^2 - 2x + 1)}{\cancel{3x}}$

$= 3x^2 - 2x + 1$ $(x \neq 0)$

Common Error

Note that in Example 4b,

$$\dfrac{9x^3 - 6x^2 + 3x}{3x} \neq 3x^2 - 2x.$$

The fundamental principle of fractions enables us to obtain an equivalent expression by dividing out any nonzero *factors* that appear in both the numerator and the denominator of a fraction. *The fundamental principle of fractions does not apply to common terms.* For example,

$$\frac{2xy}{3y} = \frac{2x}{3} \quad (y \neq 0)$$

because y is a common factor in the numerator and denominator of the first fraction. However,

$$\frac{2x + y}{3 + y} \neq \frac{2x}{3}$$

because y is a common term but *is not a common factor* of the numerator and denominator. Furthermore,

$$\frac{5x + 3}{5y} \neq \frac{x + 3}{y}$$

because 5 *is not* a factor of the *entire* numerator.

It is often necessary to factor polynomials in both the numerator and the denominator in order to see the common factors.

EXAMPLE 5

a. $\dfrac{x^2 - 7x + 6}{x^2 - 36} = \dfrac{\cancel{(x - 6)}(x - 1)}{\cancel{(x - 6)}(x + 6)}$

$= \dfrac{x - 1}{x + 6}$

$(x \neq 6, -6)$

b. $\dfrac{a - b}{b^2 - a^2} = \dfrac{-1\cancel{(b - a)}}{(b + a)\cancel{(b - a)}}$

$= \dfrac{-1}{b + a}$

$(a \neq b, -b)$

Notice that in (b) we factored $a - b$ in the numerator as $-1(b - a)$ so that the common factor $b - a$ could be "divided out."

c. $\dfrac{27x^3 - 1}{9x^2 - 1}$

$= \dfrac{\cancel{(3x - 1)}(9x^2 + 3x + 1)}{\cancel{(3x - 1)}(3x + 1)}$

$= \dfrac{9x^2 + 3x + 1}{3x + 1}$

$\left(x \neq \dfrac{1}{3}, -\dfrac{1}{3}\right)$

d. $\dfrac{bx - 3ax + 2b - 6a}{b^2 - 9a^2}$

$= \dfrac{x(b - 3a) + 2(b - 3a)}{(b - 3a)(b + 3a)}$

$= \dfrac{\cancel{(b - 3a)}(x + 2)}{\cancel{(b - 3a)}(b + 3a)}$

$= \dfrac{x + 2}{b + 3a} \quad (b \neq 3a, -3a)$

We summarize the procedure for reducing algebraic fractions as follows.

> **To Reduce a Fraction:**
>
> 1. Factor the numerator and denominator.
> 2. Divide the numerator and denominator by any common factors.

EXAMPLE 6 a. $\dfrac{4x + 2}{4}$ b. $\dfrac{9x^2 + 3}{6x + 3}$

Solutions a. $\dfrac{4x + 2}{4} = \dfrac{\cancel{2}(2x + 1)}{\cancel{2}(2)}$ b. $\dfrac{9x^2 + 3}{6x + 3} = \dfrac{\cancel{3}(3x^2 + 1)}{\cancel{3}(2x + 1)}$

$\qquad\qquad = \dfrac{2x + 1}{2}$ $\qquad\qquad = \dfrac{3x^2 + 1}{2x + 1} \quad \left(x \neq -\dfrac{1}{2}\right)$

Common Errors Note that in Example 6a above,

$$\dfrac{4x + 2}{4} \neq x + 2,$$

and in Example 6b,

$$\dfrac{9x^2 + 3}{6x + 3} \neq \dfrac{9x^2}{6x}.$$

Modeling Rational expressions arise in a number of applied situations.

EXAMPLE 7 In computing their travel time, pilots must take into account the prevailing winds. A tail wind adds to the plane's ground speed, while a head wind decreases the ground speed. Skyhigh Airlines is setting up a shuttle service from Dallas to Phoenix, a distance of 800 miles.

a. Express the time needed for a one-way trip, without wind, in terms of the speed of the plane.
b. Assume that there is a prevailing wind of 30 miles per hour blowing from the west. Write expressions for the flying times from Dallas to Phoenix and from Phoenix to Dallas.
c. Write an expression for the round trip flying time, excluding stops, with a 30-mile-per-hour wind from the west.

Solutions a. Recall that

$$\text{distance} = \text{rate} \times \text{time},$$

or, equivalently,

$$\text{time} = \frac{\text{distance}}{\text{rate}}.$$

Let r represent the speed of the plane. Then the time required for a one-way trip from Dallas to Phoenix is

$$\frac{800}{r}.$$

b. On the trip from Dallas to Phoenix the plane encounters a head wind of 30 miles per hour, so its actual ground speed is $r - 30$, where r represents the air speed of the plane. On the return trip the plane enjoys a tail wind of 30 miles per hour, so its actual ground speed is $r + 30$. Therefore, the flying times are

$$\text{Dallas to Phoenix:} \quad \frac{800}{r - 30}$$

and

$$\text{Phoenix to Dallas:} \quad \frac{800}{r + 30}.$$

c. The round-trip flying time from Dallas to Phoenix and back is

$$\frac{800}{r - 30} + \frac{800}{r + 30}.$$

EXERCISE 2.1

A

■ Simplify, and specify any values of the variables for which the fraction is undefined. See Example 1.

1. $-\dfrac{3}{-5}$
2. $-\dfrac{-3}{4}$
3. $-\dfrac{-3}{-7}$
4. $-\dfrac{-4}{-5}$
5. $-\dfrac{-3x}{4y}$
6. $-\dfrac{x}{-2y}$
7. $\dfrac{x+1}{-x}$
8. $\dfrac{x+3}{-x}$
9. $-\dfrac{7-y}{3y+2}$
10. $-\dfrac{3y-2}{4y-1}$
11. $-\dfrac{4-3x}{48-3x^2}$
12. $-\dfrac{3y^2+8}{12y^2-3}$

■ Write each fraction on the left as an equal fraction with the denominator shown on the right. (Assume that no denominator is 0.) See Example 2.

13. $-\dfrac{4}{3-y}; \dfrac{}{y-3}$
14. $\dfrac{-3}{2-x}; \dfrac{}{x-2}$
15. $\dfrac{x+1}{x-y}; \dfrac{}{y-x}$
16. $\dfrac{x+3}{y-x}; \dfrac{}{x-y}$
17. $-\dfrac{x-2}{x-y}; \dfrac{}{y-x}$
18. $-\dfrac{x-4}{x-2y}; \dfrac{}{2y-x}$
19. $\dfrac{-a+1}{-3a-b}; \dfrac{}{3a+b}$
20. $\dfrac{-a-1}{2b-3a}; \dfrac{}{3a-2b}$

■ Reduce each fraction to lowest terms. (Assume that no denominator is 0.) See Example 3.

21. $\dfrac{14cd}{-7c^2d^3}$
22. $\dfrac{100mn}{-5m^2n^3}$
23. $\dfrac{-12r^2st}{-6rst^2}$
24. $\dfrac{-15xy^3z}{-3y^2z^4}$
25. $\dfrac{2x(x-2)^2}{6x^3(x-2)}$
26. $\dfrac{3a^3(2a-1)}{9a^2(2a-1)^2}$
27. $\dfrac{5u(3u-5)}{u^5(5-3u)}$
28. $\dfrac{v^4(4v-1)}{4v(1-4v)}$

■ See Example 4.

29. $\dfrac{4x+6}{6}$
30. $\dfrac{2y-8}{8}$
31. $\dfrac{6a^3-4a^2+2a}{-2a}$
32. $\dfrac{3x^3-6x^2+3x}{-3x}$
33. $\dfrac{6x^2y^3-9x^4y^2}{9x^2y}$
34. $\dfrac{5x^3y^5+10x^4y^2}{5x^2y^2}$

■ See Example 5.

35. $\dfrac{6-6t^2}{(t-1)^2}$
36. $\dfrac{4-4x^2}{(x+1)^2}$
37. $\dfrac{2y^2-8}{2y+4}$
38. $\dfrac{5y^2-20}{2y-4}$
39. $\dfrac{6-2y}{y^3-27}$
40. $\dfrac{4-2y}{y^3-8}$
41. $\dfrac{4x^3+36x}{6x^2+18x}$
42. $\dfrac{5x^2+10x}{5x^3+20x}$

43. $\dfrac{y^2 - 9x^2}{(3x - y)^2}$

44. $\dfrac{(2x - y)^2}{y^2 - 4x^2}$

45. $\dfrac{2x^2 + x - 6}{x^2 + x - 2}$

46. $\dfrac{6x^2 - x - 1}{2x^2 + 9x - 5}$

47. $\dfrac{x - 12 + 6x^2}{17x - 12 - 6x^2}$

48. $\dfrac{2x - 30 + 4x^2}{15 - 16x + 4x^2}$

49. $\dfrac{8y^3 - 27}{4y^2 - 9}$

50. $\dfrac{8y^3 - 1}{4y^2 - 1}$

51. $\dfrac{6x^2y^2 + 7xy - 3}{4x^2y^2 + 4xy - 3}$

52. $\dfrac{8x^2y^2 - 18xy + 7}{6x^2y^2 + 7xy - 5}$

53. $\dfrac{2a^2 + 3ab - 2b^2}{a^2 - 4b^2}$

54. $\dfrac{a^2 + 4ab + 4b^2}{2a^2 + 5ab + 2b^2}$

55. $\dfrac{6x^6 - 30x^4 - 36x^2}{2x^6 + 11x^4 + 12x^2}$

56. $\dfrac{6t^6 + 10t^4 - 4t^2}{6t^6 - 6t^4 - 36t^2}$

57. $\dfrac{12p^3q^3 - 10p^2q^2 - 8pq}{18p^4q^2 - 18p^3q - 8p^2}$

58. $\dfrac{12b^4c^3 - 16b^3c^2 - 3b^2c}{8b^3c^4 - 16b^2c^3 + 6bc^2}$

59. $\dfrac{x^2 + ax + xy + ay}{2x + 2a}$

60. $\dfrac{ax^2 + x + a^2x + a}{3x + 3a}$

61. $\dfrac{ax - 2bx + ay - 2by}{a^2 - 4b^2}$

62. $\dfrac{2ax - 4ay + bx - 2by}{x^2 - 4y^2}$

63. $\dfrac{x^3 + 2x + 3x^2 + 6}{2x^3 + 4x + x^2 + 2}$

64. $\dfrac{3x^3 - 9x - x^2 + 3}{3x^3 + 3x - x^2 - 1}$

65. $\dfrac{6x^2(3x - 4) - 2x(3x - 4)^2}{x^4}$

66. $\dfrac{2(x + 2)(x^2 - 1)^3 - 6x(x^2 - 1)^2(x + 2)^2}{(x^2 - 1)^6}$

67. $\dfrac{(x + 1)^3(x - 4) - 4(x + 1)(x - 4)}{4x^2 - 4x - 48}$

68. $\dfrac{(2x - 3)^4(x^4 - 1) - x^3(2x - 3)(x^4 - 1)}{x^4 - x^3 + x^2 - x}$

■ *For Problems 69 to 72 see Example 6.*

69. Which of the following fractions are equivalent to 2a?

    a. $\dfrac{2a + 4}{4}$
    b. $\dfrac{4a^2 - 2a}{2a - 1}$
    c. $\dfrac{4a^2 - 2a}{2a}$
    d. $\dfrac{a + 3}{2a^2 + 6a}$

70. Which of the following fractions are equivalent to 3b?

    a. $\dfrac{9b^2 - 3b}{3b}$
    b. $\dfrac{b + 2}{3b^2 + 6b}$
    c. $\dfrac{3b - 9}{9}$
    d. $\dfrac{9b^2 - 3b}{3b - 1}$

71. Which of the following fractions are equivalent to −1?

    a. $\dfrac{2a + b}{2a - b}$
    b. $\dfrac{-(a + b)}{b - a}$
    c. $\dfrac{2a^2 - 1}{2a}$
    d. $\dfrac{-a^2 + 3}{a^2 + 3}$

72. Which of the following fractions are equivalent to −1?

    a. $\dfrac{2a - b}{b - 2a}$
    b. $\dfrac{-b^2 - 2}{b^2 + 2}$
    c. $\dfrac{3b^2 - 1}{3b^2 + 1}$
    d. $\dfrac{b - 1}{b}$

■ *Solve. See Example 7.*

73. River Queen Tours offers a 50-mile round-trip excursion on the Mississippi River by paddle wheel. The current in the Mississippi is 8 miles per hour.

    a. Write an expression for the time required for the downstream journey in terms of the speed of the paddle wheel.
    b. Write an expression for the time required for the return trip upstream.
    c. Write an expression for the time needed for the round trip.

74. A rowing team can maintain a speed of 15 miles per hour in still water. The team's daily training session includes a 5-mile run up the Red Cedar River and the return downstream.
   a. Write an expression for the team's time on the upstream leg in terms of the speed of the current.
   b. Write an expression for the team's time on the downstream leg.
   c. Write an expression for the total time for the training run.

75. Two pilots for the Flying Express parcel service receive packages simultaneously. Orville leaves Boston for Chicago at the same time Wilbur leaves Chicago for Boston. Each selects an air speed of 400 miles per hour for the 900-mile trip. The prevailing winds blow from east to west.
   a. Write an expression for Orville's flying time in terms of the wind speed.
   b. Write an expression for Wilbur's flying time.
   c. Who reaches his destination first? By how much time (in terms of the wind speed)?

76. On New Year's Day a blimp leaves its berth in Carson, California, and heads north for the Rose Bowl, 23 miles away. There is a breeze from the north at 6 miles per hour.
   a. Write an expression for the time required for the trip, in terms of the blimp's speed.
   b. Write an expression for the time needed for the return trip.
   c. Which trip takes longer? By how much time (in terms of the speed of the blimp)?

77. To grow enough hay for her horses a farmer needs to set aside a rectangular field of area 600 square yards.
   a. Write an expression for the length of the field in terms of its width.
   b. Write an expression for the perimeter of the field in terms of its width. Do all fields of area 600 square yards have the same perimeter?
   c. If fencing costs $10 per yard, write an expression for the cost of fencing the field.

78. A rectangular solar panel must have an area of 200 square feet in order to collect enough energy to heat a hot water tank.
   a. Write an expression for the width of the panel in terms of its length.
   b. Write an expression for the perimeter of the panel in terms of its length. Will the solar panel have the same perimeter for any dimensions that give an area of 200 square feet?
   c. Write an expression for the cost of the frame for the solar panel if the frame costs $5 per foot.

79. According to the building code, sleeping rooms in public residences such as prisons and dormitories must have a volume of 900 cubic feet. Assume the length of a dorm room is 2 feet longer than its width.
   a. Write an expression for the height of the room in terms of its width.
   b. Write an expression for the surface area of the room that must be painted. (Include the walls and ceiling but not the floor. Neglect doors and windows.)

80. You are to design a large freezer for residential use with a volume of 30 cubic feet. The width of the freezer should be 2 feet less than its length.
    a. Write an expression for the height of the freezer in terms of its length.
    b. Write an expression for the surface area of the freezer, which must be insulated.

2.2

PRODUCTS AND QUOTIENTS

Recall that the product of two fractions equals the product of their numerators divided by the product of their denominators:

$$\frac{a}{b} \cdot \frac{c}{d} = \frac{ac}{bd} \quad (b, d \neq 0). \tag{1}$$

For example,

$$\frac{6x^2}{y} \cdot \frac{xy}{2} = \frac{6x^2 \cdot xy}{y \cdot 2} = \frac{6x^3y}{2y}$$
$$= \frac{3x^3(2y)}{1(2y)} = 3x^3 \quad (y \neq 0).$$

We can arrive at the same result more directly by using slash lines:

$$\frac{6x^2}{y} \cdot \frac{xy}{2} = \frac{\overset{3}{\cancel{6}}x^2}{\cancel{y}} \cdot \frac{x\cancel{y}}{\cancel{2}} = 3x^3 \quad (y \neq 0).$$

If any of the numerators or denominators of the fractions have negative signs, it is best to proceed as if all the signs were positive and then attach the correct sign to the simplified product. If there is an even number of negative factors, the product is positive; if there is an odd number of negative factors, the product is negative.

CHAPTER 2 ■ RATIONAL EXPRESSIONS

EXAMPLE 1

a. $\dfrac{-x}{y^2} \cdot \dfrac{-2y^2}{x^2} = \dfrac{-\cancel{x}}{\cancel{y^2}} \cdot \dfrac{-2\cancel{y^2}}{\underset{x}{\cancel{x^2}}} = \dfrac{2}{x}$ $\quad (x, y \neq 0)$

b. $\dfrac{-4x^2}{3y} \cdot \dfrac{y}{2x} \cdot \dfrac{3}{5} = \dfrac{\overset{-2x}{\cancel{-4x^2}}}{\cancel{3y}} \cdot \dfrac{\cancel{y}}{\cancel{2x}} \cdot \dfrac{\cancel{3}}{5} = \dfrac{-2x}{5}$ $\quad (x, y \neq 0)$

c. $\dfrac{-4}{9}x^2 \cdot \dfrac{3}{4}x = \dfrac{\overset{-1}{\cancel{-4}}}{\underset{3}{\cancel{9}}}x^2 \cdot \dfrac{\cancel{3}}{\cancel{4}}x = \dfrac{-x^3}{3}$

d. $-\dfrac{2}{5}y \cdot \dfrac{5}{8}y^2 = -\dfrac{\cancel{2}}{\cancel{5}}y \cdot \dfrac{\cancel{5}}{\underset{4}{\cancel{8}}}y^2 = \dfrac{-y^3}{4}$

We can simplify the multiplication process by first factoring each numerator and denominator and then dividing out any common factors before applying Property (1).

EXAMPLE 2

a. $\dfrac{x^2 - 5x + 4}{3x} \cdot \dfrac{x}{x^2 - 1}$

$= \dfrac{(x-4)\cancel{(x-1)}}{3\cancel{x}} \cdot \dfrac{\cancel{x}}{(x+1)\cancel{(x-1)}}$

$= \dfrac{x-4}{3(x+1)}$ $\quad (x \neq -1, 0, 1)$

b. $\dfrac{4y^2 - 1}{4 - y^2} \cdot \dfrac{y^2 - 2y}{4y + 2}$

$= \dfrac{(2y-1)\cancel{(2y+1)}}{\cancel{(2-y)}(2+y)} \cdot \dfrac{\overset{-1}{y\cancel{(y-2)}}}{2\cancel{(2y+1)}}$

$= \dfrac{-y(2y-1)}{2(y+2)}$ $\quad \left(y \neq -2, -\dfrac{1}{2}, 2\right)$

2.2 ■ PRODUCTS AND QUOTIENTS

c. $\dfrac{x^3 - 8}{x^2 + 4x + 4} \cdot \dfrac{x^3 + 2x^2}{x^2 - 2x}$

$= \dfrac{\cancel{(x-2)}(x^2+2x+4)}{\cancel{(x+2)}(x+2)} \cdot \dfrac{\overset{x}{\cancel{x^2}}\cancel{(x+2)}}{\cancel{x}\cancel{(x-2)}}$

$= \dfrac{x(x^2 + 2x + 4)}{x + 2} \qquad (x \neq -2, 0, 2)$

d. $\dfrac{2xy + 4x + y + 2}{2x^2 - x - 1} \cdot \dfrac{(x-1)^2}{y^2 - 4}$

$= \dfrac{2x(y+2) + (y+2)}{(2x+1)(x-1)} \cdot \dfrac{(x-1)(x-1)}{(y-2)(y+2)}$

$= \dfrac{\cancel{(2x+1)}\cancel{(y+2)}}{\cancel{(2x+1)}\cancel{(x-1)}} \cdot \dfrac{(x-1)\cancel{(x-1)}}{(y-2)\cancel{(y+2)}}$

$= \dfrac{x-1}{y-2} \qquad \left(x \neq -\dfrac{1}{2}, 1; \ y \neq -2, 2\right)$

Fractions may occur in products that require the use of the distributive law.

EXAMPLE 3 a. $\dfrac{2}{3}x\left(\dfrac{1}{2}x - 9\right)$

$= \dfrac{\cancel{2}}{3}x\left(\dfrac{1}{\cancel{2}}x\right) - \dfrac{2}{\cancel{3}}x(\overset{3}{\cancel{9}})$

$= \dfrac{1}{3}x^2 - 6x$

b. $\left(x - \dfrac{1}{2}\right)^2 = \left(x - \dfrac{1}{2}\right)\left(x - \dfrac{1}{2}\right)$

$= x^2 - \dfrac{1}{2}x - \dfrac{1}{2}x + \dfrac{1}{4}$

$= x^2 - x + \dfrac{1}{4}$

CHAPTER 2 ■ RATIONAL EXPRESSIONS

Quotients of Fractions

To divide two fractions we multiply the first fraction by the reciprocal of the second fraction:

$$\frac{a}{b} \div \frac{c}{d} = \frac{a}{b} \cdot \frac{d}{c} \quad (b, c, d \neq 0). \tag{2}$$

EXAMPLE 4

a. $\dfrac{2x^3}{3y} \div \dfrac{4x}{5y^2}$

$= \dfrac{\cancel{2x^3}^{x^2}}{\cancel{3y}} \cdot \dfrac{\cancel{5y^2}^{y}}{\cancel{4x}_{2}}$

$= \dfrac{5x^2 y}{6} \quad (x, y \neq 0)$

b. $\dfrac{x^2 - 1}{x + 3} \div \dfrac{x^2 - x - 2}{x^2 + 5x + 6}$

$= \dfrac{(x - 1)\cancel{(x+1)}}{\cancel{x+3}} \cdot \dfrac{\cancel{(x+3)}(x + 2)}{\cancel{(x+1)}(x - 2)}$

$= \dfrac{(x - 1)(x + 2)}{x - 2}$

$= \dfrac{x^2 + x - 2}{x - 2} \quad (x \neq -3, -2, -1, 2)$

c. $\dfrac{4x - 6}{2x^2 - x - 1} \div \dfrac{8x^3 - 27}{4x^2 + 4x + 1}$

$= \dfrac{2\cancel{(2x-3)}}{\cancel{(2x+1)}(x - 1)} \cdot \dfrac{\cancel{(2x+1)}(2x + 1)}{\cancel{(2x-3)}(4x^2 + 6x + 9)}$

$= \dfrac{2(2x + 1)}{(x - 1)(4x^2 + 6x + 9)} \quad \left(x \neq -\dfrac{1}{2}, 1, \dfrac{3}{2}\right)$

d. $\dfrac{xy - 2y + x - 2}{y^2 + y} \div \dfrac{x^2 - 4x + 4}{4y^2 - 2y}$

$= \dfrac{y(x - 2) + (x - 2)}{y(y + 1)} \cdot \dfrac{2y(2y - 1)}{(x - 2)(x - 2)}$

$= \dfrac{\cancel{(x - 2)}\cancel{(y + 1)}}{\cancel{y}\cancel{(y + 1)}} \cdot \dfrac{2\cancel{y}(2y - 1)}{\cancel{(x - 2)}(x - 2)}$

$= \dfrac{2(2y - 1)}{x - 2} \quad \left(x \ne 2;\ y \ne -1, 0, \dfrac{1}{2}\right)$

As special cases of Property (2) observe that

$$a \div \dfrac{c}{d} = \dfrac{a}{1} \cdot \dfrac{d}{c} = \dfrac{ad}{c}, \qquad \dfrac{a}{b} \div c = \dfrac{a}{b} \cdot \dfrac{1}{c} = \dfrac{a}{bc},$$

and

$$1 \div \dfrac{a}{b} = 1 \cdot \dfrac{b}{a} = \dfrac{b}{a}.$$

EXAMPLE 5   a. $2x \div \dfrac{x}{y}$ b. $\dfrac{x}{y} \div 2x$ c. $1 \div \dfrac{x}{y}$

$= 2\cancel{x} \cdot \dfrac{y}{\cancel{x}} \qquad\qquad\ \ = \dfrac{\cancel{x}}{y} \cdot \dfrac{1}{2\cancel{x}} \qquad\quad = 1 \cdot \dfrac{y}{x}$

$= 2y \qquad\qquad\qquad\ \ = \dfrac{1}{2y} \qquad\qquad\qquad = \dfrac{y}{x}$

$(x, y \ne 0) \qquad\qquad\quad (x, y \ne 0) \qquad\qquad (x, y \ne 0)$

Polynomial Division

If a quotient of two polynomials cannot be reduced, we can simplify the expression by treating it as a division. We consider two cases: when the divisor is a monomial and when the divisor contains more than one term.

If the divisor is a monomial, we divide the monomial into each term of the numerator, that is,

$$\frac{a+b+c}{d} = \frac{a}{d} + \frac{b}{d} + \frac{c}{d}.$$

We then reduce each fraction that can be reduced.

EXAMPLE 6   a. $\dfrac{6y^2 + 4y + 1}{2}$

$= \dfrac{6y^2}{2} + \dfrac{4y}{2} + \dfrac{1}{2}$

$= 3y^2 + 2y + \dfrac{1}{2}$

b. $\dfrac{9x^3 - 6x^2 + 4}{3x}$

$= \dfrac{9x^3}{3x} - \dfrac{6x^2}{3x} + \dfrac{4}{3x}$

$= 3x^2 - 2x + \dfrac{4}{3x}$ $(x \neq 0)$

If the denominator is not a monomial, we can use a method similar to the long division algorithm used in arithmetic.

EXAMPLE 7   Divide $\dfrac{2x^2 + x - 7}{x + 3}$.

Solution   First, write

$$x + 3 \,\overline{\smash{)}\, 2x^2 + x - 7},$$

and divide $2x^2$ (the first term of $2x^2 + x - 7$) by x (the first term of $x + 3$) to obtain $2x$. (It may be helpful to write down the division: $2x^2/x = 2x$.) Write $2x$ as the first term of the quotient. Next, multiply $x + 3$ by $2x$ to obtain $2x^2 + 6x$, and subtract this product from $2x^2 + x - 7$:

$$\begin{array}{r} 2x \\ x + 3 \,\overline{\smash{)}\, 2x^2 + x - 7} \\ -\underline{(2x^2 + 6x)} \\ -5x - 7 \end{array}$$

Repeating this process, divide $-5x$ by x to obtain -5. Write -5 as the second term of the quotient. Next, multiply $x + 3$ by -5 to obtain $-5x - 15$, and subtract:

$$\begin{array}{r}
2x - 5 \\
x + 3 \overline{\smash{)}\, 2x^2 + x - 7} \\
-\underline{(2x^2 + 6x)} \\
-5x - 7 \\
-\underline{(-5x - 15)} \\
8
\end{array}$$

Since the degree of 8 is less than the degree of $x + 3$, the division is finished. The quotient is $2x - 5$, with a remainder of 8. We write the remainder as a fraction to obtain

$$\frac{2x^2 + x - 7}{x + 3} = 2x - 5 + \frac{8}{x + 3} \quad (x \neq 3).$$

When using long division it helps to write the polynomials in descending powers of the variable. If the polynomial being divided has no term of a certain power, we can insert a term with zero coefficient so that like terms will be aligned.

EXAMPLE 8 Divide $\dfrac{3x - 1 + 4x^3}{2x - 1}$.

Solution First, write $3x - 1 + 4x^3$ in descending powers as $4x^3 + 3x - 1$. Then insert $0x^2$ between $4x^3$ and $3x$ and divide as follows:

$$\begin{array}{r}
2x^2 + x + 2 \\
2x - 1 \overline{\smash{)}\, 4x^3 + 0x^2 + 3x - 1} \\
-\underline{(4x^3 - 2x^2)} \\
2x^2 + 3x \\
-\underline{(2x^2 - x)} \\
4x - 1 \\
-\underline{(4x - 2)} \\
1
\end{array}$$

Thus,

$$\frac{3x - 1 + 4x^3}{2x - 1} = 2x^2 + x + 2 + \frac{1}{2x - 1} \quad \left(x \neq \frac{1}{2}\right).$$

The following example illustrates polynomial division by a quadratic divisor.

EXAMPLE 9 Divide $\dfrac{z^4 - 3z^3 + 2z^2 - 3z + 1}{z^2 + 2z - 1}$.

Solution

$$\begin{array}{r}
z^2 - 5z + 13 \\
z^2 + 2z - 1 \overline{\smash{)}\, z^4 - 3z^3 + 2z^2 - 3z + 1} \\
-\underline{(z^4 + 2z^3 - z^2)} \\
-5z^3 + 3z^2 - 3z \\
-\underline{(-5z^3 - 10z^2 + 5z)} \\
13z^2 - 8z + 1 \\
-\underline{(13z^2 + 26z - 13)} \\
-34z + 14
\end{array}$$

Hence,

$$\frac{z^4 - 3z^3 + 2z^2 - 3z + 1}{z^2 + 2z - 1} = z^2 - 5z + 13 + \frac{-34z + 14}{z^2 + 2z - 1}.$$

EXERCISE 2.2

A

■ *Write each product as a single fraction in lowest terms. See Example 1.*

1. $\dfrac{24}{3} \cdot \dfrac{20}{36} \cdot \dfrac{3}{4}$

2. $\dfrac{3}{10} \cdot \dfrac{16}{27} \cdot \dfrac{30}{36}$

3. $\dfrac{15n^2}{3p} \cdot \dfrac{5p^2}{n^3}$

4. $\dfrac{21t^2}{5s} \cdot \dfrac{15s^3}{7st}$

5. $\dfrac{-4}{3np} \cdot \dfrac{6n^2p^3}{16}$

6. $\dfrac{14a^3b}{3b} \cdot \dfrac{-6}{7a^2}$

7. $\dfrac{1}{3}x^2 \cdot \dfrac{6}{7}x^3$

8. $\dfrac{2}{3}y \cdot \dfrac{9}{10}y^2$

9. $\dfrac{3}{4}x^2y \cdot \dfrac{2}{3}xy^2$

10. $\dfrac{1}{4}x^3y \cdot \dfrac{2}{5}xy$

11. $-\dfrac{1}{2}xyz^2 \cdot \dfrac{2}{3}x^2y$

12. $-\dfrac{3}{5}x^2y \cdot \dfrac{5}{6}xy^2z$

13. $\dfrac{-12a^2b}{5c} \cdot \dfrac{10b^2c}{24a^3b}$

14. $\dfrac{a^2}{xy} \cdot \dfrac{3x^3y}{4a}$

15. $\dfrac{-2ab}{7c} \cdot \dfrac{3c^2}{4a^3} \cdot \dfrac{-6a}{15b^2}$

16. $\dfrac{10x}{12y} \cdot \dfrac{3x^2z}{5x^3z} \cdot \dfrac{6y^2x}{3yz}$

17. $5a^2b^2 \cdot \dfrac{1}{a^3b^3}$

18. $15x^2y \cdot \dfrac{3}{45xy^2}$

2.2 ■ PRODUCTS AND QUOTIENTS

■ See Example 2.

19. $\dfrac{5x + 25}{2x} \cdot \dfrac{4x}{2x + 10}$

20. $\dfrac{3y}{4xy - 6y^2} \cdot \dfrac{2x - 3y}{12x}$

21. $\dfrac{4a^2 - 1}{a^2 - 16} \cdot \dfrac{a^2 - 4a}{2a + 1}$

22. $\dfrac{9x^2 - 25}{2x - 2} \cdot \dfrac{x^2 - 1}{6x - 10}$

23. $\dfrac{2x^2 - x - 6}{3x^2 - 4x + 1} \cdot \dfrac{3x^2 + 7x + 2}{2x^2 + 7x + 6}$

24. $\dfrac{3x^2 - 7x - 6}{2x^2 - x - 1} \cdot \dfrac{2x^2 - 9x - 5}{3x^2 - 13x - 10}$

25. $\dfrac{7a + 14}{14a - 28} \cdot \dfrac{4 + 2a - 2a^2}{2a^2 + 2a - 4} \cdot \dfrac{4a - 12}{7a + 7}$

26. $\dfrac{5x^2 - 5x}{10x - 2} \cdot \dfrac{x^2 - 9x - 10}{4x - 40} \cdot \dfrac{x^2 - 2x}{2 - 2x^2}$

27. $\dfrac{3x^4 - 48}{x^4 - 4x^2 - 32} \cdot \dfrac{4x^4 - 8x^3 + 4x^2}{2x^4 + 16x}$

28. $\dfrac{x^4 - 3x^3}{x^4 + 6x^2 - 27} \cdot \dfrac{x^4 - 81}{3x^4 - 81x}$

29. $\dfrac{4u^2 - 16v^2}{u^2 + 2uv - 2u - 4v} \cdot \dfrac{2u^3v^2 - 8uv^2}{u^3v - u^2v^2 - 6uv^3}$

30. $\dfrac{a^2 + 3ax - 3a - 9x}{3a^4 - 27a^2x^2} \cdot \dfrac{3a^4 - 11a^3x + 6a^2x^2}{3a^2x^2 - 9ax^2}$

■ Write each product as a polynomial. See Example 3.

31. $\dfrac{1}{2}x\left(\dfrac{2}{5}x - 6\right)$

32. $\dfrac{3}{4}y\left(\dfrac{1}{6}y + 8\right)$

33. $\left(x + \dfrac{1}{3}\right)\left(x + \dfrac{1}{3}\right)$

34. $\left(y - \dfrac{1}{3}\right)\left(y - \dfrac{1}{3}\right)$

35. $\left(y - \dfrac{1}{4}\right)^2$

36. $\left(y + \dfrac{1}{4}\right)^2$

■ Write each quotient as a single fraction in lowest terms. See Example 4.

37. $\dfrac{4x - 8}{3y} \div \dfrac{6x - 12}{y}$

38. $\dfrac{6y - 27}{5x} \div \dfrac{4y - 18}{x}$

39. $\dfrac{a^2 - a - 6}{a^2 + 2a - 15} \div \dfrac{a^2 - 4}{a^2 + 6a + 5}$

40. $\dfrac{a^2 + 2a - 15}{a^2 + 3a - 10} \div \dfrac{a^2 - 9}{a^2 - 9a + 14}$

41. $\dfrac{10x^2 - 13x - 3}{2x^2 - x - 3} \div \dfrac{5x^2 - 9x - 2}{3x^2 + 2x - 1}$

42. $\dfrac{9x^2 + 3x - 2}{12x^2 + 5x - 2} \div \dfrac{9x^2 - 6x + 1}{8x^2 + 10x - 3}$

43. $\dfrac{x^3 + y^3}{x} \div \dfrac{x + y}{3x}$

44. $\dfrac{8x^3 - y^3}{x + y} \div \dfrac{2x - y}{x^2 - y^2}$

45. $\dfrac{xy - 3x + y - 3}{x^2 - x - 2} \div \dfrac{x^2 - 2x - 3}{x^2 - 4}$

46. $\dfrac{2xy + 4x + 3y + 6}{2x^2 + x - 3} \div \dfrac{y^2 + 4y + 4}{y^2 - 3y + 2}$

■ See Example 5.

47. $1 \div \dfrac{x^2 - 1}{x + 2}$

48. $1 \div \dfrac{x^2 + 3x + 1}{x - 2}$

49. $(x^2 - 5x + 4) \div \dfrac{x^2 - 1}{x^2}$

50. $(x^2 - 9) \div \dfrac{x^2 - 6x + 9}{3x}$

51. $\dfrac{x^2 + 3x}{2y} \div 3x$

52. $\dfrac{2y^2 + y}{3x} \div 2y$

■ Divide. See Example 6.

53. $\dfrac{18r^2s^2 - 15rs + 6}{3rs}$

54. $\dfrac{8a^2x^2 - 4ax^2 + ax}{2ax}$

55. $\dfrac{15s^{10} - 21s^5 + 6}{-3s^2}$

56. $\dfrac{25m^6 - 15m^4 + 7}{-5m^3}$ 57. $\dfrac{9a^2b^2 + 3ab^2 + 4a^2b}{ab^2}$ 58. $\dfrac{36s^4t^5 + 24s^3t^3 - s^2t}{12st^2}$

■ See Examples 7 and 8.

59. $\dfrac{4y^2 + 12y + 7}{2y + 1}$ 60. $\dfrac{4t^2 - 4t - 5}{2t - 1}$

61. $\dfrac{x^3 + 2x^2 + x + 1}{x - 2}$ 62. $\dfrac{2x^3 - 3x^2 - 2x + 4}{x + 1}$

63. $\dfrac{4z^2 + 5z + 8z^4 + 3}{2z + 1}$ 64. $\dfrac{7 - 3t^3 - 23t^2 + 10t^4}{2t + 3}$

65. $\dfrac{x^4 - 1}{x - 2}$ 66. $\dfrac{y^5 + 1}{y - 1}$

B

■ Divide. See Example 9.

67. $\dfrac{x^3 - 3x^2 + 2x + 5}{x^2 - 2x + 7}$ 68. $\dfrac{2y^3 + 5y^2 - 3y + 2}{y^2 - y - 3}$ 69. $\dfrac{4a^4 + 3a^3 - 2a + 1}{a^2 + 3a - 1}$

70. $\dfrac{2b^4 - 3b^2 + b + 2}{b^2 + b - 3}$ 71. $\dfrac{t^4 - 3t^3 + 2t^2 - 2t + 1}{t^3 - 2t^2 + t + 2}$ 72. $\dfrac{r^4 + r^3 - 2r^2 + r + 5}{r^3 + 2r + 3}$

73. Determine k so that the polynomial $x^3 - 3x + k$ has $x - 2$ as a factor.
74. Determine k so that the polynomial $x^3 + 2x^2 + k$ has $x + 3$ as a factor.

2.3

SUMS AND DIFFERENCES

Fractions with Like Denominators

To add or subtract fractions with the same denominator, combine their numerators and use the same denominator in the sum or difference:

$$\frac{a}{c} + \frac{b}{c} = \frac{a + b}{c}$$

and

$$\frac{a}{c} - \frac{b}{c} = \frac{a - b}{c} \qquad (c \neq 0). \tag{1}$$

2.3 ■ SUMS AND DIFFERENCES

EXAMPLE 1 a. $\dfrac{2x}{9} + \dfrac{5x}{9} = \dfrac{2x + 5x}{9}$

$= \dfrac{7x}{9}$

b. $\dfrac{a + 1}{b} - \dfrac{a - 1}{b} = \dfrac{a + 1 - (a - 1)}{b}$

$= \dfrac{a + 1 - a + 1}{b} = \dfrac{2}{b} \quad (b \neq 0)$

Common Error Note that in Example 1b $-(a - 1) \neq -a - 1$.

Building Fractions If the fractions in a sum or difference have different denominators we must first *build* the fractions to equivalent fractions with a common denominator. We can then combine them as above.

Building fractions is the opposite of reducing fractions in the sense that we multiply, rather than divide, the numerator and denominator by an appropriate factor. To find the appropriate **building factor** we compare the factors of the original denominator with the factors of the desired denominator.

EXAMPLE 2 Build each fraction to an equivalent fraction with the given denominator.

a. $\dfrac{5x}{3y} = \dfrac{?}{12y^2}$ b. $\dfrac{2}{y - 1} = \dfrac{?}{y^2 - 1}$

Solutions a. We want a building factor (BF) such that

$$3y \cdot (\text{BF}) = 12y^2.$$

The building factor is $12y^2 \div 3y = 4y.$ Hence,

$$\dfrac{5x}{3y} = \dfrac{5x(4y)}{3y(4y)} = \dfrac{20xy}{12y^2} \quad (y \neq 0).$$

b. First, factor $y^2 - 1$ as $(y - 1)(y + 1)$. We want a building factor (BF) such that

$$(y - 1) \cdot (\text{BF}) = (y - 1)(y + 1).$$

(continued)

By inspection we see that the building factor is $y + 1$ Hence,

$$\frac{2}{y-1} = \frac{2(y+1)}{(y-1)(y+1)} = \frac{2y+2}{y^2-1} \qquad (y \neq 1, -1).$$

Least Common Denominator

Recall that to add arithmetic fractions we use as a common denominator the smallest natural number that is exactly divisible by each of the given denominators. For example, to add the fractions ⅙ and ⅜ we use 24 as the common denominator, because 24 is the smallest natural number that is exactly divisible by both 6 and 8.

We define the **least common denominator (LCD)** of two or more algebraic fractions as the polynomial of least degree that is exactly divisible by each of the given denominators.

To Find the LCD of Algebraic Fractions:

1. Factor each denominator completely.
2. Include as factors in the LCD each different factor the greatest number of times it occurs in any *one* of the given denominators.

EXAMPLE 3 Find the LCD of $\dfrac{1}{x^2-9}$ and $\dfrac{1}{x^2-6x+9}$.

Solution Factor each denominator:

$$x^2 - 9 = (x+3)(x-3);$$
$$x^2 - 6x + 9 = (x-3)(x-3).$$

Since the factor $(x + 3)$ occurs once in the first denominator, we include one factor $(x + 3)$ in the LCD. Since $(x - 3)$ occurs as a factor twice in the second denominator, we include two factors $(x - 3)$ in the LCD. Thus, the LCD is $(x - 3)^2(x + 3)$.

2.3 ■ SUMS AND DIFFERENCES

Fractions with Unlike Denominators

We can now add or subtract fractions with unlike denominators. We will do this in four steps.

To Add or Subtract Fractions with Unlike Denominators

1. Find the LCD for the given fractions.
2. Build each fraction to an equivalent fraction with the LCD as its denominator.
3. Add or subtract the numerators of the resulting like fractions. Use the LCD as the denominator of the sum.
4. Reduce the answer, if possible.

EXAMPLE 4 $\frac{1}{3}x - \frac{3}{8}x + \frac{1}{2}x$

Solution First, find the LCD: $3 \cdot 2 \cdot 2 \cdot 2 = 24$. Next, build each fraction to an equivalent fraction with this denominator:

$$\frac{8}{8} \cdot \frac{1}{3}x = \frac{8}{24}x; \quad \frac{3}{3} \cdot \frac{3}{8}x = \frac{9}{24}x; \quad \frac{12}{12} \cdot \frac{1}{2}x = \frac{12}{24}x.$$

Finally, combine the fractions to get

$$\frac{8}{24}x - \frac{9}{24}x + \frac{12}{24}x = \frac{8 - 9 + 12}{24}x = \frac{11}{24}x.$$

EXAMPLE 5 $\frac{3x}{x + 2} - \frac{2x}{x - 3}$

Solution By inspection, the LCD is $(x + 2)(x - 3)$. Build each fraction to an equivalent fraction with this denominator:

$$\frac{3x}{x + 2} = \frac{3x(x - 3)}{(x + 2)(x - 3)} = \frac{3x^2 - 9x}{(x - 3)(x + 2)};$$

$$\frac{2x}{x - 3} = \frac{2x(x + 2)}{(x - 3)(x + 2)} = \frac{2x^2 + 4x}{(x - 3)(x + 2)}.$$

(continued)

Combine the fractions to obtain

$$\frac{3x}{x+2} - \frac{2x}{x-3} = \frac{3x^2 - 9x - (2x^2 + 4x)}{(x-3)(x+2)}$$

$$= \frac{3x^2 - 9x - 2x^2 - 4x}{(x-3)(x+2)}$$

$$= \frac{x^2 - 13x}{(x-3)(x+2)} \quad (x \neq -2, 3).$$

The following example illustrates the addition of fractions whose denominators must be factored in order to find their LCD.

EXAMPLE 6 $\quad \dfrac{3}{x^2 - 4} - \dfrac{1}{x^2 - 5x + 6} + \dfrac{1}{x - 3}$

Solution In order to find the LCD, first factor each denominator:

$$x^2 - 4 = (x - 2)(x + 2);$$
$$x^2 - 5x + 6 = (x - 3)(x - 2);$$
$$x - 3 = x - 3.$$

Hence, the LCD is $(x - 3)(x - 2)(x + 2)$. Next, build each fraction to an equivalent fraction with this denominator:

$$\frac{3}{x^2 - 4} = \frac{3(x - 3)}{(x + 2)(x - 2)(x - 3)} = \frac{3x - 9}{(x - 3)(x - 2)(x + 2)};$$

$$\frac{1}{(x - 2)(x - 3)} = \frac{1(x + 2)}{(x - 2)(x - 3)(x + 2)} = \frac{x + 2}{(x - 3)(x - 2)(x + 2)};$$

$$\frac{1}{x - 3} = \frac{1(x - 2)(x + 2)}{(x - 3)(x - 2)(x + 2)} = \frac{x^2 - 4}{(x - 3)(x - 2)(x + 2)}.$$

Finally, combine the fractions to obtain

$$\frac{3}{x^2 - 4} - \frac{1}{x^2 - 5x + 6} + \frac{1}{x - 3}$$

$$= \frac{3x - 9 - (x + 2) + (x^2 - 4)}{(x - 3)(x - 2)(x + 2)}$$

$$= \frac{3x - 9 - x - 2 + x^2 - 4}{(x - 3)(x - 2)(x + 2)}$$

$$= \frac{x^2 + 2x - 15}{(x - 3)(x - 2)(x + 2)}$$

$$= \frac{(x + 5)\cancel{(x - 3)}}{\cancel{(x - 3)}(x - 2)(x + 2)}$$

$$= \frac{x + 5}{(x - 2)(x + 2)} \qquad (x \neq -2, 2, 3).$$

Note that we were able to reduce the answer in the example above; answers should be reduced whenever possible. We usually leave the denominator in factored form, in case further computation is necessary.

EXAMPLE 7 $\quad x - \dfrac{1}{x - 2} + \dfrac{1}{(x - 2)^2}$

Solution Regard the first term as $x/1$ to see that the LCD is $(x - 2)^2$. Build each fraction to an equivalent fraction with this denominator:

$$\frac{x}{1} = \frac{x(x - 2)^2}{1(x - 2)^2} = \frac{x^3 - 4x^2 + 4x}{(x - 2)^2};$$

$$\frac{1}{x - 2} = \frac{1(x - 2)}{(x - 2)(x - 2)} = \frac{(x - 2)}{(x - 2)^2};$$

$$\frac{1}{(x - 2)^2} = \frac{1}{(x - 2)^2}.$$

Combine the fractions to get

$$x - \frac{1}{x - 2} + \frac{1}{(x - 2)^2}$$

$$= \frac{x^3 - 4x^2 + 4x - (x - 2) + 1}{(x - 2)^2}$$

$$= \frac{x^3 - 4x^2 + 4x - x + 2 + 1}{(x - 2)^2}$$

$$= \frac{x^3 - 4x^2 + 3x + 3}{(x - 2)^2} \qquad (x \neq 2).$$

Combined Operations

Some fractional expressions involve more than one operation. When simplifying such expressions, we follow the order of operations on page 5.

EXAMPLE 8

$$\left(\frac{1}{x} - \frac{1}{x+1}\right) \cdot \frac{x^3 - x}{x^2 + 1} = \left[\frac{(x+1) - x}{x(x+1)}\right] \cdot \frac{x^3 - x}{x^2 + 1}$$

$$= \frac{1}{\cancel{x(x+1)}} \cdot \frac{\cancel{x(x+1)}(x-1)}{x^2 + 1}$$

$$= \frac{x - 1}{x^2 + 1} \quad (x \neq -1, 0)$$

EXERCISE 2.3

A

■ Write each sum or difference as a single fraction in lowest terms. See Example 1.

1. $\dfrac{x}{2} - \dfrac{3}{2}$

2. $\dfrac{y}{7} - \dfrac{5}{7}$

3. $\dfrac{1}{6}a + \dfrac{1}{6}b - \dfrac{5}{6}c$

4. $\dfrac{1}{3}x - \dfrac{2}{3}y + \dfrac{1}{3}z$

5. $\dfrac{x}{2y} + \dfrac{1}{2y} + \dfrac{x}{2y}$

6. $\dfrac{y+1}{b} + \dfrac{y-1}{b}$

7. $\dfrac{3}{x+2y} - \dfrac{x+3}{x+2y} - \dfrac{x-1}{x+2y}$

8. $\dfrac{2}{a-3b} - \dfrac{b-2}{a-3b} + \dfrac{b}{a-3b}$

9. $\dfrac{a+1}{a^2 - 2a + 1} - \dfrac{5 - 3a}{a^2 - 2a + 1}$

10. $\dfrac{x+4}{x^2 - x + 2} - \dfrac{2x - 3}{x^2 - x + 2}$

■ Express each fraction as an equivalent fraction with the given denominator. See Example 2.

11. $\dfrac{2}{6x} = \dfrac{?}{18x}$

12. $\dfrac{5}{3y} = \dfrac{?}{21y}$

13. $\dfrac{-a^2}{b} = \dfrac{?}{b^3}$

14. $\dfrac{-a}{b} = \dfrac{?}{ab^2}$

15. $y = \dfrac{?}{xy}$

16. $x = \dfrac{?}{xy^3}$

17. $\dfrac{3y}{y + 2} = \dfrac{?}{y^2 - y - 6}$

18. $\dfrac{5y}{y + 3} = \dfrac{?}{y^2 + y - 6}$

19. $\dfrac{3}{a - b} = \dfrac{?}{b^2 - a^2}$

20. $\dfrac{5}{2a + b} = \dfrac{?}{b^2 - 4a^2}$

■ *Find the LCD for each set of fractions. See Example 3.*

21. $\dfrac{5}{6(x+y)^2}, \dfrac{3}{4xy^2}$

22. $\dfrac{1}{8(a-b)^2}, \dfrac{5}{12a^2b^2}$

23. $\dfrac{2a}{a^2+5a+4}, \dfrac{2}{(a+1)^2}$

24. $\dfrac{3x}{x^2-3x+2}, \dfrac{3}{(x-1)^2}$

25. $\dfrac{x+2}{x^2-x}, \dfrac{x+1}{(x-1)^3}$

26. $\dfrac{y-1}{y^2+2y}, \dfrac{y-3}{(y+2)^2}$

27. $\dfrac{1}{6x^3}, \dfrac{x}{4x^2-4x}, \dfrac{x}{(x-1)^2}$

28. $\dfrac{1}{9y}, \dfrac{5y}{6y^3-6y}, \dfrac{y}{(y-1)^3}$

■ *Write each sum or difference as a single fraction in lowest terms. See Example 4.*

29. $\dfrac{x}{2}+\dfrac{2x}{3}$

30. $\dfrac{3y}{4}+\dfrac{y}{3}$

31. $\dfrac{2x}{3}-\dfrac{3x}{4}+\dfrac{x}{2}$

32. $\dfrac{y}{2}+\dfrac{2y}{3}-\dfrac{3y}{4}$

33. $\dfrac{5}{6}y-\dfrac{3}{4}y$

34. $\dfrac{3}{4}x-\dfrac{1}{6}x$

35. $\dfrac{2}{3}y-\dfrac{1}{6}y+\dfrac{1}{4}y$

36. $\dfrac{3}{4}y+\dfrac{1}{3}y-\dfrac{5}{6}y$

■ *See Example 5.*

37. $\dfrac{x+1}{2x}+\dfrac{2y-1}{3y}$

38. $\dfrac{y-2}{4y}+\dfrac{2x-3}{3x}$

39. $\dfrac{5}{x+1}+\dfrac{3}{x-1}$

40. $\dfrac{2}{y+2}+\dfrac{3}{y-2}$

41. $\dfrac{y}{2y-1}-\dfrac{2y}{y+1}$

42. $\dfrac{2x}{3x+1}-\dfrac{x}{x-2}$

43. $\dfrac{y-1}{y+1}-\dfrac{y-2}{2y-3}$

44. $\dfrac{x-2}{2x+1}-\dfrac{x+1}{x-1}$

■ *See Example 6.*

45. $\dfrac{7}{5x-10}-\dfrac{5}{3x-6}$

46. $\dfrac{2}{3y+6}-\dfrac{3}{2y+4}$

47. $\dfrac{2}{x^2-x-2}+\dfrac{2}{x^2+2x+1}$

48. $\dfrac{1}{y^2-1}+\dfrac{1}{y^2+2y+1}$

49. $\dfrac{y}{y^2-16}-\dfrac{y+1}{y^2-5y+4}$

50. $\dfrac{x}{x^2-5x+6}-\dfrac{x-1}{x^2-9}$

51. $\dfrac{y-1}{y^2-3y}-\dfrac{y+1}{y^2+2y}$

52. $\dfrac{x+1}{x^2+2x}-\dfrac{x-1}{x^2-3x}$

53. $\dfrac{2x+1}{x^2-4}-\dfrac{3x-2}{x^2-4x+4}$

54. $\dfrac{3y-1}{y^2-4y+3}-\dfrac{y+2}{(y-3)^2}$

55. $\dfrac{1}{z^2-7z+12}+\dfrac{2}{z^2-5z+6}-\dfrac{3}{z^2-6z+8}$

56. $\dfrac{4}{a^2-4} + \dfrac{2}{a^2+3a+2} + \dfrac{4}{a^2-a-2}$

57. $\dfrac{a+2}{a^2-6a+8} + \dfrac{3a-8}{a^2-5a+6} + \dfrac{2a-5}{a^2-7a+12}$

58. $\dfrac{2z+5}{z^2+5z+4} + \dfrac{z+13}{z^2-z-20} + \dfrac{z+7}{z^2-4z-5}$

■ See Example 7.

59. $x - \dfrac{1}{x}$

60. $1 + \dfrac{1}{y}$

61. $x + \dfrac{1}{x-1} - \dfrac{1}{(x-1)^2}$

62. $y - \dfrac{2}{y^2-1} + \dfrac{3}{y+1}$

63. $y - \dfrac{y^2}{y-1} + \dfrac{y^2}{y+1}$

64. $x + \dfrac{2x^2}{x+2} - \dfrac{3x^2}{x-1}$

65. $x - 1 + \dfrac{3}{x+2}$

66. $x + 3 + \dfrac{1}{x-1}$

■ Write each expression as a single fraction in lowest terms. See Example 8.

67. $\left(\dfrac{5}{x+5} - \dfrac{4}{x+4}\right) \cdot \dfrac{x+4}{x}$

68. $\left(\dfrac{x}{x^2+1} - \dfrac{1}{x+1}\right) \cdot \dfrac{x+1}{x-1}$

69. $\left(\dfrac{5}{x^2-9} + 1\right) \div \dfrac{x+2}{x-3}$

70. $\left(\dfrac{x}{x+1} + \dfrac{1}{x-1}\right) \div \dfrac{x^2+1}{x+1}$

2.4

COMPLEX FRACTIONS

A fraction that contains one or more fractions in either its numerator or its denominator or both is called a **complex fraction.** For example,

$$\dfrac{\dfrac{2}{3}}{\dfrac{5}{6}} \quad \text{and} \quad \dfrac{x + \dfrac{3}{4}}{x - \dfrac{1}{2}}$$

are complex fractions. Like simple fractions, complex fractions represent quotients. For example,

$$\dfrac{\dfrac{2}{3}}{\dfrac{5}{6}} = \dfrac{2}{3} \div \dfrac{5}{6} \tag{1}$$

2.4 ■ COMPLEX FRACTIONS

and

$$\frac{x + \frac{3}{4}}{x - \frac{1}{2}} = \left(x + \frac{3}{4}\right) \div \left(x - \frac{1}{2}\right) \qquad \left(x \neq \frac{1}{2}\right). \tag{2}$$

In cases like Equation (1), in which the denominator of the complex fraction is a single term, we can treat the fraction as a division problem and multiply the numerator by the reciprocal of the denominator.

EXAMPLE 1 a. $\dfrac{\frac{2}{3}}{\frac{5}{6}} = \dfrac{2}{3} \div \dfrac{5}{6}$

$\qquad\qquad = \dfrac{2}{\cancel{3}} \cdot \dfrac{\cancel{6}^2}{5} = \dfrac{4}{5}$

b. $\dfrac{\frac{3ab}{2c^2}}{\frac{6a}{b^2c}} = \dfrac{3ab}{2c^2} \div \dfrac{6a}{b^2c}$

$\qquad = \dfrac{\cancel{3ab}}{2\cancel{c^2}} \cdot \dfrac{b^2\cancel{c}}{\cancel{6a}} = \dfrac{b^3}{4c}$

$(a, b, c \neq 0)$

In a complex fraction of the form (2), in which the numerator or denominator contains sums or differences, it is more convenient to use the fundamental principle of fractions to simplify the expression. In fact, we can use the fundamental principle of fractions to simplify any complex fraction.

EXAMPLE 2 Simplify $\dfrac{\frac{2}{3}}{\frac{5}{6}}$ by using the fundamental principle of fractions.

Solution Multiply the numerator, $\frac{2}{3}$, and the denominator, $\frac{5}{6}$, by 6, the LCD of the two fractions. This gives

$$\dfrac{\frac{2}{3}}{\frac{5}{6}} = \dfrac{\frac{2}{\cancel{3}}(\cancel{6})^2}{\frac{5}{\cancel{6}}(\cancel{6})} = \dfrac{4}{5},$$

a simple fraction equivalent to the original complex fraction.

EXAMPLE 3 Simplify $\dfrac{x + \frac{3}{4}}{x - \frac{1}{2}}$.

Solution The LCD of ¾ and ½ is 4. Multiply the numerator and denominator by 4 to obtain

$$\frac{4\left(x + \frac{3}{4}\right)}{4\left(x - \frac{1}{2}\right)} = \frac{4(x) + 4\left(\frac{3}{4}\right)}{4(x) - 4\left(\frac{1}{2}\right)} = \frac{4x + 3}{4x - 2} \quad \left(x \neq \frac{1}{2}\right).$$

We summarize the method for simplifying complex fractions as follows.

To Simplify a Complex Fraction:

1. Find the LCD of all the fractions contained in the complex fraction.
2. Multiply the numerator and denominator of the complex fraction by the LCD.
3. Reduce the resulting simple fraction, if possible.

EXAMPLE 4 Write $\left(\dfrac{1}{z} - 1\right) \div \left(\dfrac{1}{z^2} - 1\right)$ as a complex fraction and simplify.

Solution Write the quotient as a fraction,

$$\left(\frac{1}{z} - 1\right) \div \left(\frac{1}{z^2} - 1\right) = \frac{\frac{1}{z} - 1}{\frac{1}{z^2} - 1},$$

then multiply the numerator and denominator by the LCD, z^2, to find

$$\frac{\frac{1}{z}-1}{\frac{1}{z^2}-1}\cdot\frac{z^2}{z^2}=\frac{\frac{1}{z}(z^2)-1(z^2)}{\frac{1}{z^2}(z^2)-1(z^2)}=\frac{z-z^2}{1-z^2}.$$

Finally, reduce the fraction to obtain

$$\frac{z-z^2}{1-z^2}=\frac{z(1-z)}{(1+z)(1-z)}=\frac{z}{1+z} \quad (z\neq -1,0,1).$$

Sometimes it is necessary to apply the fundamental principle more than once to simplify a complex fraction. In such cases we simplify a portion of the expression at a time.

EXAMPLE 5 Simplify $\dfrac{a}{a+\dfrac{3}{3+\dfrac{1}{2}}}$.

Solution First, simplify the second term of the denominator:

$$\frac{3}{3+\frac{1}{2}}=\frac{3}{3+\frac{1}{2}}\cdot\frac{2}{2}$$

$$=\frac{(3)\cdot 2}{\left(3+\frac{1}{2}\right)2}=\frac{6}{6+1}=\frac{6}{7}.$$

Substitute $6/7$ into the original fraction and apply the fundamental principle again to get

$$\frac{a}{a+\dfrac{3}{3+\dfrac{1}{2}}}=\frac{a}{a+\dfrac{6}{7}}\cdot\frac{7}{7}$$

$$=\frac{(a)7}{\left(a+\dfrac{6}{7}\right)7}=\frac{7a}{7a+6} \quad \left(a\neq\frac{-6}{7}\right).$$

EXERCISE 2.4

A

■ Write each complex fraction as a simple fraction in lowest terms. See Examples 1 and 2.

1. $\dfrac{\frac{2}{9}}{\frac{7}{3}}$

2. $\dfrac{\frac{5}{2}}{\frac{21}{4}}$

3. $\dfrac{\frac{2x}{5y}}{\frac{3x}{10y^2}}$

4. $\dfrac{\frac{3ab}{4}}{\frac{3b}{8a^2}}$

■ See Example 3.

5. $\dfrac{\frac{3}{4}}{4-\frac{1}{4}}$

6. $\dfrac{\frac{1}{3}}{4+\frac{2}{3}}$

7. $\dfrac{1-\frac{2}{3}}{3+\frac{1}{3}}$

8. $\dfrac{\frac{1}{2}+\frac{3}{4}}{\frac{1}{2}-\frac{3}{4}}$

9. $\dfrac{\frac{2}{a}+\frac{3}{2a}}{5+\frac{1}{a}}$

10. $\dfrac{\frac{2}{y}+\frac{1}{2y}}{y+\frac{y}{2}}$

11. $\dfrac{1+\frac{2}{a}}{1-\frac{4}{a^2}}$

12. $\dfrac{4-\frac{1}{x^2}}{2-\frac{1}{x}}$

13. $\dfrac{x+\frac{x}{y}}{1+\frac{1}{y}}$

14. $\dfrac{1+\frac{1}{x}}{1-\frac{1}{x}}$

15. $\dfrac{1}{1-\frac{1}{x}}$

16. $\dfrac{4}{\frac{2}{x}+2}$

17. $\dfrac{y-2}{y-\frac{4}{y}}$

18. $\dfrac{y+3}{\frac{9}{y}-y}$

19. $\dfrac{x+y}{\frac{1}{x}+\frac{1}{y}}$

20. $\dfrac{x-y}{\frac{x}{y}-\frac{y}{x}}$

21. $\dfrac{x-\frac{x}{y}}{y+\frac{y}{x}}$

22. $\dfrac{y+\frac{x}{y}}{x-\frac{y}{x}}$

23. $\dfrac{\frac{4}{x^2}-\frac{4}{z^2}}{\frac{2}{z}-\frac{2}{x}}$

24. $\dfrac{\frac{6}{b}-\frac{6}{a}}{\frac{3}{a^2}-\frac{3}{b^2}}$

25. $\dfrac{3a-\frac{b^2}{3a}}{1-\frac{b}{3a}}$

26. $\dfrac{1-\frac{2y}{x}}{x-\frac{4y^2}{x}}$

27. $\dfrac{\frac{3}{x}-\frac{9}{x^2z}}{\frac{6}{xz^2}-\frac{2}{z}}$

28. $\dfrac{\frac{4}{b^2c}-\frac{5}{bc^2}}{\frac{10}{b}-\frac{8}{c}}$

29. $\dfrac{8-\frac{2}{x}}{4-\frac{13}{x}+\frac{3}{x^2}}$

30. $\dfrac{6+\frac{1}{z}-\frac{2}{z^2}}{9+\frac{6}{z}}$

31. $\dfrac{\frac{1}{y+1}}{1-\frac{1}{y^2}}$

32. $\dfrac{\frac{1}{y-1}}{\frac{1}{y^2}+1}$

33. $\dfrac{a + 3 - \dfrac{8}{a+1}}{\dfrac{-6}{a+1} + a + 2}$ 34. $\dfrac{c + 1 - \dfrac{10}{c-2}}{\dfrac{-2}{c-2} + c - 3}$ 35. $\dfrac{\dfrac{1}{u+2} - \dfrac{2}{u-1}}{\dfrac{2}{u+1} - \dfrac{1}{u+2}}$ 36. $\dfrac{\dfrac{3}{v-2} - \dfrac{1}{v+2}}{\dfrac{1}{v-2} - \dfrac{3}{v-3}}$

37. $\dfrac{w - 1 - \dfrac{1}{w-1}}{w + 1 - \dfrac{1}{w-1}}$ 38. $\dfrac{\dfrac{1}{z-1} - \dfrac{z}{z+1}}{1 - \dfrac{z}{z+1}}$ 39. $\dfrac{\dfrac{x}{x-y} - \dfrac{y}{x+y}}{\dfrac{y}{x-y} + \dfrac{x}{x+y}}$ 40. $\dfrac{\dfrac{x}{x-y} - \dfrac{y}{x+y}}{x^2 - y^2}$

■ *Write each quotient first as a complex fraction and then as a simple fraction in lowest terms. See Example 4.*

41. $\left(\dfrac{1}{y^2} - \dfrac{1}{4}\right) \div \left(\dfrac{1}{y} + \dfrac{1}{2}\right)$ 42. $\left(\dfrac{4}{x} - \dfrac{1}{3}\right) \div \left(\dfrac{16}{x^2} - \dfrac{1}{9}\right)$

43. $\left(\dfrac{9}{a^2} - 1\right) \div \left(\dfrac{3}{a} + 1\right)$ 44. $\left(1 + \dfrac{1}{b^3}\right) \div \left(1 + \dfrac{1}{b}\right)$

45. $\left(x + 3 + \dfrac{10}{2x-3}\right) \div \left(x - \dfrac{2}{2x-3}\right)$ 46. $\left(y - 1 - \dfrac{4}{3y-2}\right) \div \left(y - \dfrac{1}{3y-2}\right)$

■ *Solve.*

47. The focal length of a lens is given by the formula

$$\dfrac{1}{f} = \dfrac{1}{p} + \dfrac{1}{q},$$

where f stands for the focal length, p is the distance from the object viewed to the lens, and q is the distance from the image to the lens. Suppose you estimate that the distance from a certain object to your camera lens is 60 inches greater than the distance from the lens to the mirror, where the image forms.

   a. Write an expression for $1/f$ in terms of q.
   b. Write an expression for f in terms of q.

48. If two resistors R_1 and R_2 in an electrical circuit are connected in parallel, the total resistance R in the circuit is given by

$$\dfrac{1}{R} = \dfrac{1}{R_1} + \dfrac{1}{R_2}.$$

   a. Assume that the second resistor, R_2, is 10 ohms greater than the first. Write an expression for $1/R$ in terms of the first resistor.
   b. Write an expression for R in terms of the first resistor.

49. Andy drives 300 miles to Lake Tahoe at 70 miles per hour and returns home at 50 miles per hour. What is his average speed for the round trip? (It is not 60 miles per hour!)

   a. Write expressions for the time it takes for each leg of the trip if Andy drives a distance d at speed r_1 and returns at speed r_2.

b. Write expressions for the total distance and total time for the trip.

c. Write an expression for the average speed for the entire trip. (Remember that rate = distance/time.)

d. Write the expression in (c) as a simple fraction that involves only the variables r_1 and r_2.

e. Using your formula, answer the question stated in the problem.

50. The owner of a print shop volunteers to produce flyers for his candidate's campaign. His large printing press can complete the job in 4 hours, and the smaller model can finish the flyers in 6 hours. How long will it take to print the flyers if he runs both presses simultaneously?

   a. Assume that the large press can complete a job in t_1 hours and the smaller press takes t_2 hours. Write expressions for the fraction of a job that each press can complete in 1 hour.

   b. Write an expression for the fraction of a job that can be completed in 1 hour with both presses running simultaneously.

   c. Write an expression for the amount of time needed to complete the job with both presses running.

   d. Using your formula from (c), answer the question stated in the problem.

B

■ *Write each fraction as a simple fraction in lowest terms. See Example 5.*

51. $\dfrac{1 + \dfrac{1}{1 - \dfrac{a}{b}}}{1 - \dfrac{3}{1 - \dfrac{a}{b}}}$

52. $\dfrac{1 - \dfrac{1}{\dfrac{a}{b} + 2}}{1 + \dfrac{3}{\dfrac{a}{2b} + 1}}$

53. $1 + \dfrac{2 + \dfrac{1}{x}}{x - \dfrac{1}{x}}$

54. $1 - \dfrac{1 - \dfrac{1}{x}}{x - \dfrac{1}{x}}$

55. $\dfrac{\dfrac{1}{ab} + \dfrac{2}{bc} + \dfrac{3}{ac}}{\dfrac{2a + 3b + c}{abc}}$

56. $\dfrac{\dfrac{a}{bc} - \dfrac{b}{ac} + \dfrac{c}{ab}}{\dfrac{1}{a^2 b^2} - \dfrac{1}{a^2 c^2} + \dfrac{1}{b^2 c^2}}$

2.5

INTEGER EXPONENTS

In this section we continue the study of exponents we began in Chapter 1. We will see that certain fractions and rational expressions can be treated more efficiently by using negative exponents.

2.5 ■ INTEGER EXPONENTS

Laws of Exponents

In Section 1.3 we introduced three laws for positive integer exponents:

> I. $a^m \cdot a^n = a^{m+n}$;
> II. $(a^m)^n = a^{mn}$;
> III. $(ab)^n = a^n b^n$.

Now consider the following quotients of powers:

$$\frac{x^7}{x^4} = \frac{\cancel{x}\cancel{x}\cancel{x}\cancel{x}xxx}{\cancel{x}\cancel{x}\cancel{x}\cancel{x}} = x^3;$$

$$\frac{x^4}{x^7} = \frac{\cancel{x}\cancel{x}\cancel{x}\cancel{x}}{\cancel{x}\cancel{x}\cancel{x}\cancel{x}xxx} = \frac{1}{x^3}.$$

These examples suggest that to divide one power by another (with the same base) we should subtract the exponents, according to the following law.

> IVa. If $m > n$, then $\dfrac{a^m}{a^n} = a^{m-n}$ $(a \neq 0)$.
>
> IVb. If $m < n$, then $\dfrac{a^m}{a^n} = \dfrac{1}{a^{n-m}}$ $(a \neq 0)$.

EXAMPLE 1

a. $\dfrac{x^4 y^6}{x^2 y} = x^{4-2} y^{6-1}$

$= x^2 y^5 \quad (x, y \neq 0)$

b. $\dfrac{x^2 y}{x^3 y^2} = \dfrac{1}{x^{3-2} y^{2-1}}$

$= \dfrac{1}{xy} \quad (x, y \neq 0)$

Note that the fourth law of exponents is consistent with the process of reducing fractions by using the fundamental principle of fractions. The same results would be obtained in the examples above by dividing out common factors.

We also have the following law for a power of a quotient:

> V. $\left(\dfrac{a}{b}\right)^n = \dfrac{a^n}{b^n}$ $(b \neq 0)$.

EXAMPLE 2   a. $\left(\dfrac{x}{y}\right)^3 = \dfrac{x^3}{y^3}$ $(y \neq 0)$ b. $\left(\dfrac{2x^2}{y}\right)^4 = \dfrac{(2x^2)^4}{y^4}$ $(y \neq 0)$

Ordinarily, two or more of the laws of exponents may be required to simplify expressions containing exponents. In Example 2b we can further simplify the result as follows: from Law (III),

$$\dfrac{(2x^2)^4}{y^4} = \dfrac{2^4(x^2)^4}{y^4} \quad (y \neq 0),$$

and from Law (II),

$$\dfrac{2^4(x^2)^4}{y^4} = \dfrac{16x^8}{y^4} \quad (y \neq 0).$$

EXAMPLE 3   a. $\left(\dfrac{2x^3}{y}\right)^2 = \dfrac{2^2 \cdot (x^3)^2}{y^2} = \dfrac{4x^6}{y^2}$ $(y \neq 0)$

b. $\left(\dfrac{x^2}{(x+2)^3}\right)^2 \left(\dfrac{x+2}{x}\right)^3 = \dfrac{x^4}{(x+2)^6} \cdot \dfrac{(x+2)^3}{x^3}$

$= \dfrac{x}{(x+2)^3}$ $(x \neq 0, -2)$

Common Error   Note that in Example 3b $(x+2)^3 \neq x^3 + 2^3$;

$$(x+2)^3 = (x+2)(x+2)(x+2)$$
$$= x^3 + 6x^2 + 12x + 8.$$

Zero and Negative Integer Exponents

If the fourth law of exponents is to hold for the case in which $m = n$ we must have that

$$\frac{a^n}{a^n} = a^{n-n} = a^0 \quad (a \neq 0).$$

But, by the definition of a quotient,

$$\frac{a^n}{a^n} = 1 \quad (a \neq 0).$$

So, to be consistent, we define a^0 as equal to 1:

$$a^0 = 1 \quad (a \neq 0).$$

Thus, for example, $3^0 = 1$ and $(-4)^0 = 1$. However, $0^0 \neq 1$; the symbol 0^0 is undefined. (This follows from the fact that % is undefined.)

We would also like the laws of exponents to hold for negative exponents. Observe that for $a \neq 0$,

$$a^n \cdot a^{-n} = a^{n-n} = a^0 = 1.$$

But, by the reciprocal property,

$$a^n \cdot \frac{1}{a^n} = 1 \quad (a \neq 0).$$

Again, for consistency we make the following definition:

$$a^{-n} = \frac{1}{a^n} \quad (a \neq 0).$$

Note that a^{-n} is simply another way to write the fraction $1/a^n$. Consequently, we also see that

$$\frac{1}{a^{-n}} = \frac{1}{\frac{1}{a^n}} = a^n \quad (a \neq 0).$$

EXAMPLE 4   a. $2^{-3} = \dfrac{1}{2^3} = \dfrac{1}{8}$ b. $9x^{-2} = 9 \cdot \dfrac{1}{x^2} = \dfrac{9}{x^2}$ $(x \neq 0)$

            c. $\dfrac{3}{2^{-3}} = 3 \cdot 2^3 = 24$ d. $\dfrac{x^{-2}}{y^{-3}} = \dfrac{y^3}{x^2}$ $(x, y \neq 0)$

Common Error

Note that in Example 4b $9x^{-2}$ equals $9/x^2$ and *not* $1/9x^2$. The exponent applies only to the base, x.

With these definitions the five laws of exponents discussed above apply to *all* integer powers and can be used to simplify expressions that involve negative exponents. In addition, by using negative exponents we can replace the two cases of Law (IV) by a single law that applies to all cases:

> IV. $\dfrac{a^m}{a^n} = a^{m-n}$ $(a \neq 0)$.

EXAMPLE 5   a. $x^{-3} \cdot x^5 = x^{-3+5}$ b. $(x^2 y^{-3})^{-1} = x^{-2} y^3$

                    $= x^2$ $(x \neq 0)$ $= \dfrac{y^3}{x^2}$ $(x, y \neq 0)$

            c. $\dfrac{8x^{-2}}{4x^{-6}} = \dfrac{8}{4} x^{-2-(-6)}$ d. $\left(\dfrac{a^2 c^{-3}}{x^0 y^{-2}}\right)^{-3} \left(\dfrac{xy^2}{a^{-1} c^3}\right)^{-2}$

                    $= 2x^4$ $(x \neq 0)$ $= \dfrac{a^{-6} c^9}{x^0 y^6} \cdot \dfrac{x^{-2} y^{-4}}{a^2 c^{-6}}$

                                                                     $= a^{-6-2} c^{9+6} x^{-2+0} y^{-4-6}$

                                                                     $= a^{-8} c^{15} x^{-2} y^{-10}$

                                                                     $= \dfrac{c^{15}}{a^8 x^2 y^{10}}$ $(a, c, x, y \neq 0)$

2.5 ■ INTEGER EXPONENTS

Finally, observe that

$$\left(\frac{a}{b}\right)^{-n} = \frac{a^{-n}}{b^{-n}} = \frac{1}{a^n} \div \frac{1}{b^n}$$
$$= \frac{1}{a^n} \cdot \frac{b^n}{1} = \frac{b^n}{a^n} = \left(\frac{b}{a}\right)^n.$$

Hence, we have the following result:

$$\left(\frac{a}{b}\right)^{-n} = \left(\frac{b}{a}\right)^n \qquad (a, b \neq 0).$$

EXAMPLE 6   a. $\left(\dfrac{x^3}{y^2}\right)^{-3} = \left(\dfrac{y^2}{x^3}\right)^3$

$= \dfrac{(y^2)^3}{(x^3)^3} = \dfrac{y^6}{x^9} \quad (x, y \neq 0)$

b. $\left(\dfrac{2}{x-y}\right)^{-4} = \left(\dfrac{x-y}{2}\right)^4$

$= \dfrac{(x-y)^4}{16} \quad (x \neq y)$

Products and Factors

The distributive law can be applied to rewrite products involving integer exponents or to factor common factors from such expressions.

EXAMPLE 7   a. $x^{-2}(x^3y + xy^2)$
$= x^{-2+3}y + x^{-2+1}y^2$
$= xy + x^{-1}y^2 \quad (x \neq 0)$

b. $x^{-3}y^{-1}(x^3y^{-2} + x^{-1}y^3)$
$= x^{-3+3}y^{-1-2} + x^{-3-1}y^{-1+3}$
$= y^{-3} + x^{-4}y^2 \quad (x, y \neq 0)$

c. $x^3(x+1)^{-2}[x^{-2} - x^{-1}(x+1)]$
$= x^{3-2}(x+1)^{-2} - x^{3-1}(x+1)^{-2+1}$
$= x(x+1)^{-2} - x^2(x+1)^{-1} \quad (x \neq -1, 0)$

In the following example we factor the powers with the smallest exponents from expressions containing more than one term. This will allow us to write the expressions as fractions that include positive exponents only.

EXAMPLE 8

a. $6x^{-2} - 3x^3 = 3x^{-2}(? - ?)$

$\qquad = 3x^{-2}(2 - x^5)$

$\qquad = \dfrac{3(2 - x^5)}{x^2} \qquad (x \neq 0)$

b. $x^{-2}y + x^2y^{-3} = x^{-2}y^{-3}(? + ?)$

$\qquad = x^{-2}y^{-3}(y^4 + x^4)$

$\qquad = \dfrac{y^4 + x^4}{x^2 y^3} \qquad (x, y \neq 0)$

c. $(x + 2)^{-1}(x - 1)^{-3} - (x + 2)^{-2}(x - 1)^{-2} = (x + 2)^{-2}(x - 1)^{-3}(? - ?)$

$\qquad = (x + 2)^{-2}(x - 1)^{-3}[(x + 2) - (x - 1)]$

$\qquad = (x + 2)^{-2}(x - 1)^{-3}(x + 2 - x + 1)$

$\qquad = 3(x + 2)^{-2}(x - 1)^{-3}$

$\qquad = \dfrac{3}{(x + 2)^2(x - 1)^3} \qquad (x \neq -2, 1)$

Sums of Powers

Note that the laws of exponents apply only to products and quotients. Special care must be taken when simplifying expressions that involve sums and differences. In many cases it is helpful to write all powers with positive exponents first and then treat the expressions as algebraic fractions.

EXAMPLE 9 Write each expression as a single fraction involving positive exponents only.

a. $x^{-1} + y^{-2}$ 　　　　　　　　b. $(x^{-1} + x^{-2})^{-1}$

Solutions a. $x^{-1} + y^{-2}$

$$= \frac{1}{x} + \frac{1}{y^2}$$

$$= \frac{(y^2)1}{(y^2)x} + \frac{1(x)}{y^2(x)}$$

$$= \frac{y^2 + x}{xy^2} \quad (x, y \neq 0)$$

b. $(x^{-1} + x^{-2})^{-1}$

$$= \left(\frac{1}{x} + \frac{1}{x^2}\right)^{-1}$$

$$= \left[\frac{(x)1}{(x)x} + \frac{1}{x^2}\right]^{-1}$$

$$= \left(\frac{x + 1}{x^2}\right)^{-1}$$

$$= \frac{x^2}{x + 1} \quad (x \neq 0, -1)$$

Common Error

Note in Example 9b that $(x^{-1} + x^{-2})^{-1} \neq (x^{-1})^{-1} + (x^{-2})^{-1}$.

Scientific Notation

An exponential form of notation is often useful in applications of mathematics that involve very large or very small quantities. For example, the mass of the earth is approximately

$$5{,}980{,}000{,}000{,}000{,}000{,}000{,}000{,}000{,}000 = 5.98 \times 10^{27} \text{ grams,}$$

and the mass of a hydrogen atom is approximately

$$0.00000000000000000000000167 = 1.67 \times 10^{-24} \text{ gram.}$$

In each case we have represented the number in **scientific notation**, that is, as the product of a number between 1 and 10 and a power of 10. The examples above suggest the following procedure.

To Write a Number in Scientific Notation:

1. Locate the decimal point so that there is exactly one nonzero digit to its left.
2. Count the number of places the decimal point was moved; this number determines the power of 10. The exponent is positive if the original number is greater than 10 and negative if the original number is less than 1.

EXAMPLE 10 a. $478{,}000 = 4.\underset{\text{5 places}}{78000} \times 10^5$ b. $0.00032 = \underset{\text{4 places}}{00003.2} \times 10^{-4}$

$\qquad\qquad\qquad\qquad = 4.78 \times 10^5 \qquad\qquad\qquad\qquad\qquad = 3.2 \times 10^{-4}$

A number written in scientific notation can be written in **standard form** by reversing the procedure above.

> **To Convert from Scientific Notation to Decimal Notation:**
>
> Move the decimal point the number of places indicated by the exponent on 10 — to the right if the exponent is positive and to the left if it is negative.

EXAMPLE 11 a. $3.75 \times 10^4 = 3\underset{\text{4 places}}{7500}$ b. $2.03 \times 10^{-3} = .\underset{\text{3 places}}{00203}$

$\qquad\qquad\qquad\qquad = 37{,}500 \qquad\qquad\qquad\qquad\qquad = 0.00203$

Using Calculators for Scientific Notation

Your calculator may use scientific notation to display very large or very small numbers. For example, 3.26×10^{-18} would appear on the display as

$$\boxed{3.26 \quad -18},$$

where the two digits on the right indicate the power of 10. To enter 3.26×10^{-18} in scientific form, first enter 3.26, then press the $\boxed{\text{EXP}}$ key, followed by 18 and finally by the $\boxed{+/-}$ key.

For some problems the "magnitude" or approximate size of the answer can be as important as the exact figure, especially if the data used in the computations are approximations or estimates. Scientific notation can be helpful in making such approximations. In the following example decimals are rounded off as necessary to make mental calculation possible.

EXAMPLE 12 a. Estimate the result of each computation.

i. $\dfrac{247{,}000}{0.0124}$

$= \dfrac{2.47 \times 10^5}{1.24 \times 10^{-2}}$

$\approx \dfrac{2.5 \times 10^5}{1.25 \times 10^{-2}}$

$= 2 \times 10^7$

ii. $\dfrac{0.00024 \times 0.000073}{0.00000021}$

$= \dfrac{2.4 \times 10^{-4} \times 7.3 \times 10^{-5}}{2.1 \times 10^{-7}}$

$\approx \dfrac{2 \times 10^{-4} \times 7 \times 10^{-5}}{2 \times 10^{-7}}$

$= 7 \times 10^{-2}$

b. Compute with the aid of a calculator.

i. $\dfrac{2.47 \times 10^5}{1.24 \times 10^{-2}}$

$\approx 1.99 \times 10^7$

ii. $\dfrac{2.4 \times 10^{-4} \times 7.3 \times 10^{-5}}{2.1 \times 10^{-7}}$

$\approx 8.3 \times 10^{-2}$

EXERCISE 2.5

A

■ *Using one or more of the laws of exponents, write each expression as a product or quotient in which each factor occurs only once and all exponents are positive. See Examples 1–3.*

1. $\dfrac{x^5}{x^3}$
2. $\dfrac{y^2}{y^6}$
3. $\dfrac{x^2 y^4}{x y^8}$
4. $\dfrac{x^4 y^6}{x^2 y}$
5. $\left(\dfrac{x}{y^2}\right)^3$
6. $\left(\dfrac{y^2}{z^3}\right)^2$
7. $\left(\dfrac{-2x}{3y^2}\right)^3$
8. $\left(\dfrac{-x^2}{2y}\right)^4$
9. $\dfrac{(4x)^3}{(-2x^2)^2}$
10. $\dfrac{(5x)^2}{(-3x^2)^3}$
11. $\dfrac{(xy)^2(-x^2y)^3}{(x^2y^2)^2}$
12. $\dfrac{(-x)^2(-x^2)^4}{(x^2)^3}$
13. $\left(\dfrac{-2x}{y^2}\right)^3 \left(\dfrac{y^2}{3x}\right)^2$
14. $\left(\dfrac{x^2 z}{2}\right)^2 \left(-\dfrac{2}{x^2 z}\right)^3$
15. $\left(\dfrac{-3}{x+y}\right)^2 \left(\dfrac{x+y}{x^2}\right)^3$
16. $\left(\dfrac{2x-y}{y}\right)^2 \left(\dfrac{-3}{2x-y}\right)^3$

■ *Write each expression without negative exponents and simplify. See Example 4.*

17. 2^{-1}
18. 3^{-2}
19. $\dfrac{1}{3^{-1}}$
20. $\dfrac{3}{4^{-2}}$

21. $(-2)^{-3}$
22. $(-5)^{-2}$
23. $\dfrac{1}{(-3)^{-2}}$
24. $\dfrac{1}{(-3)^{-3}}$
25. $\left(\dfrac{3}{5}\right)^{-1}$
26. $\left(\dfrac{1}{3}\right)^{-2}$
27. $\dfrac{5^{-1}}{3^{-2}}$
28. $\dfrac{3^{-3}}{6^{-2}}$
29. $3^{-2} + 3^2$
30. $5^{-1} + 25^0$
31. $4^{-1} - 4^{-2}$
32. $8^{-2} - 2^0$
33. $-2x^{-3}$
34. $-3x^{-2}$
35. $\dfrac{-5y^{-2}}{x^{-5}}$
36. $\dfrac{-6y^{-3}}{x^{-3}}$
37. $(x - y)^{-2}$
38. $(x + y)^{-3}$
39. $\dfrac{(b + c)^{-2}}{(b - c)^{-1}}$
40. $\dfrac{(a - 3)^{-2}}{(a + 3)^{-3}}$

■ Write each expression as a product or quotient of powers in which each factor occurs only once and all exponents are positive. See Examples 5 and 6.

41. $x^{-7} \cdot x^3$
42. $x^{-5} \cdot x^2$
43. $\dfrac{x^3}{x^{-4}}$
44. $\dfrac{x^5}{x^{-5}}$
45. $(3x^{-2}y^3)^{-2}$
46. $(2x^3y^{-4})^{-3}$
47. $\dfrac{x^0 y^{-3}}{4x^{-2}y^{-1}}$
48. $\dfrac{x^{-3}y^{-2}}{6x^{-5}y^0}$
49. $\left(\dfrac{a^{-3}}{b^2}\right)^{-2}$
50. $\left(\dfrac{a^4}{b^{-5}}\right)^{-3}$
51. $\dfrac{(a + b)^{-3}}{(a + b)^{-2}}$
52. $\dfrac{(a - b)^{-5}}{(a - b)^{-7}}$

■ Write each product as a sum of powers. See Example 7.

53. $x^{-1}(x + y)$
54. $y^{-2}(x^2 + y)$
55. $x^{-1}y^{-2}(x^2y + xy^2)$
56. $x^{-2}y^{-3}(x^3 + y^3)$

■ Factor as indicated. See Example 8.

57. $x^{-3}y^2 + 4xy^{-1} = x^{-3}y^{-1}(? + ?)$
58. $xy^{-1} + x^{-1}y^{-2} = x^{-1}y^{-2}(? + ?)$
59. $3x^{-3}y^{-4} + 4x^{-4}y^{-3} = x^{-4}y^{-4}(? + ?)$
60. $4y^{-1} + x^2y^{-3} = y^{-3}(? + ?)$

■ Write each expression as a single fraction involving positive exponents only. See Example 9.

61. $x^{-2} + y^{-2}$
62. $x^{-1} - y^{-3}$
63. $\dfrac{x}{y^{-1}} + \dfrac{x^{-1}}{y}$
64. $\dfrac{x^{-1}}{y^{-1}} + \dfrac{y}{x}$
65. $x^{-1}y - xy^{-1}$
66. $xy^{-1} + x^{-1}y$
67. $\dfrac{x^{-1} - y}{x^{-1}}$
68. $\dfrac{x + y^{-1}}{y^{-1}}$
69. $\dfrac{x^{-1} + y^{-1}}{(xy)^{-1}}$
70. $\dfrac{x^{-2} - y^{-2}}{(xy)^{-1}}$
71. $(x^{-1} + y^{-1})^{-1}$
72. $\dfrac{(x + y)^{-1}}{x^{-1} + y^{-1}}$

■ Express each number using scientific notation. See Example 10.

73. 285
74. 68,742
75. 8,372,000
76. 481,000
77. 0.024
78. 0.421
79. 0.000523
80. 0.000004

■ Express each number using standard form. See Example 11.

81. 2.4×10^2
82. 4.8×10^3
83. 6.87×10^5
84. 8.31×10^4
85. 5.0×10^{-3}
86. 8.0×10^{-1}
87. 2.02×10^{-4}
88. 4.31×10^{-5}

■ a. *Estimate the result of each computation using scientific notation.*
b. *Compute with the aid of a calculator. See Example 12.*

89. $\dfrac{0.6 \times 0.00082 \times 0.091}{0.00019 \times 0.00028}$

90. $\dfrac{0.0054 \times 0.05 \times 300}{0.0016 \times 0.27 \times 8200}$

91. $\dfrac{4{,}200{,}000 \times 0.0017 \times 61{,}000}{0.0028 \times 12{,}000{,}000 \times 23}$

92. $\dfrac{0.004 \times 27{,}000 \times 620{,}000}{2700 \times 0.0001 \times 0.009}$

93. The speed of light is approximately 300,000,000 meters per second.
   a. Express this number in scientific notation.
   b. Express the speed of light in inches per second (1 inch equals 2.54 centimeters, and 1 meter equals 100 centimeters).

94. In 1985 the public debt of the United States was $1,823,103,000,000.
   a. Express this number in scientific notation.
   b. If the population of the United States in 1985 was 238,631,000, what was the per capita debt for that year?

95. One light-year is the number of miles traveled by light in 1 year (365 days), and the speed of light is approximately 186,000 miles per second.
   a. Express in scientific notation the number of miles in 1 light-year.
   b. The star nearest to the sun is Proxima Centauri, at a distance of 4.3 light-years. How long would it take the *Pioneer 10* (the first space vehicle to achieve escape velocity from the solar system), traveling at 32,114 miles per hour, to reach Proxima Centauri?

96. Lake Superior has an area of 31,700 square miles and an average depth of 483 feet.
   a. Find the approximate volume of Lake Superior in cubic feet.
   b. If 1 cubic foot of water is equivalent to 7.48 gallons, how many gallons of water are in Lake Superior?

97. On November 6, 1923 the circulation of Reichsbank marks in Germany was 400,338,326,350,700,000,000.
   a. Express this number in scientific notation.
   b. Assume that each note is approximately 15 square inches in area. How large an area, in square feet, would that many one-mark notes cover?
   c. The total surface area of the earth is 196,937,400 square miles. How many times could you paper the earth with that many one-mark notes?

98. The mass of the earth is 6,585,600,000,000,000,000,000 tons, and its volume is 259,875,300,000 cubic miles.
   a. Express these numbers in scientific notation.
   b. Find the average density of the earth in tons per cubic mile. (Density is defined to be mass per unit volume.)
   c. Find the density of the earth in pounds per cubic foot.

B

■ *Simplify.*

99. $\left[\left(\dfrac{r^2s^3t}{xy}\right)^3\left(\dfrac{x^2y}{r^3st^2}\right)^2\right]^2$

100. $\left[\left(\dfrac{a^3bc}{x^2y}\right)^4\left(\dfrac{x^2yz}{ab^2c^3}\right)^2\right]^2$

101. $\left(\dfrac{x^2}{a^2b}\right)^2\left(-\dfrac{ab}{x^3}\right)^3\left(\dfrac{x}{ab}\right)^2$

102. $\left(\dfrac{m^3n^2p}{r^2s}\right)^2\left(\dfrac{rs}{mn^2p^2}\right)^3\left(-\dfrac{mnp}{rs}\right)^2$

103. $\left(\dfrac{3x^{-1}y^3}{2x^0y^{-5}}\right)^{-2}$

104. $\left(\dfrac{2x^{-3}z^0}{5x^{-4}z^{-2}}\right)^{-3}$

105. $\dfrac{(2^{-2}x^2y^{-1})^{-3}}{(4x^{-3}y^2)^{-2}}$

106. $\dfrac{(3y^3z^{-2})^{-1}}{(2^{-3}y^{-2}z)^{-2}}$

107. $\left(\dfrac{6x^{-2}y^2}{4z^{-1}}\right)^{-1}\cdot\left(\dfrac{3x^{-1}y^0}{z}\right)^{-2}$

108. $\left(\dfrac{2y^{-3}x}{3z^2}\right)^{-2}\left(\dfrac{2x^4}{9y^{-2}z^{-2}}\right)^{-1}$

■ *Factor.*

109. $2(x+3)^{-2} - 2(x+3)^{-3}(2x-3) = 2(x+3)^{-3}(?)$

110. $3(x-1)^{-3} - 3(3x+4)(x-1)^{-4} = 3(x-1)^{-4}(?)$

111. $(2x+1)^{-2}(3x-2)^{-2} + 3(3x-2)^{-3}(2x+1)^{-1} = (2x+1)^{-2}(3x-2)^{-3}(?)$

112. $-3(x+7)^{-4}(2x-3)^{-3} - 6(2x-3)^{-4}(x+7)^{-3} = -3(x+7)^{-4}(2x-3)^{-4}(?)$

113. Prove Laws (IV) and (IVa) for quotients of powers.
114. Prove Law (V) for powers of quotients.

CHAPTER REVIEW

A

[2.1]

■ *Reduce each fraction to lowest terms.*

1. $\dfrac{2a^2(a-1)^2}{4a(a-1)^3}$

2. $\dfrac{a^2(2a-1)}{4a(1-2a)}$

3. $\dfrac{4y-6}{6}$

4. $\dfrac{2x^2y^3 - 4x^3y}{4x^2y}$

5. $\dfrac{2x^2+6x}{2(x+3)^2}$

6. $\dfrac{(x-2y)^2}{4y^2-x^2}$

7. $\dfrac{a^2-6a+9}{2a^2-18}$

8. $\dfrac{4x^2y^2+4xy+1}{4x^2y^2-1}$

9. $\dfrac{xy+2x+y+2}{x^3+x^2+x+1}$

10. $\dfrac{(a+1)^2(a-2) - 2(a+1)(a-2)}{a^2-a-2}$

[2.2]
■ Write each expression as a single fraction in lowest terms.

11. $\dfrac{2a^2}{3b} \cdot \dfrac{15b^2}{4a}$

12. $-\dfrac{1}{3}ab^2 \cdot \dfrac{3}{4}a^3b$

13. $\dfrac{4x+6}{2x} \cdot \dfrac{6x^2}{(2x+3)^2}$

14. $\dfrac{4x^2-9}{3x-3} \cdot \dfrac{x^2-1}{4x-6}$

15. $\dfrac{x^2+x-6}{2x^2-4x+2} \cdot \dfrac{4x^2-4}{x^2+6x+9}$

16. $\dfrac{xy-x+3y-3}{x^4+4x^2+4} \cdot \dfrac{x^3-x^2+2x-2}{4x^2-36}$

17. $\dfrac{a^2-a-2}{a^2-4} \div \dfrac{a^2+2a+1}{a^2-2a}$

18. $\dfrac{a^3-8b^3}{a^2b} \div \dfrac{a^2-4ab+4b^2}{ab^2}$

19. $1 \div \dfrac{4x^2-1}{2x+1}$

20. $\dfrac{y^2+2y}{3x} \div 4y$

■ Divide.

21. $\dfrac{12y^2z^2+6yz-3}{3yz}$

22. $\dfrac{36x^6-28x^4+16x^2-4}{4x^4}$

23. $\dfrac{y^3+3y^2-2y-4}{y+1}$

24. $\dfrac{x^3-4x^2+2x+3}{x-2}$

25. $\dfrac{x^2+2x^3-1}{2x-1}$

26. $\dfrac{4-y+3y^3}{3y+1}$

27. $\dfrac{2x^3+x-5}{x^2-1}$

28. $\dfrac{y^7-1}{y^3-1}$

[2.3]
■ Write each expression as a single fraction in lowest terms.

29. $\dfrac{x+2}{3x} - \dfrac{x-4}{3x}$

30. $\dfrac{y-1}{y+3} - \dfrac{y+1}{y+3} + \dfrac{y}{y+3}$

31. $\dfrac{1}{2}a - \dfrac{2}{3}a$

32. $\dfrac{5}{6}b - \dfrac{1}{3}b + \dfrac{3}{4}b$

33. $\dfrac{3}{2x-6} - \dfrac{4}{x^2-9}$

34. $\dfrac{1}{y^2+4y+4} + \dfrac{3}{y^2-4}$

35. $\dfrac{2a+1}{a-3} - \dfrac{a-2}{a^2-4a+3}$

36. $a - \dfrac{1}{a^2+2a+1} + \dfrac{3}{a^2-1}$

[2.4]
■ Write each complex fraction as a simple fraction in lowest terms.

37. $\dfrac{\frac{2}{3}}{3-\frac{1}{3}}$

38. $\dfrac{\frac{3}{4}-\frac{1}{2}}{\frac{3}{4}+\frac{1}{2}}$

39. $\dfrac{y-\frac{2y}{x}}{1+\frac{2}{x}}$

40. $\dfrac{x-4}{x-\dfrac{16}{x}}$

41. $\dfrac{\dfrac{1}{x-1}}{1-\dfrac{1}{x^2}}$

42. $\dfrac{\dfrac{2}{x-1}-\dfrac{1}{x+2}}{\dfrac{3}{x+2}-\dfrac{2}{x-1}}$

[2.5]

■ *Write each expression as a product or quotient in which each variable occurs only once and all exponents are positive.*

43. $\dfrac{(2x)^3}{(-3x^2)^2}$

44. $\dfrac{-2}{(x-y)^2}\left(\dfrac{x-y}{2}\right)^3$

45. $\dfrac{-4x^{-2}}{6y^{-2}}$

46. $\dfrac{(x-y)^{-3}}{x-y}$

47. $(2a^{-2}b^3)^{-3}$

48. $\left(\dfrac{a^{-2}}{b^3}\right)^{-2}$

■ *Write each product as a sum of powers.*

49. $a^{-3}(a^3-a)$

50. $a^{-2}b^{-1}(a^2b^2+ab^3)$

■ *Factor as indicated.*

51. $x^{-2}y+2xy^{-2}=x^{-2}y^{-2}(?+?)$

52. $2x^{-2}y^{-1}-x^{-3}y^2=x^{-3}y^{-1}(?-?)$

■ *Write each expression as a single fraction involving positive exponents only.*

53. $x^{-3}+y^{-1}$

54. $\dfrac{x^{-1}}{y}-\dfrac{x}{y-1}$

55. $\dfrac{x^{-1}-y}{y-1}$

56. $\dfrac{x^{-1}+y^{-1}}{x^{-1}}$

57. $\dfrac{x^{-1}-y^{-1}}{(x-y)^{-1}}$

58. $\dfrac{(xy)^{-1}}{x^{-1}-y^{-1}}$

■ *Estimate the result of each computation using scientific notation.*

59. $\dfrac{0.04\times 0.00049\times 0.0025}{0.007\times 0.5\times 0.00002}$

60. $\dfrac{6{,}400{,}000\times 0.0015\times 2100}{0.0007\times 1600\times 450{,}000}$

61. Norm travels by motorboat to an island 30 miles upstream and returns in the afternoon. The current in the river is 4 miles per hour. Write an expression for the time required for the round trip in terms of the speed of Norm's boat.

62. Professor Marvel makes a round trip between Topeka and Kansas City, 65 miles to the east, by hot air balloon. Write an expression for the time required for the trip in terms of the speed of the balloon if the prevailing wind blows from the west at 12 miles per hour.

63. To light its 30-foot by 40-foot reading room the public library needs enough skylights to cover 15% of the floor space. Express the perimeter of each skylight in terms of its length if the library installs six rectangular skylights of equal size.

64. A tool and die company, needing an additional 6000 cubic feet of storage space, builds a rectangular storage shed with a 10-foot ceiling. Express the width of the shed in terms of its length.

65. Kathy and Allen drive through the sunny Southwest on their summer vacation. Kathy drives for 3 hours, and after lunch Allen drives for 3 hours at a speed 15 miles per hour slower than Kathy drove. Write an expression for their average speed in terms of Kathy's speed.

66. It takes the main inlet pipe 30 hours to fill the dolphin tank at the zoo, and the auxiliary pipe takes 45 hours. How long will it take both pipes running together to fill the tank?

67. The land area of the earth is 57,267,400 square miles. In the year 2000 the population of the earth will be 6,100,000,000.
   a. Express both of these numbers in scientific notation.
   b. In the year 2000, how many people will there be for each square mile of the earth's surface?

68. The average distance from the sun to the earth is 92,956,000 miles. Sunlight travels at 186,000 miles per second.
   a. Express each of these numbers in scientific notation.
   b. How long does it take sunlight to reach the earth?

B

■ *Divide.*

69. $\dfrac{x^3 - 2x^2 + x + 1}{x^2 - x + 2}$

70. $\dfrac{x^4 - x^2 - 2}{x^2 + 4}$

■ *Simplify.*

71. $1 + \dfrac{1 - \dfrac{1}{2}}{1 + \dfrac{1}{1 + \dfrac{1}{2}}}$

72. $\dfrac{3 - \dfrac{1}{\dfrac{x}{y} - 2}}{2 + \dfrac{3}{\dfrac{x}{2y} - 1}}$

73. $\dfrac{\dfrac{1}{a-b}}{\dfrac{a}{a^2-b^2}} + \dfrac{ab - b^2}{ab}$

74. $\dfrac{a^2 - b^2}{\dfrac{a-b}{b}} - \dfrac{a-b}{\dfrac{1}{b}}$

■ *Simplify. Write your answer using positive exponents only.*

75. $\dfrac{(xy^{-1})^{-1}(x^2y^2)^{-2}(x^{-1}y)^0}{(x^{-3}y^2)^{-1}(x^{-1}y^{-3})^{-1}}$

76. $\dfrac{(x^2y^3)^{-2}(xy^{-3})^0(x^3y^2)^{-1}}{(x^5y^{-2})^{-2}(x^4y)^2}$

■ *Factor. Write your answer using positive exponents only.*

77. $x^2(x + 1)^{-3} - 4(x + 1)^{-1}$

78. $4(x + 2)^{-2}(x - 3) + 6(x + 2)^{-1}(x - 3)^{-1}$

3 Exponents, Roots, and Radicals

3.1

nth ROOTS AND IRRATIONAL NUMBERS

If a certain square has an area of 64 square feet, how long is each side of the square? Since the area of a square is given by s^2, where s is the length of its side, we need a number whose *square* is 64. Because $8^2 = 64$, the length of the side must be 8 feet. We say that 8 is the **square root** of 64 and write $8 = \sqrt{64}$.

In general, the square root of a is a number s whose square is a.

> For $s \geq 0$,
> $$s = \sqrt{a} \quad \text{if} \quad a = s^2.$$

For example,
$$5 = \sqrt{25} \quad \text{since} \quad 25 = 5^2,$$
and
$$12 = \sqrt{144} \quad \text{since} \quad 144 = 12^2.$$

Now suppose that we want to construct a cubical box with a volume of 125 cubic inches. How long should each side of the box be? Since the volume of a cube is given by s^3, where s is the length of a side, we need a number whose *cube* is 125. Because $5^3 = 125$, the length of the side must be 5 inches. We say that 5 is the **cube root** of 125 and write $5 = \sqrt[3]{125}$.

3.1 ■ nth ROOTS AND IRRATIONAL NUMBERS

In general, the cube root of a is a number s whose cube is a:

$$s = \sqrt[3]{a} \quad \text{if} \quad a = s^3.$$

For example,

$$4 = \sqrt[3]{64} \quad \text{since} \quad 4^3 = 64,$$

and

$$9 = \sqrt[3]{729} \quad \text{since} \quad 9^3 = 729.$$

In a similar way we can look for the fourth, fifth, or sixth root of a number. For instance, the fourth root of a is a number s whose fourth power is a. In general, the nth root of a, written $\sqrt[n]{a}$, is a number whose nth power is a.

An expression of the form $\sqrt[n]{a}$ is called a **radical**, a is called the **radicand**, and n is called the **index** of the radical.

EXAMPLE 1
a. $\sqrt[4]{81} = 3$ because $3^4 = 81$.
b. $\sqrt[5]{32} = 2$ because $2^5 = 32$.
c. $\sqrt[6]{64} = 2$ because $2^6 = 64$.
d. $\sqrt[5]{100,000} = 10$ because $10^5 = 100,000$.

Does every real number have an nth root for every natural number n? That is, given a number a, will we always be able to find its square root, cube root, or nth root for any n? It turns out that the answer depends on whether n is even or odd.

Even nth Roots   First, suppose that n is an even natural number. We have, for example,

$$\sqrt{25} = 5 \quad \text{because} \quad 5^2 = 25$$

and

$$\sqrt[4]{16} = 2 \quad \text{because} \quad 2^4 = 16.$$

However, note that $(-5)^2$ also equals 25, and $(-2)^4$ also equals 16. In fact, every positive number has *two* nth roots for any even natural number n, a positive root and a negative root. To avoid confusion we use the symbol $\sqrt[n]{a}$ to denote the positive, or **principal**, nth root of a. If we want to indicate the negative nth root of a, we write $-\sqrt[n]{a}$. Thus,

$$-\sqrt{25} = -5 \quad \text{and} \quad -\sqrt[4]{16} = -2.$$

Do not confuse the symbol $-\sqrt[n]{a}$ with $\sqrt[n]{-a}$. $-\sqrt{25} = -5$, but $\sqrt{-25}$ is not a real number; that is, there is no real number whose square is -25. Similarly, there is no real number whose fourth power is -16, so $\sqrt[4]{-16}$ is not a real number. In general, *even roots of negative numbers are not real numbers*.

EXAMPLE 2
a. $\sqrt{16} = 4$ because $4^2 = 16$.
b. $-\sqrt{16} = -4$.
c. $\sqrt{-16}$ does not exist because there is no real number s for which $s^2 = -16$.
d. $\sqrt[4]{81} = 3$ because $3^4 = 81$.
e. $-\sqrt[4]{81} = -3$.
f. $\sqrt[4]{-81}$ does not exist because there is no real number s for which $s^4 = -81$.

Common Error Note that $\sqrt{16} \neq -4$ and $\sqrt[4]{81} \neq -3$ since $\sqrt[n]{a}$ stands for the *nonnegative* nth root of a when n is even.

Odd nth Roots Now consider the case in which n is an odd natural number. We have, for example,

$$\sqrt[3]{8} = 2 \quad \text{because} \quad 2^3 = 8$$

and

$$\sqrt[5]{243} = 3 \quad \text{because} \quad 3^5 = 243.$$

Note that odd roots differ from even roots in that we *can* take an *odd* root of a negative number. For example,

$$\sqrt[3]{-8} = -2 \quad \text{because} \quad (-2)^3 = -8$$

and
$$\sqrt[5]{-243} = -3 \quad \text{because} \quad (-3)^5 = -243.$$

Note also that there is only *one* odd root for each real number.

EXAMPLE 3
a. $\sqrt[3]{27} = 3$ because $3^3 = 27$.
b. $\sqrt[3]{-27} = -3$ because $(-3)^3 = -27$.
c. $-\sqrt[3]{27} = -(\sqrt[3]{27}) = -3$.
d. $\sqrt[5]{32} = 2$ because $2^5 = 32$.
e. $\sqrt[5]{-32} = -2$ because $(-2)^5 = -32$.
f. $-\sqrt[5]{32} = -(\sqrt[5]{32}) = -2$.

Exponential Notation

In Section 2.5 we extended the laws of exponents to include negative integer exponents, but we have not yet considered what fractional exponents might represent. Let us first consider powers of the form $a^{1/n}$, where n is a natural number. If we extend Law (II) of exponents (page 85) to include fractional powers, then

$$(a^{1/n})^n = a^{n/n} = a.$$

Thus, $a^{1/n}$ is a number whose nth power is a. But this means that $a^{1/n}$ is an nth root of a.

> For any natural number $n \geq 2$,
> $$a^{1/n} = \sqrt[n]{a}.$$

Recall that when n is even the symbol $\sqrt[n]{a}$ represents the principal, or positive, nth root of a. Similarly, the symbol $a^{1/n}$ indicates the *positive* root only. Likewise, $a^{1/n}$ does not represent a real number when n is even and a is negative. For example, $(-64)^{1/6}$ is not a real number. On the other hand, when n is odd, $a^{1/n}$ always represents a unique real number.

EXAMPLE 4
a. $81^{1/4} = 3$
b. $125^{1/3} = 5$
c. $(-32)^{1/5} = -2$
d. $-64^{1/6} = -2$

Common Error

In Example 4d $-64^{1/6} \neq (-64)^{1/6}$; $(-64)^{1/6}$ is not a real number. Remember that an exponent applies only to its base, so that

$$-64^{1/6} = -(64)^{1/6} = -2.$$

The use of fractional exponents greatly simplifies many calculations involving radicals. You should learn to convert easily between exponential and radical notation.

EXAMPLE 5 a. $5^{1/2} = \sqrt{5}$ b. $x^{1/5} = \sqrt[5]{x}$
c. $2x^{1/3} = 2\sqrt[3]{x}$ d. $\sqrt[4]{2y} = (2y)^{1/4}$

Common Error

In Example 5c $2x^{1/3} \neq (2x)^{1/3} = \sqrt[3]{2x}$, since the exponent $\frac{1}{3}$ applies only to its base, x.

Irrational Numbers

An nth root need not be an integer as in the examples above, or even a rational number. For example, there is no rational number r such that $r^2 = 5$, so $\sqrt{5}$ is not a rational number. $\sqrt{5}$ is nevertheless a real number; but it is irrational.

Recall from Section 1.1 that any number that can be expressed as the quotient of two integers is called a rational number, and any real number that cannot be so expressed is called an irrational number. All rational numbers have decimal representations that either terminate or repeat a pattern.

EXAMPLE 6 Find decimal representations for $\frac{3}{4}$ and for $\frac{9}{37}$.

Solution By using a calculator or by performing the divisions longhand, we find that

$$\frac{3}{4} = 3 \div 4 = 0.75, \quad \text{a terminating decimal,}$$

and

$$\frac{9}{37} = 9 \div 37 = 0.243243243\ldots,$$

where the pattern of digits 243 is repeated endlessly.

The decimal form of an irrational number never terminates, and it does not follow a repeating pattern; consequently, it is impossible to write an exact decimal equivalent for an irrational number. However, we can obtain decimal *approximations* correct to any desired degree of accuracy by rounding off. For example, the decimal representation for π, an irrational number that appears in many geometric formulas, is given by a calculator with a 10-digit display as 3.141592654. This is not the *exact* value of π, but for most calculations it is quite sufficient.

Some nth roots are rational numbers and some are irrational numbers. The radical $\sqrt[n]{a}$ represents a rational number if and only if a is the nth power of a rational number. Thus,

$$\sqrt{49}, \quad \sqrt[3]{\left(-\frac{27}{8}\right)}, \quad \text{and} \quad 81^{1/4} \quad \text{are rational numbers}$$

that can be written as 7, $-\frac{3}{2}$, and 3, respectively, but

$$\sqrt{2}, \quad \sqrt[3]{56}, \quad \text{and} \quad 7^{1/5} \quad \text{are irrational numbers.}$$

Calculators equipped with a square root key and a $\boxed{\sqrt[x]{y}}$ or radical key can be used to obtain approximations for irrational radicals.

EXAMPLE 7 Obtain decimal approximations to three decimal places for $\sqrt{7}$ and for $\sqrt[3]{56}$.

Solution To approximate $\sqrt{7}$ use the square root key:

$$7 \; \boxed{\sqrt{}} \; = 2.645751311.$$

Thus, $\sqrt{7} \approx 2.646$, where the symbol $\approx$ is read "is approximately equal to."

To approximate $\sqrt[3]{56}$, use the radical key:

$$56 \; \boxed{\sqrt[x]{y}} \; 3 \; \boxed{=} \; 3.825862366.$$

Thus, $\sqrt[3]{56} \approx 3.826$.

Since a radical can also be expressed as a fractional power, we can also use the $\boxed{y^x}$ key to evaluate radicals.

EXAMPLE 8 Use the $\boxed{y^x}$ key to approximate $\sqrt[3]{56}$.

Solution $\sqrt[3]{56} = 56^{1/3}$, so we press

$$56 \;\boxed{y^x}\; 3 \;\boxed{\tfrac{1}{x}}\; \boxed{=} \; 3.825862366.$$

Thus, $\sqrt[3]{56} \approx 3.826$.

From the definitions and properties above it is apparent that taking an nth root is the opposite, or *inverse,* of raising to a power. Raising to the power n "undoes" the result of taking an nth root:

$$(\sqrt[n]{a})^n = (a^{1/n})^n = a.$$

EXAMPLE 9
a. $(\sqrt{16})^2 = 16$
b. $(\sqrt[3]{343})^3 = 343$
c. $(\sqrt[3]{7})^3 = 7$
d. $(\sqrt[4]{x})^4 = x$

Estimating Roots

Occasionally it is useful to be able to *estimate* the value of an irrational radical without the aid of a calculator. Thus, to estimate $\sqrt[3]{47}$ we note that

$$\sqrt[3]{27} < \sqrt[3]{47} < \sqrt[3]{64}$$

(27 and 64 were chosen because they have cube roots that are integers), so

$$3 < \sqrt[3]{47} < 4.$$

EXAMPLE 10
a. $8 < \sqrt{73} < 9$, since $\sqrt{64} < \sqrt{73} < \sqrt{81}$.
b. $2 < \sqrt[4]{52} < 3$, since $\sqrt[4]{16} < \sqrt[4]{52} < \sqrt[4]{81}$.

EXERCISE 3.1

A

■ *Find the indicated root if the given expression is a real number. If it is not a real number, so state. See Examples 1–3.*

1. $\sqrt{121}$
2. $-\sqrt{169}$
3. $\sqrt[3]{-27}$
4. $\sqrt[3]{64}$
5. $\sqrt[4]{-625}$
6. $-\sqrt[4]{81}$
7. $-\sqrt[5]{\dfrac{32}{243}}$
8. $\sqrt[5]{-\dfrac{100{,}000}{32}}$
9. $\sqrt[4]{\dfrac{16}{81}}$
10. $\sqrt[4]{-\dfrac{1296}{625}}$
11. $\sqrt[3]{\dfrac{8}{729}}$
12. $-\sqrt[3]{\dfrac{343}{125}}$

■ *Find the indicated power if the given expression is a real number. If it is not a real number, so state. See Example 4.*

13. $9^{1/2}$
14. $25^{1/2}$
15. $(-81)^{1/4}$
16. $-81^{1/4}$
17. $-64^{1/6}$
18. $(-64)^{1/6}$
19. $(-32)^{1/5}$
20. $(-27)^{1/3}$
21. $(-8)^{-1/3}$
22. $(-243)^{-1/5}$
23. $\left(\dfrac{25}{64}\right)^{1/2}$
24. $\left(\dfrac{49}{81}\right)^{1/2}$

■ *Write each expression in radical form. See Example 5.*

25. $3^{1/2}$
26. $7^{1/2}$
27. $4x^{1/3}$
28. $3x^{1/4}$
29. $(4x)^{1/3}$
30. $(3x)^{1/4}$
31. $8^{-1/4}$
32. $6^{-1/3}$
33. $3(xy)^{-1/3}$
34. $y(5x)^{-1/2}$
35. $(x-2)^{1/4}$
36. $(y+2)^{1/3}$

■ *Write each expression in exponential form. See Example 5.*

37. $\sqrt{7}$
38. $\sqrt{5}$
39. $\sqrt[3]{2x}$
40. $\sqrt[3]{4y}$
41. $\dfrac{3}{\sqrt[4]{6}}$
42. $\dfrac{2}{\sqrt[5]{3}}$
43. $\dfrac{5\sqrt[3]{x}}{\sqrt[4]{yz}}$
44. $\dfrac{2\sqrt[5]{z}}{\sqrt[3]{xy}}$
45. $\sqrt[3]{x} - 3\sqrt{y}$
46. $\sqrt{x} - 2\sqrt[3]{y}$
47. $\dfrac{1}{\sqrt[3]{x-2y}}$
48. $\dfrac{1}{\sqrt[4]{3x+2y}}$

■ *Find a decimal representation for each fraction. Does the decimal terminate or repeat a pattern? See Example 6.*

49. $\dfrac{3}{8}$
50. $\dfrac{7}{16}$
51. $\dfrac{5}{6}$
52. $\dfrac{5}{12}$
53. $\dfrac{2}{7}$
54. $\dfrac{11}{13}$
55. $\dfrac{43}{11}$
56. $\dfrac{25}{6}$

■ *Use a calculator to approximate each irrational number to the nearest thousandth. See Examples 7 and 8.*

57. $\sqrt{2}$
58. $\sqrt{3}$
59. $\sqrt[3]{75}$
60. $\sqrt[4]{60}$
61. $\sqrt[5]{-43}$
62. $\sqrt[5]{-87}$
63. $\sqrt[4]{1.6}$
64. $\sqrt[3]{1.4}$

- *Simplify each expression. See Example 9.*

65. $(\sqrt[3]{125})^3$
66. $(\sqrt[4]{16})^4$
67. $(\sqrt[4]{2})^4$
68. $(\sqrt[3]{6})^3$
69. $(\sqrt[5]{-8})^5$
70. $(-\sqrt[4]{7})^4$
71. $(-\sqrt{x})^2$
72. $(-\sqrt[5]{y})^5$

- *Approximate each irrational number between two integers without the aid of a calculator. See Example 10.*

73. $\sqrt{175}$
74. $\sqrt{380}$
75. $\sqrt[3]{423}$
76. $\sqrt[3]{217}$
77. $\sqrt[3]{-84.6}$
78. $\sqrt[3]{-52.3}$
79. $\sqrt[3]{129}$
80. $\sqrt[3]{306}$

81. Find the length of the side of a square whose area is
    a. 169 square inches.
    b. 15 square feet.
    c. 2 square miles.
    d. 0.2304 square centimeter.

82. Find the radius of a circle whose area is
    a. 225π square meters.
    b. 18π square inches.
    c. 3π square yards.
    d. 0.5476π square feet.

83. Find the side of a cube whose volume is
    a. 1728 cubic millimeters.
    b. 400 cubic yards.
    c. 57 cubic feet.
    d. 15.625 cubic inches.

84. If the radius and height of a right circular cylinder are both equal to a, the volume of the cylinder is given by πa^3. Find the height of such a cylinder if its volume is
    a. 1331π cubic feet.
    b. 260π cubic meters.
    c. 23π cubic inches.
    d. 3.375π cubic centimeters.

85. The time in seconds for a pendulum of length L feet to complete one full swing is given by $2\pi\sqrt{L/g}$, where $g = 32$. How long does it take a pendulum of length 4 feet to complete one swing?

86. When a car brakes suddenly its speed in miles per hour can be estimated from the length of its skid marks, d, by using the formula $\sqrt{30kd}$. The constant k is determined by the friction of the road surface against the tires and has the value 0.8 for a dry concrete road. If a car leaves skid marks 160 feet long, how fast was it traveling when it braked?

87. The radius of a sphere of volume V is given by $\sqrt[3]{3V/4\pi}$. Find the radius of a spherical hot air balloon that holds 17,157 cubic feet of air.

88. A windmill will generate P watts of power when the wind velocity, in miles per hour, is $\sqrt[3]{P/0.015}$. Find the wind velocity necessary to generate 500 watts of power.

89. In order for a vehicle to be launched into space it must achieve a certain minimum speed called escape velocity, given by $r\sqrt{2g/(r + h)}$, where $g = 7.89 \times 10^4$ represents the gravitational pull of the earth, $r = 3960$ miles is the radius of the earth, and h is the height of the launch site above the earth's surface.

a. What escape velocity, in miles per hour, is needed to launch a satellite from the ground?
b. What escape velocity is needed to launch a satellite from a space station orbiting at a distance of 2.23×10^4 miles above the earth's surface?

90. In order to stay in orbit about the earth a satellite must maintain a certain orbital velocity given by $r\sqrt{g/(r + h)}$, where $g = 7.89 \times 10^4$ represents the gravitational pull of the earth, $r = 3960$ miles is the radius of the earth, and h is the height of the satellite above the surface of the earth.
    a. A satellite is in a geosynchronous orbit if it remains in the same position when viewed by observers on the earth. This occurs when the satellite orbits at a height of 2.23×10^4 miles above the earth. What orbital velocity, in miles per hour, will maintain a satellite in geosynchronous orbit?
    b. The moon orbits the earth at a distance of 234,000 miles. What is its orbital velocity (relative to the earth)?

91. The following formula for the flow of liquid through a tube can be used to describe blood flow through an artery. The radius in centimeters of an artery large enough to permit a blood flow of B cubic centimeters per second is given by $\sqrt[4]{8LB/k}$, where L is the length of the artery in centimeters and k is a constant determined by the blood pressure and viscosity. Find the radius of an artery that allows a blood flow of 5 liters per minute if the artery is 1 meter long and $k = 1.04 \times 10^7$.

92. The temperature of the sun, in degrees Kelvin, can be computed from the following law, which gives the temperature as $\sqrt[4]{L/4\pi R^2 s}$, where $L = 3.9 \times 10^{33}$ is the total luminosity of the sun, $R = 6.96 \times 10^{10}$ centimeters is the radius of the sun, and $s = 5.7 \times 10^{-5}$ is a constant governing radiation. Calculate the temperature of the sun.

B

93. Which is greater, $16^{1/4}$ or $16^{1/2}$? Make a conjecture about the order of $a^{1/n}$ and $a^{1/m}$ when $n > m$ and $a > 1$.
94. Which is greater, $(\frac{1}{16})^{1/4}$ or $(\frac{1}{16})^{1/2}$? Make a conjecture about the order of $a^{1/n}$ and $a^{1/m}$ when $n > m$ and $a < 1$.
95. Use your calculator to estimate each of the following to five decimal places: $2^{1/2}$, $2^{1/3}$, $2^{1/4}$, ..., $2^{1/10}$, $2^{1/100}$, $2^{1/1000}$. Make a conjecture about the value of $2^{1/n}$ when n is very large.
96. Use your calculator to estimate each of the following to five decimal places: $(\frac{1}{2})^{1/2}$, $(\frac{1}{2})^{1/3}$, $(\frac{1}{2})^{1/4}$, ..., $(\frac{1}{2})^{1/10}$, $(\frac{1}{2})^{1/100}$, $(\frac{1}{2})^{1/1000}$. Make a conjecture about the value of $(\frac{1}{2})^{1/n}$ when n is very large.
97. Use the laws of exponents to show that $\sqrt[4]{x} = \sqrt{\sqrt{x}}$.
98. Use the laws of exponents to show that $\sqrt[6]{x} = \sqrt{\sqrt[3]{x}}$.

3.2 RATIONAL EXPONENTS

To extend our definition of fractional powers to exponents of the form m/n, where m is an integer and n is a natural number, we again appeal to Law (II) of exponents on page 85:

$$a^{m/n} = (a^{1/n})^m = (a^m)^{1/n} \quad \text{(if } a^{1/n} \text{ is a real number).} \quad (1)$$

Thus, we can look at $a^{m/n}$ either as the mth power of the nth root of a or as the nth root of the mth power of a.

EXAMPLE 1

a. $8^{2/3} = (8^{1/3})^2$
$= (2)^2 = 4$

b. $8^{2/3} = (8^2)^{1/3}$
$= (64)^{1/3} = 4$

c. $27^{-2/3} = \dfrac{1}{27^{2/3}}$
$= \dfrac{1}{(27^{1/3})^2}$
$= \dfrac{1}{3^2} = \dfrac{1}{9}$

d. $(-8)^{-5/3} = \dfrac{1}{(-8)^{5/3}}$
$= \dfrac{1}{[(-8)^{1/3}]^5}$
$= \dfrac{1}{(-2)^5} = -\dfrac{1}{32}$

Since $a^{1/n} = \sqrt[n]{a}$, we may write any power with a fractional exponent in radical form as follows:

$$a^{m/n} = \sqrt[n]{a^m} = (\sqrt[n]{a})^m \quad \text{(if } \sqrt[n]{a} \text{ is a real number).} \quad (2)$$

EXAMPLE 2

a. $x^{2/3} = \sqrt[3]{x^2}$
or
$x^{2/3} = (\sqrt[3]{x})^2$

b. $x^{2/5}y^{3/5} = \sqrt[5]{x^2}\sqrt[5]{y^3}$
or
$x^{2/5}y^{3/5} = (\sqrt[5]{x})^2(\sqrt[5]{y})^3$

c. $6x^{-3/4} = \dfrac{6}{x^{3/4}}$

$= \dfrac{6}{\sqrt[4]{x^3}}$

d. $(x + 2)^{-2/3} = \dfrac{1}{(x + 2)^{2/3}}$

$= \dfrac{1}{\sqrt[3]{(x + 2)^2}}$

We may also write radical expressions as powers with fractional exponents.

EXAMPLE 3   a. $\sqrt{x^5} = x^{5/2}$

b. $\sqrt[5]{2y^2} = (2y^2)^{1/5} = 2^{1/5}y^{2/5}$

c. $\dfrac{3}{2\sqrt[3]{(4y)^5}} = \dfrac{3}{2}(4y)^{-5/3}$

d. $\sqrt{x^2 + y^2} = (x^2 + y^2)^{1/2}$

Equation (2) on page 110 shows three ways to view a radical expression. When simplifying radicals we may use whichever form is most convenient.

EXAMPLE 4   a. $\sqrt[3]{-8^2} = \sqrt[3]{-64}$
$= -4$

b. $\sqrt[4]{\left(\dfrac{16}{81}\right)^3} = \left(\sqrt[4]{\dfrac{16}{81}}\right)^3$

$= \left(\dfrac{2}{3}\right)^3 = \dfrac{8}{27}$

c. $\sqrt{x^{14}} = x^{14/2}$
$= x^7$

d. $\sqrt[3]{27x^{12}y^6} = (27x^{12}y^6)^{1/3}$
$= 27^{1/3}x^{12/3}y^{6/3}$
$= 3x^4y^2$

Using a Calculator   These observations are also helpful in using a calculator to evaluate radicals and fractional powers.

EXAMPLE 5 Use a calculator to evaluate the following.

a. $\sqrt[4]{16^3}$
b. $7^{5/3}$

Solutions a. Viewing $\sqrt[4]{16^3}$ as $(16^3)^{1/4}$ we press

$$16 \;\boxed{y^x}\; 3 \;\boxed{y^x}\; 4 \;\boxed{\tfrac{1}{x}}\; \boxed{=} \; 8.$$

Alternatively, we can express $\sqrt[4]{16^3}$ as $(16^{1/4})^3$ and press

$$16 \;\boxed{y^x}\; 4 \;\boxed{\tfrac{1}{x}}\; \boxed{y^x}\; 3 \;\boxed{=} \; 8.$$

If the calculator has parentheses, we can press

$$16 \;\boxed{y^x}\; \boxed{(}\; 3 \;\boxed{\div}\; 4 \;\boxed{)}\; \boxed{=} \; 8.$$

If the calculator has a $\boxed{\sqrt[x]{y}}$ key, we can press

$$16 \;\boxed{\sqrt[x]{y}}\; 4 \;\boxed{y^x}\; 3 \;\boxed{=} \; 8.$$

In each case we find that $\sqrt[4]{16^3} = 8$.

b. $7^{5/3} = (7^{1/3})^5$, so we press

$$7 \;\boxed{y^x}\; 3 \;\boxed{\tfrac{1}{x}}\; \boxed{y^x}\; 5 \;\boxed{=} \; 25.61513997.$$

Or, alternatively, $7^{5/3} = (\sqrt[3]{7})^5$, so we press

$$7 \;\boxed{\sqrt[x]{y}}\; 3 \;\boxed{y^x}\; 5 \;\boxed{=} \; 25.61513997.$$

Thus, $7^{5/3} \approx 25.615$.

Decimal Form for Rational Exponents

Fractional exponents may also be expressed in decimal form. For example, $3^{5/2} = 3^{2.5}$, since $5/2 = 2.5$. Powers whose exponents are expressed in decimal form are particularly convenient to evaluate with a calculator, since only the $\boxed{y^x}$ key need be used. Thus, to evaluate $3^{2.5}$ we can press

$$3 \;\boxed{y^x}\; 2.5 \;\boxed{=} \; 15.58845727,$$

so $3^{5/2} = 3^{2.5} \approx 15.588$.

3.2 ■ RATIONAL EXPONENTS 113

EXAMPLE 6 Use a calculator to evaluate $1.6^{0.7}$.

Solution We press

$$1.6 \boxed{y^x} \; 0.7 \boxed{=} \; 1.389581386,$$

so $1.6^{0.7} \approx 1.390$.

Operations with Rational Exponents

With the definitions we have made it can be shown that powers with rational exponents—positive, negative, or zero—obey the laws of exponents developed in Section 2.5.

EXAMPLE 7

a. $y^{5/6} y^{-2/3} = y^{5/6 + (-4/6)}$
$= y^{1/6}$

b. $\dfrac{7^{0.75}}{7^{0.5}} = 7^{0.75 - 0.5}$
$= 7^{0.25}$

c. $(x^8)^{0.5} = x^{8(0.5)}$
$= x^4$

d. $\dfrac{(5^{1/2} y^2)^2}{(5^{2/3} y)^3} = \dfrac{5 y^4}{5^2 y^3}$
$= \dfrac{y^{4-3}}{5^{2-1}} = \dfrac{y}{5}$

Products and Factors

We can use the distributive law to multiply factors that involve more than one term.

EXAMPLE 8

a. $y^{1/3}(y + y^{2/3})$
$= y^{1/3 + 1} + y^{1/3 + 2/3}$
$= y^{4/3} + y$

b. $(x^{1/2} - 3x^{-1/2})^2$
$= x^{1/2 + 1/2} - 6x^{1/2 - 1/2} + 9x^{-1/2 - 1/2}$
$= x - 6 + 9x^{-1}$

c. $\dfrac{5x + x^{4/3}}{4x^{2/3}}$
$= \dfrac{1}{4} x^{-2/3}(5x + x^{4/3})$
$= \dfrac{5}{4} x^{1/3} + \dfrac{1}{4} x^{2/3}$

d. $(x^{-1.2} + 2)(3x^{2.6} - 2)$
$= 3x^{-1.2 + 2.6} - 2x^{-1.2} + 6x^{2.6} - 4$
$= 3x^{1.4} - 2x^{-1.2} + 6x^{2.6} - 4$

We can also factor expressions involving powers with fractional exponents. In the following examples we factor the power with the smallest exponent from an expression with more than one term.

EXAMPLE 9

a. $3x + x^{1/2} = x^{1/2}(3x^{1/2} + 1)$
because
$x^{1/2}(3x^{1/2} + 1) = 3x + x^{1/2}$.

b. $x^{-1/2} + 1 = x^{-1/2}(1 + x^{1/2})$
because
$x^{-1/2}(1 + x^{1/2}) = x^{-1/2} + 1$.

c. $(x + 3)^{-1/3} - 3(x + 3)^{-2/3} = (x + 3)^{-2/3}[(x + 3)^{1/3} - 3]$
because
$(x + 3)^{-2/3}[(x + 3)^{1/3} - 3] = (x + 3)^{-1/3} - 3(x + 3)^{-2/3}$.

d. $x^{1/4}(x + 2)^{-1/4} + x^{-3/4}(x + 2)^{1/2} = x^{-3/4}(x + 2)^{-1/4}[x + (x + 2)^{3/4}]$
because
$x^{-3/4}(x + 2)^{-1/4}[x + (x + 2)^{3/4}] = x^{1/4}(x + 2)^{-1/4} + x^{-3/4}(x + 2)^{1/2}$.

Note that the factored expression in Example 9b can now be written as

$$\left(\frac{1}{x^{1/2}}\right)(1 + x^{1/2}) = \frac{1 + x^{1/2}}{x^{1/2}},$$

an expression involving only positive exponents.

EXAMPLE 10 Write as a single fraction involving only positive exponents.

a. $2x^{-1/2} - 4x^{1/2}$

b. $3x^{-1/3} + 2$

Solutions

a. $2x^{-1/2} - 4x^{1/2} = 2x^{-1/2}(? - ?)$
$= 2x^{-1/2}(1 - 2x)$
$= \dfrac{2(1 - 2x)}{x^{1/2}}$

b. $3x^{-1/3} + 2 = x^{-1/3}(? + ?)$
$= x^{-1/3}(3 + 2x^{1/3})$
$= \dfrac{3 + 2x^{1/3}}{x^{1/3}}$

EXERCISE 3.2

A

- In this exercise assume that all variables and radicands are positive.
- Find the indicated power. See Example 1.

1. $81^{3/4}$
2. $125^{2/3}$
3. $(-8)^{4/3}$
4. $(-64)^{2/3}$
5. $16^{-1/2}$
6. $8^{-1/3}$
7. $(-125)^{-4/3}$
8. $(-32)^{-3/5}$
9. $-\left(\dfrac{81}{625}\right)^{3/4}$
10. $\left(\dfrac{-243}{32}\right)^{2/5}$
11. $\left(\dfrac{-64}{125}\right)^{2/3}$
12. $-\left(\dfrac{289}{100}\right)^{5/2}$

- Write each expression in radical form. See Example 2.

13. $x^{4/5}$
14. $y^{3/4}$
15. $3x^{2/5}$
16. $5y^{2/3}$
17. $y^{-5/6}$
18. $x^{-2/7}$
19. $(xy)^{-2/3}$
20. $(xy)^{-3/5}$
21. $3y^{-2/3}$
22. $4x^{-3/2}$
23. $-2x^{1/4}y^{3/4}$
24. $-3x^{2/5}y^{3/5}$
25. $6(x + 2y)^{3/2}$
26. $9(x - 2y)^{2/3}$
27. $(x^2 - 4)^{-5/2}$
28. $(x^3 - 8)^{-5/3}$

- Write each expression with fractional exponents. See Example 3.

29. $\sqrt[3]{x^2}$
30. $\sqrt{y^3}$
31. $\sqrt[3]{(ab)^2}$
32. $\sqrt[3]{ab^2}$
33. $2\sqrt[5]{ab^3}$
34. $6\sqrt[5]{(ab)^3}$
35. $\dfrac{-4y}{\sqrt[4]{x^3}}$
36. $\dfrac{-2x}{\sqrt[3]{y^2}}$
37. $-\sqrt[4]{(a^2 + b)^3}$
38. $-\sqrt[3]{(2a - b^2)^2}$
39. $\dfrac{ab}{\sqrt[5]{5(a - b^2)^2}}$
40. $\dfrac{b^2}{\sqrt[4]{3(a^3 + b)^3}}$

- Find the indicated root. See Example 4.

41. $\sqrt[5]{32^3}$
42. $\sqrt[4]{16^5}$
43. $\sqrt[3]{-\left(\dfrac{27}{8}\right)^4}$
44. $\sqrt[3]{\left(\dfrac{-125}{64}\right)^2}$
45. $\sqrt[4]{16y^{12}}$
46. $\sqrt[5]{243x^{10}}$
47. $-\sqrt{a^8b^{16}}$
48. $-\sqrt{a^{10}b^{36}}$
49. $\sqrt[3]{\dfrac{-8}{125}x^9y^{27}}$
50. $-\sqrt[3]{\dfrac{64}{27}x^6y^{18}}$
51. $-\sqrt[4]{81a^8b^{12}}$
52. $\sqrt[5]{-32x^{25}y^5}$

- Use a calculator to approximate each power or root to the nearest thousandth. See Examples 5 and 6.

53. $12^{5/6}$
54. $20^{5/4}$
55. $\sqrt[3]{-6^4}$
56. $\sqrt[5]{-8^3}$
57. $(37.8)^{-2/3}$
58. $123^{-3/2}$
59. $4.7^{2.3}$
60. $16.1^{0.29}$

- Write each expression as a product or quotient of powers in which each variable occurs only once and all exponents are positive. See Example 7.

61. $2x^{3/2}x^{1/4}$
62. $4x^{1/3}x^{5/6}$
63. $\dfrac{xy}{3x^{1/3}}$

64. $\dfrac{xy^2}{5x^{1/4}}$

65. $(6x^{-2/3})\left(\dfrac{1}{3}x^{5/3}\right)$

66. $\left(\dfrac{1}{2}x^{-3/4}\right)(4x^{1/4})$

67. $(-27y^{-1/2})^{2/3}$

68. $(-32y^{-2/3})^{3/5}$

69. $\left(\dfrac{16a^{-8}}{b^{2/3}}\right)^{-3/4}$

70. $\left(\dfrac{a^{9/2}}{8b^{-6}}\right)^{-4/3}$

71. $\left(\dfrac{a^{0.2}b^3}{b^{1.7}}\right)^{-1.5}$

72. $\left(\dfrac{a^{2.1}b^{1.1}}{a^3}\right)^{-0.6}$

73. $\dfrac{(4x^{-6}y^2)^{3/2}}{6(x^{5/2}y)^{-2}}$

74. $\dfrac{(2x^{-1/2}y)^{-3}}{8(x^{-3/4}y^6)^{4/3}}$

75. $\left(\dfrac{a^{1/2}}{b^{5/6}}\right)^{2/5}\left(\dfrac{b^{2/3}}{a^{-5}}\right)^{3/5}$

76. $\left(\dfrac{a^3}{b^{4/3}}\right)^{1/4}\left(\dfrac{b^{-2}}{a^2}\right)^{3/4}$

■ *Write each product or quotient as a sum of terms. See Example 8.*

77. $2x^{1/2}(x - x^{1/2})$

78. $x^{1/3}(2x^{2/3} - x^{1/3})$

79. $3y^{-3/8}\left(\dfrac{1}{4}y^{-1/4} + y^{3/4}\right)$

80. $\dfrac{1}{2}y^{-1/3}(y^{2/3} + 3y^{-5/6})$

81. $(2x^{1/3} - 1)(x^{1/3} + 1)$

82. $(2x^{1/4} + 1)(x^{1/4} - 1)$

83. $(a^{3/4} - 2)^2$

84. $(a^{2/3} + 3)^2$

85. $\dfrac{b^{3/5} - \dfrac{1}{4}b^{-2/5}}{2b^{2/5}}$

86. $\dfrac{3b^{-1/4} + \dfrac{1}{2}b^{5/4}}{2b^{3/4}}$

87. $x^{1.3}(x - x^{2.1})$

88. $x^{-3.1}(2x^{1.2} + x^{0.4})$

89. $(x^{-1.2} - 3)(x^{1.4} + 2)$

90. $(2y^{0.2} + 1)(3y^{-1.8} - 2)$

91. During a flu epidemic in a small town health officials estimate that the number of people infected t days after the first case was discovered is given by $50t^{3/5}$. How many people have the flu after 1 week?

92. The research division of an advertising firm estimates that the number of people who have seen their ads t days after the campaign begins is given by $2000t^{5/4}$. How many people have seen the ads after 30 days?

93. The period of revolution of a planet can be calculated if its average distance a from the sun is known (Kepler's law). The period is given in years by $K^{1/2}a^{3/2}$, where $K = 1.243 \times 10^{-24}$ and a is measured in miles. Find the period of Mars if its average distance from the sun is 1.417×10^8 miles.

94. If the period of a planet is known, its average distance from the sun in miles is given by Kepler's law as $K^{1/3}p^{2/3}$, where $K = 1.243 \times 10^{-24}$ and the period is given in years. Find the distance from Venus to the sun if its period is 0.615 year.

95. The intensity of sound is measured by comparing it to the minimum intensity detectable by the human ear. This ratio is given by $10^{0.1D}$, where D is the loudness of the sound in decibels. A jet airliner taking off produces about 125 decibels. How many times as intense as the minimum detectable sound is that?

96. Earthquakes are measured by comparing their intensity to a fixed minimum intensity. This ratio is given by 10^R, where R is the magnitude of the earthquake on the Richter scale. How many times as intense as the minimum earthquake was the San Francisco earthquake of 1906, which measured 8.3 on the Richter scale?

97. A brewery wants to replace its old vats with new ones that hold 1.8 times as much as the old vats. To estimate the cost of the new equipment the accountant uses the 0.6 rule for industrial costs, which states that the cost of a new container is approximately $Cr^{0.6}$, where C is the cost of the old container and r is the ratio of the capacity of the new container to the old. If an old vat cost \$5000, how much should the accountant budget for a new one?

98. If a quantity of air expands without changing temperature, its pressure in pounds per square inch is given by $kV^{-1.4}$, where V is the volume of the air in cubic inches and $k = 2.79 \times 10^4$ is a constant. Find the air pressure of a sample when its volume is 50 cubic inches.

99. The population of a beehive increases according to the formula $20 \cdot 3^{t/8}$, where t is in days. What is the population of the beehive after 2 weeks?

100. The value of a motorcycle decreases according to the formula $2500 \cdot 2^{-t/3}$, where t is in years. How much is the motorcycle worth after 2 years?

■ *For Problems* 101 *and* 102 *use the following formula for inflation rate. An item that originally cost P dollars will cost*

$$P(1 + r)^t$$

dollars after t years if the inflation rate is 100r% annually.

101. How much will a \$25 shirt cost 30 months from now if the annual inflation rate is 5%?

102. How much will a \$1500 stereo system cost 18 months from now if the annual inflation rate is 6.5%?

B

■ *Factor. Write your answers as expressions involving positive exponents only. See Example 9 and 10.*

103. $x^{3/2} + x$

104. $y - y^{2/3}$

105. $x^{-3/2} + x^{-1/2}$

106. $y^{3/4} - y^{-1/4}$

107. $(x + 1)^{1/2} - (x + 1)^{-1/2}$

108. $(y + 2)^{1/5} - (y + 2)^{-4/5}$

109. $y(y - 2)^{1/2} + y^{3/2}(y - 2)^{-1/2}$

110. $x^{1/2}(x - 3)^{-1/2} + x^{-1/2}(x - 3)^{-3/2}$

111. $\dfrac{2}{3}(2z + 1)^{-2/3}z^{2/3} + (2z + 1)^{1/3}z^{-1/3}$

112. $-z^{-4/3}(z + 2)^{4/3} + \dfrac{4}{3}z^{-1/3}(z + 2)^{1/3}$

3.3

SIMPLIFYING RADICALS

Although we can write decimal approximations for radical expressions, in many situations we prefer to leave such expressions in radical form. In these cases it is often helpful to simplify the radicals algebraically. From the definition of a radical and the laws of exponents we can derive two important relationships that are useful for this purpose. One such relationship involves products:

For $a, b \geq 0$ and n a natural number,

$$\sqrt[n]{ab} = \sqrt[n]{a}\sqrt[n]{b}. \qquad (1)$$

This property follows from the fact that

$$\sqrt[n]{ab} = (ab)^{1/n}$$
$$= a^{1/n}b^{1/n} = \sqrt[n]{a}\sqrt[n]{b}.$$

Property (1) can be used to simplify a radical by reducing the powers of the radicand. In Example 1 each radicand is first factored into two factors, one of which consists of factors raised to multiples of the index. These factors are then removed from the radicand.

EXAMPLE 1

a. $\sqrt{18} = \sqrt{3^2}\sqrt{2}$
$= 3\sqrt{2}$

b. $\sqrt[3]{x^7} = \sqrt[3]{x^6}\sqrt[3]{x}$
$= x^2\sqrt[3]{x}$

c. $\sqrt[3]{16x^3y^5} = \sqrt[3]{2^3x^3y^3}\sqrt[3]{2y^2}$
$= 2xy\sqrt[3]{2y^2}$

d. $\sqrt[4]{6x^2y^3}\sqrt[4]{8x^3y^3} = \sqrt[4]{48x^5y^6}$
$= \sqrt[4]{16x^4y^4}\sqrt[4]{3xy^2}$
$= 2xy\sqrt[4]{3xy^2}$

Common Error

Property (1) above applies to *products* under the radical and *not* to sums. Thus, note that

$$\sqrt{4 \cdot 9} = \sqrt{4}\sqrt{9} = 2 \cdot 3, \quad \text{but} \quad \sqrt{4 + 9} \neq \sqrt{4} + \sqrt{9};$$

and

$$\sqrt[3]{x^3y^6} = \sqrt[3]{x^3}\sqrt[3]{y^6} = xy^2, \quad \text{but} \quad \sqrt[3]{x^3 + y^6} \neq \sqrt[3]{x^3} + \sqrt[3]{y^6}.$$

A second important relationship involves quotients:

For $a \geq 0$, $b > 0$ and n a natural number,
$$\sqrt[n]{\frac{a}{b}} = \frac{\sqrt[n]{a}}{\sqrt[n]{b}}. \qquad (2)$$

This property follows from the fact that
$$\sqrt[n]{\frac{a}{b}} = \left(\frac{a}{b}\right)^{1/n}$$
$$= \frac{a^{1/n}}{b^{1/n}} = \frac{\sqrt[n]{a}}{\sqrt[n]{b}}.$$

We can use Property (2) to write a radical so that the radicand contains no fractions.

EXAMPLE 2 a. $\sqrt{\dfrac{3}{4}} = \dfrac{\sqrt{3}}{\sqrt{4}} = \dfrac{\sqrt{3}}{2}$ b. $\sqrt[3]{\dfrac{5}{8}} = \dfrac{\sqrt[3]{5}}{\sqrt[3]{8}} = \dfrac{\sqrt[3]{5}}{2}$

Rationalizing Denominators

If the denominator of a fraction contains radicals, we can use the fundamental principle of fractions to obtain an equivalent form in which the denominator is free of radicals. For square roots we multiply the numerator and denominator of the fraction by the radical in the denominator.

EXAMPLE 3 a. $\sqrt{\dfrac{1}{3}} = \dfrac{\sqrt{1}}{\sqrt{3}}$ b. $\sqrt{\dfrac{2}{5x}} = \dfrac{\sqrt{2}}{\sqrt{5x}}$

$\phantom{\text{EXAMPLE 3 a.}} = \dfrac{1 \cdot \sqrt{3}}{\sqrt{3} \cdot \sqrt{3}} = \dfrac{\sqrt{2} \cdot \sqrt{5x}}{\sqrt{5x} \cdot \sqrt{5x}}$

$\phantom{\text{EXAMPLE 3 a.}} = \dfrac{\sqrt{3}}{3} = \dfrac{\sqrt{10x}}{5x}$

The process shown in the example above is called *rationalizing the denominator* of a fraction. The next example illustrates the technique for radicals of index greater than two.

EXAMPLE 4 Rationalize the denominator.

a. $\dfrac{1}{\sqrt[3]{2x}}$
b. $\sqrt[5]{\dfrac{6}{16x^3}}$

Solutions a. We need a *third* power, $(2x)^3$, beneath the radical in the denominator in order to rewrite it without a radical. Therefore, we must multiply $\sqrt[3]{2x}$ by *two* additional factors of $\sqrt[3]{2x}$. Thus, using the fundamental principle of fractions we obtain

$$\frac{1}{\sqrt[3]{2x}} = \frac{1 \cdot \sqrt[3]{2x} \cdot \sqrt[3]{2x}}{\sqrt[3]{2x} \cdot \sqrt[3]{2x} \cdot \sqrt[3]{2x}}$$
$$= \frac{\sqrt[3]{(2x)^2}}{\sqrt[3]{(2x)^3}} = \frac{\sqrt[3]{4x^2}}{2x}.$$

b. In this instance we have

$$\sqrt[5]{\frac{6}{16x^3}} = \sqrt[5]{\frac{6}{2^4 x^3}}.$$

Since we want a fifth power, namely $(2x)^5$ or $32x^5$, in the denominator, we multiply the numerator and denominator by the factor $\sqrt[5]{2x^2}$. Thus,

$$\sqrt[5]{\frac{6}{16x^3}} = \frac{\sqrt[5]{6} \cdot \sqrt[5]{2x^2}}{\sqrt[5]{16x^3} \cdot \sqrt[5]{2x^2}}$$
$$= \frac{\sqrt[5]{12x^2}}{\sqrt[5]{32x^5}} = \frac{\sqrt[5]{12x^2}}{2x}.$$

It is not always necessary to rationalize a denominator that contains radicals. Often radicals in the denominator can be eliminated by applying Properties (1) and (2) above.

EXAMPLE 5 a. $\dfrac{\sqrt{a}\sqrt{ab^3}}{\sqrt{b}} = \sqrt{\dfrac{a^2b^3}{b}}$
$= \sqrt{a^2b^2}$
$= ab$

b. $\dfrac{\sqrt[3]{16y^4}}{\sqrt[3]{y}} = \sqrt[3]{\dfrac{16y^4}{y}}$
$= \sqrt[3]{2^4 y^3}$
$= 2y\sqrt[3]{2}$

Reducing Indices

Sometimes a radical can be simplified to an equivalent radical with a lower index. This can be accomplished by first writing the radical in exponential form.

EXAMPLE 6 a. $\sqrt[4]{5^2} = 5^{2/4}$
$= 5^{1/2}$
$= \sqrt{5}$

b. $\sqrt[6]{9} = (3^2)^{1/6}$
$= 3^{1/3}$
$= \sqrt[3]{3}$

c. $\sqrt[8]{x^2} = x^{2/8}$
$= x^{1/4}$
$= \sqrt[4]{x}$

The techniques illustrated above can be used to rewrite radicals in various ways. Most often we will want to simplify radical expressions according to the following guidelines.

A Radical Expression Is in Simplest Form If:

1. The radicand contains no factor raised to a power equal to or greater than the index of the radical.
2. The radicand contains no fractions.
3. No radical expressions are contained in the denominators of fractions.
4. The index of the radical cannot be reduced.

EXAMPLE 7

a. $\dfrac{\sqrt{2x^2y}\,\sqrt{6x^3y}}{\sqrt{3y^3}} = \sqrt{\dfrac{4x^5}{y}}$

$= \dfrac{2x^2\sqrt{x}\cdot\sqrt{y}}{\sqrt{y}\cdot\sqrt{y}}$

$= \dfrac{2x^2\sqrt{xy}}{y}$

b. $\sqrt[9]{a^2b^5}\,\sqrt[9]{a^4b^4} = \sqrt[9]{a^6b^9}$
$= b\sqrt[9]{a^6} = ba^{6/9}$
$= ba^{2/3} = b\sqrt[3]{a^2}$

c. $\sqrt[3]{\dfrac{3x^4 + 2x^3}{y^2}}$

$= \dfrac{\sqrt[3]{x^3(3x+2)}\cdot\sqrt[3]{y}}{\sqrt[3]{y^2}\cdot\sqrt[3]{y}}$

$= \dfrac{\sqrt[3]{x^3}\,\sqrt[3]{(3x+2)y}}{\sqrt[3]{y^3}}$

$= \dfrac{x}{y}\sqrt[3]{(3x+2)y}$

d. $\sqrt{5a(2a+1)}\,\sqrt{10a^3(2a+1)}$
$= \sqrt{50a^4(2a+1)^2}$
$= \sqrt{25a^4(2a+1)^2}\,\sqrt{2}$
$= 5a^2(2a+1)\sqrt{2}$

Simplifying $\sqrt[n]{a^n}$

In Section 3.1 we saw that raising to powers and extracting roots are "inverse" operations; that is,

$$(\sqrt[n]{a})^n = a.$$

For example, if $a = 16$ and $n = 4$, then

$$(\sqrt[4]{16})^4 = 2^4 = 16.$$

Now consider the power and root operations in the opposite order; that is, consider

$$\sqrt[n]{a^n}.$$

We treat the two cases "n odd" and "n even" separately.
If n is an odd number, then

$$\sqrt[n]{a^n} = a \quad (n \text{ odd}).$$

For example,

$$\sqrt[3]{2^3} = \sqrt[3]{8} = 2 \quad \text{and} \quad \sqrt[3]{(-2)^3} = \sqrt[3]{-8} = -2.$$

However, because we defined a radical with an *even* index to be the principal, or nonnegative, root, we have the following special relationship for *n* even:

$$\sqrt[n]{a^n} = |a| \quad (n \text{ even}).$$

In particular,

$$\sqrt{a^2} = |a|.$$

(See the Review of Elementary Topics, page xiii, for the definition of absolute value.) For example,

$$\sqrt{3^2} = |3| = 3 \quad \text{and} \quad \sqrt{(-3)^2} = |-3| = 3.$$

Common Error Note that $\sqrt{(-3)^2} \neq -3$, since the symbol $\sqrt{}$ represents the principal, or positive, square root.

If a radical of *even* index involves variables, we must use absolute-value notation when removing factors from the radicand.

EXAMPLE 8 a. $\sqrt{16x^2} = 4|x|$

b. $\sqrt{x^2 - 2xy + y^2} = \sqrt{(x-y)^2}$
$= |x - y|$

EXERCISE 3.3

A

■ *Assume that all variables and radicands in this exercise denote positive real numbers.*

■ *Simplify. See Examples 1 and 2.*

1. $\sqrt{18}$
2. $\sqrt{50}$
3. $\sqrt[3]{24}$
4. $\sqrt[3]{54}$
5. $-\sqrt[4]{64}$
6. $-\sqrt[4]{162}$
7. $\sqrt{60{,}000}$
8. $\sqrt{800{,}000}$

9. $\sqrt[3]{900{,}000}$ 10. $\sqrt[3]{24{,}000}$ 11. $\sqrt[3]{\dfrac{-40}{27}}$ 12. $\sqrt[4]{\dfrac{80}{625}}$

13. $\sqrt[3]{x^{10}}$ 14. $\sqrt[3]{y^{16}}$ 15. $\sqrt{27z^3}$ 16. $\sqrt{12t^5}$

17. $\sqrt[4]{48a^9b^{12}}$ 18. $\sqrt[3]{81a^{12}b^8}$ 19. $-\sqrt[5]{-96p^7q^9r^{11}}$ 20. $-\sqrt[6]{256k^7u^{12}v^{15}}$

21. $-\sqrt{18s}\,\sqrt{2s^3}$ 22. $\sqrt[3]{3w^3}\,\sqrt[3]{27w^3}$ 23. $\sqrt[3]{7b^2}\,\sqrt[3]{-49b}$ 24. $-\sqrt[4]{2m^3}\,\sqrt[4]{8m}$

25. $\sqrt{16-4x^2}$ 26. $\sqrt{9y^2+18}$ 27. $\sqrt[3]{8a^3+a^6}$ 28. $\sqrt[3]{b^9-27b^{27}}$

29. $-\sqrt{\dfrac{p^{13}v^8}{225a^4}}$ 30. $\sqrt{\dfrac{-16c^{15}b^7}{-169n^6}}$ 31. $\sqrt[5]{\dfrac{-81a^{12}}{32b^{15}}}$ 32. $\sqrt[6]{\dfrac{192k^3}{m^{18}}}$

■ *Rationalize each denominator. See Example 3.*

33. $\dfrac{6}{\sqrt{3}}$ 34. $\dfrac{10}{\sqrt{5}}$ 35. $\dfrac{-\sqrt{3}}{\sqrt{7}}$ 36. $\dfrac{-\sqrt{5}}{\sqrt{6}}$

37. $\sqrt{\dfrac{7x}{18}}$ 38. $\sqrt{\dfrac{27x}{20}}$ 39. $\sqrt{\dfrac{2a}{b}}$ 40. $\sqrt{\dfrac{5p}{q}}$

41. $\dfrac{2\sqrt{3}}{\sqrt{2k}}$ 42. $\dfrac{6\sqrt{2}}{\sqrt{3v}}$ 43. $\dfrac{-9x^2\sqrt{5x^3}}{2\sqrt{6z}}$ 44. $\dfrac{-8y\sqrt{21y^5}}{3\sqrt{10t}}$

■ *See Example 4.*

45. $\dfrac{1}{\sqrt[3]{x^2}}$ 46. $\dfrac{1}{\sqrt[4]{y^3}}$ 47. $\sqrt[3]{\dfrac{2}{3y}}$ 48. $\sqrt[4]{\dfrac{2}{3x}}$

49. $\sqrt[3]{\dfrac{x}{4y^2}}$ 50. $\sqrt[4]{\dfrac{x}{8y^3}}$ 51. $\sqrt[5]{\dfrac{3}{2x^3}}$ 52. $\sqrt[5]{\dfrac{2}{9y^2}}$

53. $\dfrac{9x^3}{\sqrt[4]{27x}}$ 54. $\dfrac{15x^4}{\sqrt[3]{5x}}$ 55. $\dfrac{xy}{\sqrt[3]{(x+y)^2}}$ 56. $\dfrac{x+y}{\sqrt[4]{(x+y)^2}}$

■ *Simplify. See Example 5.*

57. $\dfrac{\sqrt{a^5b^3}}{\sqrt{ab}}$ 58. $\dfrac{\sqrt{x}\,\sqrt{xy^3}}{\sqrt{y}}$ 59. $\dfrac{\sqrt{98x^2y^3}}{\sqrt{xy}}$ 60. $\dfrac{\sqrt{45x^3}\,\sqrt{y^3}}{\sqrt{5y}}$

61. $\dfrac{\sqrt[3]{8b^4}}{\sqrt[3]{a^6}}$ 62. $\dfrac{\sqrt[3]{16r^4}}{\sqrt[3]{4t^3}}$ 63. $\dfrac{\sqrt[5]{a}\,\sqrt[5]{b^2}}{\sqrt[5]{ab}}$ 64. $\dfrac{\sqrt[5]{x^2}\,\sqrt[5]{y^3}}{\sqrt[5]{xy^2}}$

■ *Reduce the order of each radical. See Example 6.*

65. $\sqrt[4]{3^2}$ 66. $\sqrt[6]{2^2}$ 67. $\sqrt[6]{3^3}$ 68. $\sqrt[8]{5^2}$

69. $\sqrt[6]{81}$ 70. $\sqrt[10]{32}$ 71. $\sqrt[6]{x^3}$ 72. $\sqrt[9]{y^3}$

■ *Simplify. See Example 7.*

73. $\dfrac{\sqrt[3]{4a^5b^2}\,\sqrt[3]{6ab}}{\sqrt[3]{3b^4}}$ 74. $\dfrac{\sqrt[4]{8ab^3}\,\sqrt[4]{8a^2b^3}}{\sqrt[4]{2a^5}}$ 75. $\dfrac{18x^2y}{\sqrt[3]{3x^2y^3}\,\sqrt[4]{3x^9y^3}}$

76. $\dfrac{6xy^3}{\sqrt[4]{4x^3y^4}\,\sqrt[3]{2x^6y^3}}$ 77. $\sqrt[6]{2x^7z^5}\,\sqrt[6]{4x^8z^4}$ 78. $\sqrt[8]{81x^6z^5}\,\sqrt[8]{x^4z^{15}}$

79. $\sqrt{18x(x^2+1)}\,\sqrt{2x^3(x^2+1)}$ 80. $\sqrt{15y(2y-3)}\,\sqrt{30y(2y-3)^3}$

81. $\sqrt[3]{6a^2(a-b)^2}\,\sqrt[3]{4ab^2(a-b)^2}$ 82. $\sqrt[3]{9ab(a^2+3)^2}\,\sqrt[3]{12a^2b^4(a^2+3)^2}$

B

■ *Simplify. Assume that all variables and radicands represent positive numbers.*

83. $\sqrt{8(x-1)^3}$
84. $\sqrt{12(x+2)^3}$
85. $\sqrt{x^3(y-2)^5}$
86. $\sqrt{y^5(x-1)^3}$
87. $\sqrt{\dfrac{(y-3)^3}{xy^3}}$
88. $\sqrt{\dfrac{(x+2)^5}{x^3y}}$
89. $\sqrt[3]{\dfrac{(x-1)^4}{xy^2}}$
90. $\sqrt[3]{\dfrac{(y+1)^4}{x^2y}}$
91. $\sqrt[3]{(4x^5 - x^3)(x+1)^4}$
92. $\sqrt[3]{(2y^4 - y^3)(y+2)^5}$
93. $\dfrac{\sqrt{9(x-1)^2}}{\sqrt{3x}\sqrt{x^3 - x^2}}$
94. $\dfrac{\sqrt{4(y+2)^2}}{\sqrt{4y}\sqrt{y^3 + 2y^2}}$

■ *In the preceding problems variables and radicands were restricted to represent positive numbers. In Problems 95–100 consider the variables to represent any real numbers. Simplify, using absolute-value notation as needed. See Example 8.*

95. $\sqrt{4x^2}$
96. $\sqrt{9x^2y^4}$
97. $\sqrt{x^2 + 2x + 1}$
98. $\sqrt{4x^2 - 4x + 1}$
99. $\dfrac{2}{\sqrt{x^2 + 2xy + y^2}}$
100. $\dfrac{3}{\sqrt{x^4 + 2x^2y^2 + y^4}}$

3.4

OPERATIONS ON RADICAL EXPRESSIONS

Sums and Differences

Recall that sums or differences of like terms can be simplified by using the distributive law:

$$3xy + 5xy = (3 + 5)xy = 8xy.$$

"Like radicals," that is, radicals of the same index and radicand, can be combined in the same way.

EXAMPLE 1

a. $3\sqrt{3} + 4\sqrt{3} = (3 + 4)\sqrt{3}$
$= 7\sqrt{3}$

b. $7\sqrt{x} - 2\sqrt{x} = (7 - 2)\sqrt{x}$
$= 5\sqrt{x}$

c. $4\sqrt[3]{2y} - 6\sqrt[3]{2y} = (4 - 6)\sqrt[3]{2y}$
$= -2\sqrt[3]{2y}$

d. $3x\sqrt[4]{2} + x\sqrt[4]{2} = (3 + 1)x\sqrt[4]{2}$
$= 4x\sqrt[4]{2}$

Common Error

Note that in Example 1a $3\sqrt{3} + 4\sqrt{3} \neq 7\sqrt{6}$. Only the coefficients are added, in the same way that we add $3x + 4x$ to get $7x$.

It may be possible to simplify the radical expressions in a sum or difference so that two or more terms contain like radicals. Those terms can then be combined into a single term.

EXAMPLE 2

a. $\sqrt{32} + \sqrt{2} - \sqrt{18}$
$= 4\sqrt{2} + \sqrt{2} - 3\sqrt{2}$
$= 2\sqrt{2}$

b. $\sqrt[3]{24x^4} - x\sqrt[3]{81x}$
$= 2x\sqrt[3]{3x} - x \cdot 3\sqrt[3]{3x}$
$= -x\sqrt[3]{3x}$

c. $x\sqrt{4x} + 2x\sqrt[3]{x} - \sqrt{9x^3}$
$= 2x\sqrt{x} + 2x\sqrt[3]{x} - 3x\sqrt{x}$
$= -x\sqrt{x} + 2x\sqrt[3]{x}$

d. $\dfrac{\sqrt{x}}{2} + \sqrt{9x}$
$= \dfrac{1}{2}\sqrt{x} + 3\sqrt{x}$
$= \dfrac{7}{2}\sqrt{x}$

Common Error

Note that sums of radicals with different radicands or different indices *cannot* be combined. Thus,

$$\sqrt{11} + \sqrt{5} \neq \sqrt{16},$$
$$\sqrt[3]{10x} - \sqrt[3]{2x} \neq \sqrt[3]{8x},$$

and

$$\sqrt[3]{7} + \sqrt{7} \neq \sqrt[5]{7}.$$

None of the expressions above can be simplified.

Products and Factors

Radicals of the same index can be multiplied together according to Property (1) on page 118:

$$\sqrt[n]{a}\,\sqrt[n]{b} = \sqrt[n]{ab} \quad (a, b \geq 0).$$

3.4 ■ OPERATIONS ON RADICAL EXPRESSIONS

(Note that the radicals in a product need not have the same *radicands* in order to be multiplied.) Using this fact and the distributive law we can rewrite a variety products involving radicals.

EXAMPLE 3

a. $\sqrt{3}(\sqrt{2x} + \sqrt{6})$
$= \sqrt{6x} + \sqrt{18}$
$= \sqrt{6x} + 3\sqrt{2}$

c. $\sqrt[3]{x}(\sqrt[3]{2x^2} - \sqrt[3]{x})$
$= \sqrt[3]{2x^3} - \sqrt[3]{x^2}$
$= x\sqrt[3]{2} - \sqrt[3]{x^2}$

b. $\sqrt{x-1}(\sqrt{x} - \sqrt{x-1})$
$= \sqrt{x(x-1)} - (\sqrt{x-1})^2$
$= \sqrt{x^2 - x} - (x - 1)$

d. $(\sqrt{x} - \sqrt{y})(\sqrt{x} + \sqrt{y})$
$= \sqrt{x^2} - \sqrt{y^2}$
$= x - y$

To multiply radicals with different indices it is best to convert to exponential notation and use the laws of exponents.

EXAMPLE 4

a. $\sqrt{2}\sqrt[3]{4} = 2^{1/2}4^{1/3}$
$= 2^{3/6}4^{2/6}$
$= 8^{1/6}16^{1/6}$
$= 128^{1/6}$
$= \sqrt[6]{64}\sqrt[6]{2} = 2\sqrt[6]{2}$

b. $\sqrt[4]{x^3} \cdot \sqrt{2x} = x^{3/4} \cdot (2x)^{1/2}$
$= 2^{1/2} \cdot x^{3/4+1/2}$
$= 2^{2/4}x^{5/4}$
$= \sqrt[4]{2^2 x^5} = x\sqrt[4]{4x}$

We can also use the distributive law to factor expressions that contain radicals.

EXAMPLE 5

a. $3 + \sqrt{45} = 3 + 3\sqrt{5}$
$= 3(1 + \sqrt{5})$

c. $\dfrac{\sqrt{a} + \sqrt{ab}}{\sqrt{a}} = \dfrac{\sqrt{a}(1 + \sqrt{b})}{\sqrt{a}}$
$= 1 + \sqrt{b}$

b. $3\sqrt{2} - \sqrt{10} = 3\sqrt{2} - \sqrt{2}\sqrt{5}$
$= \sqrt{2}(3 - \sqrt{5})$

d. $\dfrac{x + 2\sqrt[4]{x}}{\sqrt[4]{x}} = \dfrac{\sqrt[4]{x}(\sqrt[4]{x^3} + 2)}{\sqrt[4]{x}}$
$= \sqrt[4]{x^3} + 2$

Quotients

Recall from Section 3.3 that in order to rationalize a fraction of the form $a/\sqrt{b}$ we multiply the numerator and denominator by $\sqrt{b}$. For example,

$$\frac{2}{\sqrt{3}} = \frac{2\sqrt{3}}{\sqrt{3}\sqrt{3}} = \frac{2\sqrt{3}}{3}$$

and

$$\frac{a}{\sqrt{b}} = \frac{a\sqrt{b}}{\sqrt{b}\sqrt{b}} = \frac{a\sqrt{b}}{b}.$$

The distributive property provides us with a means of rationalizing *binomial* denominators involving radicals in one or both terms. To accomplish this first recall that

$$(a - b)(a + b) = a^2 - b^2.$$

Note that the product contains no first-degree term. Each of the two factors $a - b$ and $a + b$ is said to be the **conjugate** of the other.

Now consider a fraction of the form

$$\frac{a}{b + \sqrt{c}} \quad (b + \sqrt{c} \neq 0).$$

If we multiply the numerator and denominator of this fraction by the conjugate of the denominator, the denominator of the resulting fraction will be free of radicals. That is,

$$\frac{a(b - \sqrt{c})}{(b + \sqrt{c})(b - \sqrt{c})} = \frac{ab - a\sqrt{c}}{b^2 - c} \quad (b^2 - c \neq 0),$$

where the denominator has been rationalized.

This process also applies to fractions of the form

$$\frac{a}{\sqrt{b} + \sqrt{c}},$$

since

$$\frac{a(\sqrt{b} - \sqrt{c})}{(\sqrt{b} + \sqrt{c})(\sqrt{b} - \sqrt{c})} = \frac{a\sqrt{b} - a\sqrt{c}}{b - c} \quad (b - c \neq 0).$$

3.4 ■ OPERATIONS ON RADICAL EXPRESSIONS

EXAMPLE 6 Rationalize each denominator.

a. $\dfrac{2}{\sqrt{3} - 1}$

b. $\dfrac{x}{\sqrt{2} + \sqrt{x}}$

Solutions

a. $\dfrac{2}{\sqrt{3} - 1}$

$= \dfrac{2(\sqrt{3} + 1)}{(\sqrt{3} - 1)(\sqrt{3} + 1)}$

$= \dfrac{2(\sqrt{3} + 1)}{3 - 1}$

$= \sqrt{3} + 1$

b. $\dfrac{x}{\sqrt{2} + \sqrt{x}}$

$= \dfrac{x(\sqrt{2} - \sqrt{x})}{(\sqrt{2} + \sqrt{x})(\sqrt{2} - \sqrt{x})}$

$= \dfrac{x(\sqrt{2} - \sqrt{x})}{2 - x}$

$= \dfrac{x\sqrt{2} - x\sqrt{x}}{2 - x}$

In some situations it is helpful to rationalize the numerator of a fraction.

EXAMPLE 7 Rationalize each numerator.

a. $\dfrac{1 + \sqrt{x}}{\sqrt{x}}$

b. $\dfrac{\sqrt{x + 1} - \sqrt{2}}{x - 2}$

Solutions

a. $\dfrac{1 + \sqrt{x}}{\sqrt{x}} = \dfrac{(1 + \sqrt{x})(1 - \sqrt{x})}{\sqrt{x}(1 - \sqrt{x})}$

$= \dfrac{1 - x}{\sqrt{x} - x}$

b. $\dfrac{\sqrt{x + 1} - \sqrt{2}}{x - 2} = \dfrac{(\sqrt{x + 1} - \sqrt{2})(\sqrt{x + 1} + \sqrt{2})}{(x - 2)(\sqrt{x + 1} + \sqrt{2})}$

$= \dfrac{(x + 1) - 2}{(x - 2)(\sqrt{x + 1} + \sqrt{2})}$

$= \dfrac{x - 1}{(x - 2)(\sqrt{x + 1} + \sqrt{2})}$

EXERCISE 3.4

A

■ Assume that all radicands and variables in this exercise are positive real numbers.

■ Simplify each sum by combining like terms. See Examples 1 and 2.

1. $3\sqrt{7} + 2\sqrt{7}$
2. $5\sqrt{2} - 3\sqrt{2}$
3. $4\sqrt{3} - \sqrt{27}$
4. $\sqrt{75} + 2\sqrt{3}$
5. $\sqrt{50x} + \sqrt{32x}$
6. $\sqrt{8y} - \sqrt{18y}$
7. $3\sqrt[3]{16} - \sqrt[3]{2} - 2\sqrt[3]{54}$
8. $\sqrt[3]{81} + 2\sqrt[3]{24} - 3\sqrt[3]{3}$
9. $4\sqrt[3]{40} + 6\sqrt[3]{80} - 5\sqrt{45} - \sqrt[3]{135}$
10. $6\sqrt[3]{32} - 3\sqrt{32} + \sqrt[3]{128} - 2\sqrt{128}$
11. $3\sqrt{4xy^2} - 4\sqrt{9xy^2} + 2\sqrt{4x^2y}$
12. $2\sqrt{8y^2z} - 3\sqrt{9yz^2} + 3\sqrt{32y^2z}$
13. $xy\sqrt[4]{2x^2} - \sqrt[4]{32x^6y^4} - \dfrac{3}{y}\sqrt[4]{x^6y^8}$
14. $x^2\sqrt[4]{48xy} - \dfrac{3x}{y}\sqrt[4]{3x^5y^5} - 7\sqrt[4]{3x^9y}$
15. $\dfrac{\sqrt{3}}{2} + \sqrt{27}$
16. $\dfrac{\sqrt{5}}{3} - \sqrt{80}$
17. $\dfrac{3}{2}\sqrt[3]{2x} - \sqrt[3]{16x}$
18. $\dfrac{2}{3}\sqrt[3]{3y} + \sqrt[3]{24y}$
19. $\dfrac{2a}{\sqrt{5}} - \dfrac{a\sqrt{5}}{2}$
20. $\dfrac{ab\sqrt{2}}{3} + \dfrac{3ab}{\sqrt{2}}$

■ Write each expression without parentheses and simplify. See Example 3.

21. $2(3 - \sqrt{5})$
22. $5(2 - \sqrt{7})$
23. $\sqrt{2}(\sqrt{6} + \sqrt{10})$
24. $\sqrt{3}(\sqrt{12} - \sqrt{15})$
25. $\sqrt[3]{2}(\sqrt[3]{20} - 2\sqrt[3]{12})$
26. $\sqrt[3]{3}(2\sqrt[3]{18} + \sqrt[3]{36})$
27. $2\sqrt{x}(\sqrt{24x} + \sqrt{12})$
28. $\sqrt{3y}(\sqrt{6y} - \sqrt{18})$
29. $\sqrt{x-3}(\sqrt{x} - \sqrt{x+3})$
30. $\sqrt{4-x}(\sqrt{4x} + \sqrt{1-x})$
31. $\sqrt[3]{x+1}(\sqrt[3]{(x+1)^2} - \sqrt[3]{x+1})$
32. $\sqrt[3]{(x+2)^2}(\sqrt[3]{(x+2)^2} + \sqrt[3]{2x})$
33. $(\sqrt{x} - 3)(\sqrt{x} + 3)$
34. $(2 + \sqrt{x})(2 - \sqrt{x})$
35. $(\sqrt{2} - \sqrt{3})(\sqrt{2} + 2\sqrt{3})$
36. $(\sqrt{3} - \sqrt{5})(2\sqrt{3} + \sqrt{5})$
37. $(\sqrt{5} - \sqrt{2})^2$
38. $(\sqrt{2} - 2\sqrt{3})^2$
39. $(3\sqrt{x} + \sqrt{2y})(2\sqrt{x} - 3\sqrt{2y})$
40. $(\sqrt{5x} - 2\sqrt{y})(2\sqrt{5x} - 3\sqrt{y})$
41. $(\sqrt{a} - 2\sqrt{b})^2$
42. $(\sqrt{2a} - 2\sqrt{b})(\sqrt{2a} + 2\sqrt{b})$

■ Simplify. See Example 4.

43. $\sqrt{7}\sqrt[3]{7}$
44. $\sqrt[3]{6}\sqrt{6}$
45. $\sqrt{5}\sqrt[3]{2}$
46. $\sqrt[3]{6}\sqrt{3}$
47. $\sqrt[4]{4x}\sqrt[3]{4x}$
48. $\sqrt[3]{2x^2}\sqrt[5]{x^2}$
49. $\sqrt[4]{xy^2}\sqrt{xy}$
50. $\sqrt{xy}\sqrt[3]{x^2y^3}$

■ Change each expression to the form indicated. See Example 5.

51. $2 + 2\sqrt{3} = 2(? + ?)$
52. $5 + 10\sqrt{2} = 5(? + ?)$
53. $2\sqrt{27} + 6 = 6(? + ?)$
54. $5\sqrt{5} - \sqrt{25} = 5(? - ?)$
55. $4 + \sqrt{16y} = 4(? + ?)$
56. $3 + \sqrt{18x} = 3(? + ?)$
57. $\sqrt{2} - \sqrt{6} = \sqrt{2}(? - ?)$
58. $\sqrt{12} - 2\sqrt{6} = 2\sqrt{3}(? - ?)$
59. $2y\sqrt{x} + 3\sqrt{xy} = \sqrt{x}(? + ?)$
60. $a\sqrt{5b} - \sqrt{3ab} = \sqrt{b}(? - ?)$
61. $4x - \sqrt{12x} = 2\sqrt{x}(? - ?)$
62. $6y + \sqrt{18y} = 3\sqrt{y}(? + ?)$

3.4 ■ OPERATIONS ON RADICAL EXPRESSIONS

■ *Reduce each fraction to lowest terms after factoring the numerator. See Example 5.*

63. $\dfrac{2 + 2\sqrt{3}}{2}$
64. $\dfrac{6 + 2\sqrt{5}}{2}$
65. $\dfrac{6 + 2\sqrt{18}}{6}$
66. $\dfrac{8 - 2\sqrt{12}}{4}$

67. $\dfrac{x - \sqrt{x^3}}{x}$
68. $\dfrac{xy - x\sqrt{xy^2}}{xy}$
69. $\dfrac{x\sqrt{y} - \sqrt{y^3}}{\sqrt{y}}$
70. $\dfrac{\sqrt{x} - y\sqrt{x^3}}{\sqrt{x}}$

71. $\dfrac{\sqrt{8ab} + 8a}{4\sqrt{a}}$
72. $\dfrac{12b - \sqrt{12ab}}{6\sqrt{b}}$
73. $\dfrac{\sqrt[3]{x^2} - 3x}{\sqrt[3]{x}}$
74. $\dfrac{4x + \sqrt[4]{x^3}}{\sqrt[4]{x}}$

■ *Rationalize each denominator. See Example 6.*

75. $\dfrac{4}{1 + \sqrt{3}}$
76. $\dfrac{1}{2 - \sqrt{2}}$
77. $\dfrac{2}{\sqrt{7} - 2}$
78. $\dfrac{2}{4 - \sqrt{5}}$

79. $\dfrac{x}{\sqrt{x} - 3}$
80. $\dfrac{y}{\sqrt{3} - y}$
81. $\dfrac{\sqrt{6} - 3}{2 - \sqrt{6}}$
82. $\dfrac{\sqrt{x} + \sqrt{y}}{\sqrt{x} - \sqrt{y}}$

83. $\dfrac{\sqrt{5}}{5\sqrt{3} + 3\sqrt{5}}$
84. $\dfrac{\sqrt{3}}{2\sqrt{3} - 3\sqrt{2}}$
85. $\dfrac{2\sqrt{a} - \sqrt{3b}}{\sqrt{3a} + 2\sqrt{b}}$
86. $\dfrac{\sqrt{6b} + 6\sqrt{a}}{3\sqrt{b} - \sqrt{3a}}$

■ *Rationalize each numerator. See Example 7.*

87. $\dfrac{\sqrt{x-1} - 1}{\sqrt{x} - 1}$
88. $\dfrac{1 - \sqrt{x+1}}{\sqrt{x+1}}$
89. $\dfrac{\sqrt{x+1} - \sqrt{x}}{\sqrt{x+1} + \sqrt{x}}$
90. $\dfrac{\sqrt{x-1} + \sqrt{x}}{\sqrt{x-1} - \sqrt{x}}$

B

■ *Write each expression as a single fraction in which the denominator is rationalized.*

91. $\sqrt{x} + \dfrac{1}{\sqrt{x}}$
92. $\dfrac{\sqrt{x}}{x} - \dfrac{x}{\sqrt{x}}$
93. $\sqrt{x+1} - \dfrac{x}{\sqrt{x+1}}$

94. $\sqrt{x^2 - 2} - \dfrac{x^2 + 1}{\sqrt{x^2 - 2}}$
95. $\dfrac{x}{\sqrt{x^2 + 1}} - \dfrac{\sqrt{x^2 + 1}}{x}$
96. $\dfrac{x}{\sqrt{x^2 - 1}} + \dfrac{\sqrt{x^2 - 1}}{x}$

■ *Write each expression as a simple fraction in which the denominator is rationalized.*

97. $\dfrac{1 - \dfrac{1}{\sqrt{x}}}{1 + \dfrac{1}{\sqrt{x}}}$
98. $\dfrac{\sqrt{x} + \dfrac{1}{\sqrt{x}}}{x + 1}$
99. $\dfrac{\dfrac{1}{\sqrt{x}} - \dfrac{1}{\sqrt{x+2}}}{2}$

100. $\dfrac{\dfrac{1}{\sqrt{x+2}} + \sqrt{x}}{\dfrac{\sqrt{x}}{\sqrt{x+2}}}$
101. $\dfrac{\sqrt{x^2 - 2} + \dfrac{x}{\sqrt{x^2 - 2}}}{\dfrac{x + 2}{\sqrt{x^2 - 2}}}$
102. $\dfrac{\dfrac{x}{\sqrt{x^2 - 1}} - \sqrt{x^2 - 1}}{x^2\sqrt{x^2 - 1}}$

3.5

COMPLEX NUMBERS

In Section 3.1 we noted that the square root of a negative number is not a real number. However, for many applications it is necessary to consider such square roots. In this section we introduce a set of numbers, C, called **complex numbers,** which includes all the real numbers and also square roots of negative real numbers.

Imaginary Numbers

We begin by defining a new number, i, whose square is -1.

$$i^2 = -1 \quad \text{or} \quad i = \sqrt{-1}.$$

Furthermore, we define the square root of *any* negative real number in the following way.

For $a > 0$,
$$\sqrt{-a} = \sqrt{-1}\sqrt{a} = i\sqrt{a}.$$

EXAMPLE 1 a. $\sqrt{-4} = \sqrt{-1}\sqrt{4}$
$\phantom{\sqrt{-4}} = i\sqrt{4} = 2i$

b. $\sqrt{-3} = \sqrt{-1}\sqrt{3}$
$\phantom{\sqrt{-3}} = i\sqrt{3}$

Thus, a square root of a negative real number can be represented as the product of a real number and the number $\sqrt{-1}$ or i. For historical reasons such numbers are called **imaginary numbers.**

Note that each negative real number $-a$, for $a > 0$, has *two* imaginary square roots, $i\sqrt{a}$ and $-i\sqrt{a}$, since

$$(i\sqrt{a})^2 = i^2(\sqrt{a})^2 = i^2 a = -a$$

and

$$(-i\sqrt{a})^2 = i^2(-\sqrt{a})^2 = i^2 a = -a.$$

3.5 ■ COMPLEX NUMBERS   133

For example, the two square roots of -9 are

$$\sqrt{-9} = \sqrt{-1}\sqrt{9} = 3i \quad \text{and} \quad -\sqrt{-9} = -\sqrt{-1}\sqrt{9} = -3i.$$

Complex Numbers

Now consider all possible expressions of the form $a + bi$, where a and b are real numbers and $i = \sqrt{-1}$. Such an expression represents a **complex number**, that is, a number in the set C. Here a is called the **real part** of the number and b is the **imaginary part**. If $b = 0$, then $a + bi = a$, and it is evident that the set R of real numbers is contained in the set C of complex numbers. The relationships among the subsets of C are shown in Figure 3.1.

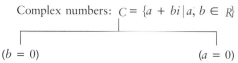

FIGURE 3.1

Complex numbers: $C = \{a + bi \mid a, b \in R\}$

$(b = 0)$ Real numbers: $a + bi = a$*

$(a = 0)$ Imaginary numbers: $a + bi = bi$

EXAMPLE 2 Write each expression in the form $a + bi$.

a. $3\sqrt{-18}$

b. $2 - 3\sqrt{-16}$

Solutions a. $3\sqrt{-18} = 3\sqrt{-1 \cdot 9 \cdot 2}$
$= 3\sqrt{-1}\sqrt{9}\sqrt{2}$
$= 3i(3)\sqrt{2}$
$= 9i\sqrt{2}$

b. $2 - 3\sqrt{-16} = 2 - 3\sqrt{-1 \cdot 16}$
$= 2 - 3\sqrt{-1}\sqrt{16}$
$= 2 - 3i(4)$
$= 2 - 12i$

Operations on Complex Numbers

To add or subtract complex numbers we simply add or subtract their real parts and their imaginary parts.

EXAMPLE 3 a. $(2 + 3i) + (5 - 4i)$
$= (2 + 5) + (3 - 4)i$
$= 7 - i$

b. $(2 + 3i) - (5 - 4i)$
$= (2 - 5) + [3 - (-4)]i$
$= -3 + 7i$

---

*See Figure 1.1, page 4, for sets of numbers that are contained in the set of real numbers.

To multiply complex numbers we treat them as though they were binomials and replace i^2 with -1.

EXAMPLE 4

a. $(2 - 1)(1 + 3i)$
$= 2 + 6i - i - 3i^2$
$= 2 + 6i - i - 3(-1)$
$= 2 + 6i - i + 3$
$= 5 + 5i$

b. $(3 - i)^2 = (3 - i)(3 - i)$
$= 9 - 3i - 3i + i^2$
$= 9 - 6i + (-1)$
$= 8 - 6i$

The quotient of two complex numbers can be found by using the following property, which is analogous to the fundamental principle of fractions in the set of real numbers. First, recall from Section 3.4 that for $b > 0$ the conjugate of $a + \sqrt{b}$ is $a - \sqrt{b}$. Similarly, the conjugate of $a + bi$ is $a - bi$.

EXAMPLE 5

a. The conjugate of $2 + 3i$ is $2 - 3i$.
b. The conjugate of $-3 - i$ is $-3 + i$.
c. The conjugate of $2i$ is $-2i$.
d. The conjugate of $-4 + i$ is $-4 - i$.

The quotient

$$\frac{a + bi}{c + di}$$

of two complex numbers can be simplified by multiplying the numerator and denominator by $c - di$, the conjugate of the denominator. That is,

$$\frac{a + bi}{c + di} = \frac{(a + bi)(c - di)}{(c + di)(c - di)}.$$

If the divisor is of the form bi, we need only multiply the numerator and denominator by i.

EXAMPLE 6

a. $\dfrac{4-i}{-2i} = \dfrac{(4-i)i}{-2i \cdot i}$

$= \dfrac{4i - i^2}{-2i^2}$

$= \dfrac{4i - (-1)}{-2(-1)}$

$= \dfrac{4i + 1}{2}$

$= \dfrac{1}{2} + 2i$

b. $\dfrac{4+i}{2+3i} = \dfrac{(4+i)(2-3i)}{(2+3i)(2-3i)}$

$= \dfrac{8 - 10i - 3i^2}{4 - 9i^2}$

$= \dfrac{8 - 10i - 3(-1)}{4 - 9(-1)}$

$= \dfrac{8 - 10i + 3}{4 + 9}$

$= \dfrac{11}{13} - \dfrac{10}{13}i$

Using Radical Notation

The symbol $\sqrt{-b}$ $(b > 0)$ should be used with care, since certain properties involving square root symbols are valid for real numbers but are *not* valid when the symbols do not represent real numbers.

Common Error

$\sqrt{-2}\sqrt{-3} \neq \sqrt{(-2)(-3)} = \sqrt{6}$;
$\sqrt{-2}\sqrt{-3} = (i\sqrt{2})(i\sqrt{3}) = i^2\sqrt{6} = -\sqrt{6}$.

To avoid difficulty with this point,

rewrite all expressions of the form $\sqrt{-b}$ $(b > 0)$ in the form $i\sqrt{b}$ before performing any computations.

EXAMPLE 7

a. $\sqrt{-2}(3 - \sqrt{-5})$
$= i\sqrt{2}(3 - i\sqrt{5})$
$= 3i\sqrt{2} - i^2\sqrt{10}$
$= 3i\sqrt{2} - (-1)\sqrt{10}$
$= \sqrt{10} + 3i\sqrt{2}$

b. $(2 + \sqrt{-3})(2 - \sqrt{-3})$
$= (2 + i\sqrt{3})(2 - i\sqrt{3})$
$= 4 - 3i^2$
$= 4 - 3(-1)$
$= 7$

c. $\dfrac{2}{\sqrt{-3}}$

$= \dfrac{2}{i\sqrt{3}} \cdot \dfrac{i\sqrt{3}}{i\sqrt{3}}$

$= \dfrac{2i\sqrt{3}}{3i^2}$

$= \dfrac{2i\sqrt{3}}{3(-1)} = \dfrac{-2i\sqrt{3}}{3}$

d. $\dfrac{1}{3 - \sqrt{-1}}$

$= \dfrac{1}{3 - i}$

$= \dfrac{1 \cdot (3 + i)}{(3 - i)(3 + i)}$

$= \dfrac{3 + i}{9 - i^2}$

$= \dfrac{3 + i}{9 - (-1)} = \dfrac{3}{10} + \dfrac{1}{10}i$

EXERCISE 3.5

A

■ Write each expression in the form $a + bi$ or $a + ib$. See Examples 1 and 2.

1. $\sqrt{-4}$
2. $\sqrt{-9}$
3. $\sqrt{-32}$
4. $\sqrt{-50}$
5. $3\sqrt{-8}$
6. $4\sqrt{-18}$
7. $3\sqrt{-24}$
8. $2\sqrt{-40}$
9. $5\sqrt{-64}$
10. $7\sqrt{-81}$
11. $-2\sqrt{-12}$
12. $-3\sqrt{-75}$
13. $4 + 2\sqrt{-1}$
14. $5 - 3\sqrt{-1}$
15. $3\sqrt{-50} + 2$
16. $5\sqrt{-12} - 1$
17. $\sqrt{4} + \sqrt{-4}$
18. $\sqrt{20} - \sqrt{-20}$

■ See Example 3.

19. $(2 + 4i) + (3 + i)$
20. $(2 - i) + (3 - 2i)$
21. $(4 - i) - (6 - 2i)$
22. $(2 + i) - (4 - 2i)$
23. $3 - (4 + 2i)$
24. $(2 - 6i) - 3$

■ See Example 4.

25. $(2 - i)(3 + 2i)$
26. $(1 - 3i)(4 - 5i)$
27. $(3 + 2i)(5 + i)$
28. $(-3 - i)(2 - 3i)$
29. $(6 - 3i)(4 - i)$
30. $(7 + 3i)(-2 - 3i)$
31. $(2 - i)^2$
32. $(2 + 3i)^2$
33. $(2 - i)(2 + i)$
34. $(1 - 2i)(1 + 2i)$

■ See Examples 5 and 6.

35. $\dfrac{1}{3i}$
36. $\dfrac{-2}{5i}$
37. $\dfrac{3 - i}{5i}$
38. $\dfrac{4 + 2i}{3i}$
39. $\dfrac{2}{1 - i}$
40. $\dfrac{-3}{2 + i}$
41. $\dfrac{2 + i}{1 + 3i}$
42. $\dfrac{3 - i}{1 + i}$

43. $\dfrac{2-3i}{3-2i}$
44. $\dfrac{6+i}{2-5i}$
45. $\dfrac{3+2i}{5-3i}$
46. $\dfrac{-4-3i}{2+7i}$

■ *See Example 7.*

47. $\sqrt{-4}\,(1-\sqrt{-4})$
48. $\sqrt{-9}\,(3+\sqrt{-16})$
49. $(2+\sqrt{-9})(3-\sqrt{-9})$
50. $(4-\sqrt{-2})(3+\sqrt{-2})$
51. $\dfrac{3}{\sqrt{-4}}$
52. $\dfrac{-1}{\sqrt{-25}}$
53. $\dfrac{2-\sqrt{-1}}{2+\sqrt{-1}}$
54. $\dfrac{1+\sqrt{-2}}{3-\sqrt{-2}}$

55. For what values of x will $\sqrt{x-5}$ be real? Imaginary?
56. For what values of x will $\sqrt{x+3}$ be real? Imaginary?
57. Simplify. (*Hint:* $i^2=-1$ and $i^4=1$.)
   a. i^6 b. i^{12} c. i^{15} d. i^{102}
58. Express with a positive exponent and simplify.
   a. i^{-1} b. i^{-2} c. i^{-3} d. i^{-6}
59. Evaluate x^2+2x+3 for $x=1+i$.
60. Evaluate $2y^2-y+2$ for $y=2-i$.

CHAPTER REVIEW

A

[3.1]

■ *Find each root or power if it is a real number. If it is not a real number, say so.*

1. $\sqrt[3]{-125}$
2. $-\sqrt[4]{16}$
3. $\sqrt[4]{-16}$
4. $\sqrt[3]{\dfrac{1}{8}}$
5. $-27^{1/3}$
6. $-16^{1/2}$
7. $(-16)^{1/2}$
8. $\left(\dfrac{8}{27}\right)^{1/3}$

■ *Write each expression in radical form.*

9. $3x^{1/2}$
10. $(3x)^{1/2}$
11. $4x^{-1/3}$
12. $(4x)^{-1/3}$

■ *Write each expression in exponential form.*

13. $\sqrt[3]{5x}$
14. $\dfrac{5}{\sqrt[3]{x}}$
15. $3\sqrt{xy}$
16. $\sqrt{x}-\sqrt[3]{y}$

[3.2]

■ *Write each expression in radical form.*

17. $4x^{-3/4}$
18. $-2x^{1/3}y^{2/3}$
19. $4(x-1)^{3/2}$
20. $2(x^2-1)^{-3/2}$

■ Write each expression in exponential form.

21. $4\sqrt{x^3}$
22. $2\sqrt[3]{x^2y}$
23. $\dfrac{-2x}{\sqrt[3]{y^2}}$
24. $\dfrac{y}{\sqrt[3]{(x-y)^2}}$

■ Write each expression so that each variable occurs only once and all exponents are positive.

25. $3x^{1/2}x^{3/4}$
26. $\dfrac{x^2y}{2x^{1/2}}$
27. $x^{-1/2}(2x^{3/4})$
28. $(-8y^{-1/2})^{2/3}$
29. $\left(\dfrac{4x^{-3}}{y^{1/2}}\right)^{-1/2}$
30. $\left(\dfrac{8y^{2/3}}{x^{-1/3}}\right)^{-3}$
31. $\dfrac{(x^{-1}y^{-1/2})^{2/3}}{(x^{5/2}y)^{-2}}$
32. $\left(\dfrac{x^2}{y^{2/3}}\right)^3\left(\dfrac{y^{-1}}{x^3}\right)^2$

■ Write each expression as a sum of terms.

33. $y^{-1/2}(y + y^{1/2})$
34. $x^{1/3}(x^{-2/3} - x^{1/3})$
35. $(3x^{1/2} + 2)(x^{1/2} - 3)$
36. $(x^{-1} + 2)^2$

[3.3]
■ Simplify. Assume that all variables and radicands denote positive real numbers.

37. $\sqrt[3]{16x^4y^3}$
38. $\sqrt{2x^3}\sqrt{8x^3y}$
39. $\sqrt{8a^4 + 4a^2}$
40. $\sqrt[5]{\dfrac{y^6}{32x^{10}}}$

■ Simplify. Rationalize denominators as necessary.

41. $\dfrac{3}{\sqrt[2]{x^3}}$
42. $\sqrt[3]{\dfrac{x}{2y}}$
43. $\dfrac{\sqrt{4x}\sqrt{9y^3}}{\sqrt{6y}}$
44. $\sqrt[4]{x^6}$
45. $\sqrt{3x(2x-1)}\sqrt{6x(2x-1)^3}$
46. $\sqrt[3]{4y(y+1)^2}\sqrt[3]{12y^4(y+1)^2}$

[3.4]
■ Simplify each sum by combining like terms.

47. $2\sqrt{2} - \sqrt{8} + 4\sqrt{18}$
48. $3\sqrt{x^3y} + 2x\sqrt{xy}$
49. $\dfrac{1}{2}\sqrt[3]{2x} + \dfrac{3}{4}\sqrt[3]{16x}$
50. $\dfrac{x\sqrt{3}}{2} - \dfrac{2x}{\sqrt{3}}$

■ Write each expression without parentheses and simplify.

51. $\sqrt{2y}(\sqrt{8y} - \sqrt{6y^3})$
52. $\sqrt{x-1}\left(\sqrt{x-1} - \dfrac{1}{\sqrt{x-1}}\right)$
53. $(\sqrt{a} + \sqrt{2b})(2\sqrt{a} - 3\sqrt{2b})$
54. $(\sqrt{2a} - \sqrt{b})^2$

■ Simplify.

55. $\sqrt[3]{x^2y}\sqrt{xy}$
56. $\sqrt[3]{2y^2}\sqrt[4]{4y}$

■ Reduce each fraction.

57. $\dfrac{x\sqrt{xy^4} - xy^2}{xy}$
58. $\dfrac{2y - \sqrt[3]{y^2}}{\sqrt[3]{y}}$

■ *Rationalize each denominator.*

59. $\dfrac{x}{\sqrt{x}+2}$

60. $\dfrac{\sqrt{x}-2\sqrt{y}}{\sqrt{x}+2\sqrt{y}}$

■ *Rationalize each numerator.*

61. $\dfrac{1+\sqrt{y-1}}{\sqrt{y-1}}$

62. $\dfrac{\sqrt{y}+\sqrt{y-1}}{\sqrt{y}}$

[3.5]
■ *Write each expression in the form $a + bi$ or $a + ib$.*

63. $(3 - 2i) + (4 + i)$

64. $(1 + 3i) - (2 - 4i)$

65. $(3 - i)(4 + 3i)$

66. $(4 + i)^2$

67. $\dfrac{2}{5i}$

68. $\dfrac{-3}{4 + i}$

69. $\dfrac{2 - i}{3 + i}$

70. $\dfrac{5 + i}{2 - 3i}$

71. $\sqrt{-3}(1 + 2\sqrt{-3})$

72. $(3 - \sqrt{-2})(5 + \sqrt{-2})$

73. $\dfrac{5}{\sqrt{-8}}$

74. $\dfrac{3 - \sqrt{-3}}{2 + \sqrt{-3}}$

75. According to the theory of relativity the mass of an object traveling at velocity v is given by

$$\dfrac{M}{\sqrt{1 - \dfrac{v^2}{c^2}}},$$

where M is the mass of the object at rest and c is the speed of light. Find the mass of a 180-pound man traveling at a velocity of $0.7c$.

76. The cylinder of the smallest surface area for a given volume has a radius and height both equal to $\sqrt[3]{V/\pi}$. Find the dimensions of the tin can of smallest surface area with volume 60 cubic inches.

77. Membership in the Wildlife Society has grown according to the formula $30t^{3/4}$, where t is the number of years since its founding in 1970. What was the society's membership in 1990?

78. The heron population in Saltmarsh Refuge is estimated by conservationists at $36t^{-2/3}$, where t is the number of years since the refuge was established in 1980. How many heron were there in 1985?

79. The population of Dry Gulch has been declining according to the formula $3800 \cdot 2^{-t/20}$, where t is the number of years since the town's heyday in 1910. What was the population of Dry Gulch in 1990?

80. The number of compact discs produced each year by Delta Discs is given by $8000 \cdot 3^{t/2}$, where t is the number of years since discs were introduced in 1980. How many discs did Delta produce in 1989?

81. How much will a $90 camera cost 10 months from now if the inflation rate is 6% annually?

82. How much will a $1200 sofa cost 20 months from now if the inflation rate is 8% annually?

B

- *Factor. Write your answer using only positive exponents.*

83. $x^{-1/3} - x^{1/3}$

84. $x^{1/4}(x+2)^{3/4} + x^{1/2}(x+2)^{-3/4}$

- *Simplify, using absolute-value notation as needed.*

85. $\sqrt[3]{(x^6 - x^3)(x+1)^4}$

86. $\sqrt{x^2 - 4x + 4}$

- *Write as a single fraction and simplify.*

87. $\dfrac{x}{\sqrt{1-x}} - \sqrt{1-x}$

88. $\dfrac{1}{\sqrt{x}} + \dfrac{1}{\sqrt{x+1}}$

- *Simplify.*

89. $\dfrac{x + \dfrac{1}{\sqrt{x}}}{\dfrac{1}{\sqrt{x}} - x}$

90. $\dfrac{1}{1 - \dfrac{1}{1 + \sqrt{x}}}$

4 Equations in One Variable

In this chapter we review some procedures for solving a variety of equations in one variable.

Recall that any number that satisfies an equation is called a **solution** of the equation. The process of finding solutions of equations involves generating simpler equations that have the same solutions. Equations that have identical solutions are called **equivalent equations**.

There are several ways to form equivalent equations. One way is to simplify one or both sides of the equation by combining like terms. For example, the solution of $9x - 8x = 12$ becomes evident when it is written in the equivalent form $x = 12$.

The following properties also enable us to generate equivalent equations.

---

1. The addition of the same expression to each side of an equation produces an equivalent equation.
2. The multiplication of each side of an equation by the same nonzero number produces an equivalent equation.

The application of Properties (1) and (2) above enables us to transform an equation whose solution may not be obvious, through a series of equivalent equations, to an equation whose solution is obvious.

4.1

LINEAR EQUATIONS

The discussion above applies to a wide variety of equations. In this section we give our attention to first-degree, or **linear**, equations.

EXAMPLE 1 Solve $2x + 5 = 11 - x$.

Solution First, add $x - 5$ to each side to obtain

$$3x = 6.$$

Then multiply each side by $\frac{1}{3}$ (or divide by 3) to get

$$\frac{1}{3}(3x) = \frac{1}{3}(6)$$
$$x = 2.$$

The solution is 2.

We can always check an apparent solution by substituting the suggested value into the original equation and verifying that the resulting statement is true. If the equations generated at each step of the solution process are obtained by application of Properties (1) and (2) above, the sole purpose of such a check is to detect arithmetic errors.

Equations Containing Parentheses

When an equation contains grouping symbols, it may be necessary to use the distributive law to simplify one or both sides of the equation.

EXAMPLE 2 Solve $0.03(x + 20) - 0.02(x + 100) = 8$.

Solution Multiply each side of the equation by 100 to clear the decimals, yielding

$$100[0.03(x + 20) - 0.02(x + 100)] = 100(8)$$
$$100[0.03(x + 20)] - 100[0.02(x + 100)] = 100(8)$$
$$3(x + 20) - 2(x + 100) = 800.$$

Then apply the distributive law to obtain

$$3x + 60 - 2x - 200 = 800,$$

from which

$$x - 140 = 800$$
$$x = 940.$$

Hence, the solution is 940.

Common Error

Note that in Example 2,

$$100[0.03(x + 20)] \neq 100(0.03) \cdot 100(x + 20),$$

since $0.03(x + 20)$ is a single term. Each *term* in the equation is multiplied once by 100.

If parentheses are nested within a second set of grouping symbols, we may need to apply the distributive law more than once.

EXAMPLE 3 Solve $4[x - 3(x + 2) - 4] = 8$.

Solution First, simplify $-3(x + 2)$ to obtain

$$4[x - 3x - 6 - 4] = 8$$
$$4[-2x - 10] = 8.$$

Applying the distributive law a second time yields

$$-8x - 40 = 8$$
$$-8x = 48$$
$$x = -6.$$

Hence, the solution is -6.

Solving Formulas

An equation containing more than one variable can be solved for one of the variables in terms of the others. In general, we apply Properties (1) and (2) on page 141 until the desired symbol stands alone on one side of the equation. The following suggestions may be helpful.

> **To Solve a First-degree Equation for a Specified Variable:**
>
> 1. Transform the equation into a form in which all terms containing the specified variable are on one side and all terms not containing that variable are on the other side.
> 2. Combine like terms on each side.
> 3. If unlike terms containing the specified variable remain, factor that variable from the terms.
> 4. Divide each side by the coefficient of the specified variable.

EXAMPLE 4 Solve the formula $S = lw + lh + 2wh$ for w.

Solution First, obtain

$$S - lh = lw + 2wh,$$

in which all terms containing w are on one side. Then factor w from $lw + 2wh$ to obtain

$$S - lh = w(l + 2h),$$

and finally divide each side by $(l + 2h)$, the coefficient of w, to get

$$\frac{S - lh}{l + 2h} = w \quad \text{or} \quad w = \frac{S - lh}{l + 2h}.$$

Solving Problems

In previous chapters we wrote mathematical expressions for word phrases and relationships between variables. Now we shall use equations to model a variety of problems that require numerical solutions. In the examples below we begin by representing the unknown quantity or quantities by algebraic expressions in one variable. Then, using the conditions given in the problem, we write an equation that models the problem. Finally, we solve the equation and interpret the solution in terms of the original problem.

EXAMPLE 5 The Little Theater auditorium holds 200 seats. Reserved seats are generally sold for $16 and student rush seats for $5. If the theater department hopes to bring in $2760 on each performance of its production of *Macbeth*, how many seats should be designated as reserved? How many as rush seats?

Solution Write expressions for the number of reserved seats and the number of rush seats in terms of a single variable:

$$\text{number of reserved seats:} \quad x\,;$$
$$\text{number of rush seats:} \quad 200 - x.$$

The total income, $2760, is the sum of the incomes from the reserved seats and the rush seats:

$$\underbrace{16x}_{\text{Income from reserved seats}} + \underbrace{5(200 - x)}_{\text{Income from rush seats}} = 2760.$$

Solve the equation:

$$16x + 1000 - 5x = 2760$$
$$11x = 1760$$
$$x = 160.$$

The theater department should sell 160 reserved seats and 200 − 160, or 40, student rush seats.

EXAMPLE 6 The price of housing in urban areas has been increasing at a rate of 4% per year. If a house costs $100,000 today, what was its cost 2 years ago?

Solution The cost of the house 2 years ago: x.
After 1 year the price of the house increased by 4%, or $0.04x$. Thus, 1 year ago the house cost

$$\underbrace{x}_{\text{Original cost}} + \underbrace{0.04x}_{\text{Price increase}} = x(1 + 0.04) = 1.04x.$$

This year the price increased by 4% of *last year's* cost, or $0.04(1.04x)$. The cost of the house is now

$$\underbrace{1.04x}_{\text{Cost last year}} + \underbrace{0.04(1.04x)}_{\text{Price increase}} = 1.04x(1 + 0.04) = (1.04)^2 x.$$

This expression is equal to the present price of the home:

$$(1.04)^2 x = 100,000.$$

Solve the equation:

$$x = \frac{100,000}{(1.04)^2}$$
$$= 92,455.62.$$

Two years ago the house cost $92,456.

Weighted Averages

The notion of *weighted average* has application to a variety of problems. Recall that the average of a set of numbers $\{x_1, x_2, \ldots, x_n\}$ is given by

$$\bar{x} = \frac{x_1 + x_2 + \cdots + x_n}{n}.$$

In a **weighted average** the numbers being averaged occur with different frequencies, or are "weighted" differently in their contribution to the average value. For instance, suppose a biology class of 12 students takes a 10-point quiz. Two students receive 10s, three receive 9s, five receive 8s, and 1 receives a score of 6. The average score on the quiz is then

$$\bar{x} = \frac{2(10) + 3(9) + 5(8) + 1(6)}{12} = 7.75.$$

The numbers in color are the weights—in this example the number of times each score occurred. Note that n, the total number of scores, is equal to the sum of the weights:

$$12 = 2 + 3 + 5 + 1.$$

EXAMPLE 7   Kwan's grade in his accounting class will be computed as follows: tests count for 50% of the grade, homework counts for 20%, and the final exam counts for 30%. If Kwan has an average of 84 on tests and 92 on homework, what score does he need on the final exam to earn a grade of 90?

Solution   Score on final exam: x.
Kwan's grade is the weighted average of his test, homework, and final exam scores:

$$\frac{0.50(84) + 0.20(92) + 0.30x}{1.00} = 90, \tag{1}$$

or

$$0.50(84) + 0.20(92) + 0.30x = 1.00(90). \tag{2}$$

Solve the equation for x:

$$60.4 + 0.30x = 90$$
$$x \approx 98.7.$$

Kwan needs a score of 98.7 on the final exam to earn a grade of 90.

Note that in Equation (1) of Example 7 we *divided by the sum of the weights*, $0.50 + 0.20 + 0.30 = 1.00$, to find the weighted average. In Equation (2) we multiplied both sides of Equation (1) by the sum of the weights to obtain an equivalent form of the equation in which *the sum of the weighted scores equals the average times the sum of the weights*. This form is particularly useful for problem solving.

EXAMPLE 8 A farmer plans to buy two types of seed corn to plant: a standard variety that yields 400 bushels per acre and a new hybrid that yields 500 bushels per acre. If he plants 200 acres of the standard variety, how many acres of the hybrid should he plant so that his total average yield will be 420 bushels per acre?

Solution Number of acres to be planted with hybrid seed: x.

The total yield is the sum of the yields from the two varieties (the "weights" in this problem are the number of acres planted in each variety of corn):

$$\underbrace{(400)(200)}_{\text{Yield from standard variety}} + \underbrace{500x}_{\text{Yield from hybrid variety}} = \underbrace{420}_{\text{Average}} \underbrace{(x + 200)}_{\text{Sum of weights}}.$$

Solve the equation:

$$80{,}000 + 500x = 420x + 84{,}000$$
$$80x = 4000$$
$$x = 50.$$

The farmer should plant 50 acres of the hybrid corn.

EXAMPLE 9 A chemistry student wants to produce a 50% acid solution by adding some 10% solution of the acid to 20 liters of a 60% solution. How many liters of the 10% solution are needed?

Solution Number of liters of 10% solution: x.

The *amount* of pure acid in the mixture must be equal to the sum of the *amounts* of pure acid in the 10% and 60% solutions. A figure may be helpful in visualizing these relationships. The amount of pure acid in each solution is shown in color; the remainder is water.

Write an equation relating the amounts of pure acid (the "weights" in this problem are the amounts of each concentration of acid):

$$\underbrace{0.10\,x}_{\substack{\text{Liters of pure acid}\\\text{in 10\% solution}}} + \underbrace{0.60(20)}_{\substack{\text{Liters of pure acid}\\\text{in 60\% solution}}} = \underbrace{0.50(x+20)}_{\substack{\text{Liters of pure acid}\\\text{in 50\% solution}}}.$$

Solve the equation. First, multiply each side by 100 to clear the decimals:

$$10x + 60(20) = 50(x + 20)$$
$$10x + 1200 = 50x + 1000$$
$$-40x = -200$$
$$x = 5.$$

Five liters of 10% acid solution are needed.

Linear equations sometimes arise in problems involving rate, time, and distance.

EXAMPLE 10 Leo helped his sister drive her car to college and then flew home. They averaged 50 miles per hour on the road, and Leo's flight home traveled at 300 miles per hour. If Leo's round trip took 15 hours and his flight left 1 hour after Leo and his sister arrived at the school, how far is it from their home to the school?

Solution For this problem it is convenient first to find the time taken for Leo to drive to school. Using this information, the distance from home to school can be found:

time driving to school: t.

Leo's round trip of 15 hours was spent driving, waiting for his flight, and flying home, so

$$\begin{aligned}\text{Time flying home} &= \underbrace{\text{total time}}_{15} - \underbrace{\text{time driving}}_{t} - \underbrace{\text{time waiting}}_{1}\\ &= 14 - t\end{aligned}$$

It is helpful to organize the information into a chart.

	Rate	Time	Distance
Driving	50 mph	t hr	$50t$ mi
Flying	300 mph	$(14 - t)$ hr	$300(14 - t)$ mi

The distances in the third column are obtained by using the fact that distance = rate × time $(d = rt)$. Since the distance Leo traveled on each leg of the journey was the same,

$$50t = 300(14 - t).$$

Solve the equation:

$$50t = 4200 - 300t$$
$$350t = 4200$$
$$t = 12.$$

Leo and his sister spent 12 hours driving, so they traveled a distance of (12 hours) (50 miles per hour), or 600 miles.

EXERCISE 4.1

A

- Solve each equation. See Examples 1, 2, and 3.

1. $3x + 5 = 26$
2. $2 + 5x = 37$
3. $3(z + 2) = 14$
4. $2(z - 3) = 15$
5. $3y - (y - 4) = 12$
6. $5y - (y + 1) = 14$
7. $2[x - 3(x + 2) - 4] = 6$
8. $3[3x - 2(x - 3) + 1] = 8$
9. $-2[y - (y + 1)] = 3(y - 2)$
10. $-3[2y - (y - 2)] = 2(y + 3)$
11. $(z - 3)^2 - z^2 = 6$
12. $(z + 2)^2 - z^2 = 8$
13. $4 + (x - 1)(x + 2) = x^2 + 8$
14. $2 + (x - 2)(x + 3) = x^2 + 12$
15. $3[x + 2(x + 2)^2 - x^2] = 3x^2 + 3$
16. $2[2x + 3(x + 1)^2 - x^2] = 4x^2 + 1$
17. $4(x - 1)^2 + 2x = (2x + 1)^2 + 2x$
18. $2(x + 1)^2 + 2x(x + 1) = 2 + (2x - 1)^2$
19. $0.40(y - 4) = 2.80$
20. $0.60(y + 2) = 3.60$
21. $0.25y + 0.10(y + 32) = 11.60$
22. $0.12y + 0.08(y + 10,000) = 12,000$

23. $0.10x + 0.10(20) = 0.50(x + 20)$
24. $0.10x + 0.12(x + 4000) = 920$
25. $0.8x - 2.6 = 1.4x + 0.3$
26. $4.8 - 1.3x = 0.7x + 2.1$
27. $0.8843x + 2.3214 = 1.7652 - 3.1621x$
28. $6.4872x - 0.2183 = 2.1847x + 3.4629$
29. $5.532x - 12.078 = 6.157 - 2.323(x - 0.012)$
30. $1.187x + 4.296 = 11.091 + 6.215(x - 7.825)$

■ *Solve each formula for the specified variable. See Example 4.*

31. $I = prt$, for p
32. $V = lwh$, for h
33. $P = 2l + 2w$, for w
34. $S = 3\pi d + \pi a$, for d
35. $v = k + gt$, for g
36. $v = k + gt$, for t
37. $A = P(1 + rt)$, for r
38. $A = P(1 + rt)$, for t
39. $R = 2d + h(a + b)$, for b
40. $S = a + (n - 1)d$, for n
41. $A = \pi rh + \pi r^2$, for h
42. $A = 2w^2 + 4lw$, for l
43. $S = 2(lw + lh + wh)$, for h
44. $S = 2(ab + bc + ac)$, for a

■ *Solve each problem. See Example 5.*

45. A math contest exam has 40 questions. A contestant scores 5 points for each correct answer but loses 2 points for each wrong answer. If Lupe answered all the questions and her score was 102, how many questions did she answer correctly?

46. A game show contestant wins $25 for each correct answer he gives but loses $10 for each incorrect response. Roger answered 24 questions and won $355. How many answers did he get right?

47. The reprographics department has a choice of two new copying machines. One costs $20,000 and $0.02 per copy to operate. The other sells for $17,000, but operating costs are $0.025 per copy. How many copies must the repro department make before the more expensive copier pays for itself?

48. Annie needs a new refrigerator and can choose between two models of the same size. One model costs $525 and $0.08 per hour to run. The more energy efficient model costs $700 but runs for $0.05 per hour. If Annie buys the more expensive model, how long will it be before she starts saving money?

49. A noted illustrator receives a fee of $10,000 to create illustrations for a children's book. She will also receive royalties of 2% of the price of each book. If the books sell for $12.95, how many copies must be sold before the illustrator makes $20,000 for her work?

50. Willie makes $21,000 per year as a vacuum cleaner salesman. He is offered a job by a competing company that would pay him $19,000 a year plus a 6% commission on his sales. If the vacuum cleaners cost $320, how many would Willie have to sell to top his present salary?

■ *See Example 6.*

51. The population of Midland has been growing at a rate of 8% over the past 5 years. Its present population is 135,000.
    a. What was the population of Midland last year?
    b. What was the population 2 years ago?

52. For the past 3 years the inflation rate has been 6%. This year a steak dinner costs $12.
    a. What did a steak dinner cost last year?
    b. What did a steak dinner cost 2 years ago?
53. Virginia took a 7% pay cut when she changed jobs last year.
    a. What percent pay increase must she receive this year in order to match her old salary of $24,000?
    b. Can you answer the question in (a) without knowing Virginia's old salary?
54. Clarence W. Networth took a 16% loss in the stock market last year.
    a. What percent gain must he realize this year in order to restore his original holdings of $85,000?
    b. Can you answer the question in (a) without knowing the original value of Clarence's holdings?

■ *See Example 7.*

55. Delbert's test average in algebra is 77%. If the final exam counts for 30% of the grade and the test average counts for 70%, what must Delbert score on the final to have a term average of 80%?
56. Harold's batting average for the first 8 weeks of the baseball season is .385. What batting average must he maintain over the last 18 weeks so that his season average will be .350?

■ *See Example 8.*

57. In 1985 a survey found that the average weekly income for 317 lawyers was $717. If the average weekly income for male lawyers was $776 and the average weekly income for female lawyers was $558, how many female lawyers were represented in the survey?
58. Two hundred and ninety-four juniors and seniors in a large high school took the Scholastic Aptitude Test last spring. Their average score on the math portion was 520. The average score for the seniors was 526, and the average score for the juniors was 512. How many juniors took the exam?

■ *See Example 9.*

59. A sculptor wants to cast a bronze statue from an alloy that is 60% copper. He has 30 pounds of a 45% alloy. How much 80% copper alloy should he mix with it to obtain the 60% copper alloy?
60. A horticulturist needs a fertilizer that contains 8% potash, but she can find only fertilizers that contain 6% and 15% potash. How much of each should she mix to obtain 10 pounds of 8% potash fertilizer?
61. Barbara wants to earn $500 a year by investing $5000 in two accounts, a savings plan that pays 8% annual interest and a high-risk option that pays 13.5% interest. How much should she invest in each account?

62. An investment broker promises his client a 12% return on her funds. If the broker invests $3000 in bonds paying 8% interest, how much must he invest in stocks paying 15% interest to keep his promise?

■ *See Example 10.*

63. A cyclist sets out on a 3-day trip at an average speed of 16 miles per hour. Six hours later his wife finds his patch kit on the dining room table.
    a. If she heads after him in the car at 45 miles per hour, how long will it be before she catches him?
    b. How far from home will they be when she catches him?
64. Kate and Julie set out in their sailboat on a straight course at 9 miles per hour. Two hours later their mother becomes worried and sends their father after them in the speedboat.
    a. If their father travels at 24 miles per hour, how long will it be before he catches them?
    b. How far from shore will they be when he catches them?
65. Patrick jogs to the gym at 6 miles per hour, works out for an hour, and walks home at 4 miles per hour. If he is gone for 1 hour and 50 minutes, how far does he live from the gym?
66. Erin drives to the market at 40 miles per hour, shops for an hour and a half, and drives home in rush hour traffic at 30 miles per hour. If she is gone for 2 hours and 5 minutes, how far from the market does she live?

B

■ *Solve each problem.*

67. Larry wants to have a swimming pool built in his backyard. A pool that is 6 feet deep will cost $45 in materials for each square foot of swimming area. The contractor charges $3000 plus 40% of the cost of the materials to pay for labor. If Larry has $12,000 to spend on the pool, what is the largest pool he can afford?
68. Martin and Trudy are in the market for a new car. They can afford payments of $220 per month, and they have $1200 for a down payment. If the dealer will arrange a 3-year payment plan at 8.5% annual simple interest, what is the most expensive car that Martin and Trudy can consider?
69. Don makes $13 per hour for 40 hours per week, 50 weeks per year. His income tax comes to 18% of his gross annual income. How many hours of overtime must Don work, at time and a half, to pay for his taxes? (Assume that his overtime will also be taxed at the same rate.)
70. Rosa makes $8 per hour for a 40-hour work week. The payroll office deducts 13% of her weekly paycheck for taxes and Social Security, 0.5% for union dues, and $25 for a company savings plan. If Rosa would like to take home $300 a week, how many hours of overtime, at time and a half, must she work? (Assume that the deductions from her overtime are at the same rate as her regular pay.)

4.2

EQUATIONS CONTAINING FRACTIONS

To solve an equation that includes fractions we first generate an equivalent equation that does not involve fractions. To do this we multiply each side of the equation by the LCD of all the fractions that appear in the equation.

EXAMPLE 1 Solve $\dfrac{x}{3} + 2 = \dfrac{x}{2}$.

Solution Multiply each side by 6, the LCD of the denominators 3 and 2, to produce an equivalent equation free of fractions. This gives

$$6\left(\frac{x}{3} + 2\right) = 6\left(\frac{x}{2}\right)$$
$$2x + 12 = 3x.$$

Now add $-2x$ to each side to obtain

$$2x + 12 + (-2x) = 3x + (-2x)$$
$$12 = x.$$

Hence, the solution is 12.

Care must be exercised in the application of Property (2) on page 141, for we have specifically *excluded multiplication by zero*. For example, the equation $x = 3$, whose solution is 3, is *not* equivalent to $0 \cdot x = 0 \cdot 3$, for which every real number is a solution.

EXAMPLE 2 Solve $\dfrac{x}{x-3} = \dfrac{3}{x-3} + 2$. (1)

Solution Multiply each side by $(x - 3)$ to clear the equation of fractions. This gives

$$(x - 3)\frac{x}{x - 3} = (x - 3)\frac{3}{x - 3} + (x - 3)2 \qquad (2)$$

or

$$x = 3 + 2x - 6,$$

from which

$$x = 3.$$

Thus, 3 appears to be a solution of (1). But, on substituting 3 for x in (1), we have

$$\frac{3}{0} \quad \text{and} \quad \frac{3}{0} + 2,$$

where both sides of the equation are undefined. To obtain Equation (2) we multiplied each side of Equation (1) by $(x - 3)$. However, when x is 3, $(x - 3)$ is 0, so Equation (2) is *not* equivalent to Equation (1) for $x = 3$. Thus, 3 is *not* a solution of Equation (1). Equation (1) has no solution.

Common Error Note that in Example 2 *each term* of the equation must be multiplied by the LCD, including the constant term, 2.

Whenever we multiply both sides of an equation by an expression containing the variable we must check the apparent solution by substituting it into the original equation. Apparent solutions that do not satisfy the original equation are called **extraneous solutions**.

EXAMPLE 3 Solve $\dfrac{2}{x - 4} + \dfrac{4}{x + 4} = \dfrac{2x}{x^2 - 16}$.

Solution Factor the denominator on the right side as $(x - 4)(x + 4)$. Multiply each side by the LCD, $(x - 4)(x + 4)$, to clear the equation of fractions. This gives

$$(x - 4)(x + 4)\left(\frac{2}{x - 4} + \frac{4}{x + 4}\right) = (x - 4)(x + 4)\left(\frac{2x}{x^2 - 16}\right),$$

or
$$2(x + 4) + 4(x - 4) = 2x,$$
from which
$$6x - 8 = 2x,$$
or
$$x = 2.$$

Check Since we multiplied each side of the equation by $(x - 4)(x + 4)$, we must check the solution. Substitute 2 for x in the original equation. Does
$$\frac{2}{2 - 4} + \frac{4}{2 + 4} = \frac{2(2)}{2^2 - 16}?$$

Yes; $-1 + \frac{2}{3} = \frac{-4}{12}$, so 2 *is* a solution of the equation.

Proportions

A **proportion** is a special kind of fractional equation of the form $\frac{a}{b} = \frac{c}{d}$ in which each side is a ratio. If each side of the proportion is multiplied by the LCD, bd, we have
$$(bd)\frac{a}{b} = (bd)\frac{c}{d}$$
$$ad = bc.$$

This is a fundamental property of proportions:

$$\text{If } \frac{a}{b} = \frac{c}{d}, \text{ then } ad = bc. \tag{3}$$

EXAMPLE 4 Solve $\dfrac{x + 6}{3} = \dfrac{x + 1}{2}$.

Solution Using Property (3) write the equivalent equation
$$2(x + 6) = 3(x + 1),$$

from which

$$2x + 12 = 3x + 3$$
$$9 = x.$$

The solution is 9. Note that we would have obtained the same solution if we had multiplied each side of the original equation by the LCD, 6.

Common Error

Property (3) can only be used to solve a proportion. It does *not* apply to other kinds of fractional equations. For example, Property (3) does not apply to the equation

$$x + \frac{1}{2} = \frac{2x}{3},$$

which is not a proportion. To solve this equation we multiply each side by the LCD, 6, to clear the fractions.

Solving Formulas

In Section 4.1 we solved formulas for a specified variable in terms of the other variables. Formulas that involve fractions can be solved for one variable in terms of the others by applying similar techniques. It is often helpful to begin by clearing the fractions.

EXAMPLE 5 Solve the formula

$$\frac{1}{a} = \frac{1}{b} + \frac{1}{c}$$

for c.

Solution Multiply each side by the LCD, abc, to obtain

$$(abc)\frac{1}{a} = (abc)\frac{1}{b} + (abc)\frac{1}{c}$$
$$bc = ac + ab.$$

Subtracting ac from each side gives

$$bc - ac = ab.$$

4.2 ■ EQUATIONS CONTAINING FRACTIONS

Now factor c from the left side and divide each side by $(b - a)$ to obtain

$$c(b - a) = ab$$
$$c = \frac{ab}{b - a} \quad (b \neq a).$$

Problem Solving

Problems involving motion at a constant speed often lead to fractional equations.

EXAMPLE 6 Two trains travel from New Orleans to Birmingham, Alabama, a distance of 360 miles. The express train travels at twice the rate of the local and takes 4 hours less time for the trip. Find the rate of each train.

Solution Represent the rate of each train in terms of a single variable:

rate of the local train: r;
rate of the express train: $2r$.

The problem gives a relationship between the travel times for the two trains, so use the formula $d = rt$, or $t = \dfrac{d}{r}$, to write the time that each train traveled in terms of the variable, r:

travel time of local train: $\dfrac{360}{r}$

travel time of express train: $\dfrac{360}{2r}$

The travel time of the express train is 4 hours less than the travel time of the local, so subtract 4 from the local train's time and equate the difference to the express train's time:

$$\frac{360}{2r} = \frac{360}{r} - 4.$$

Solve the equation. First, multiply each side by the LCD, $2r$, to clear the fractions:

$$2r\left(\frac{360}{2r}\right) = 2r\left(\frac{360}{r} - 4\right)$$
$$360 = 720 - 8r$$
$$8r = 360$$
$$r = 45.$$

158   CHAPTER 4 ■ EQUATIONS IN ONE VARIABLE

Check that 45 is not an extraneous solution by substituting 45 for r in the original equation.

The local train's rate is 45 miles per hour, and the express train's rate is $2(45) = 90$ miles per hour.

Proportions can be used to solve a variety of problems involving the ratio of two quantities.

EXAMPLE 7 Jan's car uses 8 gallons of gas to travel 140 miles. How many gallons will be required for a trip of 450 miles?

Solution Gallons of gas needed for 450 miles: x.
Set up a ratio between gallons of gas and miles, using the given values, and set up a second ratio using the unknown quantity. Make sure that you use the same quantity (gallons in this case) in the numerator of each ratio. The two ratios can then be equated, yielding the proportion

$$\frac{8 \text{ gallons}}{140 \text{ miles}} = \frac{x \text{ gallons}}{450 \text{ miles}}.$$

Solve the proportion, using Property (3) on page 155:

$$8(450) = 140\, x$$

$$x = \frac{8(450)}{140}$$

$$= \frac{180}{7} = 25\frac{5}{7}.$$

Hence, $25\frac{5}{7}$ gallons of gas are needed to travel 450 miles.

EXAMPLE 8 Find x in the figure.

Solution Note that $\triangle ABC$ is similar to $\triangle ADE$, so the corresponding sides of the two triangles are proportional. (See the Review of Elementary Topics.) Thus,

$$\frac{3}{5} = \frac{4}{4 + x}$$

or, from Property (3) on page 155,

$$3(4 + x) = 4(5),$$

from which

$$12 + 3x = 20$$
$$x = \frac{8}{3}.$$

Check that $8/3$ is not an extraneous solution.

EXERCISE 4.2

A

■ *Solve each equation. See Example 1.*

1. $1 + \dfrac{x}{9} = \dfrac{4}{3}$
2. $4 + \dfrac{x}{5} = \dfrac{5}{3}$
3. $\dfrac{1}{5}x - \dfrac{1}{2}x = 9$
4. $\dfrac{1}{4}x = 2 - \dfrac{1}{3}x$
5. $\dfrac{x}{3} = 3 - \dfrac{x + 1}{2}$
6. $\dfrac{x - 5}{4} - \dfrac{x - 9}{12} = 1$
7. $\dfrac{2x}{3} - \dfrac{2x + 5}{6} = \dfrac{1}{2}$
8. $\dfrac{3x}{4} = \dfrac{2}{3} - \dfrac{x - 7}{6}$

■ *See Examples 2 and 3.*

9. $\dfrac{2}{x + 1} = \dfrac{x}{x + 1} + 1$
10. $\dfrac{5}{x - 3} = \dfrac{x + 2}{x - 3} + 3$
11. $\dfrac{3}{x - 2} = \dfrac{1}{2} + \dfrac{2x - 7}{2x - 4}$
12. $\dfrac{2}{x + 1} + \dfrac{1}{3x + 3} = \dfrac{1}{6}$
13. $\dfrac{4}{x + 2} - \dfrac{1}{x} = \dfrac{2x - 1}{x^2 + 2x}$
14. $\dfrac{1}{x - 1} + \dfrac{2}{x + 1} = \dfrac{x - 2}{x^2 - 1}$
15. $\dfrac{x}{x + 2} - \dfrac{3}{x - 2} = \dfrac{x^2 + 8}{x^2 - 4}$
16. $\dfrac{4}{2x - 3} + \dfrac{4x}{4x^2 - 9} = \dfrac{1}{2x + 3}$

■ *Solve each proportion by using Property (3) on page 155. See Example 4.*

17. $\dfrac{2}{3} = \dfrac{x}{x + 2}$
18. $\dfrac{7}{5} = \dfrac{x}{x - 2}$
19. $\dfrac{3}{4} = \dfrac{y + 2}{12 - y}$
20. $\dfrac{-3}{4} = \dfrac{y - 7}{y + 14}$
21. $\dfrac{50}{r} = \dfrac{75}{r + 20}$
22. $\dfrac{30}{r} = \dfrac{20}{r - 10}$

■ *Solve each formula for the specified variable. See Example 5.*

23. $F = \dfrac{9}{5}C + 32$, for C

24. $C = \dfrac{5}{9}(F - 32)$, for F

25. $A = \dfrac{h}{2}(b + c)$, for b

26. $S = \dfrac{n}{2}(a + s)$, for s

27. $E = \dfrac{1}{2}mv^2 + mgh$, for h

28. $S = vt + \dfrac{1}{2}at^2$, for a

29. $I = \dfrac{E}{R}$, for R

30. $I^2 = \dfrac{P}{R}$, for R

31. $0.8738 = \dfrac{h}{r + 10}$, for r

32. $2.483 = \dfrac{h}{r - 6}$, for r

33. $S = \dfrac{a}{1 - r}$, for r

34. $I = \dfrac{E}{r + R}$, for R

35. $\dfrac{W}{w} = \dfrac{D}{d}$, for w

36. $\dfrac{P}{T} = \dfrac{p}{t}$, for t

37. $V = C\left(1 - \dfrac{t}{n}\right)$, for n

38. $P = \dfrac{A}{1 + rt}$, for t

39. $\dfrac{1}{f} = \dfrac{1}{p} + \dfrac{1}{q}$, for q

40. $\dfrac{1}{R} = \dfrac{1}{r} + \dfrac{1}{s}$, for s

41. $H = \dfrac{2xy}{x + y}$, for x

42. $M = \dfrac{ab}{a + b}$, for b

43. $\dfrac{p}{a} = \dfrac{1 + e}{1 - e}$, for e

44. $\dfrac{p}{q} = \dfrac{r}{q + r}$, for r

■ *Solve each problem. See Example 6.*

45. Two student pilots leave the airport at the same time. They both fly at an air speed of 180 miles per hour, but one flies with the wind and the other flies against the wind. Both pilots check in with their instructor at the same time, and the first pilot has traveled 500 miles while the second pilot has gone 400 miles. Find the speed of the wind.

46. Pam's outboard motorboat travels at 20 miles per hour in still water. Pam drove 8 miles upstream to the gas station in two-thirds of the time it took her to travel 18 miles downstream to Marie's house. What is the speed of the current in the river?

47. A chartered sight-seeing flight over the Grand Canyon is scheduled to return to its departure point in 3 hours. If the plane flies at 100 miles per hour in still air and there is a head wind of 20 miles per hour on the outward journey, how far can the plane go before it must turn around?

48. The Explorer's Club plans a canoe trip up the Lazy River. Club members plan to paddle for 6 hours each day, and they take enough food for 4 days. If they paddle at 8 miles per hour in still water and the current in the Lazy River is 2 miles per hour, how far can they go upriver before they must turn around?

■ *See Example 7.*

49. If the taxes on a house worth $120,000 are $2700, what would the taxes be on a house assessed at $275,000 at the same tax rate?

50. If a typical household in the Midwest uses 83 million BTUs of electricity annually and pays $1236, how much will a household that uses 70 million BTUs annually spend for energy?

51. Your rich uncle leaves $100,000 to be divided between you and his daughter, Myrtle, in the ratio of 3 to 5. How much will you get?

52. The school district receives $320,000 to be divided between maintenance and new equipment in the ratio of 4 to 5. How much should be allocated to maintenance?

53. The scale on a map of Michigan uses 3/8 inch to represent 10 miles. If Isle Royale is 1 11/16 inches long on the map, what is the actual length of the island?

54. A photographer plans to enlarge a photograph that measures 8.3 centimeters by 11.2 centimeters to produce a poster that is 36 centimeters wide. How long will the poster be?

55. A compact car gets 32 miles to the gallon. When the car is exported to Europe its mileage must be converted to metric units. What is its mileage in kilometers per liter?

56. The orbital velocity of the earth around the sun is 29.8 kilometers per second. What is the speed in miles per hour?

57. The Forest Service tags 200 perch and releases them into Spirit Lake. One month later it captures 80 perch and finds that 18 of them are tagged. What is the Forest Service's estimate of the original perch population of the lake?

58. The Wildlife Commission tags 30 Canada Geese at one of its migratory feeding grounds. When the geese return the commission captures 45 geese, of which four are tagged. What is the commission's estimate of the number of geese that use the feeding ground?

■ *See Example 8. (See the Review of Elementary Topics for properties of similar triangles.)*

59. A 6-foot man stands 12 feet from a lamppost. His shadow is 9 feet long. How tall is the lamppost?

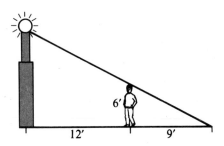

60. A rock climber estimates the height of a cliff she plans to scale as follows. She places a mirror on the ground so that she can just see the top of the cliff in the mirror while she stands straight. (The angles 1 and 2 formed by the light rays are

162 CHAPTER 4 ■ EQUATIONS IN ONE VARIABLE

equal.) She then measures the distance to the mirror (2 feet) and the distance from the mirror to the base of the cliff (56 feet). If she is 5 feet, 6 inches tall, how high is the cliff?

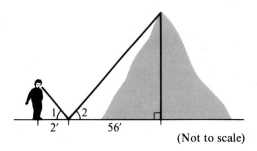

(Not to scale)

61. A conical tank is 12 feet deep and the diameter of the top is 8 feet. If the tank is filled with water to a depth of 7 feet, what is the area of the surface of the water?

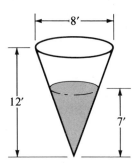

62. A florist fits a cylindrical piece of foam into a conical vase that is 10 inches high and measures 8 inches across the top. If the radius of the foam cylinder is $2\frac{1}{2}$ inches, how tall should it be just to reach the top of the vase?

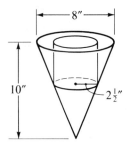

63. To measure the distance across a river, stand at A and sight across the river to a convenient landmark at B. Then measure the distances AC, CD, and DE. If AC = 20 feet, CD = 13 feet, and DE = 58 feet, how wide is the river?

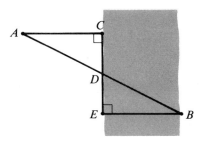

64. To measure the distance EC across the lake shown in the figure, stand at A and sight point C across the lake, then mark point B. Then sight to point E and mark point D so that DB is parallel to CE. If AD = 25 yards, AE = 60 yards, and BD = 30 yards, how wide is the lake?

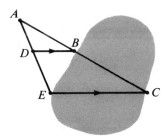

B

■ *Solve each formula for the specified variable.*

65. $v = \dfrac{1 - \dfrac{e}{c}}{\dfrac{1}{c} - e}$, for e

66. $k^2 = \dfrac{b}{1 + \dfrac{m}{M}}$, for m

67. $V^2 = Gm\left(\dfrac{2}{r} - \dfrac{1}{a}\right)$, for r

68. $\dfrac{V^2}{2} = gR^2\left(\dfrac{1}{R} - \dfrac{1}{r}\right)$, for r

69. $V = \dfrac{2}{\dfrac{1}{v_1} + \dfrac{1}{v_2}}$, for v_2

70. $V = \dfrac{v_1 + v_2}{1 + \dfrac{v_1 v_2}{c}}$, for v_1

4.3

QUADRATIC EQUATIONS

A second-degree equation in one variable is called a **quadratic equation.** We shall call

$$ax^2 + bx + c = 0$$

the *standard form* for such equations, where a, b, and c are constants representing real numbers and $a \neq 0$.

Solution by Factoring

If the left side of a quadratic equation in standard form can be factored, we can solve the equation by making use of the following principle, called the **zero-factor principle.**

> *The product of two factors equals 0 if and only if one or both of the factors equals 0.*

Thus,

$$ab = 0 \quad \text{if and only if} \quad a = 0 \quad \text{or} \quad b = 0.$$

EXAMPLE 1 $(x - 1)(x + 2) = 0$ if and only if

$$x - 1 = 0 \quad (x = 1)$$

or

$$x + 2 = 0 \quad (x = -2).$$

The word "or" in the statement above is used in the inclusive sense to mean either one *or* the other *or* both.

EXAMPLE 2 Solve $3x(x + 1) = 2x + 2$.

Solution First, write the equation in standard form as
$$3x^2 + x - 2 = 0.$$

Then factor the left side to obtain
$$(3x - 2)(x + 1) = 0$$

and apply the zero-factor principle, which gives
$$3x - 2 = 0 \quad \text{or} \quad x + 1 = 0$$
$$x = \frac{2}{3} \qquad x = -1.$$

The solutions are ⅔ and −1.

EXAMPLE 3 Solve $x^2 - \frac{17}{3}x = 2$.

Solution First, write the equation in standard form as
$$x^2 - \frac{17}{3}x - 2 = 0.$$

Multiply each side by 3 and simplify to obtain
$$3(x^2) - 3\left(\frac{17}{3}x\right) - 3(2) = 3(0)$$
$$3x^2 - 17x - 6 = 0.$$

Factor the left side and apply the zero-factor principle to obtain
$$(3x + 1)(x - 6) = 0,$$
$$3x + 1 = 0 \quad \text{or} \quad x - 6 = 0$$
$$x = -\frac{1}{3} \qquad x = 6.$$

The solutions are −⅓ and 6.

Common Error

Note that to use the zero-factor principle one side of the equation *must be zero*. Thus, in solving the equation

$$(x - 2)(x - 4) = 15$$

it is *incorrect* to set each factor equal to 15:

$$x - 2 \neq 15; \qquad x - 4 \neq 15.$$

Before applying the zero-factor principle we must simplify the left side and write the equation in standard form.

Equations containing fractions are often quadratic.

EXAMPLE 4 Solve $\dfrac{6}{x} + 1 = \dfrac{1}{x + 2}$.

Solution Multiply both sides of the equation by the LCD, $x(x + 2)$, to get

$$x(x + 2)\left(\frac{6}{x} + 1\right) = x(x + 2)\left(\frac{1}{x + 2}\right),$$

or

$$6(x + 2) + x(x + 2) = x.$$

Use the distributive law to remove parentheses and write in standard form to obtain

$$6x + 12 + x^2 + 2x = x$$
$$x^2 + 7x + 12 = 0.$$

Factor the left side to get

$$(x + 3)(x + 4) = 0$$

and apply the zero-factor principle:

$$x + 3 = 0 \quad \text{or} \quad x + 4 = 0$$
$$x = -3 \qquad\qquad x = -4.$$

Check Since we multiplied both sides of the equation by $x(x + 2)$, we must check the answers. Substitute each value for x into the original equation:

$$x = -3: \text{ does } \frac{6}{-3} + 1 = \frac{1}{-3 + 2}?$$

Yes; $-2 + 1 = -1$.

$$x = -4: \text{ does } \frac{6}{-4} + 1 = \frac{1}{-4 + 2}?$$

Yes; $\frac{-3}{2} + 1 = \frac{-1}{2}$.

(Or, note that neither value for x causes any denominator to equal zero.) The solutions are -3 and -4.

Common Error

Note that the LCD for the equation in Example 4 is *not* $x + 2$. The first denominator, x, is not a factor of $x + 2$.

Solutions of Quadratic Equations

In general, a quadratic equation has two solutions. However, if the left side is the square of a binomial, we find only one solution. For example, the quadratic equation

$$x^2 - 2x + 1 = 0,$$

or

$$(x - 1)^2 = 0,$$

has only one solution, 1. The unique solution obtained in this example is said to be of *multiplicity two*.

Notice that the solutions of the quadratic equation

$$(x - r_1)(x - r_2) = 0 \qquad (1)$$

are r_1 and r_2. Therefore, if two solutions r_1 and r_2 of a quadratic equation are known, we can reconstruct the equation directly as (1). By multiplying the factors together the equation can be written in standard form.

EXAMPLE 5 If ½ and -3 are given as the solutions of a quadratic equation, the equation is

$$\left(x - \frac{1}{2}\right)[x - (-3)] = 0.$$

To obtain an equivalent equation with integer coefficients we multiply both sides of the equation by 2 to clear the fractions (this also facilitates multiplication of the binomials):

$$2\left(x - \frac{1}{2}\right)(x + 3) = 2(0),$$

or

$$(2x - 1)(x + 3) = 0,$$

from which

$$2x^2 + 5x - 3 = 0.$$

Extraction of Roots

Quadratic equations of the form

$$x^2 = b \tag{2}$$

can be solved by a method called **extraction of roots**. From the definition of square root, x must be a square root of b. Since each nonzero real number b has two square roots (either real or imaginary), Equation (2) has two solutions, $\sqrt{b}$ and $-\sqrt{b}$. These solutions are real if $b > 0$ and imaginary if $b < 0$.

EXAMPLE 6 Solve.

    a. $2x^2 - 6 = 0$ b. $4x^2 + 3 = 0$

Solutions For each equation isolate x^2 on one side of the equation and then extract roots.

a. $$2x^2 - 6 = 0$$
$$x^2 = 3$$
$$x = \sqrt{3} \quad \text{or} \quad x = -\sqrt{3}$$
The solutions are $\sqrt{3}$ and $-\sqrt{3}$.

b.
$$4x^2 + 3 = 0$$
$$x^2 = -\frac{3}{4}$$
$$x = \sqrt{-\frac{3}{4}} \quad \text{or} \quad x = -\sqrt{-\frac{3}{4}}$$

The solutions are $\frac{1}{2}i\sqrt{3}$ and $-\frac{1}{2}i\sqrt{3}$.

Equations of the form

$$(x - p)^2 = q$$

can also be solved by extraction of roots.

EXAMPLE 7 Solve the equation $(x - 2)^2 = 16$.

Solution The equation says that $(x - 2)$ is a number whose square is 16. Hence,

$$x - 2 = 4 \quad \text{or} \quad x - 2 = -4,$$

from which we have

$$x = 6 \quad \text{or} \quad x = -2.$$

The solutions are 6 and -2.

Problem Solving

In some cases a mathematical model for a physical situation may lead to a quadratic equation and hence may have two solutions. It may be that one, but not both, of the solutions fits the physical situation.

EXAMPLE 8 A ball thrown into the air reaches a height h in feet given by the formula $h = 64t - 16t^2$, where t is the time in seconds after the throw. How long will it take the ball to reach a height of 48 feet on its way up?

Solution The model is given by the formula $h = 64t - 16t^2$, so we need only substitute 48 for h and solve for t:

$$48 = 64t - 16t^2$$
$$16t^2 - 64t + 48 = 0$$
$$16(t^2 - 4t + 3) = 0$$
$$16(t - 1)(t - 3) = 0$$
$$t = 1 \quad \text{or} \quad t = 3$$

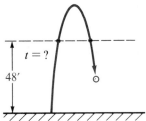

Use the figure to interpret the two solutions; it takes 1 second to reach 48 feet *on the way up*. (In 3 seconds the ball is also 48 feet high, but on the way down.)

EXAMPLE 9 The size of a rectangular computer monitor screen is the length of its diagonal. If the length of the rectangle is 3 inches greater than the width, what are the dimensions of a 15-inch monitor?

Solution Express the two dimensions of the monitor in terms of a single variable:

width: x;
length: $x + 3$.

Use the Pythagorean theorem (see the Review of Elementary Topics) to write the equation

$$x^2 + (x + 3)^2 = 15^2.$$

Solve the equation:

$$x^2 + x^2 + 6x + 9 = 225$$
$$2x^2 + 6x - 216 = 0$$
$$2(x - 9)(x + 12) = 0,$$

so

$$x - 9 = 0 \quad \text{or} \quad x + 12 = 0$$
$$x = 9 \qquad\qquad x = -12.$$

Since dimensions cannot be negative, the width is 9 inches and the length is $9 + 3 = 12$ inches.

EXAMPLE 10
Rachel built a deck of uniform width around her 20-foot by 25-foot rectangular pool. If she used 196 square feet of cedar decking, how wide was the deck?

Solution Width of deck: x.
The area of the deck can be obtained by subtracting the area of the pool from the entire enlarged area:

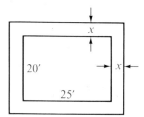

$$\underbrace{(25 + 2x)(20 + 2x)}_{\text{Enlarged area}} - \underbrace{500}_{\text{Area of pool}} = \underbrace{196}_{\text{Area of deck}}.$$

Solve the equation:

$$500 + 90x + 4x^2 - 500 = 196$$
$$4x^2 + 90x - 196 = 0$$
$$2x^2 + 45x - 98 = 0$$
$$(2x + 49)(x - 2) = 0$$
$$x = -\frac{49}{2} \quad \text{or} \quad x = 2.$$

The width cannot be negative, so the width of the deck is 2 feet.

EXERCISE 4.3

A

■ *Solve. Determine your answer by inspection or see Example 1.*

1. $(2x + 5)(x - 2) = 0$
2. $(x + 1)(3x - 1) = 0$
3. $x(2x + 1) = 0$
4. $x(3x - 7) = 0$
5. $4(x - 6)(2x + 3) = 0$
6. $5(2x - 7)(x + 1) = 0$

■ *See Example 2.*

7. $2x^2 = 6x$
8. $3z^2 = 3z$
9. $2z^2 - 18 = 0$
10. $3x^2 - 12 = 0$
11. $9x^2 = 4$
12. $25x^2 = 4$
13. $3y^2 - 6y = -3$
14. $12y^2 = 8y + 15$
15. $x(2x - 3) = -1$
16. $2x(x - 2) = x + 3$
17. $x(x - 3) = 2(x - 3)$
18. $5(t + 2) = t(t + 2)$
19. $y(3y + 2) = (y + 2)^2$
20. $(x - 1)^2 = 2x^2 + 3x - 5$

21. $x(x + 1) = 4 - (x + 2)^2$
22. $6(1 - y) = 12 - (y + 1)^2$
23. $(t + 2)(t - 5) = 8$
24. $(x + 1)(2x - 3) = 3$
25. $4x - [(x + 1)(x - 2) + 6] = 0$
26. $2x - [(x + 2)(x - 3) + 8] = 0$
27. $[(x + 1)^2 - 2x] = 10$
28. $3[(x + 2)^2 - 4x] = 15$

■ See Examples 3 and 4.

29. $\dfrac{2x^2}{3} + \dfrac{x}{3} - 2 = 0$
30. $2x - \dfrac{5}{3} = \dfrac{x^2}{3}$
31. $\dfrac{x^2}{6} + \dfrac{x}{3} = \dfrac{1}{2}$

32. $\dfrac{x}{4} - \dfrac{3}{4} = \dfrac{1}{x}$
33. $3 = \dfrac{10}{x^2} - \dfrac{7}{x}$
34. $5 = \dfrac{6}{x^2} - \dfrac{7}{x}$

35. $\dfrac{4}{3x} + \dfrac{3}{3x + 1} + 2 = 0$
36. $-3 = \dfrac{-10}{x + 2} + \dfrac{10}{x + 5}$

37. $\dfrac{9}{x^2 + x - 2} + \dfrac{1}{x^2 - x} = \dfrac{4}{x - 1}$
38. $\dfrac{2}{x^2 - 2x} + \dfrac{1}{2x} = \dfrac{-1}{x^2 + 2x}$

39. $\dfrac{x + 2}{x - 3} - \dfrac{8x + 11}{2x^2 - 5x - 3} = \dfrac{x + 1}{2x + 1}$
40. $\dfrac{2x + 1}{2x - 3} + \dfrac{x^2 + 3x - 7}{2x^2 - 7x + 6} = \dfrac{x + 1}{x - 2}$

■ Given the solutions of a quadratic equation, r_1 and r_2, write the equation in standard form with integral coefficients. See Example 5.

41. -2 and 1
42. -4 and 3
43. 0 and -5
44. 0 and 5
45. -3 and $\dfrac{1}{2}$
46. $-\dfrac{2}{3}$ and 4
47. $-\dfrac{1}{4}$ and $\dfrac{3}{2}$
48. $-\dfrac{1}{3}$ and $-\dfrac{1}{2}$

■ Solve by extraction of roots. See Example 6.

49. $9x^2 = 25$
50. $4x^2 = 9$
51. $2x^2 = 14$
52. $3x^2 = 15$
53. $4x^2 + 24 = 0$
54. $3x^2 + 9 = 0$
55. $\dfrac{2x^2}{3} = 4$
56. $\dfrac{3x^2}{5} = 6$

■ See Example 7.

57. $(x - 2)^2 = 9$
58. $(x + 3)^2 = 4$
59. $(2x - 1)^2 = 16$
60. $(3x + 1)^2 = 25$
61. $(x + 2)^2 = -3$
62. $(x - 5)^2 = -7$
63. $\left(x - \dfrac{1}{2}\right)^2 = \dfrac{3}{4}$
64. $\left(x - \dfrac{2}{3}\right)^2 = \dfrac{5}{9}$
65. $\left(x + \dfrac{1}{3}\right)^2 = \dfrac{-1}{81}$
66. $\left(x + \dfrac{1}{2}\right)^2 = \dfrac{-1}{16}$
67. $(8x - 7)^2 = 8$
68. $(5x - 12)^2 = 24$

■ *Solve each problem. See Examples 8, 9, and 10.*

69. Delbert stands at the top of a 300-foot cliff and throws his algebra book directly upward with a velocity of 20 feet per second. The height of his book above the ground t seconds later is given by $h = -16t^2 + 20t + 300$, where h is in feet.

   a. How long will it be before Delbert's book passes him on the way down?

   b. How long will it take Delbert's book to hit the ground at the bottom of the cliff?

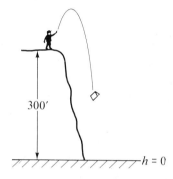

70. James Bond stands on top of a 240-foot building and throws a film canister upward to a fellow agent in a helicopter 16 feet above the building. The height of the film above the ground t seconds later is $h = -16t^2 + 32t + 240$, where h is in feet.

   a. How long will it take the film canister to reach the agent in the helicopter?

   b. If the agent misses the canister, how long will it be before the film hits the ground?

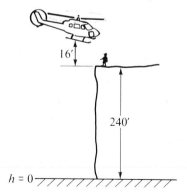

71. The area of an equilateral triangle is given by the formula $A = (\sqrt{3}/4)\, s^2$, where s is the length of the side. How long is the side of an equilateral triangle whose area is 12 square centimeters?

72. The area of the ring pictured is given by the formula $A = \pi R^2 - \pi r^2$, where R is the radius of the outer circle and r is the radius of the inner circle. If the outer radius is 4 centimeters and the area of the ring is 11π square centimeters, what is the radius of the inner circle?

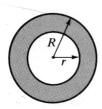

73. One end of a ladder is 10 feet from the base of a wall, and the other end reaches a window in the wall. The ladder is 2 feet longer than the height of the window. Find the height of the window.

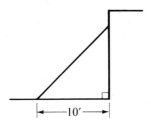

74. The diagonal of a rectangle is 20 inches. One side of the rectangle is 4 inches shorter than the other side. Find the dimensions of the rectangle.

75. What size square can be inscribed in a circle of radius 8 inches?

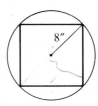

76. What size rectangle can be inscribed in a circle of radius 30 feet if the length of the rectangle must be three times its width?

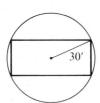

77. Rani has a rectangular rose garden that measures 30 feet by 50 feet. She wants to border the garden with a redwood chip path of uniform width. If she has enough redwood chips to cover 425 square feet, how wide can the path be?

78. Mehrdad wants to enlarge his 12-foot by 15-foot patio by paving a border of uniform width around three sides, excluding one 15-foot edge where the patio meets the house. He has enough bricks to pave 135 square feet. How wide can the border be?

79. A rancher has 360 yards of fence to enclose a rectangular pasture. If the pasture should be 8000 square yards in area, what should its dimensions be?

80. If the rancher in Problem 79 uses a riverbank to bound one side of the pasture, he can enclose 16,000 square yards with 360 yards of fence. What will the dimensions of the pasture be then?

81. A box is made from a square piece of cardboard by cutting 2-inch squares from each corner and turning up the edges. If the volume of the box is 50 cubic inches, how large should the piece of cardboard be?

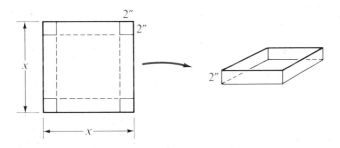

82. A length of rain gutter is made from a piece of aluminum 6 feet long and 1 foot wide. How much should be turned up along each long edge so that the gutter has a capacity of ¾ cubic feet of rainwater?

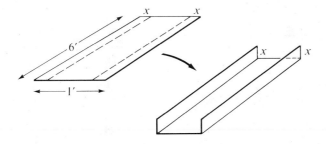

83. The owner of a motel has 60 rooms to rent. She finds that if she charges $20 per night all the rooms will be occupied. For each $2 that she increases the price, three rooms stand vacant. How much should she charge so that one night's revenue will be $1350?

84. The owner of a video store sells 80 blank tapes per week if he charges $8 per tape. For every $0.50 that he increases the price, he sells four fewer tapes per week. How much should he charge in order to bring in $648 per week from tapes?

85. The manager of Joe's Burgers discovers that he will sell $160/x$ burgers per day if the price of a burger is x dollars. On the other hand, he can afford to make $6x + 49$ burgers if he charges x dollars apiece. At what price will he sell as many burgers as he makes?

86. A florist finds that she will sell $300/x$ dozen roses per week if she charges x dollars for a dozen. Her suppliers will sell her $5x - 55$ dozen roses to be sold at x dollars per dozen. At what price will she sell all the roses she purchases?

87. A riverboat that travels at 18 miles per hour in still water can go 30 miles upriver in 1 hour's less time than it can go 63 miles downriver. What is the speed of the current in the river?

88. Gene's motorboat travels 20 miles per hour in still water. It took him 3 hours longer to go 90 miles upriver on a fishing trip than it took to return 75 miles downriver. How fast was the current in the river?

B

■ *Solve for x in terms of a, b, and c.*

89. $x^2 - (a + b)^2 = 0$

90. $(2a + x)^2 - b^2 = 0$

91. $x^2 - (a + b)x + ab = 0$

92. $a^2x^2 + 2a(b + c)x + (b + c)^2 = 0$

93. $x^2 - \dfrac{b + c}{a}x + \dfrac{bc}{a^2} = 0$

94. $x^2 - \dfrac{a^2 + b^2}{ab}x + 1 = 0$

95. If one solution of $2x^2 - bx - 21 = 0$ is 7, find the other solution.

96. If one solution of $ax^2 + 16x - 12 = 0$ is 6, find the other solution.

97. Show that the sum of the solutions of $x^2 + bx + c = 0$ is $-b$ and their product is c.

98. Show that the sum of the solutions of $ax^2 + bx + c = 0$ is $-b/a$ and their product is c/a.

4.4

COMPLETING THE SQUARE; QUADRATIC FORMULA

Completing the Square

Not every quadratic equation can be solved by factoring. For example, the trinomial $x^2 + x + 1$ cannot be factored, so the equation $x^2 + x + 1 = 0$ cannot be solved by factoring. However, we can use the method of extraction

4.4 ■ COMPLETING THE SQUARE; QUADRATIC FORMULA

of roots to solve *any* quadratic equation. In Section 4.3 we used extraction of roots to solve equations of the form

$$(x - p)^2 = q,$$

where the left side is the square of a binomial. In this section we will see how to rewrite any quadratic equation in a form that can be solved by extraction of roots.

Recall that in Section 1.5 we factored *perfect-square trinomials* as squares of binomials. For example,

$$x^2 + 10x + 25 = (x + 5)^2;$$
$$x^2 - 6x + 9 = (x - 3)^2.$$

Note that the constant term in each perfect-square trinomial is the *square of one-half the coefficient of x;* thus, in the examples given above we have $25 = [½(10)]^2$ and $9 = [½(-6)]^2$. Obtaining the constant term in this way is called **completing the square**.

EXAMPLE 1 Complete the square by adding an appropriate constant, and write the result as the square of a binomial.

a. $x^2 - 12x$ ___   b. $x^2 + 5x$ ___

Solutions a. One-half of -12 is -6, so the constant term is $(-6)^2$ or 36. Add 36 to obtain

$$x^2 - 12x + 36 = (x - 6)^2.$$

b. One-half of 5 is $5/2$, so the constant term is $(5/2)^2$, or $25/4$. Add $25/4$ to obtain

$$x^2 + 5x + \frac{25}{4} = \left(x + \frac{5}{2}\right)^2.$$

If we think of the coefficient of x as $2p$, the constant term is then p^2 and the resulting trinomial can be factored as $(x + p)^2$. In general,

$$x^2 + 2px + p^2 = (x + p)^2.$$

Now let us see how to apply the technique of completing the square to solving quadratic equations. Consider the equation

$$x^2 - 6x - 7 = 0.$$

We first rewrite the equation as

$$x^2 - 6x = 7$$

and then complete the square on the left side. Since $p = \tfrac{1}{2}(-6) = -3$ and $p^2 = (-3)^2 = 9$, we add 9 to complete the square. Thus,

$$x^2 - 6x + 9 = 7 + 9.$$

Note that we must add 9 to *both* sides of the equation! The left side of the equation is now the square of a binomial, $(x - 3)^2$, so we can write

$$(x - 3)^2 = 16.$$

Finally, we obtain the solution by extraction of roots:

$$x - 3 = 4 \quad \text{or} \quad x - 3 = -4;$$

thus,

$$x = 7 \quad \text{or} \quad x = -1.$$

The solutions of the equation are 7 and −1.

Solutions obtained by completing the square may be irrational or complex numbers.

EXAMPLE 2 Solve $x^2 - 4x + 20 = 0$.

Solution First, write the equation with the constant term on the right side:

$$x^2 - 4x = -20.$$

Now complete the square on the left side; one-half the coefficient of x is $\tfrac{1}{2}(-4) = -2$, and $(-2)^2 = 4$, so add 4 to *both* sides of the equation:

$$x^2 - 4x + 4 = -20 + 4.$$

Write the left side as the square of a binomial, and combine terms on the right side:

$$(x - 2)^2 = -16.$$

Finally, extract roots to obtain

$$x - 2 = 4i \quad \text{or} \quad x - 2 = -4i.$$

Thus,

$$x = 2 + 4i \quad \text{or} \quad x = 2 - 4i.$$

Common Error Note that in Example 2 the solution $2 + 4i$ is *not* equal to $6i$. Remember that the real and imaginary parts of a complex number cannot be combined.

The method just presented does not work if the leading coefficient is a number other than 1. For example, $2x^2 - 6x = 2$ cannot be solved by adding 9 to each side of the equation, since $2x^2 - 6x + 9$ is not a perfect square. We must first divide each side of the equation by the leading coefficient, 2. This will enable us to write the equation in the form $(x - p)^2 = q$. We can then proceed as above.

EXAMPLE 3 Solve $2x^2 - 6x - 2 = 0$:

Solution First, divide each side by 2:

$$x^2 - 3x - 1 = 0.$$

Then rewrite the equation with the constant term on the right side:

$$x^2 - 3x = 1.$$

One-half the coefficient of the first-degree term is $[\frac{1}{2}(-3)] = -\frac{3}{2}$ and $(-\frac{3}{2})^2 = \frac{9}{4}$, so add $\frac{9}{4}$ to both sides of the equation:

$$x^2 - 3x + \frac{9}{4} = 1 + \frac{9}{4}.$$

Rewrite the left side as the square of a binomial and simplify the right side to get

$$\left(x - \frac{3}{2}\right)^2 = \frac{13}{4}.$$

Finally, extract roots by setting $(x - 3/2)$ equal to each square root of $13/4$. This gives the two equations

$$x - \frac{3}{2} = \sqrt{\frac{13}{4}} \quad \text{and} \quad x - \frac{3}{2} = -\sqrt{\frac{13}{4}}$$

$$x = \frac{3}{2} + \frac{\sqrt{13}}{2} \qquad\qquad x = \frac{3}{2} - \frac{\sqrt{13}}{2}.$$

The solutions are $\dfrac{3 + \sqrt{13}}{2}$ and $\dfrac{3 - \sqrt{13}}{2}$.

We summarize the solution of quadratic equations by completing the square as follows.

> **To Solve a Quadratic Equation by Completing the Square:**
>
> 1. Write the equation in standard form.
> 2. Divide each side by the coefficient of the second-degree term, and subtract the constant term from each side of the equation.
> 3. Add the square of one-half the coefficient of the first-degree term to each side of the equation.
> 4. Write the left side as the square of a binomial. Simplify the right side.
> 5. Complete the solution of the equation by extracting roots.

The Quadratic Formula

We can solve the general quadratic equation,

$$ax^2 + bx + c = 0,$$

by completing the square to obtain the following formula for the solutions of a quadratic equation:

$$x = \frac{-b \pm \sqrt{b^2 - 4ac}}{2a}.$$

(The proof is left as an exercise.) This formula, called the **quadratic formula,** expresses the solutions of a quadratic equation in terms of its coefficients. The symbol ±, read "plus or minus," is used to condense the two equations

$$x = \frac{-b + \sqrt{b^2 - 4ac}}{2a} \quad \text{and} \quad x = \frac{-b - \sqrt{b^2 - 4ac}}{2a}$$

into a single equation. We need only substitute the coefficients a, b, and c of a given quadratic equation into the formula to find the solutions for that equation.

The quadratic formula can be used to solve any quadratic equation, although solution by factoring is often faster and simpler. The quadratic formula should be used if the quadratic trinomial cannot be factored easily.

EXAMPLE 4 Solve $x^2 = 2x - \dfrac{1}{2}$.

Solution Write the equation in standard form as

$$2x^2 - 4x + 1 = 0.$$

Then substitute 2 for a, -4 for b, and 1 for c in the quadratic formula to obtain

$$\begin{aligned} x &= \frac{-(-4) \pm \sqrt{(-4)^2 - 4(2)(1)}}{2(2)} \\ &= \frac{4 \pm \sqrt{16 - 8}}{4} \\ &= \frac{4 \pm 2\sqrt{2}}{4}. \end{aligned}$$

Reduce by factoring the numerator to get

$$\frac{2(2 \pm \sqrt{2})}{4} = \frac{2 \pm \sqrt{2}}{2}.$$

The solutions are $\dfrac{2 + \sqrt{2}}{2}$ and $\dfrac{2 - \sqrt{2}}{2}$.

Common Error

The expression in Example 4 *cannot* be reduced by "canceling" the fours; that is,

$$x \neq \frac{\cancel{4} \pm 2\sqrt{2}}{\cancel{4}} = \pm 2\sqrt{2}$$

because 4 is not a factor of the numerator.

EXAMPLE 5 Solve $x = \dfrac{-2}{2x - 1}$.

Solution Write the equation in standard form as follows:

$$(2x - 1)x = \frac{-2}{2x - 1}(2x - 1)$$
$$2x^2 - x = -2$$
$$2x^2 - x + 2 = 0.$$

Substitute 2 for a, -1 for b, and 2 for c in the quadratic formula to obtain

$$x = \frac{-(-1) \pm \sqrt{(-1)^2 - 4(2)(2)}}{2(2)}$$
$$= \frac{1 \pm \sqrt{-15}}{4} = \frac{1 \pm i\sqrt{15}}{4}.$$

Check Notice that neither value for x causes the denominator in the original equation to equal zero.

The solutions are $\dfrac{1 - i\sqrt{15}}{4}$ and $\dfrac{1 + i\sqrt{15}}{4}$.

Notice that in Example 4 we can express the solutions as $1 + \sqrt{2}/2$ and $1 - \sqrt{2}/2$, and in Example 5 we can express the solutions as $\frac{1}{4} - i\sqrt{15}/4$ and $\frac{1}{4} + i\sqrt{15}/4$. In fact, it is clear from the form

$$x = \frac{-b}{2a} \pm \frac{\sqrt{b^2 - 4ac}}{2a}$$

that if the solutions of a quadratic equation are irrational or complex they are conjugates:

$$\frac{-b}{2a} + \frac{\sqrt{b^2 - 4ac}}{2a} \quad \text{or} \quad \frac{-b}{2a} - \frac{\sqrt{b^2 - 4ac}}{2a}.$$

Solving Formulas

Quadratic equations in more than one variable can be solved for a specified variable in terms of the others by using extraction of roots or the quadratic formula.

EXAMPLE 6 Solve $V = \frac{1}{3}\pi(r + 2)^2 h$ for r.

Solution Use extraction of roots. First, multiply both sides by 3 to clear the fractions, then divide by πh so that the squared binomial is isolated on one side of the equation:

$$3V = \pi(r + 2)^2 h$$

$$\frac{3V}{\pi h} = (r + 2)^2.$$

Extracting square roots gives

$$\pm\sqrt{\frac{3V}{\pi h}} = r + 2,$$

or

$$r = -2 \pm \sqrt{\frac{3V}{\pi h}}.$$

EXAMPLE 7 Solve $x^2 - xy + y = 2$ for x in terms of y.

Solution First, write the equation in standard form as a quadratic equation in the variable x:

$$x^2 - yx + (y - 2) = 0.$$

184 CHAPTER 4 ■ EQUATIONS IN ONE VARIABLE

Expressions involving y are treated as constants with respect to x. Use the quadratic formula, substituting 1 for a, $-y$ for b, and $y - 2$ for c. Thus,

$$x = \frac{-(-y) \pm \sqrt{(-y)^2 - 4(1)(y - 2)}}{2(1)}$$

$$= \frac{y \pm \sqrt{y^2 - 4y + 8}}{2}.$$

Problem Solving

Applied problems may lead to quadratic equations that cannot be solved by factoring. Extraction of roots or the quadratic formula can be used to solve such equations.

EXAMPLE 8 The owners of a day-care center plan to enclose a divided play area against the back wall of their building as shown. They have 300 feet of picket fence and would like the total area of the playground to be 6000 square feet. Find the dimensions of the playground.

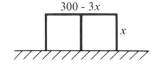

Solution Represent the dimensions in terms of a single variable:

width of the playground: x;
length of the playground: $300 - 3x$.

Since the area (length × width) should be 6000 square feet, write the equation:

$$x(300 - 3x) = 6000.$$

In standard form,

$$3x^2 - 300x + 6000 = 0,$$

or

$$x^2 - 100x + 2000 = 0.$$

The left side cannot be factored, so use the quadratic formula with $a = 1$, $b = -100$, and $c = 2000$:

$$x = \frac{-(-100) \pm \sqrt{(-100)^2 - 4(1)(2000)}}{2(1)}$$

$$= \frac{100 \pm \sqrt{10{,}000 - 8000}}{2}$$

$$= \frac{100 \pm \sqrt{2000}}{2}$$

$$\approx \frac{100 \pm 44.7}{2}.$$

Thus, $x \approx 72.35$ or $x \approx 27.65$. Both values give feasible solutions to the problem. If the width is 72.35 feet, the length is $300 - 3(72.35)$, or 82.95, feet. If the width is 27.65 feet, the length is $300 - 3(27.65)$, or 217.05, feet.

EXERCISE 4.4

A

■ *Complete the square and write the result as the square of a binomial. See Example 1.*

1. $x^2 + 8x$
2. $x^2 - 14x$
3. $x^2 - 7x$
4. $x^2 + 3x$
5. $x^2 + \dfrac{3}{2}x$
6. $x^2 - \dfrac{5}{2}x$
7. $x^2 - \dfrac{4}{5}x$
8. $x^2 + \dfrac{2}{3}x$

■ *Solve by completing the square. See Examples 2 and 3.*

9. $x^2 - 2x + 1 = 0$
10. $x^2 + 4x + 4 = 0$
11. $x^2 + 9x + 20 = 0$
12. $x^2 - x - 20 = 0$
13. $x^2 = 3 - 3x$
14. $x^2 = 5 - 5x$
15. $2x^2 + 4x - 3 = 0$
16. $3x^2 + x - 4 = 0$
17. $2x^2 - 5 = 3x$
18. $4x^2 - 3 = 2x$
19. $2x^2 + 4x = -3$
20. $3x^2 + x = -4$

Solve for x, y, or z using the quadratic formula. See Examples 4 and 5.

21. $x^2 - 5x + 4 = 0$
22. $x^2 - 4x + 4 = 0$
23. $y^2 + 3y = 4$
24. $y^2 - 5y = 6$
25. $z^2 = 3z - 1$
26. $2z^2 = 7z - 6$
27. $0 = x^2 - \frac{5}{3}x + \frac{1}{3}$
28. $0 = x^2 - \frac{1}{2}x + \frac{1}{2}$
29. $5z + 6 = 6z^2$
30. $13z + 5 = 6z^2$
31. $x^2 - 5x = 0$
32. $y^2 + 3y = 0$
33. $4y^2 + 8 = 0$
34. $2z^2 + 1 = 0$
35. $2y^2 = y - 1$
36. $x^2 + 2x = -5$
37. $y = \frac{1}{y - 3}$
38. $2z = \frac{3}{z - 2}$
39. $3y = \frac{1 + y}{y - 1}$
40. $2x = \frac{x + 1}{x - 1}$
41. $2y^2 = y - 2 - y^2$
42. $3z^2 + 2z + 2 = z$
43. $2x = \frac{-1}{2x - 1}$
44. $y = \frac{-1}{y - 1}$
45. $\frac{x}{x^2 + 2x - 2} = 1$
46. $\frac{3x + 1}{x^2 + x + 5} = 1$
47. $\frac{2x}{x - 1} - \frac{x + 1}{2} = 0$
48. $\frac{3}{2x + 1} - \frac{2x - 3}{x} = 0$
49. $\frac{2x}{x - 1} + \frac{3}{x^2 - x} = \frac{x + 1}{x}$
50. $\frac{3x - 1}{3x + 1} + \frac{2x}{2x + 1} = 1$

Solve each formula for the indicated variable. See Example 6.

51. $A = 4\pi r^2$, for r
52. $K = \frac{1}{2}mv^2$, for v
53. $F = \frac{mv^2}{r}$, for v
54. $A = \frac{\sqrt{3}}{4}s^2$, for s
55. $A = P(1 + r)^2$, for r
56. $V = \pi(r - 2)^2 h$, for r
57. $F = \frac{Gm_1 m_2}{d^2}$, for d
58. $F = \frac{kq_1 q_2}{r^2}$, for r
59. $V = \pi(r^2 + R^2)h$, for R
60. $V = 2(s^2 + t^2)w$, for t
61. $a^2 + b^2 = c^2$, for b
62. $2a^2 + 3b^2 = c^2$, for a
63. $\frac{x^2}{8} + \frac{y^2}{3} = 1$, for y
64. $\frac{x^2}{5} - \frac{y^2}{6} = 1$, for x
65. $E = \frac{1}{2}mv^2 + mgh$, for v
66. $h = \frac{1}{2}gt^2 + dl$, for t
67. $h = \frac{gr^2 - rv^2}{v^2}$, for v
68. $R^2 = \frac{E^2 - I^2(wL)^2}{I^2}$, for I
69. $a = \frac{mr}{2m - rV^2}$, for V
70. $m = \frac{4A^2}{a(1 - e^2)}$, for e

■ *See Example 7.*

71. $A = 2w^2 + 4lw$, for w
72. $A = \pi r^2 + \pi rs$, for r
73. $h = 4t - 16t^2$, for t
74. $h = 6t - 3t^2$, for t
75. $P = IE - RI^2$, for I
76. $s = vt - \frac{1}{2}at^2$, for t
77. $D = \frac{n^2 - 3n}{2}$, for n
78. $S = \frac{n^2 + n}{2}$, for n
79. $3x^2 + xy + y^2 = 2$, for y
80. $y^2 - 3xy + x^2 = 3$, for x

■ *Solve each problem. See Example 8.*

81. A car traveling at s miles per hour on a dry road surface will require approximately d feet to stop, where d is given by $d = s^2/24 + s/2$. If a car must be able to stop in 50 feet, what is the maximum speed it can safely travel?

82. A car traveling at s miles per hour on a wet road surface will require approximately d feet to stop, where d is given by $d = s^2/12 + s/2$. If a car failed to stop in 100 feet, what speed did it exceed?

83. A skydiver jumps out of an airplane at 11,000 feet. Her altitude in feet t seconds after jumping is given by $h = -16t^2 - 16t + 11{,}000$.
    a. If she must open her parachute at 1000 feet, how long can she free-fall?
    b. If the skydiver drops a marker just before she opens her parachute, how long will it take the marker to hit the ground?

84. A high diver jumps from the 10-meter springboard. His height in meters above the water t seconds after leaving the board is given by $h = -9.8t^2 + 8t + 10$.
    a. How long is it before the diver passes the board on the way down?
    b. How long is it before the diver hits the water?

85. Cyril plans to invest $5000 in a money market account paying interest compounded annually. If he would like to have $6250 in 2 years, what interest rate must Cyril obtain?

86. Two years ago Carol's living expenses were $1200 per month. This year the same items cost Carol $1400 per month. What was the annual inflation rate for the past 2 years?

87. An animal trainer has 100 meters of chain link fence with which to enclose 250 square meters in three pens of equal size as shown. Find the dimensions of each pen.

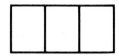

88. An architect wants to include in his plans a rectangular window topped by a semicircle. In order to admit enough light the window should have an area of 120 square feet. The rectangular portion of the window should be 2 feet longer than it is wide. Find the dimensions of the window.

188 CHAPTER 4 ■ EQUATIONS IN ONE VARIABLE

89. When a person looks down from a height, his line of sight is tangent to the earth at the horizon.

   a. If the radius of the earth is 3960 miles, use the Pythagorean theorem to calculate how far a person can see on a clear day from the top of the World Trade Center in New York, 1350 feet high.

   b. How high would a person have to be to see 100 miles?

90. a. If the radius of the earth is 6370 kilometers, how far can a person see from an airplane at an altitude of 10,000 meters? (See Problem 89.)

   b. How high would a person have to be to see 10 kilometers?

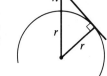

B

■ *Given one solution of a quadratic equation with rational coefficients, find the other. Write a quadratic equation with these solutions.*

91. $2 + \sqrt{5}$ 92. $3 - \sqrt{2}$ 93. $4 - 3i$ 94. $5 + i$

95. Complete the square to find the solutions of

$$x^2 + bx + c = 0$$

in terms of b and c.

96. Complete the square to find the solutions of

$$ax^2 + bx + c = 0$$

in terms of a, b, and c.

■ *The **discriminant** of a quadratic equation is the expression*

$$D = b^2 - 4ac.$$

(Note that the discriminant is the expression under the radical in the quadratic formula.) It can be used to determine the nature of the solutions of a quadratic equation:

   if $D > 0$, there are two distinct real solutions;
   if $D = 0$, there is one real solution (of multiplicity two);
   if $D < 0$, there are two complex (conjugate) solutions.

Use the discriminant to classify the solutions of the following quadratic equations.

97. $3x^2 + 26 = 17x$ 98. $4x^2 + 23x = 19$
99. $16x^2 - 712x + 7921 = 0$ 100. $121x^2 + 1254x + 3249 = 0$
101. $65.2x = 13.2x^2 + 41.7$ 102. $0.03x^2 = 0.05x - 0.12$

4.5 OTHER NONLINEAR EQUATIONS

Equations Involving Radicals

To solve equations containing radicals we make use of the following property.

> If each side of an equation is raised to the same natural-number power, the solutions of the original equation are also solutions of the new equation.

Raising both sides of an equation to a power does not always result in an equivalent equation; the new equation may have additional solutions that are not solutions of the original equation. These are called **extraneous solutions**. For example, by squaring each side of

$$x = 3 \tag{1}$$

we obtain

$$x^2 = 9. \tag{2}$$

The solutions of Equation (2) are 3 and -3. In this instance -3 is an extraneous solution of Equation (1), since -3 does not satisfy the original equation, $x = 3$.

If each side of an equation is raised to an *odd* power, extraneous solutions will not be introduced. However, if an *even* power is used, each solution obtained *must* be checked in the original equation to verify its validity. The check is part of the solution process.

EXAMPLE 1 Solve the equation $\sqrt{x + 2} + 4 = x$.

Solution First, obtain the radical expression $\sqrt{x + 2}$ as the only term on one side, yielding

$$\sqrt{x + 2} = x - 4.$$

Then square each side to get

$$(\sqrt{x + 2})^2 = (x - 4)^2$$
$$x + 2 = x^2 - 8x + 16.$$

Solve the quadratic equation by factoring:

$$x^2 - 9x + 14 = 0$$
$$(x - 2)(x - 7) = 0$$
$$x = 2 \quad \text{or} \quad x = 7.$$

Check Does $\sqrt{2 + 2} + 4 = 2$? No; 2 is not a solution.
Does $\sqrt{7 + 2} + 4 = 7$? Yes. The solution is 7.

Sometimes it is necessary to square both sides of an equation more than once in order to eliminate the radicals.

EXAMPLE 2 Solve $\sqrt{x - 7} + \sqrt{x} = 7$.

Solution First, add $-\sqrt{x}$ to each side to obtain

$$\sqrt{x - 7} = 7 - \sqrt{x},$$

which has only one term with a radical on each side. Square each side to remove one radical:

$$(\sqrt{x - 7})^2 = (7 - \sqrt{x})^2$$
$$x - 7 = 49 - 14\sqrt{x} + x.$$

Simplify the result, and isolate the radical on one side of the equation:

$$-56 = -14\sqrt{x}$$
$$4 = \sqrt{x}.$$

Now square each side again to obtain

$$(4)^2 = (\sqrt{x})^2$$
$$16 = x.$$

Check Does $\sqrt{16 - 7} + \sqrt{16} = 7$? Yes. The solution is 16.

Common Error

Note that we cannot solve the equation in Example 2 by squaring each term. Thus, it would be *incorrect* to write

$$(\sqrt{x-7})^2 + (\sqrt{x})^2 = 7^2.$$

The property on page 189 can also be used to solve equations that involve roots of index higher than two.

EXAMPLE 3 Solve $\sqrt[3]{x^2 + 12x} = 4$.

Solution Cube each side of the equation to obtain

$$(\sqrt[3]{x^2 + 12x})^3 = (4)^3$$
$$x^2 + 12x = 64.$$

Solve the resulting quadratic equation by factoring:

$$x^2 + 12x - 64 = 0$$
$$(x + 16)(x - 4) = 0$$
$$x = -16 \quad \text{or} \quad x = 4.$$

Note that we do not need to check the solutions, since we raised each side of the equation to an odd power. The solutions are -16 and 4.

EXAMPLE 4 Solve $(2x + 1)^{2/3} = 25$.

Solution First, cube each side of the equation to obtain

$$[(2x + 1)^{2/3}]^3 = 25^3,$$

or

$$(2x + 1)^2 = 25^3.$$

Solve this quadratic equation by extracting roots:

$$2x + 1 = \pm 25^{3/2},$$

from which

$$2x + 1 = 125 \quad \text{or} \quad 2x + 1 = -125.$$

Solving for x yields

$$x = 62 \quad \text{or} \quad x = -63.$$

We do not need to check the solutions, since we *cubed* each side and then extracted square roots.

We can also use the property on page 189 to solve formulas that involve radicals for a specified variable.

EXAMPLE 5 Solve $t = \sqrt{\dfrac{1 + s^2}{g}}$ for s.

Solution Square each side to get

$$t^2 = \frac{1 + s^2}{g},$$

from which

$$gt^2 = 1 + s^2$$
$$s^2 = gt^2 - 1.$$

Thus,

$$s = \pm\sqrt{gt^2 - 1}.$$

Equations Quadratic in Form

Some equations that are not quadratic equations are nevertheless quadratic in form—that is, they are of the form

$$au^2 + bu + c = 0, \qquad (3)$$

where u represents some expression in terms of another variable. For example,

$$x^4 - 10x^2 + 9 = 0 \quad \text{or} \quad (x^2)^2 - 10(x^2) + 9 = 0$$

is quadratic in the variable x^2, and

$$y^{2/3} - 5y^{1/3} + 4 = 0 \quad \text{or} \quad (y^{1/3})^2 - 5(y^{1/3}) + 4 = 0$$

is quadratic in the variable $y^{1/3}$.

4.5 ■ OTHER NONLINEAR EQUATIONS

To solve equations that are quadratic in form we use a technique called *substitution of variables*.

EXAMPLE 6 Solve $x^4 - 10x^2 + 9 = 0$.

Solution Make the substitution $x^2 = u$. Then $x^4 = u^2$, and the equation reduces to

$$u^2 - 10u + 9 = 0.$$

Solve this equation by factoring to obtain

$$(u - 9)(u - 1) = 0$$
$$u = 9 \quad \text{or} \quad u = 1.$$

But, since $u = x^2$, we have

$$x^2 = 9 \quad \text{or} \quad x^2 = 1,$$

from which we obtain the solutions 3, −3, 1, −1.

The equation in Example 6 could also have been solved by factoring the left side to obtain

$$(x^2 - 9)(x^2 - 1) = 0,$$

from which

$$x^2 - 9 = 0 \quad \text{or} \quad x^2 - 1 = 0$$
$$x = \pm 3 \qquad \qquad x = \pm 1$$

so that the solutions are the same as above.

EXAMPLE 7 Solve $y^{2/3} - 5y^{1/3} + 4 = 0$ by substitution of variables.

Solution Make the substitution $y^{1/3} = u$. Then $y^{2/3} = u^2$, and the equation becomes

$$u^2 - 5u + 4 = 0.$$

Solve for u by factoring:

$$(u - 4)(u - 1) = 0$$
$$u = 4 \quad \text{or} \quad u = 1.$$

Finally, replace u with $y^{1/3}$ and solve for y:

$$y^{1/3} = 4 \quad \text{or} \quad y^{1/3} = 1$$
$$y = 64 \qquad\qquad y = 1.$$

The solutions are 1 and 64.

Solving Other Equations by Factoring

The zero-factor principle, introduced on page 164, can sometimes be used to solve equations that are not necessarily quadratic.

EXAMPLE 8 Solve $2x(x - 1)^{-1} - 2x^2(2x - 1)^{-2} = 0$.

Solution Factor $2x(2x - 1)^{-2}$ from the left side to obtain

$$2x(2x - 1)^{-2}[(2x - 1) - x] = 0$$
$$2x(2x - 1)^{-2}(x - 1) = 0,$$

or

$$\frac{2x(x - 1)}{(2x - 1)^2} = 0.$$

Since a fraction equals zero only if its numerator is zero, set the numerator equal to zero to get

$$2x = 0 \quad \text{or} \quad x - 1 = 0$$
$$x = 0 \qquad\qquad x = 1.$$

The solutions are 0 and 1.

Common Error

Note that in Example 8 we did *not* find a solution by setting the denominator equal to zero. The left side is undefined when $x = \frac{1}{2}$.

4.5 ■ OTHER NONLINEAR EQUATIONS

Problem Solving

Equations that involve radicals can arise in the solution of applied problems.

EXAMPLE 9 Two oil derricks are located the same distance offshore from a straight section of coast. A supply boat returns from the derrick at point A to the harbor at C (see the figure), reloads, and sails to the derrick at B, traveling a total distance of 102 miles. How far are the derricks from the coast?

Solution Distance from each derrick to coast: x.

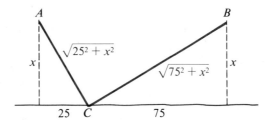

Use the Pythagorean theorem to write an expression for the distance traveled by the supply boat on each leg of its journey. The sum of these distances is 102 miles:

$$\sqrt{x^2 + (25)^2} + \sqrt{x^2 + (75)^2} = 102.$$

Solve the equation. First, rewrite the equation with one radical on each side, and then square each side:

$$(\sqrt{x^2 + 625})^2 = (102 - \sqrt{x^2 + 5625})^2$$
$$x^2 + 625 = 10{,}404 - 204\sqrt{x^2 + 5625} + x^2 + 5625.$$

Collect like terms and square again:

$$204\sqrt{x^2 + 5625} = 15{,}404$$
$$\sqrt{x^2 + 5625} \approx 75.5$$
$$x^2 + 5625 \approx 5700.25$$
$$x \approx \pm 8.7.$$

Since x must be greater than 0, we find that $x \approx 8.7$. The derricks are about 8.7 miles offshore.

EXERCISE 4.5

A

■ Solve and check. If there is no solution, so state. See Example 1.

1. $\sqrt{x} - 5 = 3$
2. $\sqrt{x} - 4 = 1$
3. $\sqrt{y + 6} = 2$
4. $\sqrt{y - 3} = 5$
5. $3z + 4 = \sqrt{3z + 10}$
6. $2z - 3 = \sqrt{7z - 3}$
7. $2x + 1 = \sqrt{10x + 5}$
8. $4x + 5 = \sqrt{3x + 4}$
9. $\sqrt{y + 4} = y - 8$
10. $4\sqrt{x - 4} = x$
11. $\sqrt{2y - 1} = \sqrt{3y - 6}$
12. $\sqrt{4y + 1} = \sqrt{6y - 3}$
13. $\sqrt{x - 3}\sqrt{x} = 2$
14. $\sqrt{x}\sqrt{x - 5} = 6$

■ See Example 2.

15. $\sqrt{y + 4} = \sqrt{y + 20} - 2$
16. $4\sqrt{y} + \sqrt{1 + 16y} = 5$
17. $\sqrt{x} + \sqrt{2} = \sqrt{x + 2}$
18. $\sqrt{4x + 17} = 4 - \sqrt{x + 1}$
19. $(5 + x)^{1/2} + x^{1/2} = 5$
20. $(y + 7)^{1/2} + (y + 4)^{1/2} = 3$
21. $(y^2 - 3y + 5)^{1/2} - (y + 2)^{1/2} = 0$
22. $(z - 3)^{1/2} + (z + 5)^{1/2} = 4$

■ See Examples 3 and 4.

23. $\sqrt[3]{x} = -3$
24. $\sqrt[3]{x} = -4$
25. $\sqrt[4]{x - 1} = 2$
26. $\sqrt[4]{x - 1} = 3$
27. $x^{2/3} - 1 = 15$
28. $x^{3/4} + 3 = 11$
29. $(3x - 4)^{3/2} = 27$
30. $(6x - 2)^{5/3} = -32$
31. $x^{-2/5} = 9$
32. $x^{-3/2} = 8$
33. $(2x - 3)^{-1/4} = \dfrac{1}{2}$
34. $(5x + 2)^{-1/3} = \dfrac{1}{4}$
35. $\sqrt[3]{x^2 - 3} = 3$
36. $\sqrt[4]{x^3 - 7} = 3$
37. $\sqrt[3]{2x^2 - 15x} = 5$
38. $\sqrt[3]{2x^2 - 11x} = 6$

■ Solve each formula for the specified variable. See Example 5.

39. $T = 2\pi\sqrt{\dfrac{L}{g}}$, for L
40. $T = 2\pi\sqrt{\dfrac{m}{k}}$, for m
41. $S = r\sqrt{\dfrac{g}{r + h}}$, for h
42. $S = r\sqrt{\dfrac{2g}{r + h}}$, for h
43. $r = \sqrt[3]{\dfrac{3V}{4\pi}}$, for V
44. $d = \sqrt[3]{\dfrac{16Mr^2}{m}}$, for M
45. $R = \sqrt[4]{\dfrac{8Lvf}{\pi p}}$, for p
46. $T = \sqrt[4]{\dfrac{E}{SA}}$, for A
47. $r = \sqrt{t^2 - s^2}$, for s
48. $c = \sqrt{a^2 - b^2}$, for b

49. $A = B + C\sqrt{D + E^2}$, for E
50. $A = B - C\sqrt{D - E^2}$, for E
51. $A = \dfrac{\pi a^2 \sqrt{1 - e^2}}{p}$, for e
52. $f = \dfrac{30}{\pi}\sqrt{\dfrac{g}{l^2 - r^2}}$, for l
53. $T = \pi\sqrt{\dfrac{(r + R)^3}{8m}}$, for r
54. $k = \dfrac{2\pi}{p}\sqrt{\dfrac{(a + R)^3}{1 + m}}$, for a
55. $m = \dfrac{M}{\sqrt{1 - \dfrac{v^2}{c^2}}}$, for v
56. $V = \sqrt{k\left(\dfrac{2}{r} - \dfrac{1}{a}\right)}$, for r

■ Solve. See Examples 6 and 7.

57. $y^4 - 4y^2 + 3 = 0$
58. $y^4 - 6y^2 + 5 = 0$
59. $2z^6 - 15z^3 = 8$
60. $3z^6 - 7z^3 = 6$
61. $x - 2\sqrt{x} = 15$
62. $x + 3\sqrt{x} = 10$
63. $\sqrt{x} - 6\sqrt[4]{x} + 8 = 0$
64. $\sqrt[3]{x^2} - 12\sqrt[3]{x} + 20 = 0$
65. $\sqrt[3]{(x + 1)^2} - 3\sqrt[3]{x + 1} - 4 = 0$
66. $\sqrt{x - 6} + 3\sqrt[4]{x - 6} = 18$
67. $z^3 + 7z^{3/2} - 8 = 0$
68. $27z^3 + 26z^{3/2} - 1 = 0$
69. $x^{2/3} - 3x^{1/3} = 4$
70. $2y^{2/3} + 5y^{1/3} = 3$
71. $2x - 9x^{1/2} = -4$
72. $8x^{1/2} + 7x^{1/4} = 1$
73. $y^{-2} - y^{-1} - 12 = 0$
74. $z^{-2} + 9z^{-1} - 10 = 0$
75. $(x - 1)^{1/2} - 2(x - 1)^{1/4} - 15 = 0$
76. $(x - 2)^{1/2} - 11(x - 2)^{1/4} + 18 = 0$

■ Solve each problem. See Example 9.

77. Two highways intersect at right angles as shown. At the instant when a car heading east at 50 miles per hour passes the intersection a car traveling north at 40 miles per hour is already 5 miles north of the intersection. When will the cars be 200 miles apart?

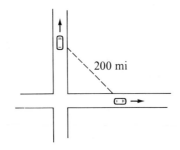

78. A plane flying north at 100 miles per hour passes over St. Louis at 10 A.M. A plane flying east at 200 miles per hour passes over St. Louis at 11 A.M. When will the planes be 700 miles apart?

79. Porterville and Perrysburg are located on opposite sides of a river as shown. The road joining the towns crosses a bridge over the river at point B. If the road from Porterville to Perrysburg is 50 miles long, how far is the bridge from point A?

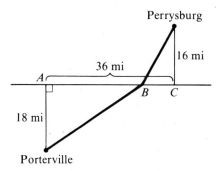

80. Two radio antennae are 75 feet apart. They are supported by a guy wire attached to the first antenna at a height of 20 feet, anchored to the ground between the antennae, and attached to the second antenna at a height of 25 feet. If the wire is 90 feet long, how far from the base of the first antenna should it be anchored to the ground?

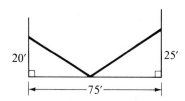

81. A cable television service wants to run a cable from its station at A to an island at C. It costs $150 per mile to run the cable underground and $500 per mile to run the cable underwater. If it costs $15,200 to run the cable to the island, how far is point P from the station?

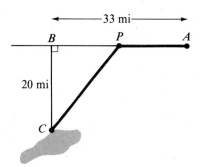

82. Henry lives on an island 4 miles from shore. There is a town at B, 7 miles from point A. On Saturdays Henry rows to point P and then walks the rest of the way to town. He can row 2 miles per hour and walk 5 miles per hour. If it takes him 3 hours and 18 minutes to get to town, how far does he walk?

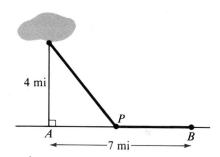

B

■ *Solve. See Example 8.*

83. $3x(2x - 1) - 4(2x - 1)^2 = 0$
84. $x(2 - 3x) + (3x^2 + 4)(2 - 3x) = 0$
85. $(x + 4)^3 - 36(x + 4) = 0$
86. $(x - 3)^2 - (x - 3)(x^2 - 5x) = 0$
87. $3x^2(x + 3)^{-2} - 2x^3(x + 3)^{-3} = 0$
88. $2x(3x + 4)^{-3} - 9x^2(3x + 4)^{-4} = 0$
89. $2(x - 3)^{5/4} + x(x - 3)^{1/4} = 0$
90. $(6x + 2)^{4/3} - 2x(6x + 2)^{1/3} = 0$
91. $\frac{4}{3}x^{1/3} + \frac{1}{3}x^{-2/3} = 0$
92. $\frac{1}{2}x^{-1/2} + \frac{3}{2}x^{1/2} = 0$
93. $(x^2 - 1)^{1/2} + x^2(x^2 - 1)^{-1/2} = 0$
94. $2(x^3 - 1)^{1/3} + x^3(x^3 - 1)^{-2/3} = 0$

CHAPTER REVIEW

A

[4.1]
■ *Solve.*

1. $2x - 6 = 4x - 8$
2. $0.40x = 240$
3. $2(x - 3) + 4(x + 2) = 6$
4. $2x - (x + 3) = 2(x + 1)$
5. $2[2x - (x + 3) + 4] = 12$
6. $-[x - (3x - 2) - 2] = 16$
7. $-[x^2 - (x - 3)^2] = 0$
8. $-[2(x + 1)^2 - 2x^2] = 0$
9. $0.30(y + 2) = 2.10$
10. $0.06y + 0.04(y + 1000) = 60$
11. Solve $3N = 5t - 3c$ for t.
12. Solve $C = 10 + 2p - 2t$ for p.

13. Solve $l = a + nd - d$ for n.
14. Solve $2s = at + k$ for a.
15. Solve $S = 2\pi(R - r)$ for R.
16. Solve $9C = 5(F - 32)$ for F.

[4.2]
- Solve.

17. $7 + \dfrac{5x}{3} = x - 2$
18. $\dfrac{2x - 1}{5} - \dfrac{x + 1}{2} = 0$
19. $\dfrac{x}{x - 2} = \dfrac{2}{x - 2} + 7$
20. $\dfrac{x}{x - 3} + \dfrac{9}{x + 3} = 1$

- Solve each proportion.

21. $\dfrac{y + 3}{y + 5} = \dfrac{1}{3}$
22. $\dfrac{y}{6 - y} = \dfrac{1}{2}$

- Solve each formula for the indicated variable.

23. $S = \dfrac{n}{2}(a + s)$, for a
24. $V = C\left(1 - \dfrac{t}{n}\right)$, for t
25. $\dfrac{1}{f} = \dfrac{1}{p} + \dfrac{1}{q}$, for p
26. $\dfrac{p}{q} = \dfrac{r}{q + r}$, for q

[4.3]
- Solve by factoring.

27. $x(3x + 2) = (x + 2)^2$
28. $6y = (y + 1)^2 + 3$
29. $4x - [(x + 1)(x - 2) + 6] = 0$
30. $3[(x + 2)^2 - 4x] = 15$
31. $\dfrac{1}{x} + 2x = \dfrac{33}{4}$
32. $\dfrac{1}{x} + \dfrac{2}{x - 2} = \dfrac{2}{3}$

- Solve by extraction of roots.

33. $5x^2 = 30$
34. $3x^2 + 6 = 0$
35. $(2x - 5)^2 = 9$
36. $(7x - 1)^2 = -15$

[4.4]
- Solve by completing the square.

37. $x^2 - 4x - 6 = 0$
38. $x^2 + 3x = 3$
39. $x^2 - x + 2 = 0$
40. $2x^2 = 2x - 3$

- Solve by using the quadratic formula.

41. $\dfrac{1}{2}x^2 + 1 = \dfrac{3}{2}x$
42. $x^2 - 3x + 1 = 0$
43. $x^2 - x + 2 = 0$
44. $2x^2 + 3x + 2 = 0$
45. $\dfrac{2}{x - 1} - \dfrac{x + 2}{x} = 0$
46. $\dfrac{3x}{x + 1} - \dfrac{2}{x^2 + x} = \dfrac{4}{x}$

■ *Solve each formula for the indicated variable.*

47. $\dfrac{x^2}{4} - \dfrac{y^2}{9} = 36$, for y

48. $p = \sqrt{\dfrac{1 - 2t^2}{s}}$, for t

49. $2x^2 - kx + 1 = 0$, for x

50. $2x^2 + xy + y^2 = 0$, for y

[4.5]
■ *Solve.*

51. $x - 3\sqrt{x} + 2 = 0$

52. $\sqrt{x + 1} + \sqrt{x + 8} = 7$

53. $(x + 7)^{1/2} + x^{1/2} = 7$

54. $(y - 3)^{1/2} + (y + 4)^{1/2} = 7$

55. $\sqrt[3]{x + 1} = 2$

56. $x^{2/3} + 2 = 6$

57. $(x - 1)^{-3/2} = \dfrac{1}{8}$

58. $(2x + 1)^{-1/2} = \dfrac{1}{3}$

■ *Solve each formula for the indicated variable.*

59. $t = \sqrt{\dfrac{2v}{g}}$, for g

60. $q - 1 = 2\sqrt{\dfrac{r^2 - 1}{3}}$, for r

61. $R = \dfrac{1 + \sqrt{p^2 + 1}}{2}$, for p

62. $q = \sqrt[3]{\dfrac{1 + r^2}{2}}$, for r

■ *Solve.*

63. $x^4 - 5x^2 + 4 = 0$

64. $y^4 - 6y^2 + 5 = 0$

65. $y^2 + 7 - \sqrt{y^2 + 7} = 12$

66. $y^2 - 1 - \sqrt{y^2 - 1} = 6$

67. $y^{2/3} - 2y^{1/3} = 8$

68. $x^{2/3} - 2x^{1/3} = 35$

69. $(x + 1)^3 - 12(x + 1) = 0$

70. $(y^2 - 4)^{1/2} - (y^2 - 4)^{1/4} = 0$

71. Two retired tennis pros go into business to market their own brand of tennis racket. Their initial outlay is $35,000, and the rackets cost $22 to manufacture.
    a. If they sell the rackets for $40 each, how many must they sell before they break even?
    b. How many must they sell in order to clear $10,000 apiece in the first year?

72. A cosmetics manufacturer expands her product line to include bubble bath. She spends $16,000 on start-up costs and markets the bubble bath for $8 per bottle.
    a. If the bubble bath costs $5 per bottle to produce, how many bottles must the manufacturer sell to break even?
    b. If the manufacturer can afford a loss no greater than $10,000 on the new product, how many bottles must she sell?

73. A solar energy firm experienced a 12% increase in sales this year on top of a 9% increase last year. What is its total percent increase in sales over 2 years ago?

74. A suburban school district experienced a 15% increase in enrollment last year and anticipates an 8% increase this year. What is the total percent increase in enrollment over 2 years ago?

75. Olga keeps up a 6-minute pace on the first 5 miles of a 12-mile run. What pace must she maintain on the last 7 miles so that her average pace will be 7 minutes per mile?

76. Bert's new car gets 28 miles to the gallon in highway driving and 21 in the city. If he drove 315 miles on a 14-gallon tank of gas, how many miles of highway driving did he do?

77. A recipe for applesauce bread calls for 1½ cups of flour and ⅓ cup of honey. How much honey should be used if the flour is increased to 2½ cups?

78. The instructions for planting an azalea call for ¾ pound of peat moss mixed with 1⅓ pounds of nitrohumus. How many pounds of nitrohumus will you need to make azalea mix with a 7-pound bag of peat moss?

79. Roscoe, who is exactly 6 feet tall, estimates the height of a bridge as follows. He stands at one end of the bridge and sights to the base of the supports at the opposite end. He then paces off the distance to point P (28 feet) and the distance from P to the other end of the bridge (92 feet). How tall is the bridge?

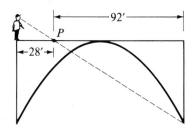

80. Find the size of the largest square that can be inscribed in an isosceles triangle of base 8 inches and height 12 inches. (See figure at right.)

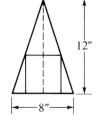

81. In a tennis tournament among n competitors $\dfrac{n(n-1)}{2}$ matches must be played. If the organizers can schedule 36 matches, how many players should they invite?

82. The formula $S = \dfrac{n(n+1)}{2}$ gives the sum of the first n positive integers. How many integers must be added to make a sum of 91?

83. A wooden packing crate has a square base and top and is 4 feet high. If the wood for the crate costs 10 cents per square foot, how large a crate can be built for $16.80?

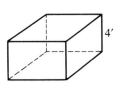

84. A redwood planter has a square cross section and a length of 40 inches. If the redwood costs ½ cent per square inch, how large a planter can be built for $5.44?

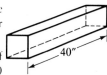

85. A travel agent offers a group rate of $2400 per person for a week in London if 15 people sign up for the tour. The price per person is reduced by another $100 each time an additional person signs up. How many people must sign up for the tour in order for the agent to collect $38,000?

86. The credit union divides $12,600 equally among its members each year as a dividend. This year there are six fewer members than last year, and each person receives $5 more. How many members are there this year?

87. Norm takes a commuter train 10 miles to his job in the city. The evening train returns him home at a rate 10 miles per hour faster than the morning train takes him to work. If Norm spends a total of 50 minutes per day commuting, what is the rate of each train?

88. Kristen drove 50 miles to her sister's house, traveling 10 miles in heavy traffic to get out of the city and then 40 miles in less congested traffic. Her average speed in the city was 20 miles per hour less than her speed in light traffic. What was each rate if her trip took 1 hour and 30 minutes?

89. Lewis invested $2000 in an account that compounds interest annually. Two years later he closed the account, withdrawing $2464.20. What interest rate did he earn?

90. Earl borrowed $5500 from his uncle for 2 years with interest compounded annually. At the end of 2 years he owed his uncle $6474.74. What was the interest on the loan?

B

■ Solve each formula for the indicated variable.

91. $w = \left(\dfrac{r}{r+h}\right)^2 s,\ \ \text{for } h$

92. $P = \dfrac{E^2 r}{(R+r)^2},\ \ \text{for } R$

93. $r = \dfrac{2gR^2}{2gR - V^2},\ \ \text{for } R$

94. $\dfrac{V^2}{2} = gR^2\left(\dfrac{1}{R} - \dfrac{1}{r}\right),\ \ \text{for } R$

95. $\dfrac{D}{L} = \dfrac{\sqrt{1 + \dfrac{v}{c}}}{\sqrt{1 - \dfrac{v}{c}}} - 1,\ \ \text{for } v$

96. $f = \dfrac{1}{2\pi}\sqrt{\dfrac{\dfrac{1}{C} + \dfrac{1}{D}}{L}},\ \ \text{for } C$

■ Solve.

97. $(x^2 - 2)(2x + 5)^{-3} - 2(2x + 5)^{-2} = 0$
98. $2x^2(3x + 2)^{-2/3} - (3x + 2)^{1/3} = 0$
99. $x^{-1/2}(x + 2)^{1/2} - x^{3/2}(x + 2)^{-1/2} = 0$
100. $3(x^2 - 2)(x^2 - 4) - 2(x^2 - 4)^2 = 0$

Summary for Part I

The section in which the symbol or property is first used is shown in parentheses.

Symbols

a^n: nth power of a, or a to the nth power:
$$a^n = a \cdot a \cdot a \cdots a \quad (n \text{ factors})$$
(1.1)

$a^0 = 1 \quad (a \neq 0)$ (2.5)

$a^{-n} = \dfrac{1}{a^n} \quad (a \neq 0)$ (2.5)

$a^{1/n}$: nth root of a (3.1)

$(a^{1/n})^n = a \quad (a \geq 0 \text{ if } n \text{ is even})$ (3.1)

$a^{m/n} = (a^m)^{1/n} = (a^{1/n})^m \quad (a > 0)$ (3.1)

$\sqrt[n]{a} = a^{1/n} \quad (n \geq 2)$ (3.1)

C: set of complex numbers   (3.5)

i: imaginary unit:
$$i = \sqrt{-1}; \quad i^2 = -1$$
(3.5)

$\sqrt{-b} = \sqrt{-1}\sqrt{b} = i\sqrt{b} \quad (b > 0)$ (3.5)

$a + bi$:   complex number   (3.5)

bi: imaginary number   (3.5)

Subsets of the Set of Complex Numbers

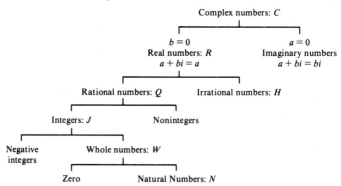

Properties

Special Products and Factors

$(x + a)(x + b) = x^2 + (a + b)x + ab$ (1.3)
$(x + a)^2 = x^2 + 2ax + a^2$ (1.5)
$(x - a)^2 = x^2 - 2ax + a^2$ (1.5)
$(x + a)(x - a) = x^2 - a^2$ (1.5)
$(x + a)(x^2 - ax + a^2) = x^3 + a^3$ (1.5)
$(x - a)(x^2 + ax + a^2) = x^3 - a^3$ (1.5)

Fractions (Denominators do not equal zero)

$\dfrac{a}{b} = \dfrac{a \cdot c}{b \cdot c}$ (2.1)

$-\dfrac{a}{b} = \dfrac{-a}{b} = \dfrac{a}{-b}$ (2.1)

$\dfrac{a}{b} = \dfrac{-a}{-b}$ (2.1)

$\dfrac{a}{b} \cdot \dfrac{c}{d} = \dfrac{a \cdot c}{b \cdot d}; \quad \dfrac{a}{b} \div \dfrac{c}{d} = \dfrac{a \cdot d}{b \cdot c}$ (2.2)

$\dfrac{a}{c} + \dfrac{b}{c} = \dfrac{a + b}{c}; \quad \dfrac{a}{c} - \dfrac{b}{c} = \dfrac{a - b}{c}$ (2.3)

Exponents

I. $a^m \cdot a^n = a^{m+n}$ (1.3)
II. $(a^m)^n = a^{mn}$ (1.3)
III. $(ab)^m = a^m b^m$ (1.3)
IV. $\dfrac{a^m}{a^n} = a^{m-n} = \dfrac{1}{a^{n-m}}$ $(a \neq 0)$ (2.5)
V. $\left(\dfrac{a}{b}\right)^n = \dfrac{a^n}{b^n}$ $(b \neq 0)$ (2.5)
VI. $\left(\dfrac{a}{b}\right)^{-n} = \left(\dfrac{b}{a}\right)^n$ $(a, b \neq 0)$ (2.5)

Radicals

$\sqrt[n]{a^m} = (\sqrt[n]{a})^m = a^{m/n}$ $(\sqrt[n]{a}\text{ real})$ (3.2)
$\sqrt[n]{a^n} = |a|$ for n even   (3.3)
$\sqrt[n]{a^n} = a$ for n odd   (3.3)
$\sqrt[n]{ab} = \sqrt[n]{a}\sqrt[n]{b}$ $(a > 0,\ b > 0)$ (3.3)
$\sqrt[n]{\dfrac{a}{b}} = \dfrac{\sqrt[n]{a}}{\sqrt[n]{b}}$ $(a \geq 0,\ b > 0)$ (3.3)

Nonlinear Equations

$ab = 0$ if a equals zero, b equals zero, or a and b equal zero   (4.3)

$x^2 = b$ is equivalent to
$\quad x = \sqrt{b}$ or $x = -\sqrt{b}$ (4.3)

$ax^2 + bx + c = 0$ is equivalent to
$$x = \dfrac{-b \pm \sqrt{b^2 - 4ac}}{2a}$$ (4.4)

If each side of an equation is raised to the same natural-number power, the solutions of the original equation are also solutions of the resulting equation.   (4.5)

Cumulative Review Exercises for Part I

■ The numbers in brackets refer to the sections in which such problems are first considered.

A

■ Simplify.

1. $\dfrac{-3^2 - 3(-2)^3}{3 - 8} - \dfrac{1 - 2^4(-2)^2}{(4 - 7)^2}$ [1.1]

2. $3[x - 2(x + 1) - 3]$ [1.3]

3. $\dfrac{4x^2 - 1}{1 - 2x}$ [2.1]

4. $\left(\dfrac{x^2 y^{-2}}{x^{-2} y}\right)^{-2}$ [2.5]

5. $\dfrac{xy - \sqrt{x^3 y^2}}{xy}$ [3.3]

6. $\dfrac{x^2}{y - x} \div \dfrac{x^3 - x^2}{x - y}$ [2.2]

7. $\left(\dfrac{x^{2/3} y^{1/2}}{x^{1/3}}\right)^{-6}$ [3.2]

8. $\dfrac{(4 \times 10^3)(6 \times 10^{-4})}{3 \times 10^{-2}}$ [2.5]

9. $\dfrac{(2x + 1)^{-1}}{(2x)^{-1} + 1^{-1}}$ [2.5]

10. $3\sqrt{2x} - \sqrt{32x} + 4\sqrt{50x}$ [3.4]

11. $2x - (x^2 - 4x) - [(1 - 2x^2) - (3x - 4)]$ [1.2]

12. $(-2b)^3(a^2 b)^2 + ab(3a^2 b)(-ab^3) - a^2(ab^2)^2$ [1.3]

■ Solve.

13. $0.08x + 0.12(x - 3000) = 1240$ [4.1]

14. $\dfrac{x + 1}{2x + 1} = \dfrac{x - 3}{x - 2} - \dfrac{1}{2}$ [4.2]

15. $x + \dfrac{3}{x} = 4$ [4.2]

16. $\dfrac{3}{y - 1} + \dfrac{2}{3y - 3} = \dfrac{11}{9}$ [4.2]

17. $3x^2 = 3x - 1$ [4.4]

18. $y - 2\sqrt{y} - 8 = 0$ [4.5]

19. $2x^2 - x + 3 = 0$ [4.4]

20. $\dfrac{x + 2}{x - 3} = \dfrac{3}{4}$ [4.2]

21. $R = \dfrac{1 + \sqrt{p^2 + 1}}{2}$, for p [4.5]
22. $S = 2\pi r(r + h)$, for h [4.1]

- Write each expression as a single fraction in lowest terms.

23. $\dfrac{y^3 - y}{2y + 1} \cdot \dfrac{4y + 2}{y^2 + 2y + 1}$ [2.2]

24. $\dfrac{x^2 - y^2}{2y^2 - 2x^2}$ [2.1]

25. $\dfrac{y}{y^2 - 16} - \dfrac{y + 1}{y^2 - 5y + 4}$ [2.3]

26. $\dfrac{2y - 6}{y + 2x} \div \dfrac{4y - 12}{2y + 4x}$ [2.2]

27. $\dfrac{x^3 - y^3}{x^2 + xy} \cdot \dfrac{x + y}{x^2 - xy}$ [2.2]

28. $\dfrac{a - 3 - \dfrac{2}{a - 1}}{a + 2 - \dfrac{3}{a - 1}}$ [2.4]

29. Divide $\dfrac{x^3 - 2x^2 + x + 1}{x^2 - x + 2}$.   [2.2]

30. Factor $y^2 - 4x^2 + 4x - 1$.   [1.5]

31. Evaluate $ar^{n-1} - br^{n+1}$ for $a = 2$, $b = -3$, $r = 2$, and $n = 3$.   [1.1]

32. Write as a single fraction in lowest terms:

$$\dfrac{2x}{x - 1} - \dfrac{x + 2}{x + 1} \div \left(\dfrac{1}{x} - \dfrac{2}{x + 1}\right).$$ [2.3]

33. Write in scientific notation.
    a. 0.000000023   b. 402,000,000,000   [2.5]

34. Rationalize the denominator: $\dfrac{y}{2\sqrt{y} - 3}$.   [3.4]

35. Solve $x - x^{1/2} = 12$.   [4.5]

36. Divide $\dfrac{x^5 - 3x^3 - 1}{x - 2}$.   [2.2]

37. Factor completely: $3x^2(x - 3)^3(x + 2) - 12x(x + 2)^2(x - 3)^2$.   [1.4]

38. Find the value of $x^2 - 2xy^2 - xy + y^4$ for $x = 3$ and $y = -1$.   [1.1]

39. Rationalize the numerator: $\dfrac{1 - \sqrt{x - 1}}{x\sqrt{x - 1}}$.   [3.4]

40. Solve $(x - 1)^{1/2} - (x - 1)^{1/4} - 6 = 0$.   [4.5]

41. Solve $2[(x - 1)^2 + 3x] = 14$.   [4.3]

42. Write each expression in exponential form.
    a. $\dfrac{\sqrt{x}}{\sqrt[3]{y}}$ b. $2\sqrt[3]{x^2 y}$ [3.1]

43. Write each expression in the form $a + bi$.
    a. $\dfrac{-2}{3 - i}$ b. $(2 + \sqrt{-3})(3 - \sqrt{-3})$ [3.5]

44. Simplify $[y - (x + y)] - [3y - (2x - y) + 3x]$ [1.2]
45. Evaluate $y^5 - y^3$ for $y = -3$.   [1.2]
46. Simplify $\dfrac{\frac{x-1}{x}}{1 - \frac{1}{x^2}}$.   [2.4]
47. Reduce $\dfrac{xy - 3x - 2y + 6}{x^3 - 2x^2 + x - 2}$ to lowest terms.   [2.1]
48. Factor $a^3 - (a - 1)^3$ and simplify.   [1.5]
49. Simplify $\sqrt[3]{2x(x-1)^2}\sqrt[3]{12x^4(x-1)}$.   [3.3]
50. Solve $0.40(x - 2) + 0.20x = 3.40$.   [4.1]

■ *Solve.*

51. Dependable Insurance Co. offers an employee benefit plan to small companies with 50 or fewer employees. For companies with 20 or fewer employees the annual cost is $100 per employee. For each additional employee, the cost is reduced by $2 per employee. Write a polynomial expressing the total annual cost of the plan for a company with $20 + x$ employees.

52. The Amazon River discharges 4,200,000 cubic feet of water per second into the Atlantic Ocean. How many gallons per day reach the ocean from the Amazon? (One cubic foot of water is equivalent to 7.48 gallons.)

53. Roger heads north on his bicycle at an average speed of 15 miles per hour, riding for 7 hours per day. Three days later Carol sets out on the same route in the car. She averages 50 miles per hour for 7 hours per day. How long will it be before she overtakes Roger?

54. A truck driver wants to average 50 miles per hour on his cross-country hauls. One morning he runs into traffic outside of Chicago and travels only 60 miles in his first 2 hours. If he drives 6 more hours that day, what speed must he drive in order to achieve the 50-mile-per-hour average for the day?

55. The population of Bellevue has been growing at a rate of $9\frac{1}{2}\%$ over the past 8 years. Its present population is 85,000 people. What was its population 3 years ago?

56. The number (in millions) of pairs of specialty sports shoes sold annually in the United States has been growing according to the formula $3t^{3/2}$, where t is the number of years since 1975. How many pairs of shoes were sold in 1987?

57. If the interest on an account is computed every 3 months and added to the principal, the interest is said to be compounded quarterly. At the end of n years the amount of money in the account is given by

$$P\left(1 + \frac{r}{4}\right)^{4n},$$

where P is the original principal and r is the interest rate.

   a. Write polynomials for the amount of money in the account after 6 months and after 1 year.

b. If $1200 is invested in the account at 5% interest, how much is the account worth after 1 year?

58. The image of a tall building just fills the screen in a pinhole camera. If the camera is 10 inches long by 8 inches high and is placed 200 feet from the building, how tall is the building?

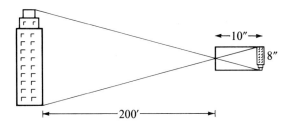

59. On an illustrated map of Paris 4.5 centimeters represents 1 kilometer. If it is about 10 centimeters on the map from your hotel to the Louvre, how long a walk will it be?

60. Irene wants to enclose two adjacent chicken coops of equal size against the henhouse wall. She has 66 feet of chicken wire fencing and would like the total area of the two coops to be 360 square feet. What should the dimensions of the chicken coops be?

PART II

Functions and Graphs

Many applied problems involve relationships among two or more variables. For example, we might want to relate the size of a wheat harvest to the amount of rainfall in the spring or the speed of a jet airliner to its fuel consumption. A small manufacturing company may be interested in how its profit is related to the amount it spends on advertising or to its volume of sales. To study such relationships we look for a mathematical description, or model, for the situation.

There are two aspects to mathematical modeling: first to find a model that describes a given situation (usually an equation or set of equations), and then to interpret the model to explain known results or predict new ones. In this chapter and the following ones we will study both aspects of modeling.

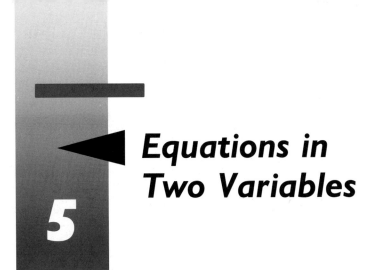

Equations in Two Variables

5.1

GRAPHS

Graphing is one of the most useful tools in the process of modeling. For example, the data in Table 5.1 show the atmospheric pressure at different altitudes above the surface of the earth on a certain day. Meteorologists regularly collect such data by attaching to a weather balloon a device called a radiosonde, which is equipped with a barometer and a radio transmitter. Altitudes are given in feet, and atmospheric pressures are given in inches of mercury.

Altitude	0	5000	10,000	20,000	30,000	40,000	50,000
Pressure	29.7	24.8	20.5	14.6	10.6	8.5	7.3

TABLE 5.1

We observe a generally decreasing trend in pressure as the altitude increases, but it is difficult to say anything more precise about the relationship between pressure and altitude. A clearer picture emerges if we plot the data on a graph. To do this we use two perpendicular number lines called **axes**; we use the horizontal axis for the values of the first variable, altitude, and the vertical axis for the values of the second variable, pressure.

The entries in Table 5.1 are called **ordered pairs** in which the **first component** is the altitude and the **second component** is the atmospheric pressure measured at that altitude. For example, the first two entries can be represented by $(0, 29.7)$ and $(5000, 24.8)$. We can then plot the points whose **coordinates** are given by the ordered pairs, as shown in Figure 5.1a.

213

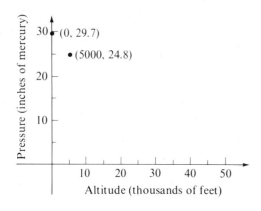

FIGURE 5.1a

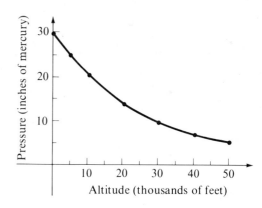

FIGURE 5.1b

If we plot all the ordered pairs from the table, we can connect them with a smooth curve as shown in Figure 5.1b. In doing this we are really "estimating" the pressures that correspond to altitudes between those given, for example, altitudes of 15,000 feet or 37,000 feet. However, for many physical situations variables are related so that one changes "smoothly" with respect to the other. We will assume that this is the case in most of the modeling we do.

EXAMPLE 1 From the graph in Figure 5.1b, estimate

a. the atmospheric pressure measured at an altitude of 15,000 feet;
b. the altitude at which the pressure is 12 inches of mercury.

Solutions a. Note that the point on the graph with first coordinate 15,000 has a second coordinate of approximately 17.4. Hence, we estimate the pressure at 15,000 feet to be 17.4 inches of mercury.

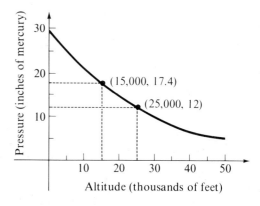

b. Note that the point on the graph with second coordinate 12 has a first coordinate of approximately 25,000, so an atmospheric pressure of 12 inches of mercury occurs at about 25,000 feet.

By using the graph of the data in Figure 5.1b we can obtain information about the relationship between altitude and pressure that would be difficult or impossible to obtain from the data alone.

EXAMPLE 2 a. For what altitudes is the pressure less than 18 inches of mercury?
b. How much does the pressure decrease as the altitude increases from 15,000 to 25,000 feet?
c. For which 10,000-foot increase in altitude does the pressure change most rapidly?

Solutions a. From the graph in Figure 5.1b we see that the pressure has dropped to 18 inches of mercury at about 14,000 feet and that it continues to decrease as the altitude increases. Therefore, the pressure is less than 18 inches of mercury for altitudes greater than 14,000 feet.
b. At 15,000 feet the pressure is approximately 17.4 inches of mercury, and at 25,000 feet it is 12.3 inches. This represents a decrease in pressure of $17.4 - 12.3$, or 5.1, inches of mercury.
c. By studying the graph we see that the pressure decreases most rapidly for low altitudes, so we conclude that the greatest drop in pressure occurs between 0 and 10,000 feet.

Equations in Two Variables

In the example above we used a graph to illustrate a collection of data given in a table. Graphs can also help us to analyze models given in the form of equations. We first review some facts about solutions of equations in two variables.

An equation in two variables, such as $y = 2x + 3$, is said to be *satisfied* if the variables are replaced by a pair of numbers that make the statement true. The pair of numbers is called a **solution** of the equation and is usually written as an ordered pair (x, y) because it is understood that the numbers are considered in a particular order, x first and y second. (Of course, any two convenient letters may be used to name the variables in an equation.)

Unlike the equations in one variable that we have studied, equations in two variables may have infinitely many solutions. To find ordered pairs that are solutions of a given equation, we assign any number to one of the variables and then determine the related value (if one exists) for the second variable. For example, to obtain solutions to

$$y - x = 1$$

we might substitute the values 2, 3, and 4 for x to get

$$y - (2) = 1, \quad \text{from which} \quad y = 3;$$
$$y - (3) = 1, \quad \text{from which} \quad y = 4;$$
$$y - (4) = 1, \quad \text{from which} \quad y = 5.$$

Thus, $(2, 3)$, $(3, 4)$, and $(4, 5)$ are three solutions of $y - x = 1$. We could have used any values for x and obtained corresponding values for y, so there are infinitely many solutions of this equation.

EXAMPLE 3 Find the missing components so that each ordered pair (x, y) is a solution of $y - 2x = 4$.

a. $(0, ?)$ b. $(?, 0)$ c. $(3, ?)$

Solutions

a. $y - 2x = 4$
$y - 2(0) = 4$
$y = 4$
$(0, 4)$

b. $y - 2x = 4$
$(0) - 2x = 4$
$x = -2$
$(-2, 0)$

c. $y - 2x = 4$
$y - 2(3) = 4$
$y = 10$
$(3, 10)$

EXAMPLE 4 List the ordered pairs (t, s) that satisfy the equation $s = 3t + 1$ and have t-components 1, 2, and 3.

Solution

For $t = 1$,
$s = 3(1) + 1$
$= 4.$

For $t = 2$,
$s = 3(2) + 1$
$= 7.$

For $t = 3$,
$s = 3(3) + 1$
$= 10.$

Thus, the desired solutions are $(1, 4)$, $(2, 7)$, and $(3, 10)$.

For many equations it is helpful to express one variable in terms of the other before looking for solutions. For example, we can transform the equation

$$y - 3x = -2 \tag{1}$$

into the equivalent equation

$$y = 3x - 2. \tag{2}$$

Both equations have the same solution set. In (1) the variables x and y are **implicitly** related, while in (2) y is said to be expressed **explicitly** in terms of x.

EXAMPLE 5 Transform each equation into one in which y is expressed explicitly in terms of x, then find values of y for the given values of x.

a. $3y - xy = 4$; $\quad x = 2, 5$ $\qquad$ b. $x^2 + 4y^2 = 5$; $\quad x = 0, 1$

Solutions a. $3y - xy = 4$
$$y(3 - x) = 4$$
$$y = \frac{4}{3 - x}$$
When $x = 2$, $y = \dfrac{4}{3 - (2)} = 4$;

when $x = 5$, $y = \dfrac{4}{3 - (5)} = -2$.

b. $x^2 + 4y^2 = 5$
$$4y^2 = 5 - x^2$$
$$y^2 = \frac{1}{4}(5 - x^2)$$
$$y = \pm\frac{1}{2}\sqrt{5 - x^2}$$
When $x = 0$,
$$y = \pm\frac{1}{2}\sqrt{5 - 0^2} = \pm\frac{1}{2}\sqrt{5};$$
when $x = 1$,
$$y = \pm\frac{1}{2}\sqrt{5 - 1^2} = \pm 1.$$

Graphs of Equations in Two Variables

Since an equation in two variables often has infinitely many solutions, we cannot list them all. However, we can display the solutions on a graph. We first establish coordinate axes labeled with the variables they represent. Most commonly, the horizontal axis is called the **x-axis**, the vertical axis is called the **y-axis**, and their point of intersection is called the **origin**. The axes divide the plane into four regions called **quadrants**, which are referred to by Roman numerals, as illustrated in Figure 5.2.

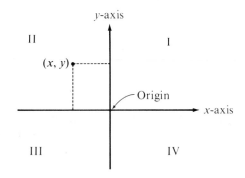

FIGURE 5.2

The system described above is called the **Cartesian (or rectangular) coordinate system.** Notice that each point in the plane can be located by its coordinates and every pair of coordinates corresponds to a unique point in the plane.

The **graph** of an equation is the graph of all the solutions of the equation. Thus, a particular point is included in the graph of an equation if the coordinates of the point satisfy the equation. If the coordinates of a point do not satisfy the equation, then the point is not part of the graph. We can think of the graph of an equation as a picture of the solutions of the equation.

Most of the graphs we will study can be obtained by plotting a few solutions of the equation and then connecting these points by a smooth curve. For example, to graph

$$y = 3x - 2 \tag{2}$$

we first find a number of solutions of the equation by choosing values for x and solving for the corresponding y-values.

For $x = -2$, $y = 3(-2) - 2 = -8$.
For $x = -1$, $y = 3(-1) - 2 = -5$.
For $x = 0$, $y = 3(0) - 2 = -2$.
For $x = 1$, $y = 3(1) - 2 = 1$.
For $x = 2$, $y = 3(2) - 2 = 4$.

We then tabulate these values and plot the ordered pairs.

x	y	Solutions
-2	-8	$(-2, -8)$
-1	-5	$(-1, -5)$
0	-2	$(0, -2)$
1	1	$(1, 1)$
2	4	$(2, 4)$

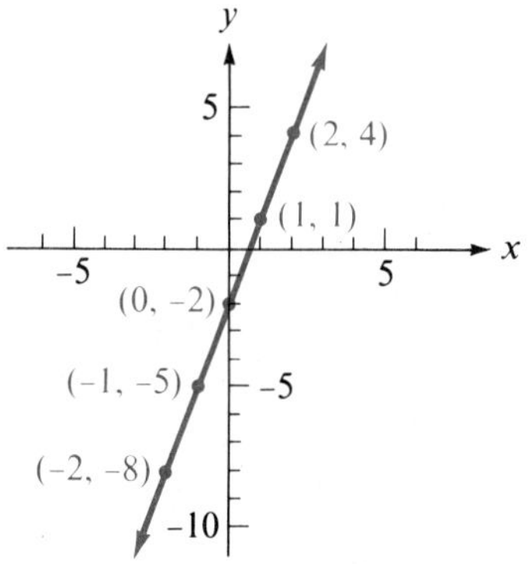

FIGURE 5.3

By connecting the plotted points we obtain the graph in Figure 5.3. The graph does not display *all* the solutions of Equation (2). Since there is a solution corresponding to every real number x, the graph extends infinitely in either direction, as indicated by the arrows.

5.1 ■ GRAPHS

EXAMPLE 6 Graph each equation.

a. $y = x^2 - 2$
b. $y = \dfrac{1}{x}$

Solutions Choose values for x and solve for the corresponding values of y. Plot the solutions and connect them with a smooth curve.

a.

x	y	Solutions
-2	2	$(-2, 2)$
-1	-1	$(-1, -1)$
0	-2	$(0, -2)$
1	-1	$(1, -1)$
2	2	$(2, 2)$

b.

x	y	Solutions
-2	$-\dfrac{1}{2}$	$\left(-2, -\dfrac{1}{2}\right)$
-1	-1	$(-1, -1)$
$-\dfrac{1}{2}$	-2	$\left(-\dfrac{1}{2}, -2\right)$
0	—	—
$\dfrac{1}{2}$	2	$\left(\dfrac{1}{2}, 2\right)$
1	1	$(1, 1)$
2	$\dfrac{1}{2}$	$\left(2, \dfrac{1}{2}\right)$

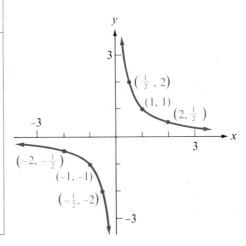

Notice that the graph in (b) has no point whose x-coordinate is zero. This is because the equation $y = 1/x$ has no solution for $x = 0$.

As before, we use arrows to show that for larger or smaller values of x the graphs continue in the directions indicated. However, we must choose enough x-values to show all the major features of the graph. As you learn more about graphing you will be better able to choose an appropriate set of x-values.

EXERCISE 5.1

A

■ *For Problems 1–6, see Examples 1 and 2.*

1. The figure shows a graph of the temperature recorded during a winter day in Billings, Montana.

   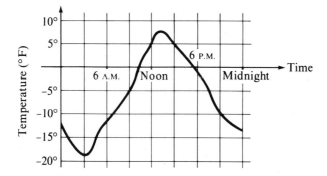

   a. What were the high and low temperatures recorded during the day?
   b. During what time intervals is the temperature above 5°F? Below −5°F?
   c. Estimate the temperatures at 7 A.M. and 2 P.M. At what time(s) is the temperature approximately 0°F? Approximately −12°F?
   d. How much did the temperature increase between 3 A.M. and 6 A.M.? Between 9 A.M. and noon? How much did the temperature decrease between 6 P.M. and 9 P.M.?
   e. During which 3-hour interval did the temperature increase most rapidly? Decrease most rapidly?

2. The figure shows a graph of the altitude of a commercial jetliner during its flight from Denver to Los Angeles.

   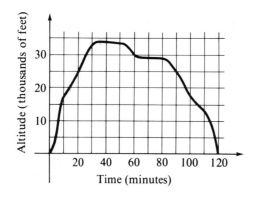

a. What was the highest altitude the jet achieved? At what time(s) was this altitude recorded?
b. During what time intervals was the altitude greater than 10,000 feet? Below 20,000 feet?
c. Estimate the altitudes 15 minutes into the flight and 35 minutes into the flight. At what time(s) was the altitude approximately 16,000 feet? 32,000 feet?
d. How many feet did the jet climb during the first 10 minutes of flight? Between 20 minutes and 30 minutes? How many feet did the jet descend between 100 minutes and 120 minutes?
e. During which 10-minute interval did the jet ascend most rapidly? Descend most rapidly?

3. The graph shows the gas mileage achieved by an experimental model automobile at different speeds.

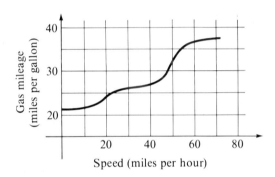

a. Estimate the gas mileage achieved at 43 miles per hour.
b. Estimate the speed at which a gas mileage of 34 miles per gallon is achieved.
c. At what speed is the best gas mileage achieved? Do you think that the gas mileage will continue to improve as the speed increases? Why or why not?
d. The data illustrated by the graph were collected under ideal test conditions. What factors might affect the gas mileage if the car were driven under more realistic conditions?

4. The graph on page 222 shows the average height of young women aged 0 to 18 years.
a. Estimate the average height of 5-year-old girls.
b. Estimate the age at which the average young woman is 50 inches tall.
c. At what age does the average woman achieve her adult height? Do you think that the height will continue to increase as age increases? Why or why not?
d. The data recorded in the graph reflect the average heights for young women at given ages. What factors might affect the data for specific individuals?

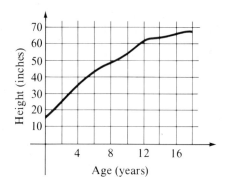

5. The graph shows the speed of a car during an hour-long journey.

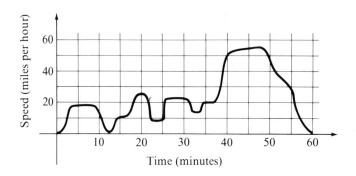

   a. When did the car stop at a traffic signal?
   b. During what time interval did the car drive in "stop-and-go" city traffic?
   c. During what time interval did the car travel on the freeway?

6. The graph shows the fish population of a popular fishing pond.

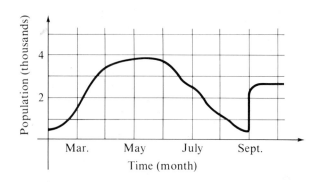

   a. During what months do the young fish hatch?
   b. During what months is fishing allowed?
   c. When does the park service restock the pond?

■ *Find the missing components so that each ordered pair is a solution of the given equation. See Example 3.*

7. $y = x + 7$; a. $(0, ?)$, b. $(2, ?)$, c. $(-2, ?)$
8. $y = 6 - 2x$; a. $(0, ?)$, b. $(?, 0)$, c. $(-1, ?)$
9. $3x - 4y = 6$; a. $(0, ?)$, b. $(?, 0)$, c. $(-5, ?)$
10. $x + 2y = 5$; a. $(0, ?)$, b. $(5, ?)$, c. $(-3, ?)$

■ *Find the ordered pairs that satisfy the given equation and have the given x-components. See Example 4.*

11. $y = x - 4$; $-3, 0, 3$
12. $y = 2x + 6$; $-2, 0, 2$
13. $y = \dfrac{3}{x + 2}$; $1, 2, 3$
14. $y = \dfrac{4x}{x^2 - 1}$; $0, 2, 4$
15. $y = \sqrt{x^2 - 1}$; $1, 3, 5$
16. $y = \dfrac{1}{2}\sqrt{4 - x^2}$; $0, 1, 2$

■ *Transform each equation into one in which y is expressed explicitly in terms of x. Find points with the given x-components that lie on the graph of the equation. See Example 5.*

17. $2x + y = 6$; $2, 4$
18. $4x - y = 2$; $-2, -4$
19. $xy - x = 2$; $-2, 2$
20. $3x - xy = 6$; $1, 3$
21. $xy - y = 4$; $4, 8$
22. $x^2y - xy = -5$; $2, 4$
23. $x^2y - 4y = xy + 2$; $-1, 1$
24. $x^2y - xy + 3 = 5y$; $-2, 2$
25. $4 = \dfrac{x}{y^2 - 2}$; $-1, 4$
26. $3 = \dfrac{x}{y^2 + 1}$; $-2, 1$
27. $3x^2 - 4y^2 = 4$; $2, 3$
28. $5x^2 - 4y^2 = 2$; $2, 4$

■ *Graph each equation. Use the suggested values for the variables. See Example 6.*

29. $y = 2x + 4$; $x = -3, -2, -1, 0, 1, 2$
30. $y = x - 4$; $x = -2, -1, 0, 1, 2, 3, 4, 5$
31. $s = -2t + 4$; $t = -2, -1, 0, 1, 2, 3$
32. $s = -t - 4$; $t = -5, -4, -3, -2, -1, 0, 1$
33. $w = v^2 + 2$; $v = -3, -2, -1, 0, 1, 2, 3$
34. $w = v^2 - 4$; $v = -3, -2, -1, 0, 1, 2, 3$
35. $p = \dfrac{1}{m - 1}$; $m = -2, -1, 0, \dfrac{1}{2}, \dfrac{3}{4}, \dfrac{5}{4}, \dfrac{3}{2}, 2, 3$
36. $p = \dfrac{1}{m + 1}$; $m = -3, -2, -\dfrac{3}{2}, -\dfrac{5}{4}, -\dfrac{3}{4}, -\dfrac{1}{2}, 0, 1, 2$
37. $h = \sqrt{z}$; $z = 0, 1, 2, 3, 4, 9$
38. $h = \sqrt[3]{z}$; $z = -8, -2, -1, 0, 1, 2, 8$
39. $q = |s|$; $s = -3, -2, -1, 0, 1, 2, 3$

40. $q = |2s|$; $\quad s = -3, -2, -1, 0, 1, 2, 3$

41. $y = x^3$; $\quad x = -2, -1, -\dfrac{1}{2}, 0, \dfrac{1}{2}, 1, 2$

42. $y = \dfrac{1}{x^2}$; $\quad x = -2, -1, -\dfrac{1}{2}, -\dfrac{1}{3}, \dfrac{1}{3}, \dfrac{1}{2}, 1, 2$

B

43. Which graph best illustrates the following?
    a. The stopping distances for cars traveling at various speeds

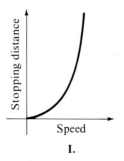
I.
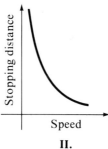
II.

b. Your pulse rate during an aerobics class

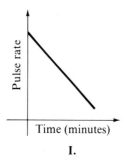
I.
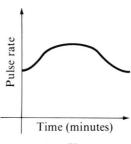
II.

44. Which graph best illustrates the following?
    a. Your income in terms of the number of hours you worked

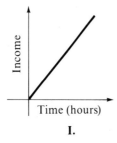
I.
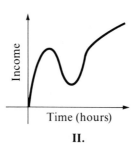
II.

b. Your temperature during an illness

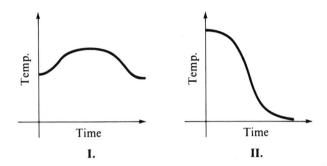

I. II.

■ *Sketch graphs to illustrate the following situations.*

45. The height above the ground of your head during a ride on a Ferris wheel.
46. The height above the ground of a rubber ball dropped from the top of a 10-foot ladder.
47. The time it takes to drive 200 miles at various speeds.
48. The length of a rectangle whose area is 24 square feet, for various widths.

■ *Answer the following questions with a sentence or two.*

49. What is a solution of an equation in two variables?
50. Give examples of equations in two variables in which the variables are related **(a)** explicitly and **(b)** implicitly.
51. What is the graph of an equation in two variables?
52. How can you determine whether a particular point lies on the graph of an equation?

5.2

GRAPHS OF LINEAR EQUATIONS

Any first-degree equation in two variables, that is, any equation that can be written equivalently in the form

$$ax + by = c \quad (a \text{ and } b \text{ not both } 0), \tag{1}$$

has a graph that is a straight line. For this reason such equations are called **linear equations**. Equation (1) is called the standard form for a linear equation. Linear equations and their graphs can be used to model a wide variety of applied problems.

EXAMPLE 1 Yumiko's long-distance telephone company charges a $3 access fee for a call to Tokyo and $2 for each minute of the call. (A fraction of a minute is charged as the corresponding fraction of $2.)

a. Write an equation that expresses the cost of a call to Tokyo in terms of the length of the call.
b. Graph the equation in (a).

Solutions a. Let t represent the length of the call in minutes and let C represent the cost of the call. Then

$$C = 3 + 2t \qquad (t \geq 0).$$

b. Choose several values for t and calculate the corresponding values for C to obtain the following ordered pairs.

t	C	(t, C)
0	3	$(0, 3)$
1	5	$(1, 5)$
2	7	$(2, 7)$
3	9	$(3, 9)$

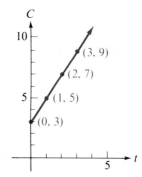

The graphs of these points lie on a straight line, as shown. Note that the line extends infinitely in only one direction, since negative values of t do not make sense here.

In order to draw a graph that is useful in analyzing a problem we must choose scales for the axes that reflect the magnitudes of the variables involved.

EXAMPLE 2 In 1960 a three-bedroom house in Midville cost $30,000. The price of a home has increased by an average of $4000 per year since then.

a. Write an equation that expresses the price of a three-bedroom house in Midville in terms of the number of years since 1960.

b. Graph the equation in (a).
c. Estimate the cost of a three-bedroom house in 1990.

Solutions a. Let P represent the price of the house t years after 1960. Then

$$P = 30{,}000 + 4000t \qquad (t \geq 0). \qquad (2)$$

b. Choose several values for t and calculate the corresponding values for P to obtain the following ordered pairs.

t	P	(t, P)
0	30,000	(0, 30,000)
5	50,000	(5, 50,000)
10	70,000	(10, 70,000)
20	110,000	(20, 110,000)

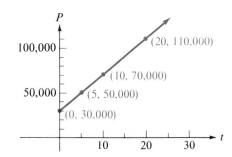

To graph the equation we scale the horizontal axis, or t-axis, in 5-year intervals and the vertical axis, or P-axis, in intervals of $10,000, then plot the points found above.

c. In 1990 $t = 30$, so we substitute $t = 30$ into Equation (2) to obtain

$$P = 30{,}000 + 4000(30)$$
$$= 150{,}000.$$

A three-bedroom house cost $150,000 in 1990.

Intercept Method of Graphing

Since any two distinct points determine a straight line, only two solutions of a linear equation are needed to determine its graph. Two solutions that are easy to find are $(x_1, 0)$ and $(0, y_1)$. These points, called the *x*- and *y*-intercepts of the graph, are also easy to locate because they are the points where the graph intersects the *x*- and *y*-axes, respectively. For simplicity, we often refer to the *numbers* x_1 and y_1 as the *x*- and *y*-intercepts also. To find the *x*-intercept we substitute 0 for y in the equation and solve for x; to find the *y*-intercept we substitute 0 for x and solve for y.

EXAMPLE 3 Graph $3x + 4y = 12$.

Solution If $y = 0$, then

$$3x + 4(0) = 12$$
$$3x = 12$$
$$x = 4.$$

The x-intercept is 4, and the point $(4, 0)$ lies on the graph. If $x = 0$, then

$$3(0) + 4y = 12$$
$$4y = 12$$
$$y = 3.$$

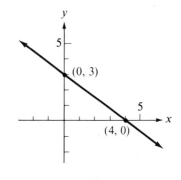

The y-intercept is 3, and the point $(0, 3)$ lies on the graph. Thus, the graph of the equation appears as shown.

EXAMPLE 4 Phil and Ernie buy a used photocopier for $800 and set up a copy service on their campus. For each hour that the copier runs continuously, Phil and Ernie make $40.

a. Write an equation that relates Phil and Ernie's profit (or loss) to the number of hours they run the copier.
b. Find the intercepts and sketch the graph.
c. What is the significance of the intercepts to the problem?

Solutions a. Let t represent the number of hours that Phil and Ernie run the copier, and let P represent their net profit or loss. Then

$$P = -800 + 40t \quad (t \geq 0).$$

b. If $t = 0$, then

$$P = -800 + 40(0)$$
$$P = -800.$$

If $P = 0$, then

$$0 = -800 + 40t$$
$$-40t = -800$$
$$t = 20.$$

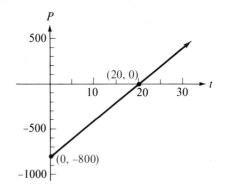

The P-intercept is -800 and the t-intercept is 20. Graph the points $(0, -800)$ and $(20, 0)$ and draw a line through them. Note that we do not include points with negative t-coordinates in the graph.

c. The P-intercept, -800, is the value of P when $t = 0$, that is, the initial, or starting, value of the profit variable. Phil and Ernie start out $800 in debt.

The t-intercept, 20, is the value of t when $P = 0$, the number of hours of operation required for Phil and Ernie to break even.

If a graph intersects both axes at the origin or if the intercepts are too close to the origin to be helpful in drawing the graph, it is necessary to plot *at least* one other point to establish the line with accuracy.

EXAMPLE 5 Graph $y = 3x$.

Solution If $x = 0$, then $y = 0$, and both intercepts of the graph are at the origin, $(0, 0)$. Assigning any other replacement for x, say 2, we obtain a second solution, $(2, 6)$. We first graph the ordered pairs $(0, 0)$ and $(2, 6)$ and then complete the graph as shown.

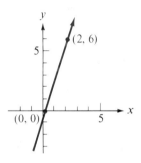

Horizontal and Vertical Lines

There are two special cases of linear equations worth noting. First, an equation such as

$$y = 4$$

can be considered an equation in two variables,

$$0x + y = 4.$$

For each value of x this equation assigns the value 4 to y. That is, any ordered pair of the form $(x, 4)$ is a solution of the equation. For example,

$$(-1, 4) \quad (2, 4), \quad \text{and} \quad (4, 4)$$

are all solutions of the equation. If we draw a straight line through these points, we obtain the graph shown in Figure 5.4.

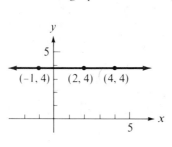

FIGURE 5.4   **FIGURE 5.5**

The other special case of a linear equation is of the type

$$x = 3,$$

which may be considered an equation in two variables,

$$x + 0y = 3.$$

Here, only one value is permissible for x, namely, 3, while any value may be assigned to y. Thus, any ordered pair of the form $(3, y)$ is a solution of this equation. If we choose two solutions, say $(3, 1)$ and $(3, 3)$, and draw a straight line through these two points, we have the graph shown in Figure 5.5. In general, we have the following.

The graph of $y = k$ (k a constant) is a horizontal line.

The graph of $x = k$ (k a constant) is a vertical line.

EXAMPLE 6   a. Graph $y = 2$.   b. Graph $x = -4$.

Solutions   a.   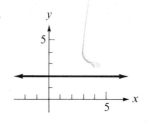   b.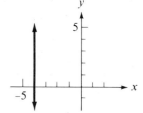

EXERCISE 5.2
A

- *For Problems 1–12, see Examples 1 and 2.*

1. Frank plants a dozen corn seedlings, each 6 inches tall. With plenty of water and sunlight they will grow approximately 2 inches per day.
   a. Write an equation that expresses the height of the seedlings in terms of the number of days since they were planted.
   b. Graph the equation.
   c. How tall is the corn after 3 weeks?
   d. How long will it be before the corn is 6 feet tall?

2. In the desert the temperature at 6 A.M., just before sunrise, was 65°F. The temperature rose about 5 degrees every hour until it reached its maximum value at about 5 P.M.
   a. Write an equation that expresses the temperature in the desert in terms of the number of hours since 6 A.M.
   b. Graph the equation.
   c. How hot is it at noon?
   d. When will the temperature be 110°F?

3. On October 31 Betty and Paul fill their 250-gallon heating fuel oil tank. Beginning in November they use an average of 15 gallons per week in heating fuel oil.
   a. Write an equation that relates the amount of oil in the tank to the number of weeks since October 31.
   b. Graph the equation.
   c. How much fuel oil is left in the tank 8 weeks later?
   d. When will there be only 70 gallons of fuel oil left?

4. Leon's camper has a 20-gallon gas tank, and he gets 12 miles to the gallon.
   a. Write an equation that relates the amount of gasoline in Leon's fuel tank to the number of miles he has driven.
   b. Graph the equation.
   c. How much gas does Leon have after driving 100 miles?
   d. If Leon's gas gauge registers one-fourth of a tank, how far has he driven?

5. Sandra works for the post office. She is assigned to sort and stack mail for 2 hours in the morning. After that, she delivers mail to 15 houses per hour.
   a. Write an equation that expresses the number of hours in Sandra's shift in terms of the number of houses on her route.
   b. Graph the equation.
   c. How long is Sandra's shift if there are 60 houses on her route?
   d. How many houses can she cover in an 8-hour workday?

6. Jim works at the blood bank. It takes him 2½ hours in the morning to collect samples from the patients and to set up the chemicals for the tests. He can then process an average of 20 samples per hour.
   a. Write an equation that expresses the number of samples Jim processes in terms of the number of hours he works.
   b. Graph the equation.
   c. How many samples can Jim process on a 7-hour shift?
   d. How long will it take Jim to process 125 samples?
7. Here is a formula for calculating the daily calorie intake for a woman that will result in a weight loss of 1 pound per week. First, determine the woman's ideal weight (lean weight at age 22). For each pound of ideal weight, allow 15 calories per day. Now subtract 500 calories.
   a. Write an equation that relates a woman's daily calorie allowance to her ideal weight.
   b. Graph the equation.
   c. What is the calorie intake recommended for a woman whose ideal weight is 125 pounds?
   d. What is the ideal weight of a woman who should consume 1300 calories per day on this diet?
8. Repeat Problem 7 to determine a diet program for a man. A man's ideal weight is his lean weight at age 25, and he is allowed 17 calories for each pound of ideal weight.
   a. Write an equation that relates a man's daily calorie allowance to his ideal weight.
   b. Graph the equation.
   c. What calorie intake is recommended for a man whose ideal weight is 175 pounds?
   d. What is the ideal weight of a man who should consume 2200 calories per day on this diet?
9. The owner of a gas station has $4800 to spend on unleaded gasoline this month. Regular unleaded costs him $0.60 per gallon, and premium unleaded costs $0.80 per gallon.
   a. Write an equation that relates the amount of regular unleaded gasoline he can buy to the amount of premium unleaded he can buy.
   b. Graph the equation.
   c. If the owner buys 3000 gallons of premium unleaded, how many gallons of regular unleaded can he buy?
10. Five pounds of body fat is equivalent to 16,000 calories. Carol can burn 600 calories per hour bicycling and 400 calories per hour swimming.
    a. Write an equation that relates the number of hours of each activity Carol needs to perform in order to lose 5 pounds.
    b. Graph the equation.
    c. If Carol bicycles for 10 hours, how many hours must she swim?

11. A real estate agent receives a salary of $10,000 plus 3% of her total sales for the year.
    a. Write an equation that relates the agent's salary to her total annual sales.
    b. Graph the equation.
    c. If the agent sells $500,000 worth of property, what will her salary be?
    d. How much property must the agent sell to make a yearly salary of $40,000?
12. Under a proposed graduated income tax system, a single taxpayer whose taxable income is between $13,920 and $16,190 would pay taxes of $1706.30 plus 20% of the amount of his income over $13,920.
    a. Write an equation that relates the taxes owed to the amount of income over $13,920.
    b. Graph the equation.
    c. How much does a single taxpayer with a taxable income of $15,000 pay in taxes?
    d. If Everett paid $1800 in taxes last year, what was his taxable income?

■ *Graph each equation by the intercept method. See Example 3.*

13. $x + 2y = 8$
14. $2x - y = 6$
15. $3x - 4y = 12$
16. $2x + 6y = 6$
17. $\dfrac{x}{9} - \dfrac{y}{4} = 1$
18. $\dfrac{x}{5} + \dfrac{y}{8} = 1$
19. $\dfrac{2x}{3} + \dfrac{3y}{11} = 1$
20. $\dfrac{8x}{7} - \dfrac{2y}{7} = 1$
21. $20x = 30y - 45,000$
22. $30x = 45y + 60,000$
23. $0.4x + 1.2y = 4.8$
24. $3.2x - 0.8y = 12.8$

■ *For Problems 25–36, find the intercepts of the graphs in Problems 1–12. In each case, explain what the intercepts represent in the terms of the problem. Do the intercepts make sense in the context of the problem? See Example 4.*

■ *Graph each equation. See Example 5.*

37. $x + y = 0$
38. $x - y = 0$
39. $2x - y = 0$
40. $2x + y = 0$
41. $x = 3y$
42. $x = -3y$

43. A freight train travels at a constant speed of 50 miles per hour.
    a. Write an equation that expresses the distance the train has traveled in terms of the number of hours elapsed.
    b. Graph the equation.
44. A computer programmer is paid $12 per hour.
    a. Write an equation that expresses the programmer's wages in terms of the number of hours she works.
    b. Graph the equation.

■ *Graph each equation. See Example 6.*

45. $y = -3$
46. $x = -2$
47. $2x = 8$
48. $3y = 15$
49. $x = 0$
50. $y = 0$

B

51. Find an equation of the form $ax + by = 12$ for the line whose x-intercept is 6 and whose y-intercept is 4.
52. Find an equation of the form $x/a + y/b = 1$ for the line whose x-intercept is 6 and whose y-intercept is 3.
53. Find the x- and y-intercepts of the line whose equation is $Mx + Ny = P$.
54. Find the x- and y-intercepts of the line whose equation is $x/P + y/Q = 1$.

■ *For each equation, graph the line and find the area of the triangle formed by the line and the coordinate axes.*

55. $3x + 2y = 18$
56. $3x - 5y = 15$

■ *Answer the following questions with a sentence or two.*

57. Why is an equation of the form $ax + by = c$ called linear?
58. Describe how you would graph a line using the intercept method.
59. What kinds of lines cannot be graphed by the intercept method?
60. Why is $x = k$, where k is a constant, the equation of a vertical line?

5.3

POINT-SLOPE FORMULA FOR LINES

Slope

The graphs of all linear equations share a common property that distinguishes them from other types of equations and makes linear equations particularly convenient to use as models. This property involves the notion of *slope*.

Consider the graph in Example 1 on page 226, which shows the cost of a long-distance call to Tokyo. If we start at any point on the graph in Figure 5.6 and increase the horizontal or t-coordinate by one unit, we must increase the vertical or C-coordinate by two units to reach another point on the graph. In

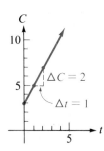

FIGURE 5.6

other words, if we increase the length of the call by 1 minute, the cost increases by $2. We may think of this in terms of a ratio: the cost of the call increases by $2 for each 1-minute increase in the length of the call. Thus,

$$\frac{\text{change in } C}{\text{change in } t} = \frac{\Delta C}{\Delta t} = \frac{2}{1},$$

where the symbol Δ stands for the change in the value of a variable.

The ratio $\Delta C/\Delta t$ is called the **slope** of the line. It represents the rate of change of the cost C with respect to the length t of the call: $2 per minute.

EXAMPLE 1 A driver for a cross-country trucking firm travels at a constant speed of 50 miles per hour.

a. Write an equation that expresses the distance traveled in terms of the number of hours driven.
b. Graph the equation.
c. Find the slope of the graph and interpret it as a rate of change.

Solutions a. Let t represent the number of hours driven and D the distance traveled. Then

$$D = 50t \qquad (t \geq 0).$$

b. Choose several values for t and calculate the corresponding values for D to obtain the graph.

c. Note that if we move from one point to another on the graph, then

$$\frac{\Delta D}{\Delta t} = \frac{50 \text{ miles}}{1 \text{ hour}},$$

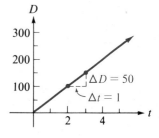

so the slope is 50. It represents the change in distance with respect to time traveled, or the speed of the truck in miles per hour.

Slope Formula In the preceding examples we have seen that the slope of a line is the ratio of the *vertical* change to the *horizontal* change required to move from one point on the line to another. We can obtain a formula for the slope of a line as

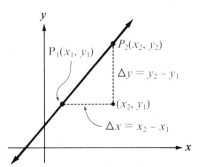

FIGURE 5.7

follows. Let P_1 and P_2 be points on the line with coordinates (x_1, y_1) and (x_2, y_2), respectively, as shown in Figure 5.7. As we move from P_1 to P_2 the change in y value, or vertical change, is $y_2 - y_1$, and the change in x value, or horizontal change, is $x_2 - x_1$. Thus, using the letter m to designate slope, we have the following formula:

$$m = \frac{\Delta y}{\Delta x} = \frac{y_2 - y_1}{x_2 - x_1} \quad (x_2 \neq x_1).$$

EXAMPLE 2 Find the slope of the line segment joining the points $(2, -1)$ and $(4, 3)$.

Solution Let $(2, -1)$ be (x_1, y_1) and $(4, 3)$ be (x_2, y_2). Then

$$m = \frac{y_2 - y_1}{x_2 - x_1} = \frac{3 - (-1)}{4 - 2}$$

$$= \frac{4}{2} = 2.$$

Notice that the order in which the points are chosen does not matter, since

$$\frac{y_1 - y_2}{x_1 - x_2} = \frac{-(y_2 - y_1)}{-(x_2 - x_1)} = \frac{y_2 - y_1}{x_2 - x_1}.$$

5.3 ■ POINT-SLOPE FORMULA FOR LINES

In the example above, if we had chosen $(4, 3)$ for (x_1, y_1) and $(2, -1)$ for (x_2, y_2), then

$$m = \frac{y_2 - y_1}{x_2 - x_1} = \frac{-1 - 3}{2 - 4}$$

$$= \frac{-4}{-2} = 2,$$

the same answer as above.

It is also true that no matter which two points on the line are used to compute the slope, the result will be the same. For example, consider the line in Example 1 on page 000. If we use the points $(0, 0)$ and $(1, 50)$ to compute the slope we obtain

$$m = \frac{50 - 0}{1 - 0} = 50,$$

and using the points $(1, 50)$ and $(3, 150)$ we find

$$m = \frac{150 - 50}{3 - 1} = \frac{100}{2} = 50,$$

as before. This is the property that distinguishes *linear equations* from other equations: *their graphs have constant slope.*

EXAMPLE 3 The graph of the equation $y = \frac{3}{4}x - 3$ is shown below. If we use the points $(x_1, y_1) = (0, -3)$ and $(x_2, y_2) = (4, 0)$ to compute the slope, we find

$$m = \frac{y_2 - y_1}{x_2 - x_1} = \frac{0 - (-3)}{4 - 0} = \frac{3}{4}.$$

If we use two other points, say $(x_1, y_1) = (5, \tfrac{3}{4})$ and $(x_2, y_2) = (2, -\tfrac{3}{2})$, we find

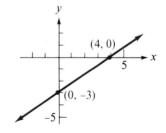

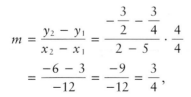

as before.

Geometric Interpretation of Slope

Consider the three graphs of linear equations shown in Figure 5.8.

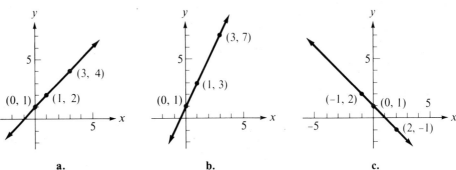

FIGURE 5.8　　a.　　　　　　　　b.　　　　　　　　c.

All three have the same y-intercept, (0, 1). The difference in these graphs is the "steepness" of the lines, or their slope. Since the slope gives the ratio of the change in y-value to the change in x-value, the greater the slope, the steeper the line. The slope of the first line, (a), is

$$m = \frac{4-2}{3-1} = \frac{2}{2} = 1,$$

or

$$\frac{\Delta y}{\Delta x} = \frac{1}{1}.$$

For each unit that the x-coordinate increases, the y-coordinate increases one unit as we move from one point to another on the line.

For the second line, (b), the slope is

$$m = \frac{7-3}{3-1} = \frac{4}{2} = 2,$$

or

$$\frac{\Delta y}{\Delta x} = \frac{2}{1}.$$

For each unit that the x-coordinate increases, the y-coordinate increases two units. The second line is "steeper" than the first line.

The slope of the third line, (c), is

$$m = \frac{-1-2}{2-(-1)} = \frac{-3}{3} = -1,$$

or

$$\frac{\Delta y}{\Delta x} = \frac{-1}{1}.$$

The y-coordinate *decreases* one unit for each unit that the x-coordinate increases. In general, a negative slope indicates a line whose graph runs from upper left to lower right; the graph of a line with positive slope runs from lower left to upper right. The greater the absolute value of the slope, the steeper the line.

EXAMPLE 4 Several lines with different slopes are shown below.

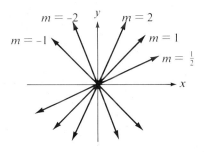

The appearance of the slope of a line is affected by the choice of scales for the axes. In Figure 5.9 all three lines have slope 2, but the line in Figure 5.9a appears "steeper" because the units on the y-axis are larger than those on the x-axis. To compare the slopes of two lines visually we must check that both lines are graphed on coordinate systems with the same scale.

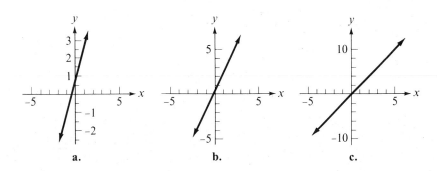

FIGURE 5.9   a.   b.   c.

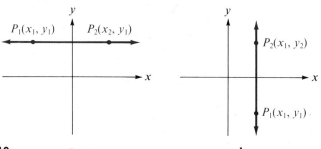

FIGURE 5.10   a.   b.

If a line is horizontal, as shown in Figure 5.10a, then

$$m = \frac{y_2 - y_1}{x_2 - x_1} = \frac{0}{x_2 - x_1} = 0,$$

and its slope is 0. If a line is vertical, as shown in Figure 5.10b,

$$m = \frac{y_2 - y_1}{x_2 - x_1} = \frac{y_2 - y_1}{0},$$

and the slope is *undefined*. In other words, a vertical line does not have a slope.

EXAMPLE 5   The line $y = 5$ is horizontal, so its slope is 0. The line $x = -3$ is vertical, so its slope is undefined.

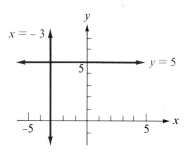

Point-Slope Form

There is only one line in the plane that has a given slope and passes through a given point. Therefore, we should be able to find the equation of a line given only its slope, m, and one point, (x_1, y_1), as shown in Figure 5.11. Since the slope of the line is constant, if (x, y) represents any other point on the line, then

$$\frac{y - y_1}{x - x_1} = m \quad \text{(if } x \neq x_1\text{)}.$$

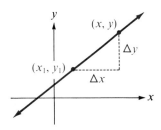

FIGURE 5.11

Thus:

$$y - y_1 = m(x - x_1).$$

This equation holds for any point on the line, so it is an equation for the line. It is called the **point-slope form** for a linear equation.

EXAMPLE 6 Find an equation for the line that passes through the point $(1, -4)$ and has slope $-\tfrac{3}{4}$.

Solution Substitute -4 for y_1, 1 for x_1, and $-\tfrac{3}{4}$ for m in the point-slope form to obtain

$$y - (-4) = -\frac{3}{4}(x - 1).$$

To change to standard form, multiply each side by 4 to obtain

$$4y + 16 = -3(x - 1)$$
$$4y + 16 = -3x + 3$$
$$3x + 4y = -13.$$

We can find an equation of a line whose graph includes two given points by first finding the slope of the line and then using the point-slope formula.

EXAMPLE 7   Find an equation for the line whose graph includes the points $(2, 2)$ and $(-4, 1)$.

Solution   First find the slope, selecting either point as (x_1, y_1) and the other as (x_2, y_2).

$$m = \frac{y_2 - y_1}{x_2 - x_1} = \frac{1 - 2}{-4 - 2}$$
$$= \frac{-1}{-6} = \frac{1}{6}.$$

Then use the point-slope formula with either point. Using $(2, 2)$ for (x_1, y_1) in the formula yields

$$y - 2 = \frac{1}{6}(x - 2),$$

from which

$$6y - 12 = x - 2$$
$$x - 6y = -10.$$

Linear equations are used to model variable relationships in which the rate of change of one variable with respect to the other is constant.

EXAMPLE 8   Ms. Randolph bought a new car in 1980. In 1982 the car was worth $9000, and in 1985 it was valued at $4500.

   a. Assuming that the depreciation is linear, that is, that the value of the car decreases by the same amount each year, write an equation that expresses the value of Ms. Randolph's car in terms of the number of years she has owned it.
   b. Interpret the slope as a rate of change.
   c. Find the value of the car when it was new.

Solutions   a. Let t represent the number of years that Ms. Randolph has owned her car, and let V represent its value after t years. Then the two ordered pairs $(2, 9000)$ and $(5, 4500)$ represent points on the graph of V versus t. To find its equation, first compute the slope:

$$m = \frac{V_2 - V_1}{t_2 - t_1} = \frac{9000 - 4500}{2 - 5}$$
$$= \frac{4500}{-3} = -1500.$$

Then use the point-slope formula with either point, say $(2, 9000)$, and $m = -1500$:

$$V - V_1 = m(t - t_1)$$
$$V - 9000 = -1500(t - 2)$$
$$V - 9000 = -1500t + 3000$$
$$V + 1500t = 12{,}000.$$

b. The slope represents the change in the value of the car per year. Thus,

$$m = \frac{\Delta V}{\Delta t} = \frac{-1500 \text{ dollars}}{1 \text{ year}},$$

so the car depreciated at a rate of $1500 per year.

c. The car was new when $t = 0$, so

$$V + 1500(0) = 12{,}000,$$

from which

$$V = 12{,}000.$$

The car was worth $12,000 when new.

EXERCISE 5.3

A

■ *For Problems 1 and 2, see Example 1.*

1. The distance covered by a cross-country competitor is given by $d = 6t$, where t is the number of hours she runs.
   a. Graph the equation.
   b. Using two points on the graph, compute the slope $\Delta d/\Delta t$ (including units).
   c. What is the significance of the slope in terms of the problem?

244 CHAPTER 5 ■ EQUATIONS IN TWO VARIABLES

2. A temporary typist's paycheck (before deductions) is given by $S = 8t$, where t is the number of hours she worked.
   a. Graph the equation.
   b. Using two points on the graph, compute the slope $\Delta S/\Delta t$ (including units).
   c. What is the significance of the slope in terms of the problem?

■ *Find the slope of the line segment joining each pair of points. See Example 2.*

3. $(-3, 2), (2, 14)$
4. $(-4, -3), (1, 9)$
5. $(2, -3), (-2, -1)$
6. $(5, -4), (-1, 1)$
7. $(-6, -3), (-6, 3)$
8. $(-2, 9), (4, 9)$
9. $(7.6, -4.2), (-3.1, 6.8)$
10. $(-9.7, -2.1), (-3.5, -6.2)$
11. $\left(\dfrac{3}{4}, -\dfrac{1}{8}\right), \left(\dfrac{5}{6}, -\dfrac{1}{2}\right)$
12. $\left(-\dfrac{7}{5}, -\dfrac{1}{3}\right), \left(\dfrac{1}{10}, -\dfrac{4}{3}\right)$

■ *Find the slope of each line.*

13.

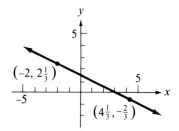

14.

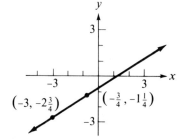

15.

16.

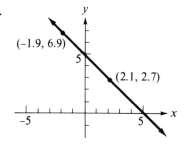

17.

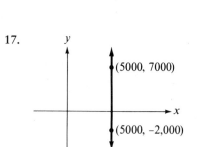

18.

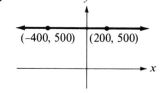

■ *For each linear equation, do the following.*
a. *Graph the equation.*
b. *Choose two points on the line and compute its slope.*
c. *Choose two more points on the line and compute the slope again. See Example 3.*

19. $y = -2x - 3$

20. $y = 3x - 4$

21. $y = \dfrac{2}{3}x - 2$

22. $y = -\dfrac{3}{2}x + 1$

■ *Which of the following tables represent variables that are related by a linear equation? (Hint: Which relationships have constant slope?) See Example 3.*

23. a.

x	y
2	12
3	17
4	22
5	27
6	32

b.

t	P
2	4
3	9
4	16
5	25
6	36

c.

h	w
−6	20
−3	18
0	16
3	14
6	12

d.

t	d
5	0
10	3
15	6
20	12
25	24

24. a.

r	E
1	5
2	$\dfrac{5}{2}$
3	$\dfrac{5}{3}$
4	$\dfrac{5}{4}$
5	1

b.

s	t
10	6.2
20	9.7
30	12.6
40	15.8
50	19.0

c.

w	A
2	−13
4	−23
6	−33
8	−43
10	−53

d.

x	C
0	0
2	5
4	10
8	20
16	40

■ *For the graphs in Problems 25 and 26 on page 246, do the following.*
a. *Determine whether the slope is positive, negative, zero, or undefined.*
b. *Arrange the lines in order of increasing slope.*

See Examples 4 and 5.

25.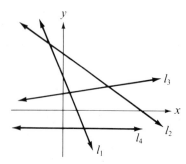
26.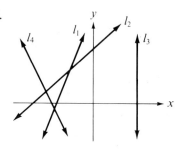

■ Find an equation for the line that passes through the given point and has the given slope. Write the equation in standard form. See Example 6.

27. $(2, -5)$; $m = -3$
28. $(-6, -1)$; $m = 4$
29. $(2, -1)$; $m = \dfrac{5}{3}$
30. $(-1, 2)$; $m = -\dfrac{3}{2}$
31. $(-6.4, -3.5)$; $m = -0.27$
32. $(7.2, -1.3)$; $m = 1.55$

■ Find an equation for the line whose graph includes the two given points. Write the equation in standard form. See Example 7.

33. $(-4, 2), (3, 3)$
34. $(5, -1), (2, -3)$
35. $(-2, -6), (-2, 5)$
36. $(-1, 3), (4, 3)$
37. $(15.3, 9.6), (-2.4, -10.8)$
38. $(11.2, -18.3), (5.1, -6.8)$

■ For Problems 39–44, do the following.
a. Find a linear equation relating the variables.
b. Graph the equation.
c. State the slope of the line, including units, and explain its meaning in the context of the problem.

See Example 8.

39. It cost a bicycle company $9000 to make 50 touring bikes in its first month of operation and $15,000 to make 125 bikes during its second month. Express its production costs in terms of the number of bicycles made.
40. Under ideal conditions Andrea's Porsche can travel 312 miles on a full tank (12 gallons of gasoline) and 130 miles on 5 gallons. Express the distance Andrea can drive in terms of the amount of gasoline she buys.
41. On an international flight a passenger may check two bags weighing 70 pounds, or 154 kilograms, each and one carry-on bag weighing 50 pounds, or 110 kilograms. Express the weight of a bag in kilograms in terms of its weight in pounds.
42. A radio station in Detroit, Michigan, reports the high and low temperatures in the Detroit/Windsor area as 59°F and 23°F, respectively. A station in Windsor, Ontario, reports the same temperatures as 15°C and −5°C. Express the Fahrenheit temperature in terms of the Celsius temperature.
43. When Harold and Nancy leave their motel at 8 A.M. on the second day of their summer vacation, they are 265 miles from Los Angeles. When they stop for lunch

at 1 P.M., they are 590 miles from Los Angeles. Express their distance from Los Angeles on the second day in terms of the time they have driven.

44. Flying lessons cost $645 for an 8-hour course and $1425 for a 20-hour course. Both prices include a fixed insurance fee. Express the cost of flying lessons in terms of the length of the course.

B

45. A line of slope -4 passes through the points $(-2, -2)$ and $(x, 19)$. Find x.
46. A line of slope $8/5$ passes through the points $(12, 10)$ and $(x, -6)$. Find x.
47. A line of slope $1/4$ passes through the points $(2, -1)$ and $(-6, y)$. Find y.
48. A line of slope -3 passes through the points $(7, -6)$ and $(2, y)$. Find y.
49. A line has slope $7/3$. Find the vertical change associated with each horizontal change along the line.
    a. 6   b. 10   c. -24
50. A line has slope $-4/5$. Find the horizontal change associated with each vertical change along the line.
    a. 2   b. -12 c. 5
51. Residential staircases are usually built with a slope of 70%, or $7/10$. If the vertical distance between stories is 10 feet, how much horizontal space does the staircase require?
52. A straight section of highway in the Midwest maintains a grade (or slope) of 4%, or $1/25$, for 12 miles. How much does your elevation change as you travel the road?

■ *Answer the following questions with a sentence or two.*

53. Explain slope. Use a diagram.
54. What property distinguishes linear equations from other equations in two variables?
55. How does slope measure the "steepness" of a line?
56. What is the slope of a horizontal line? A vertical line? Explain why.

5.4

FURTHER PROPERTIES OF LINES

Slope-Intercept Form

Every nonvertical line crosses the y-axis and thus has a y-intercept, whose coordinates we designate by $(0, b)$. If we substitute these coordinates into the point-slope formula, we get

$$y - b = m(x - 0),$$

or

$$y = mx + b.$$

This equation is called the **slope-intercept form** for a linear equation. Note that the coefficient of x is the slope, m, and the constant term, b, is the y-intercept. (See Figure 5.12.)

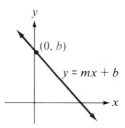

FIGURE 5.12

Recall that the point-slope formula is useful for finding the equation of a line when its slope and one point on the line are known. The slope-intercept form, on the other hand, is useful when we have an equation for a nonvertical line and want to find its slope and y-intercept. Any such equation can be put into slope-intercept form by solving explicitly for y. Once that is done, the slope and y-intercept of the line can be found without further calculation by reading off their values from the equation.

EXAMPLE 1 Write $3x + 4y = 6$ in slope-intercept form and specify the slope of the line and its y-intercept.

Solution First, solve the equation explicitly for y:

$$4y = -3x + 6$$
$$y = \frac{-3}{4}x + \frac{3}{2}.$$

Hence, the slope is $-3/4$ (the coefficient of x), and the y-intercept is $3/2$ (the constant term).

5.4 ■ FURTHER PROPERTIES OF LINES   249

We can graph a line whose equation is given in slope-intercept form by using the definition of slope, $m = \Delta y/\Delta x$.

EXAMPLE 2   Graph the line $y = \dfrac{4}{3}x - 2$.

Solution   Begin by plotting the y-intercept. The y-intercept of the line is -2, so the point $(0, -2)$ lies on the graph. Find a second point on the line by using the slope,

$$m = \frac{\Delta y}{\Delta x} = \frac{4}{3}.$$

Starting at $(0, -2)$, move four units in the y-direction and three units in the x-direction to find the point $(3, 2)$. Then draw the line through these two points.

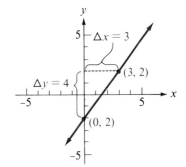

We can also determine the equation of a line from its graph by noting its slope and y-intercept.

EXAMPLE 3   Find an equation for the line below.

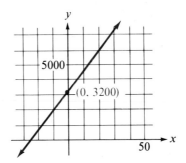

Solution The line crosses the y-axis at the point $(0, 3200)$, so the y-intercept is 3200. To calculate the slope of the line, locate another convenient point on the line, say $(20, 6000)$, and compute:

$$m = \frac{\Delta y}{\Delta x} = \frac{6000 - 3200}{20 - 0}$$

$$= \frac{2800}{20} = 140.$$

Thus the slope-intercept form of the line, with $m = 140$ and $b = 3200$, is

$$y = 140x + 3200.$$

Parallel and Perpendicular Lines

Consider the graphs of the equations

$$y = \frac{2}{3}x - 4 \tag{1}$$

$$y = \frac{2}{3}x + 2 \tag{2}$$

shown in Figure 5.13. The lines have the same slope, ⅔, but different y-intercepts. Since slope measures the "slant," or inclination, of a line, lines with the same slope are parallel.

> *Two lines with slopes m_1 and m_2 are parallel if and only if $m_1 = m_2$.*

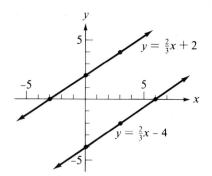

FIGURE 5.13

EXAMPLE 4 Are the graphs of the equations $3x + 6y = 6$ and $y = -\frac{1}{2}x + 5$ parallel?

Solution The lines are parallel if their slopes are equal. The slope of the first line can be found by putting its equation into slope-intercept form:

$$6y = -3x + 6$$
$$y = -\frac{3}{6}x + \frac{6}{6}$$
$$y = -\frac{1}{2}x + 1.$$

The slope of the first line is $m_1 = -\frac{1}{2}$. The equation of the second line is in slope-intercept form, and its slope is $m_2 = -\frac{1}{2}$. Thus, $m_1 = m_2$, so the lines are parallel.

Now consider the graphs of the equations

$$y = \frac{2}{3}x - 2 \qquad (3)$$

and

$$y = -\frac{3}{2}x + 3 \qquad (4)$$

shown in Figure 5.14. The lines appear to be perpendicular. The relationship between the slopes of perpendicular lines is not as easy to see as the relationship for parallel lines. However, for this example, $m_1 = \frac{2}{3}$ and $m_2 = -\frac{3}{2}$.

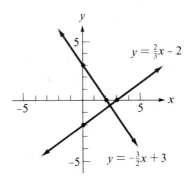

FIGURE 5.14

Note that

$$m_2 = -\frac{3}{2} = \frac{-1}{\frac{2}{3}} = \frac{-1}{m_1}.$$

This relationship holds for any two perpendicular lines with slopes m_1 and m_2, where $m_1 \neq 0$ and $m_2 \neq 0$.

> *Two lines with slopes m_1 and m_2 are perpendicular if* $m_2 = \dfrac{-1}{m_1}.$

We say that m_2 is the *negative reciprocal* of m_1. (Geometric arguments for this result and the corresponding result about parallel lines are given in Problems 39 and 40 at the end of this section.)

EXAMPLE 5 Are the graphs of $3x - 5y = 5$ and $2y = \frac{10}{3}x + 3$ perpendicular?

Solution Find the slope of each line by putting it into slope-intercept form. For the first line,

$$y = \frac{3}{5}x - 1,$$

so $m_1 = 3/5$. For the second line,

$$y = \frac{5}{3}x + \frac{3}{2},$$

so $m_2 = 5/3$. Now, the negative reciprocal of m_1 is

$$\frac{-1}{m_1} = \frac{-1}{\frac{3}{5}} = \frac{-5}{3},$$

but $m_2 = 5/3$. Thus, $m_2 \neq -1/m_1$, so the lines are not perpendicular.

EXAMPLE 6 Show that the triangle with vertices $A(0, 8)$, $B(6, 2)$, and $C(-4, 4)$ is a right triangle.

Solution Show that two of the sides of the triangle are perpendicular. The line segment $\overline{AB}$ has slope

$$m_1 = \frac{2 - 8}{6 - 0} = \frac{-6}{6} = -1,$$

and the line segment $\overline{AC}$ has slope

$$m_2 = \frac{4 - 8}{-4 - 0} = \frac{-4}{-4} = 1.$$

Since

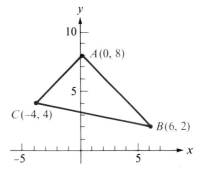

$$\frac{-1}{m_1} = \frac{-1}{-1} = 1 = m_2,$$

the sides $\overline{AB}$ and $\overline{AC}$ are perpendicular, and the triangle is a right triangle.

We can find an equation for a line parallel to a given line by first finding the slope of the given line and then using the point-slope formula.

EXAMPLE 7 Find an equation for the line that passes through the point $(1, 4)$ and is parallel to the line $4x - 2y = 6$.

Solution The slope of the given line can be found by putting its equation into slope-intercept form:

$$4x - 2y = 6$$
$$-2y = -4x + 6$$
$$y = 2x - 3.$$

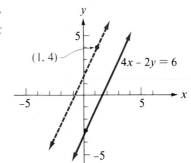

The given line has slope 2. Since the unknown line is parallel to the given line, its slope is also 2. The unknown line passes through the point (1, 4), so use the point-slope formula with $m = 2$ and $(x_1, y_1) = (1, 4)$:

$$y - y_1 = m(x - x_1)$$
$$y - 4 = 2(x - 1)$$
$$y - 4 = 2x - 2$$
$$y = 2x + 2.$$

The line that is perpendicular to both lines in Example 7 and passes through the point (1, 4) has slope $-\frac{1}{2}$, the negative reciprocal of the given slope, 2. We can find its equation by using the point-slope formula with the given point, (1, 4):

$$y - 4 = -\frac{1}{2}(x - 1)$$
$$y - 4 = -\frac{1}{2}x + \frac{1}{2}.$$

Thus, an equation for the perpendicular line is

$$y = -\frac{1}{2}x + \frac{9}{2}.$$

EXERCISE 5.4

A

■ a. *Write each equation in slope-intercept form.*
   b. *State the slope and y-intercept of the line.*
See Example 1.

1. $3x + 2y = 1$
2. $3x - y = 7$
3. $x - 3y = 2$
4. $5x - 4y = 0$
5. $\frac{1}{4}x + \frac{3}{2}y = \frac{1}{6}$
6. $\frac{7}{6}x - \frac{2}{9}y = 3$
7. $4.2x - 0.3y = 6.6$
8. $0.8x + 0.004y = 0.24$
9. $y + 29 = 0$
10. $y - 37 = 0$
11. $250x + 150y = 2450$
12. $80x - 360y = 6120$

■ a. *Graph the line with the given slope and y-intercept.*
   b. *Write an equation for the line.*
See Example 2.

13. $m = 3$ and $b = -2$
14. $m = -4$ and $b = 1$
15. $m = -2$ and $b = 4$
16. $m = 5$ and $b = -3$
17. $m = \dfrac{5}{3}$ and $b = -6$
18. $m = -\dfrac{3}{4}$ and $b = -2$
19. $m = -\dfrac{1}{2}$ and $b = 3$
20. $m = \dfrac{2}{3}$ and $b = -4$

■ **a.** *Estimate the slope and vertical intercept of each line.*
b. *Using your estimates from (a), write an equation for the line.*
See Example 3.

21.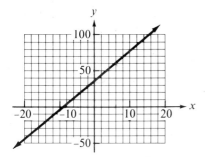

22.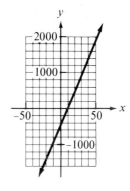

23.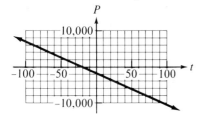

24.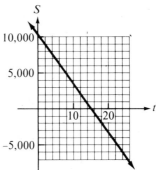

25.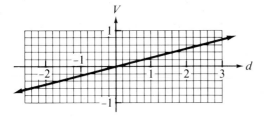

26.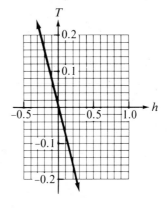

■ For Problems 27 and 28, see Examples 4 and 5.

27. Determine whether the given lines are parallel, perpendicular, or neither.
   a. $y = \frac{3}{5}x - 7$; $3x - 5y = 2$ b. $y = 4x + 3$; $y = \frac{1}{4}x - 3$
   c. $6x + 2y = 1$; $x = 1 - 3y$ d. $2y = 5$; $5y = -2$

28. Determine whether the given lines are parallel, perpendicular, or neither.
   a. $2x - 7y = 14$; $7x - 2y = 14$ b. $x + y = 6$; $x - y = 6$
   c. $x = -3$; $3y = 5$ d. $\frac{1}{4}x - \frac{3}{4}y = \frac{2}{3}$; $\frac{1}{6}x = \frac{1}{2}y + \frac{1}{3}$

■ For Problems 29–32, see Example 6.

29. Show that the triangle with vertices $A(2, 5)$, $B(5, 2)$, and $C(10, 7)$ is a right triangle.

30. Show that the triangle with vertices $P(-1, 3)$, $Q(3, 8)$, and $R(4, 5)$ is a right triangle.

31. Show that the points $P(2, 4)$, $Q(3, 8)$, $R(5, 1)$, and $S(4, -3)$ are the vertices of a parallelogram.

32. Show that the points $P(-5, 4)$, $Q(7, -11)$, $R(12, 25)$, and $S(0, 40)$ are the vertices of a parallelogram.

■ For Problems 33–36, see Example 7.

33. Write an equation of the line that is parallel to the graph of $x - 2y = 5$ and passes through the point $(2, -1)$. Sketch the graphs of both equations.

34. Write an equation of the line that is parallel to the graph of $2y - 3x = 5$ and passes through the point $(-3, 2)$. Sketch the graphs of both equations.

35. Write an equation of the line that is perpendicular to the graph of $2y - 3x = 5$ and passes through the point $(1, 4)$. Sketch the graphs of both equations.

36. Write an equation of the line that is perpendicular to the graph of $x - 2y = 5$ and passes through the point $(4, -3)$. Sketch the graphs of both equations.

B

37. Given the points $P(4, -1)$, $Q(2, 7)$, and $S(-3, 4)$, find the value of k in the ordered pair $T(5, k)$ that makes $\overline{PQ}$ parallel to $\overline{ST}$.

38. Using the points in Problem 37, find the value of k that makes $\overline{PQ}$ perpendicular to $\overline{ST}$.

39. In this exercise we will show that parallel lines have the same slope. In the figure, l_1 and l_2 are two parallel lines that are neither horizontal nor vertical. Their y-intercepts are A and B. The segments $\overline{AC}$ and $\overline{CD}$ are constructed parallel to the x- and y-axes, respectively. Explain why each of the following statements is true.
   a. Angle ACD equals angle CAB.
   b. Angle DAC equals angle ACB.
   c. Triangle ACD is similar to triangle CAB.

d. $m_1 = \dfrac{CD}{AC}$; $m_2 = \dfrac{AB}{AC}$.

e. $m_1 = m_2$.

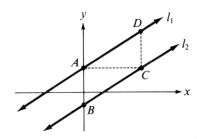

40. In this exercise we will show that for two perpendicular lines with slopes m_1 and m_2 (where neither line is vertical), m_2 is the negative reciprocal of m_1. In the figure, lines l_1 and l_2 are perpendicular. Their y-intercepts are B and C. The segment $\overline{AP}$ is constructed through the point of intersection of l_1 and l_2 parallel to the x-axis. Explain why each of the following statements is true.

   a. Angle ABC and angle ACB are complementary.
   b. Angle ABC and angle BAP are complementary.
   c. Angle BAP equals angle ACB.
   d. Angle CAP and angle ACB are complementary.
   e. Angle CAP equals angle ABC.
   f. Triangle ABP is similar to triangle CAP.
   g. $m_1 = \dfrac{BP}{AP}$; $m_2 = -\dfrac{CP}{AP}$.
   h. $m_2 = \dfrac{-1}{m_1}$.

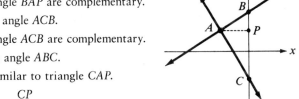

41. If the lines $ax + by = c$ and $px + qy = r$ are parallel, show that
$$aq - pb = 0.$$

42. If the lines $ax + by = c$ and $px + qy = r$ are perpendicular, show that
$$ap + bq = 0.$$

■ *Answer the following questions with a sentence or two.*

43. Explain how the definition of slope can be used to graph a line whose equation is given in slope-intercept form.

44. Give the formulas for the point-slope and slope-intercept forms of the equation of a line. Describe how each is used.

CHAPTER REVIEW

[5.1]

1. The figure shows a graph of the snow level at a ski resort in Colorado during a 2-week period in January 1989.

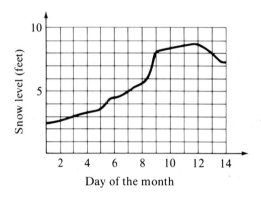

   a. What was the highest snow level recorded?
   b. On how many days was the snow level above 4 feet?
   c. By how much did the snow level increase between January 7 and January 9?
   d. During which daily interval did the snow level increase most rapidly? Decrease most rapidly?

2. The figure shows a graph of the maximum daily temperatures at the ski resort in Problem 1 over the same 2-week period.

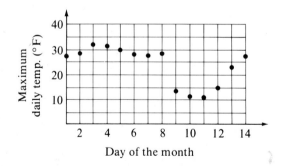

   a. Estimate the highest maximum temperature and the lowest maximum temperature.
   b. On how many days was the maximum temperature below 20°F?

c. On what day did a cold front pass over the resort?
d. By approximately how much did the maximum temperature drop on the day the cold front passed?

■ *Find the missing components so that each ordered pair is a solution of the given equation.*

3. $3x - 5y = 10$; a. $(0, ?)$, b. $(?, 0)$, c. $(2, ?)$
4. $x + 6y = 8$; a. $(0, ?)$, b. $(?, 0)$, c. $(?, -2)$

■ *Transform each equation into one in which y is expressed explicitly in terms of x. Find ordered pairs with the given x-components that satisfy the equation.*

5. $2x + xy = 4$; $2, 4$
6. $2x - xy = 6$; $-2, 2$
7. $xy - 3y = 6$; $-3, 0$
8. $2y = 8 - x^2 y$; $-4, -2$
9. $x^2 - 4y^2 = 16$; $4, 5$
10. $x^2 + 2y^2 = 8$; $-1, 0$

■ *Graph each equation. Use the suggested values for the variables.*

11. $s = 4t - 2$; $t = -2, -1, 0, 1, 2$
12. $g = t^2 - 9$; $t = -4, -3, -2, -1, 0, 1, 2, 3, 4$
13. $q = \dfrac{1}{r - 2}$; $r = -2, -1, 0, 1, 2, 3, 4$
14. $p = \sqrt{4q}$; $q = 0, 1, 4, 16$

[5.2]
■ *Graph each equation.*

15. $4x - 3y = 12$
16. $\dfrac{x}{6} - \dfrac{y}{2} = 1$
17. $50x = 40y - 20{,}000$
18. $1.4x + 2.1y = 8.4$
19. $3x - 4y = 0$
20. $x = -4y$
21. $4x = -12$
22. $2y - 6 = 0$

[5.3]
■ *Find the slope of the line segment joining each pair of points.*

23. $(-1, 4), (3, -2)$
24. $(5, 0), (2, -6)$
25. $(6.2, 1.4), (-2.1, 4.8)$
26. $(0, -6.4), (-5.6, 3.2)$

■ *Find an equation for the line that passes through the given point and has the given slope. Write the equation in standard form.*

27. $(-4, 6)$; $m = -\dfrac{2}{3}$
28. $(2, -5)$; $m = \dfrac{3}{2}$

■ *Find an equation for the line whose graph includes the two given points. Write the equation in standard form.*

29. $(3, -5), (-2, 4)$
30. $(0, 8), (4, -2)$

[5.4]

■ a. *Write each equation in slope-intercept form.*
  b. *State the slope and y-intercept of the line.*

31. $2x - 4y = 5$

32. $\frac{1}{2}x + \frac{2}{3}y = \frac{5}{6}$

33. $8.4x + 2.1y = 6.3$

34. $y - 3 = 0$

■ a. *Graph the line with the given slope and y-intercept.*
  b. *Write an equation for the line in standard form.*

35. $m = -2$ and $b = 3$

36. $m = \frac{3}{2}$ and $b = -5$

■ *Determine whether the given lines are parallel, perpendicular, or neither.*

37. $y = \frac{1}{2}x + 3$; $x - 2y = 8$

38. $4x - y = 6$; $x + 4y = -2$

39. Write an equation for the line that is *parallel* to the graph of $2x + 3y = 6$ and passes through the point $(1, 4)$.

40. Write an equation for the line that is *perpendicular* to the graph of $2x + 3y = 6$ and passes through the point $(1, 4)$.

■ *Solve.*

41. Last year Pinwheel Industries introduced a new model calculator. It cost $2000 to develop the calculator and $20 to manufacture each one.
    a. Write an equation that expresses the total costs in terms of the number of calculators produced.
    b. Graph the equation.
    c. What is the cost of producing 1000 calculators?
    d. How many calculators can be produced for $10,000?

42. Megan weighed 5 pounds at birth and gained 18 ounces per month during her first year.
    a. Write an equation that expresses Megan's weight in terms of her age.
    b. Graph the equation.
    c. How much did Megan weigh at 9 months?
    d. When did Megan weigh 9 pounds?

43. The world's oil reserves were 1660 billion barrels in 1976; total annual consumption is 20 billion barrels.
    a. Write an equation that expresses the remaining oil reserves in terms of time.
    b. Graph the equation.
    c. How much oil will be left in the year 2000?
    d. When will the world's oil reserves be completely depleted?

44. The world's copper reserves were 500 million tons in 1976; total annual consumption is 8 million tons.
   a. Write an equation that expresses the remaining copper reserves in terms of time.
   b. Graph the equation.
   c. How much copper will be left in the year 2000?
   d. When will the world's copper reserves be completely depleted?

45. The owner of a movie theater needs to bring in $1000 at each screening in order to stay in business. He sells adult tickets at $5 apiece and children's tickets at $2 each.
   a. Write an equation that relates the number of adult tickets he must sell to the number of children's tickets he must sell.
   b. Graph the equation.
   c. If the owner sells 120 adult tickets, how many children's tickets must he sell?
   d. In the context of the problem, what is the significance of the x- and y-intercepts of the graph?

46. Alida plans to spend part of her vacation in Atlantic City and part in Saint-Tropez. She estimates that after airfare her vacation will cost $60 per day in Atlantic City and $100 per day in Saint-Tropez. She has $1200 to spend after airfare.
   a. Write an equation that relates the number of days Alida can spend in Atlantic City to the number of days in Saint-Tropez.
   b. Graph the equation.
   c. If Alida spends 10 days in Atlantic City, how long can she spend in Saint-Tropez?
   d. In the context of the problem, what is the significance of the x- and y-intercepts of the graph?

47. An interior decorator charges a consulting fee of $500 plus 10% of the cost of the remodeling.
   a. Write an equation that relates the decorator's fee to the cost of the remodeling.
   b. Graph the equation.
   c. How much is the decorator's fee on a $12,000 remodeling job?

48. Auto registration fees in Connie's home state are $35 plus 2% of the value of the automobile.
   a. Write an equation that relates the registration fee to the value of the automobile.
   b. Graph the equation.
   c. How much is the registration fee on a $20,000 sports car?

■ *For Problems 49 and 50, do the following.*
  a. *Find a linear equation relating the variables.*
  b. *Graph the equation.*
  c. *State the slope of the line, including units, and explain its meaning in the context of the problem.*

49. The population of Maple Rapids was 4800 in 1972 and had grown to 6780 by 1987. Assuming a constant increase, express the population of Maple Rapids in terms of years since 1972.

50. Cicely's odometer read 112 miles when she filled up her 14-gallon gas tank and 308 when the gas gauge read half full. Express her mileage in terms of the amount of gas she used.

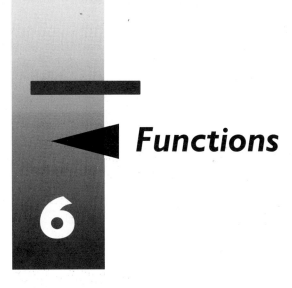

Functions

6.1

DEFINITIONS AND NOTATION

In Chapter 5 we studied a number of relationships between two variables; in particular, we considered equations that define linear relationships. For example, the equation

$$C = 30t + 800 \qquad (t \geq 0) \tag{1}$$

relates the variables t and C. To each value of $t \geq 0$ we associate the value of C given by Equation (1). Thus,

when $t = 0$, $C = 30(0) + 800 = 800$;
when $t = 4$, $C = 30(4) + 800 = 920$;
when $t = 10$, $C = 30(10) + 800 = 1100$.

We have represented an association between the values of two variables by a chart, or by ordered pairs, as follows.

t	C
0	800
4	920
10	1100

(t, C)
$(0, 800)$
$(4, 920)$
$(10, 1100)$

Notice that for the relationship defined by Equation (1) we can always determine the value of C associated with a given value of t. Such a relationship between variables is called a **function**. The variable t in Equation (1) is called the **independent** variable; C is the **dependent** variable because its values are determined by the values of t. In general, we make the following definition.

> *A* **function** *is a relationship between two variables for which a unique value of the* **dependent** *variable can be determined from a value of the* **independent** *variable.*

Note that the definition calls for a *unique value,* that is, exactly one value of the dependent variable corresponding to each value of the independent variable.

EXAMPLE 1
a. The distance d traveled by a car in 1 hour is a function of its speed r. If we know the speed of the car we can determine the distance it travels by the formula $d = rt$.

b. The price of a fill-up with unleaded gasoline is a function of the number of gallons purchased. The gas pump represents the function by displaying the values of the independent variable (number of gallons) and the dependent variable (total cost).

c. Score on the Scholastic Aptitude Test is *not* a function of score on an IQ test, since two people with the same score on an IQ test may score differently on the SAT. That is, a person's score on the SAT is not uniquely determined by his or her score on an IQ test.

A function can be described in a number of different ways. In the following examples we consider functions defined by tables or charts and by equations. In Section 6.2 we will consider functions defined by graphs.

EXAMPLE 2 The accounting office of an automobile parts manufacturer compiled the following data on sales over several years.

Year (t)	Total Sales (S)
1985	$612,000
1986	$663,000
1987	$692,000
1988	$749,000
1989	$804,000

Here the year is considered the independent variable, and we say that the total sales *S is a function of t*.

EXAMPLE 3 The table below gives the cost of sending printed material by third-class mail.

Weight in Ounces (w)	Postage (p)
$0 < w \leq 1$	$0.25
$1 < w \leq 2$	$0.45
$2 < w \leq 3$	$0.65
$3 < w \leq 4$	$0.85
$4 < w \leq 5$	$1.05
$5 < w \leq 6$	$1.25
$6 < w \leq 7$	$1.45

If the weight of the article is known, the postage required can be determined from the table. For instance, a catalog weighing 4½ ounces would require $1.05 in postage. Here w is the independent variable and p is the dependent variable. We say that *p is a function of w*.

EXAMPLE 4 The table below records the age and cholesterol count for 20 patients tested in a hospital survey.

Patient Number	Age	Cholesterol Count	Patient Number	Age	Cholesterol Count
301	53	217	332	51	209
308	48	232	336	53	241
312	55	198	339	49	186
313	56	238	340	51	216
316	51	227	343	57	208
320	52	264	347	52	248
322	53	195	356	50	214
324	47	203	359	56	271
325	48	212	362	53	193
328	50	234	370	48	172

The data do *not* define cholesterol count as a function of age, since several patients of the same age have different cholesterol levels. For example, patients 316, 332, and 340 are all 51 years old but have cholesterol counts of 227, 209, and 216, respectively. Thus, we cannot determine a *unique* value of the dependent variable (a patient's cholesterol level) from the value of the independent variable (his or her age). *There must be other factors besides age that influence a person's cholesterol level.*

The following example illustrates a function defined by an equation.

EXAMPLE 5 The Sears Tower in Chicago is the world's tallest building at 1454 feet. If an algebra book is dropped from the top of the Sears Tower, its height above the ground after t seconds is given by the equation

$$h = 1454 - 16t^2. \qquad (2)$$

Thus, after 1 second the book's height is

$$h = 1454 - 16(1)^2 = 1438 \text{ feet};$$

after 2 seconds its height is

$$h = 1454 - 16(2)^2 = 1390 \text{ feet}.$$

For this function, t is the independent variable and h is the dependent variable. Notice that for any value of t the corresponding unique value of h can be determined from Equation (2). We say that *h is a function of t*.

Function Notation

There is a convenient notation that is used to indicate a functional relationship between two variables. We may use a letter, such as f, g, or h, or F, G, or H, to name a particular function. (Any letter can be used, but these are the most popular choices.) For instance, in Example 5 we expressed the height h of an algebra book falling from the top of the Sears Tower as a function of the time t that it has been falling. We may write this as

$$f(t) = 1454 - 16t^2.$$

The symbol $f(t)$, read "f of t," is another name for the height h. It indicates that h is a function of t. Thus, we might speak of "the height $f(t)$ of the book at time t."

Note that the parentheses in the symbol $f(t)$ do *not* indicate multiplication. (It would not make sense to multiply the name of a function by a variable.)

Evaluating Functions

Think of the symbol $f(t)$ as a single entity that represents the dependent variable of a function.

The notation $f(t)$ is very useful for showing the correspondence between particular values of the variables h and t. In Example 5 we saw that when $t = 1$, $h = 1438$ and when $t = 2$, $h = 1390$. These relationships can be expressed as

$$f(1) = 1438 \quad \text{and} \quad f(2) = 1390,$$

which are read "f of 1 equals 1438" and "f of 2 equals 1390." Note that values for the independent variable, t, appear inside the parentheses and values for the dependent variable, h, appear on the other side of the equation.

EXAMPLE 6 Let g be the name of the function defined by the table in Example 3 on page 265. Find $g(1)$, $g(3)$, and $g(6.75)$.

Solution According to the table,

when $w = 1$, $p = 0.25$, so $g(1) = 0.25$;
when $w = 3$, $p = 0.65$, so $g(3) = 0.65$;
when $w = 6.75$, $p = 1.45$, so $g(6.75) = 1.45$.

Finding the value of the dependent variable that corresponds to a particular value of the independent variable is called **evaluating the function**. If a function is described by an equation, we simply substitute the given value into the equation to find the corresponding function value.

EXAMPLE 7 If $H(s) = \dfrac{\sqrt{s + 3}}{s}$, find the following.

a. $H(6)$
b. $H(-1)$

Solutions
a. $H(6) = \dfrac{\sqrt{6 + 3}}{6} = \dfrac{\sqrt{9}}{6}$
$= \dfrac{3}{6} = \dfrac{1}{2}$

b. $H(-1) = \dfrac{\sqrt{-1 + 3}}{-1}$
$= \dfrac{\sqrt{2}}{-1} = -\sqrt{2}$

Sometimes it is useful to evaluate a function for a value represented by an algebraic expression.

EXAMPLE 8 Evaluate $f(x) = 4x^2 - x + 5$ for the following.

a. $x = 3b$ b. $x = a + 3$

Solutions
a. $f(3b) = 4(3b)^2 - (3b) + 5$
$= 4(9b^2) - 3b + 5 = 36b^2 - 3b + 5$

b. $f(a + 3) = 4(a + 3)^2 - (a + 3) + 5$
$= 4(a^2 + 6a + 9) - a - 3 + 5$
$= 4a^2 + 24a + 36 - a - 3 + 5$
$= 4a^2 + 23a + 38$

Common Error

In Example 8b, note that $f(a + 3) \neq f(a) + f(3)$;

$f(a) + f(3) = [4a^2 - a + 5] + [4(3)^2 - 3 + 5] = 4a^2 - a + 43$.

EXAMPLE 9 The cost C of manufacturing a certain kind of shoe is a function of the number of pairs produced, given by the equation

$$C(x) = 3000 + 20x,$$

where x is the number of pairs produced and $C(x)$ is measured in dollars. Find the cost of producing 500 pairs of shoes.

Solution To find the value of C that corresponds to $x = 500$, find $C(500)$:

$$C(500) = 3000 + 20(500) = 13,000.$$

The cost of producing 500 pairs of shoes is $13,000.

We can perform algebraic operations on function values just as we do on any other variables. Remember that the expression $f(x)$ represents a *single* value and that the parentheses do *not* indicate multiplication.

EXAMPLE 10 If $f(x) = x^3 - 1$, find the following.

a. $f(2) + f(3)$ b. $f(2 + 3)$ c. $2f(x) + 3$

Solutions

a. $f(2) + f(3) = (2^3 - 1) + (3^3 - 1)$
$= (8 - 1) + (27 - 1)$
$= 7 + 26 = 33$

b. $f(2 + 3) = f(5)$
$= 5^3 - 1$
$= 125 - 1 = 124$

c. $2f(x) + 3 = 2(x^3 - 1) + 3$
$= 2x^3 - 2 + 3$
$= 2x^3 + 1$

EXERCISE 6.1

A

■ *For which of the following pairs is the second quantity a function of the first? (Why?) See Example 1.*

1. Price of an item; sales tax at 4%
2. Time traveled at constant speed; distance traveled
3. Number of years of education; annual income
4. Distance flown in an airplane; price of the ticket
5. Volume of a container of water; its weight
6. Amount of a paycheck; amount of Social Security tax withheld

■ *Each of the following establishes a correspondence between two variables. Suggest appropriate independent and dependent variables and decide whether the relationship is a function. See Example 1.*

7. An itemized grocery receipt
8. An inventory list
9. An index
10. A will
11. An instructor's grade book
12. An address book
13. A bathroom scale
14. A radio dial

Which of the following tables describe functions? Why or why not? See Examples 2–4.

15.

x	t
−1	2
0	9
1	−2
0	−3
−1	5

16.

y	w
0	8
1	12
3	7
5	−3
7	4

17.

x	y
−3	8
−2	3
−1	0
0	−1
1	0
2	3
3	8

18.

s	t
2	5
4	10
6	15
8	20
6	25
4	30
2	35

19.

r	−4	−2	0	2	4
v	6	6	3	6	6

20.

p	−5	−4	−3	−2	−1
d	−5	−4	−3	−2	−1

21.

Pressure (p)	Volume (v)
15	100.0
20	75.0
25	60.0
30	50.0
35	42.8
40	37.5
45	33.3
50	30.0

22.

Frequency (f)	Wavelength (w)
5	60.0
10	30.0
20	15.0
30	10.0
40	7.5
50	6.0
60	5.0
70	4.3

23.

Temperature (T)	Humidity (h)
Jan. 1 34°F	42%
Jan. 2 36°F	44%
Jan. 3 35°F	47%
Jan. 4 29°F	50%
Jan. 5 31°F	52%
Jan. 6 35°F	51%
Jan. 7 34°F	49%

24.

Inflation Rate (I)	Unemployment Rate (U)
1972 5.6%	5.1%
1973 6.2%	4.5%
1974 10.1%	4.9%
1975 9.2%	7.4%
1976 5.8%	6.7%
1977 5.6%	6.8%
1978 6.7%	7.4%

25.

Adjusted Gross Income (I)	Tax Bracket (T)
0–2479	0%
2480–3669	11%
3670–4749	12%
4750–7009	14%
7010–9169	15%
9170–11,649	16%
11,650–13,919	18%

26.

Cost of Merchandise (m)	Shipping Charge (C)
$0.01–10.00	$2.50
10.01–20.00	3.75
20.01–35.00	4.85
35.01–50.00	5.95
50.01–75.00	6.95
75.01–100.00	7.95
Over 100.00	8.95

■ For each of the relationships described below, do the following.
  a. Identify the independent and dependent variables.
  b. Make a table of ordered pairs for at least five choices of the independent variable.
  c. Decide whether the equation defines a function.

See Example 5.

27. A computer costs $28,000 and depreciates according to the formula

$$V = 28,000(1 - 0.06t),$$

where V is the value of the computer after t years.

28. In a profit-sharing plan an employee receives a salary of

$$S = 20,000 + 0.01x,$$

where x represents the company's profit for the year.

29. An advertising agency accrues revenue in thousands of dollars according to the formula

$$R = 50x^2 - 200x + 800,$$

where x represents the number of its clients.

30. A manufacturer of machine parts finds that his profit is given by the formula

$$P = 0.02x^2 - 800x - 500,000 \quad (x \geq 10,000),$$

where x is the number of parts manufactured.

31. The number of compact cars that a large dealership can sell at price p is given by

$$N = \frac{12,000,000}{p}.$$

32. A department store finds that the market value of its Christmas-related merchandise is given by

$$M = \frac{600{,}000}{t} \quad (t \leq 30),$$

where t is the number of weeks after Christmas.

33. The velocity of a car that brakes suddenly can be determined from the length of its skid marks d by

$$v = \sqrt{12d},$$

where d is in feet and v is in miles per hour.

34. The distance d in miles that a person can see on a clear day from a height h in feet is given by

$$d = 1.22\sqrt{h}.$$

■ See Example 6.

35. If the function described in Problem 21 is called g, find the following.
   a. $g(25)$ b. $g(40)$ c. x so that $g(x) = 50$
36. If the function described in Problem 22 is called h, find the following.
   a. $h(20)$ b. $h(60)$ c. x so that $h(x) = 10$
37. If the function described in Problem 25 is called T, find the following.
   a. $T(8750)$ b. $T(6249)$ c. x so that $T(x) = 15\%$
38. If the function described in Problem 26 is called C, find the following.
   a. $C(11.50)$ b. $C(47.24)$ c. x so that $C(x) = 7.95$

■ Evaluate each function for the given values. See Example 7.

39. $f(x) = 6 - 2x$
   a. $f(3)$ b. $f(-2)$ c. $f(-12.7)$ d. $f\left(\frac{2}{3}\right)$
40. $g(t) = 5t - 3$
   a. $g(1)$ b. $g(-4)$ c. $g(14.1)$ d. $g\left(\frac{3}{4}\right)$
41. $h(v) = 2v^2 - 3v + 1$
   a. $h(0)$ b. $h(-1)$ c. $h\left(\frac{1}{4}\right)$ d. $h(-6.2)$
42. $r(s) = 2s - s^2$
   a. $r(2)$ b. $r(-4)$ c. $r\left(\frac{1}{3}\right)$ d. $r(-1.3)$

43. $H(z) = \dfrac{2z - 3}{z + 2}$

    a. $H(4)$ b. $H(-3)$ c. $H\left(\dfrac{4}{3}\right)$ d. $H(4.5)$

44. $F(x) = \dfrac{1 - x}{2x - 3}$

    a. $F(0)$ b. $F(-3)$ c. $F\left(\dfrac{5}{2}\right)$ d. $F(9.8)$

45. $E(t) = \sqrt{t - 4}$

    a. $E(16)$ b. $E(4)$ c. $E(7)$ d. $E(4.2)$

46. $D(r) = \sqrt{5 - r}$

    a. $D(4)$ b. $D(-3)$ c. $D(-9)$ d. $D(4.6)$

47. $Q(y) = 4y^{5/2}$

    a. $Q(16)$ b. $Q\left(\dfrac{1}{4}\right)$ c. $Q(35)$ d. $Q(100)$

48. $T(w) = -3w^{2/3}$

    a. $T(27)$ b. $T\left(\dfrac{1}{8}\right)$ c. $T(20)$ d. $T(1000)$

■ *Evaluate the following functions. See Example 8.*

49. $G(s) = 3s^2 - 6s$

    a. $G(3a)$ b. $G(a + 2)$ c. $G(a) + 2$ d. $G(-a)$

50. $h(x) = 2x^2 + 6x - 3$

    a. $h(2a)$ b. $h(a + 3)$ c. $h(a) + 3$ d. $h(-a)$

51. $g(x) = 8$

    a. $g(2)$ b. $g(8)$ c. $g(a + 1)$ d. $g(-x)$

52. $f(t) = -3$

    a. $f(4)$ b. $f(-3)$ c. $f(b - 2)$ d. $f(-t)$

53. $P(x) = x^3 - 1$

    a. $P(2x)$ b. $2P(x)$ c. $P(x^2)$ d. $[P(x)]^2$

54. $Q(t) = 5t^3$

    a. $Q(2t)$ b. $2Q(t)$ c. $Q(t^2)$ d. $[Q(t)]^2$

■ *Solve each problem and express your answer using function notation. See Example 9.*

55. If the function described in Problem 27 is called $V(t)$, find the value of the computer after 10 years.

56. If the function described in Problem 28 is called $S(x)$, find the employee's salary if her company makes a profit of $850,000.

57. If the function described in Problem 29 is called $R(x)$, find the agency's revenue when it has 40 clients.

58. If the function described in Problem 30 is called $P(x)$, find the manufacturer's profit if 20,000 parts are made.

59. If the function described in Problem 31 is called $N(p)$, how many compact cars can be sold at a price of $6000?

60. If the function described in Problem 32 is called $M(t)$, find the value of the merchandise 12 weeks after Christmas.

61. If the function described in Problem 33 is called $v(d)$, how fast was a car traveling if it left skid marks 250 feet long?

62. If the function described in Problem 34 is called $d(h)$, how far can a person see from the top of Mount McKinley at 20,320 feet?

■ For each of the functions below, compute the following.
  a. $f(2) + f(3)$
  b. $f(2 + 3)$
  c. $f(a) + f(b)$
  d. $f(a + b)$
For which functions does $f(a + b) = f(a) + f(b)$? See Example 10.

63. $f(x) = 3x - 2$ 64. $f(x) = 1 - 4x$ 65. $f(x) = x^2 + 3$ 66. $f(x) = x^2 - 1$

67. $f(x) = \sqrt{x + 1}$ 68. $f(x) = \sqrt{6 - x}$ 69. $f(x) = \dfrac{-2}{x}$ 70. $f(x) = \dfrac{3}{x}$

■ For each pair of functions, do the following.
  a. Compute $f(0)$ and $g(0)$.
  b. Find all values of x for which $f(x) = 0$.
  c. Find all values of x for which $g(x) = 0$.
  d. Find all values of x for which $f(x) = g(x)$.

71. $f(x) = 2x^2 + 3x$, $g(x) = 5 - 6x$ 72. $f(x) = 3x^2 - 6x$, $g(x) = 8 + 4x$
73. $f(x) = \sqrt{x + 2}$, $g(x) = 3x - 4$ 74. $f(x) = \sqrt{x - 3}$, $g(x) = 2x - 7$

B

■ For each of the functions in Problems 75–82, compute the following.
  a. $f(x + h)$ b. $f(x + h) - f(x)$ c. $\dfrac{f(x + h) - f(x)}{h}$

75. $f(x) = 3x - 5$ 76. $f(x) = 2 - 4x$
77. $f(x) = 2x^2$ 78. $f(x) = 5x^2$
79. $f(x) = x^2 - 3x + 1$ 80. $f(x) = x^2 + 2x - 6$
81. $f(x) = \dfrac{3}{x + 2}$ 82. $f(x) = \dfrac{1}{x - 4}$

■ *Find an equation that describes each of the following functions.*

83.

x	f(x)
1	1
2	4
3	7
4	10
5	13

84.

x	g(x)
1	16
2	12
3	8
4	4
5	0

85.

t	G(t)
1	2
2	5
3	10
4	17
5	26

86.

t	H(t)
1	2
2	8
3	18
4	32
5	50

6.2

GRAPHS OF FUNCTIONS

In Section 6.1 we considered functions defined by tables or by equations. Functions can also be described by graphs. For example, the graph in Figure 6.1 gives the Dow-Jones Industrial Average (the average value of the stock prices of 500 major companies) for the stock market "correction" of October 1987.

In this example the Dow-Jones Industrial Average (DJIA) is given as a function of time during the 8 days from October 15 to October 22. Note that values of the independent variable, time, are displayed on the horizontal axis

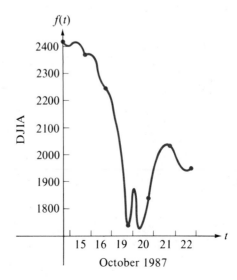

FIGURE 6.1

and values of the dependent variable, DJIA, are displayed on the vertical axis. For example, the DJIA was 2412 at noon on October 15 and 1726 at noon on October 20. In fact, each point on the graph of the function has coordinates of the form $(t, f(t))$, where $f(t)$ is simply another name for the corresponding DJIA.

EXAMPLE 1 Consider the graph of the function g given below.

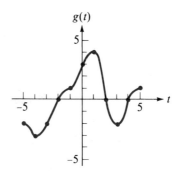

a. Find $g(-2)$, $g(0)$, and $g(5)$.
b. For what value(s) of t is $g(t) = -2$? For what value(s) of t is $g(t) = 0$?
c. What is the largest, or maximum, value of $g(t)$? For what value of t does the function take on its maximum value?

Solutions
a. The points $(-2, 0)$, $(0, 3)$, and $(5, 1)$ lie on the graph of g. Therefore, $g(-2) = 0$, $g(0) = 3$, and $g(5) = 1$.
b. Since the point $(-3, -2)$ lies on the graph, $g(-3) = -2$. (The t-value is -3.) Since the points $(-2, 0)$, $(2, 0)$, and $(4, 0)$ lie on the graph, $g(-2) = 0$, $g(2) = 0$, and $g(4) = 0$. (The t-values are -2, 2, and 4.)
c. The maximum value of $g(t)$ is the second coordinate of the highest point on the graph, $(1, 4)$. Thus, the maximum value of $g(t)$ is 4, and it occurs when $t = 1$.

Although some functions are *defined* by their graphs, we can also construct graphs for functions described by tables or by equations. This is done in much the same way that we obtained graphs from tables and equations in Chapter 5; for each value of the independent variable we form an ordered pair whose second component is the related function value.

EXAMPLE 2 Graph the function $f(x) = x^2 - 6$ $(0 \leq x \leq 4)$.

Solution Obtain several ordered pairs by evaluating the function at a number of values for x:

x	$f(x)$
0	−6
1	−5
2	−2
3	3
4	10

since $f(0) = (0)^2 - 6 = -6$;
since $f(1) = (1)^2 - 6 = -5$;
since $f(2) = (2)^2 - 6 = -2$;
since $f(3) = (3)^2 - 6 = 3$;
since $f(4) = (4)^2 - 6 = 10$.

The ordered pairs $(0, -6)$, $(1, -5)$, $(2, -2)$, $(3, 3)$, and $(4, 10)$ represent points on the graph of the function, where the horizontal axis represents the x-values and the vertical axis represents the $f(x)$-values. Plot the points and connect them with a smooth curve to obtain the graph shown below.

Domain and Range

Consider the graph of the function

$$f(t) = 1454 - 16t^2$$

given in Example 5 in Section 6.1, in which a book was dropped from a 1454-foot-tall building. We can find ordered pairs by evaluating the function

$$f(t) = h = 1454 - 16t^2$$

for choices of t:

t	$f(t)$		Ordered Pair
0	1454	since $f(0) = 1454 - 16(0)^2 = 1454$;	$(0, 1454)$
1	1438	since $f(1) = 1454 - 16(1)^2 = 1438$;	$(1, 1438)$
3	1310	since $f(3) = 1454 - 16(3)^2 = 1310$;	$(3, 1310)$
5	1054	since $f(5) = 1454 - 16(5)^2 = 1054$;	$(5, 1054)$
7	670	since $f(7) = 1454 - 16(7)^2 = 670$;	$(7, 670)$
9	158	since $f(9) = 1454 - 16(9)^2 = 158$;	$(9, 158)$
10	−146	since $f(10) = 1454 - 16(10)^2 = -146$.	$(10, -146)$

If we plot these ordered pairs, we obtain the graph shown in Figure 6.2.

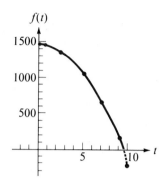

FIGURE 6.2

Notice that the function makes sense only for values of t between 0 and approximately 9.5, since we do not consider negative time values and the book hits the ground $(h = 0)$ after falling between 9 and 10 seconds. These permissible values of the independent variable make up the **domain** of the function f. All the points on the graph of $f(t)$ have t-coordinates between 0 and 9.5.

We can also see from the graph that all the points have h-coordinates between 0 and 1454. These function values that correspond to the permissible t-values make up the **range** of the function.

> The **domain** of a function is the set of permissible values for the independent variable. The **range** is the set of function values (i.e., values of the dependent variable) that correspond to the domain values.

Sometimes the domain is given as part of the definition of a function, and sometimes it can be determined from the conditions of the problem, as in the example above. The range can be determined most easily from the graph of the function.

EXAMPLE 3 Find the domain and range of the function graphed in Example 2,

$$f(x) = x^2 - 6 \quad (0 \leq x \leq 4).$$

Solution Because the domain is the set of x-values between 0 and 4, inclusive, all the points on the graph have x-values between 0 and 4. The $f(x)$-values of the points on the graph fall between -6 and 10, so the range of the function f is the set of $f(x)$-values between -6 and 10, inclusive.

Interval Notation

There is a convenient notation for specifying the domain and range of many functions. A set that consists of all the real numbers between two numbers a and b, with $a < b$, is called an **interval**. If the set includes both of the endpoints a and b, so that $a \leq x \leq b$, then the set is called a **closed interval** and is denoted by the symbol $[a, b]$. If the set does not include its endpoints, so that $a < x < b$, then it is called an **open interval** and is denoted by (a, b). An interval that includes one of its endpoints but not both is called **half-open** or **half-closed**. The domain and range of the function f in Example 3 are both closed intervals, so we may write that the domain of f is $[0, 4]$ and its range is $[-6, 10]$.

EXAMPLE 4

a. Determine the domain and range of the function h graphed below.

b. For the indicated points, show the domain values and their corresponding range values in the form of ordered pairs.

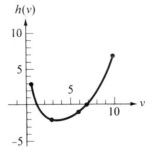

Solutions

a. All the points on the graph have v-coordinates between 1 and 10, so the domain of the function is the interval $[1, 10]$. The $h(v)$-coordinates have values between -2 and 7, inclusive, so the range of the function is the interval $[-2, 7]$.

b. Read the coordinates of the indicated points to obtain the ordered pairs $(1, 3)$, $(3, -2)$, $(7, 0)$, and $(10, 7)$.

Figure 6.3 shows the graph of the function h in Example 4 with the domain values marked on the horizontal axis and the range values marked on the vertical axis. It is sometimes helpful to imagine a rectangle whose length and width are given by those segments as shown, so that we have a "window" in the plane through which to view the graph of the function. All the points $(v, h(v))$ on the graph will lie in this window.

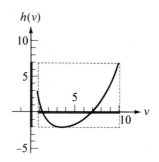

FIGURE 6.3

Not all functions have domains and ranges that are intervals.

EXAMPLE 5 a. Graph the postage function p given in Example 3 of Section 6.1.
b. Determine the domain and range of the function.

Solutions a. From the table that defines the postage function, note that articles of all weights up to 1 ounce require \$0.25 postage. That is, for all w-values greater than 0 but less than or equal to 1, the p-value is 0.25. Thus, the graph for w between 0 and 1 looks like a small piece of the horizontal line $p = 0.25$. Similarly, for all w-values greater than 1 but less than or equal to 2 the p-value is 0.45, so the graph for w-values between 1 and 2 is a small piece of the line $p = 0.45$. Continue in this way to obtain the graph shown.

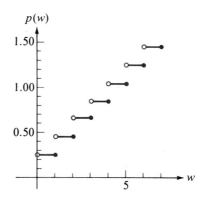

The open circles at the left endpoint of each segment indicate that that point is not included in the graph; the closed circles are points on the graph. For instance, if $w = 3$ the postage p is $0.65, not $0.85. Thus, the point $(3, 0.65)$ is part of the graph of p but the point $(3, 0.85)$ is not.

b. Postage rates are given for all weights greater than 0 ounces but less than or equal to 7 ounces, so the domain of the function is the half-open interval $(0, 7]$. On the other hand, the range of the function is *not* an interval, since the possible values for p do not include *all* the real numbers between 0.25 and 1.45. The range is the set of discrete values 0.25, 0.45, 0.65, 0.85, 1.05, 1.25, and 1.45.

We often think of the elements of the domain as the "input" values for a function and the elements of the range as the corresponding "output" values. Thus, for the function $f(x) = 3x^2 + 2$, if we choose $x = 2$ as an input value, then $f(2) = 3(2)^2 + 2 = 14$ is the output value.

The Vertical Line Test

Using the notions of domain and range, we can restate the definition of a function as follows.

A relationship between two variables is a function if each element of the domain is paired with only one element of the range.

Thus, there can be only *one* ordered pair with any given value as a first component, or, equivalently, there cannot be two different ordered pairs with the *same* first component. What does this mean in terms of the graph of the function?

Consider the graph shown in Figure 6.4a. Every vertical line intersects the graph in at most one point. Hence, the graph represents a function. In Figure 6.4b the line $x = 2$ intersects the graph at two points, $(2, 1)$ and $(2, 4)$. The domain value 2 is associated with two different range values, 1 and 4, so the graph cannot be that of a function.

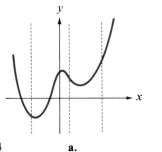 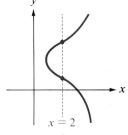

FIGURE 6.4 a. b.

We summarize these results as follows.

> **The Vertical Line Test**
>
> A graph represents a function if every vertical line intersects the graph in at most one point.

EXAMPLE 6 Use the vertical line test to determine which of the following graphs represent functions.

a.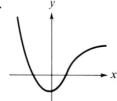

b.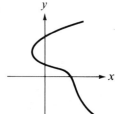

c.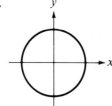

d.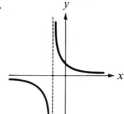

e.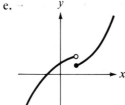

f.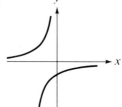

Solutions Graphs (a), (d), and (e) represent functions, since no vertical line intersects in more than one point. Graphs (b), (c), and (f) do not represent functions.

EXERCISE 6.2

A

■ *Consider the following graphs of functions. See Example 1.*

1. a. Find $h(-3)$, $h(1)$, and $h(3)$.
   b. For what value(s) of z is $h(z) = 3$?
   c. Find the x- and y-intercepts of the graph.
   d. What is the maximum value of $h(z)$?
   e. For what value(s) of z does h take on its maximum value?

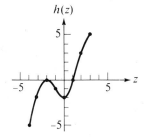

2. a. Find $G(-4)$, $G(-1)$, and $G(4)$.
   b. For what value(s) of s is $G(s) = 3$?
   c. Find the x- and y-intercepts of the graph.
   d. What is the minimum value of $G(s)$?
   e. For what value(s) of s does G take on its minimum value?

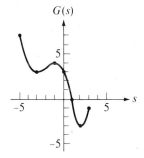

3. a. Find $R(1)$ and $R(3)$.
   b. For what value(s) of p is $R(p) = 2$?
   c. Find the x- and y-intercepts of the graph.
   d. Find the maximum and minimum values of $R(p)$.
   e. For what value(s) of p does R take on its maximum and minimum values?

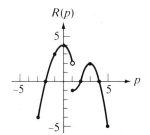

4. a. Find $f(-1)$ and $f(3)$.
   b. For what value(s) of t is $f(t) = 5$?
   c. Find the x- and y-intercepts of the graph.
   d. Find the maximum and minimum values of $f(t)$.
   e. For what value(s) of t does f take on its maximum and minimum values?

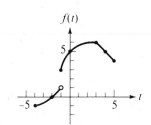

284   CHAPTER 6 ■ FUNCTIONS

5. a. Find $S(0)$, $S(\pi/6)$, and $S(-\pi)$.
    b. Estimate the value of $S(\pi/3)$ from the graph.
    c. For what value(s) of x is $S(x) = -\frac{1}{2}$?
    d. Find the maximum and minimum values of $S(x)$.
    e. For what value(s) of x does S take on its maximum and minimum values?

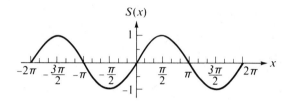

6. a. Find $C(0)$, $C(-\pi/3)$, and $C(\pi)$.
    b. Estimate the value of $C(\pi/6)$ from the graph.
    c. For what value(s) of x is $C(x) = \frac{1}{2}$?
    d. Find the maximum and minimum values of $C(x)$.
    e. For what value(s) of x does C take on its maximum and minimum values?

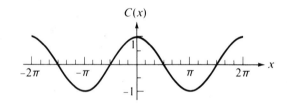

7. a. Find $F(-3)$, $F(-2)$, and $F(2)$.
    b. For what value(s) of s is $F(s) = -1$?
    c. Find the maximum and minimum values of $F(s)$.
    d. For what value(s) of s does F take on its maximum and minimum values?

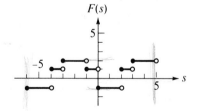

8. a. Find $P(-3)$, $P(-2)$, and $P(1)$.
    b. For what value(s) of n is $P(n) = 0$?
    c. Find the maximum and minimum values of $P(n)$.
    d. For what value(s) of n does P take on its maximum and minimum values?

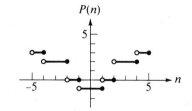

■ *Graph. See Example 2.*

9. $f(x) = x^2 - 4x$; $-2 \leq x \leq 6$
10. $g(x) = 6x - x^2$; $-1 \leq x \leq 7$
11. $g(t) = -x^2 - 2x$; $-5 \leq t \leq 3$
12. $f(t) = -t^2 - 4t$; $-6 \leq t \leq 2$
13. $h(x) = x^3 - 1$; $-2 \leq x \leq 2$
14. $q(x) = x^3 + 4$; $-3 \leq x \leq 2$
15. $F(t) = \sqrt{t+4}$; $-4 \leq t \leq 5$
16. $G(t) = \sqrt{8-t}$; $-1 \leq t \leq 8$
17. $G(x) = \dfrac{1}{3-x}$; $-\dfrac{5}{4} \leq x \leq \dfrac{11}{4}$
18. $F(x) = \dfrac{1}{x-1}$; $-\dfrac{13}{4} \leq x \leq -\dfrac{5}{4}$
19. $G(x) = \dfrac{1}{3-x}$; $3 < x \leq 6$
20. $F(x) = \dfrac{1}{x-1}$; $1 < x \leq 4$

■ *For Problems 21–40, find the domain and range of the functions in Problems 1–20. If possible, express your answers using interval notation. See Examples 3–5.*

■ *Which of the following graphs represent functions? See Example 6.*

41.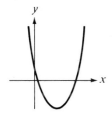

42.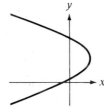

43.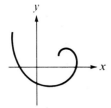

44.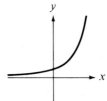

45.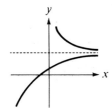

46.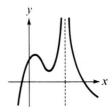

47.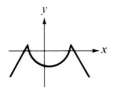

48.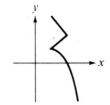

286   CHAPTER 6 ■ FUNCTIONS

49.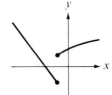

50.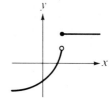

6.3

SOME BASIC GRAPHS

In Section 6.2 we graphed functions by plotting a large number of ordered pairs. In this section we develop some techniques for analyzing functions and sketching their graphs quickly, without resorting to point-plotting. First, we consider the graphs of functions defined by linear equations.

Graphs of Linear Functions

The graphs of all lines except vertical lines pass the vertical line test and hence are the graphs of functions. (See Figure 6.5.)

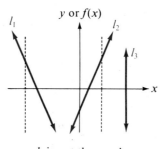

l_3 is not the graph of a function.

FIGURE 6.5

Functions of the form $f(x) = mx + b$ are therefore called **linear functions**. By letting the vertical axis, or y-axis, represent the values of $f(x)$ we can graph linear functions in the same way that we graphed linear equations of the form $y = mx + b$ in Section 5.3.

EXAMPLE 1   Graph the function $f(x) = -2x + 5$.

Solution The graph is a line of slope $m = -2$ and y-intercept $b = 5$. Plot the point $(0, 5)$ and use the definition of slope,

$$m = \frac{\Delta y}{\Delta x} = \frac{-2}{1},$$

to find a second point on the line. Draw the line through the two points.

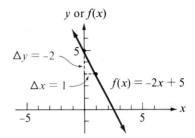

If the domain of a function is not given as part of its definition, we assume that the domain is as large as possible; that is, we include in the domain all x-values that "make sense" when substituted into the equation that defines the function. A linear function $f(x) = mx + b$ can be evaluated at any real-number value of x, so its domain is the set of all real numbers. This set is represented in interval notation as $(-\infty, \infty)$. The symbol ∞, or "infinity," is *not* a real number; it merely indicates that the set extends infinitely in either direction on the number line. There is no number on the number line called ∞, so we use parentheses, ( ), to show that $-\infty$ and ∞ are not themselves elements of the domain.

The range of the linear function $f(x) = mx + b$ $(m \neq 0)$ is also the set of all real numbers, since any number can be obtained as an output value of the function. (If $m = 0$, then the range consists of one element, b.) Thus, the graph of a linear function continues infinitely at both ends, as we noted in Section 5.1. Since we cannot show all of the graph, we show just a portion of it, usually enough to include both the x- and y-intercepts. We use arrows to indicate that the graph continues in the same direction, as in the figure for Example 1.

Some Basic Nonlinear Functions

In the examples and exercises of Section 5.1 we obtained the graphs of a number of simple functions by constructing tables of values. Several of these graphs are shown for reference in the figures below. Because these graphs are fundamental to further study of mathematics and its applications, you should become familiar with the properties of each, so that you can sketch them easily from memory.

The functions $f(x) = x^2$ and $g(x) = x^3$ are graphed in Figure 6.6. Each has as its domain the set of all real numbers, and the graphs extend infinitely as indicated by the arrows.

The graph of the quadratic function $f(x) = x^2$ is called a **parabola**. If we think of moving along the x-axis from left to right, we see that the graph is "falling" for x-values less than zero and "rising" for x-values greater than zero. On the other hand, the graph of the cubic function $g(x) = x^3$ is rising

FIGURE 6.6

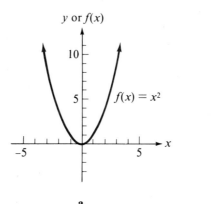
a.

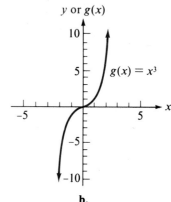
b.

for all values of x. From their graphs we see that the range of the function $f(x) = x^2$ is the interval $[0, \infty)$ and the range of $g(x) = x^3$ is the interval $(-\infty, \infty)$.

The functions $f(x) = \sqrt{x}$ and $g(x) = \sqrt[3]{x}$ graphed in Figure 6.7 are radical functions.

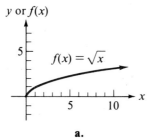
a.

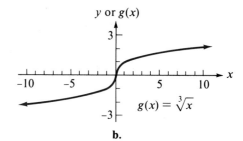
b.

FIGURE 6.7

Since the square root of a negative number is not a real number, the domain of $f(x) = \sqrt{x}$ is the interval $[0, \infty)$. As x increases the graph rises, and although the function values grow slowly, every nonnegative real number will eventually be attained as a function value. Thus, the range of f is the interval $[0, \infty)$.

Since the cube root of every real number, whether positive, negative, or zero, is a real number, the domain of $g(x) = \sqrt[3]{x}$ is the set of all real numbers. The graph rises for all values of x, and the range of $g(x) = \sqrt[3]{x}$ is the set of all real numbers.

The functions $f(x) = 1/x$ and $g(x) = 1/x^2$ graphed in Figure 6.8 are examples of rational functions. Note that neither function can be evaluated for $x = 0$, since division by zero is undefined. The domain of each function is the set of all real numbers *except* zero.

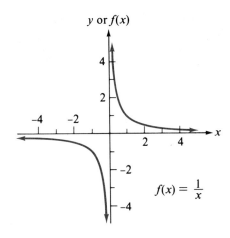
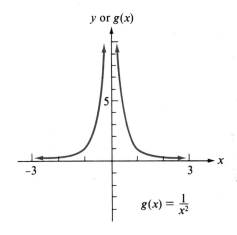

FIGURE 6.8 a. b.

Asymptotes

Because zero is not in the domain of f, there is no point on the graph of $f(x) = 1/x$ with x-coordinate 0. To understand the behavior of the graph near $x = 0$ we evaluate the function for several x-values close to zero.

x	$f(x)$
-0.5	-2
-0.1	-10
-0.01	-100
-0.001	-1000
0.5	2
0.1	10
0.01	100
0.001	1000

As we choose x-values closer and closer to zero but still greater than zero, the function values increase. As we choose x-values less than zero, but approaching zero, the function values decrease. We say that f has a **vertical asymptote** at $x = 0$; that is, the graph approaches, but never touches, the line $x = 0$.

Now consider the rest of the graph. For positive x, as the value of x increases, the value of $f(x) = 1/x$ decreases (but remains positive). (You should verify this by evaluating the function for some large values of x.) Thus, the value of $f(x)$ gets closer to zero, and the graph approaches the x-axis. For negative x-values, as x decreases, the function values increase toward zero and the graph approaches the x-axis from below. Since $1/x$ never *equals* zero for any x-value, the graph never actually touches the x-axis. We say that the x-axis (i.e., the line $y = 0$) is a **horizontal asymptote** for the graph. The range of f is the set of all real numbers except zero.

The graph of $g(x) = 1/x^2$ is similar in behavior to the graph of $f(x) = 1/x$, except that the function values of $g(x)$ are always positive. Thus, the range of g is the set $(0, \infty)$.

Functions Defined Piecewise

A function for which different equations are used to determine the function values on different portions of the domain is said to be defined "piecewise." To graph a function defined piecewise, we consider each piece of the domain separately.

EXAMPLE 2 Graph the function defined by

$$f(x) = \begin{cases} x + 1 & \text{if } x \leq 1 \\ 3 & \text{if } x > 1 \end{cases}.$$

Solution Think of the plane as divided into two regions by the vertical line $x = 1$. In the left-hand region $(x \leq 1)$, graph the line $y = x + 1$. Notice that the value $x = 1$ is included in the first region, so $f(1) = 2$, and the point $(1, 2)$ is included on the graph. We indicate this with a solid dot at the point $(1, 2)$. In the right-hand region $(x > 1)$, graph the line $y = 3$. The value $x = 1$ is *not* included in the second region, so the point $(1, 3)$ is *not* part of the graph. We indicate this with an open circle at the point $(1, 3)$.

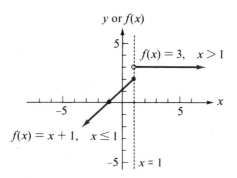

Absolute-Value Functions

The absolute-value function $f(x) = |x|$ is an example of a function that is defined piecewise. From the definition of the absolute value of x we have

$$f(x) = |x| = \begin{cases} x & \text{if } x \geq 0 \\ -x & \text{if } x < 0 \end{cases}.$$

To graph the absolute-value function, think of dividing the plane into two regions, the first region including all points with x-coordinates less than zero (to the left of the y-axis) and the second region including all points with x-coordinates greater than or equal to zero (to the right of and including the y-axis). In the first region we graph a portion of the line $y = -x$, and in the second region we graph a portion of the line $y = x$. (See Figure 6.9.)

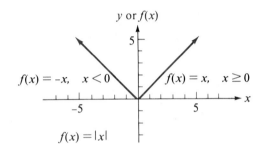

FIGURE 6.9

Notice that the two "pieces" of the graph of the absolute-value function meet at the origin, but the two pieces of the graph in Example 2 did not meet.

EXERCISE 6.3

A

■ *Graph the following linear functions. See Example 1.*

1. $f(x) = 3x - 4$
2. $g(x) = -4x + 6$
3. $h(t) = -2t + 5$
4. $r(t) = 6t - 3$
5. $G(s) = -\dfrac{5}{3}s + 50$
6. $F(s) = -\dfrac{3}{4}s + 60$
7. $R(u) = 2.6u - 120$
8. $H(v) = 1.8v - 240$

■ *Construct tables of values and verify the following basic graphs introduced in this section. State the domain and range of each function. See Figures 6.6–6.8.*

9. $y = x^2$
10. $y = x^3$
11. $y = \sqrt{x}$

12. $y = \sqrt[3]{x}$
13. $y = \dfrac{1}{x}$
14. $y = \dfrac{1}{x^2}$

■ *Graph the following functions. See Example 2.*

15. $f(x) = \begin{cases} -2 & \text{if } x \leq 1 \\ x - 3 & \text{if } x > 1 \end{cases}$

16. $h(x) = \begin{cases} -x + 2 & \text{if } x \leq -1 \\ 3 & \text{if } x > -1 \end{cases}$

17. $G(t) = \begin{cases} 3t + 9 & \text{if } t < -2 \\ -3 - \dfrac{1}{2}t & \text{if } t \geq -2 \end{cases}$

18. $F(s) = \begin{cases} \dfrac{1}{3}s + 3 & \text{if } s < 3 \\ 2s - 5 & \text{if } s \geq 3 \end{cases}$

19. $H(t) = \begin{cases} t^2 & \text{if } t \leq 1 \\ \dfrac{1}{2}t + \dfrac{1}{2} & \text{if } t > 1 \end{cases}$

20. $g(t) = \begin{cases} \dfrac{3}{2}t + 7 & \text{if } t \leq -2 \\ t^2 & \text{if } t > -2 \end{cases}$

21. $k(x) = \begin{cases} |x| & \text{if } x \leq 2 \\ \sqrt{x} & \text{if } x > 2 \end{cases}$

22. $S(x) = \begin{cases} \dfrac{1}{x} & \text{if } x < 1 \\ |x| & \text{if } x \geq 1 \end{cases}$

23. $D(x) = \begin{cases} |x| & \text{if } x < -1 \\ x^3 & \text{if } x \geq -1 \end{cases}$

24. $m(x) = \begin{cases} x^2 & \text{if } x \leq \dfrac{1}{2} \\ |x| & \text{if } x > \dfrac{1}{2} \end{cases}$

25. $P(t) = \begin{cases} t^3 & \text{if } t \leq 1 \\ \dfrac{1}{t^2} & \text{if } t > 1 \end{cases}$

26. $Q(t) = \begin{cases} t^2 & \text{if } t \leq -1 \\ \sqrt[3]{t} & \text{if } t > -1 \end{cases}$

B

■ *Evaluate the functions for a sufficient number of points in the indicated intervals in order to answer the questions.*

27. a. On the same set of axes, graph $f(x) = x^2$ and $g(x) = x^3$ for $0 \leq x \leq 1$. On the interval $(0, 1)$, which is greater, $f(x)$ or $g(x)$?
    b. On the same set of axes, graph $f(x) = x^2$ and $g(x) = x^3$ for $x \geq 1$. On the interval $(1, \infty)$, which is greater, $f(x)$ or $g(x)$?
    c. On the same set of axes, graph $f(x) = x^2$ and $g(x) = x^3$ for the interval $-\infty < x < \infty$.

28. a. On the same set of axes, graph $f(x) = \sqrt{x}$ and $g(x) = \sqrt[3]{x}$ for the interval $0 \leq x \leq 1$. On the interval $[0, 1)$, which is greater, $f(x)$ or $g(x)$?
    b. On the same set of axes, graph $f(x) = \sqrt{x}$ and $g(x) = \sqrt[3]{x}$ for $x \geq 1$. On the interval $(1, \infty]$, which is greater, $f(x)$ or $g(x)$?
    c. On the same set of axes, graph $f(x) = \sqrt{x}$ and $g(x) = \sqrt[3]{x}$ for the interval $-\infty < x < \infty$.

29. a. On the same set of axes, graph $f(x) = 1/x$ and $g(x) = 1/x^2$ for the interval $0 < x \leq 1$. On the interval $(0, 1)$, which is greater, $f(x)$ or $g(x)$?

b. On the same set of axes, graph $f(x) = 1/x$ and $g(x) = 1/x^2$ for $x \geq 1$. On the interval $(1, \infty)$, which is greater, $f(x)$ or $g(x)$?

c. On the same set of axes, graph $f(x) = 1/x$ and $g(x) = 1/x^2$ for the interval $-\infty < x < \infty$, $x \neq 0$.

30. a. On the same set of axes, graph $f(x) = x^2$, $g(x) = \sqrt{x}$, and $h(x) = 1/x^2$.

b. On the same set of axes, graph $f(x) = x^3$, $g(x) = \sqrt[3]{x}$, and $h(x) = 1/x$.

6.4

FUNCTIONS AS MATHEMATICAL MODELS

In applications of mathematics, models involving one or more functions are often used to answer questions or make predictions. In this section we consider several such models.

EXAMPLE 1 A company that produces computer chips finds that after a start-up cost of $10,000, each new chip costs $200 to manufacture.

a. Express the cost C of producing x new chips as a function of x.

b. If the company can sell the chips for $250 each, express the revenue R from the sale of x chips as a function of x.

c. Express the company's profit P from producing and selling x chips as a function of x.

Solutions a. The cost C is the sum of the start-up cost and the cost of manufacturing x chips. Since the cost of manufacturing x chips is

$$\$200 \quad \cdot \quad x$$
(cost per chip) · (number of chips)

and the start-up cost is $10,000, we have

$$C(x) = 10{,}000 + 200x.$$

Since the company cannot produce a negative number of chips, the domain of C is the interval $[0, \infty)$.

b. The total revenue R is given by the product of the selling price per chip times the number of chips sold. Thus,

$$R(x) = 250x.$$

The domain of R is the interval $[0, \infty)$. The graphs of the functions C and R are shown in Figure (a) below.

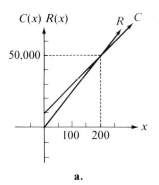

a.

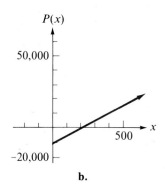

b.

c. In general,

$$\text{profit} = \text{revenue} - \text{cost},$$

so

$$\begin{aligned} P(x) &= R(x) - C(x) \\ &= 250x - (10{,}000 + 200x) \\ &= 50x - 10{,}000. \end{aligned}$$

The graph of $P(x)$ is shown in Figure (b) above.

In Example 1 the three functions $C(x)$, $R(x)$, and $P(x)$, together with their graphs, constitute a model for the company's production and marketing efforts. When $x > 200$, $R(x)$ is greater than $C(x)$, and the company earns a profit. When $x < 200$, $R(x)$ is less than $C(x)$, so the company experiences a loss. The value $x = 200$, where $R(x) = C(x)$, is called the **break-even point**.

EXAMPLE 2 A beekeeper has beehives distributed over 60 square miles of pastureland. When she places four hives per square mile, each hive produces about 32 pints of honey per year. For each additional hive per square mile, honey production drops by 4 pints per hive. Express the number of pints P of honey produced on each square mile as a function of the number x of additional hives.

Solution If x additional hives are set up, then the number of hives per square mile is $4 + x$, and each hive will produce $32 - 4x$ pints of honey. The amount of honey produced on each square mile is the product of the number of hives per square mile and the number of pints produced by each hive. Thus,

$$P(x) = (4 + x)(32 - 4x)$$
$$= 128 + 16x - 4x^2.$$

The graph of the function P is shown in Figure 6.10. Note that the domain of P is the interval $[0, 8]$, since x-values greater than 8 result in negative values for $P(x)$.

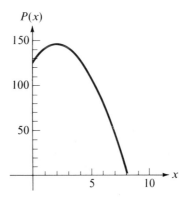

FIGURE 6.10

EXAMPLE 3 A wire 12 feet long is to be cut into two pieces of unequal length, and each piece is to be bent into a square. Express the total area A enclosed by both pieces as a function of the length x of the shorter piece.

Solution Draw and label a figure as shown:

length of short piece: x;
length of long piece: $12 - x$;
smaller area: A_1;
larger area: A_2;
total area enclosed: $A_1 + A_2$.

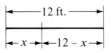
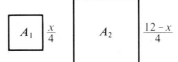

Since the perimeter of the smaller square is x, the side length of this square is $x/4$, as shown in the figure. Similarly, the side length of the larger square is $(12 - x)/4$.

Since the area of a square is the square of the length of its side,

$$A_1 = \left(\frac{x}{4}\right)^2 \quad \text{and} \quad A_2 = \left(\frac{12 - x}{4}\right)^2,$$

and the total area is given by

$$A(x) = \left(\frac{x}{4}\right)^2 + \left(\frac{12 - x}{4}\right)^2.$$

We can now find the total area enclosed for any specified length x simply by evaluating the function at the given x-value. Since x represents a length, it cannot be negative. If $x = 0$, then the wire is actually left in one piece, so we also exclude the value $x = 0$. Furthermore, since x represents the shorter of the two pieces of wire, it is less than one-half the total length of the wire. Thus, the domain of the function $A(x)$ is the interval $(0, 6)$.

Direct Variation

There are two types of functional relationships, widely used in the sciences, to which custom has assigned special names. First, the variable y is said to **vary directly** with (or be **directly proportional** to) the nth power of the variable x if

$$y = kx^n,$$

where k is a positive constant and n is a natural number.

EXAMPLE 4 a. The circumference of a circle varies directly with the radius, since

$$C = 2\pi r.$$

b. The area of a circle varies directly with the square of the radius, since

$$A = \pi r^2.$$

c. The amount of interest earned in 1 year on an account paying 7% simple interest varies directly with the principal invested, since

$$I = 0.07P.$$

In each of the examples above, as the independent variable increases through positive values the dependent variable increases also. Thus, a direct variation is an example of an **increasing** function. This is clear when we consider the graphs of some typical direct variations in Figure 6.11.

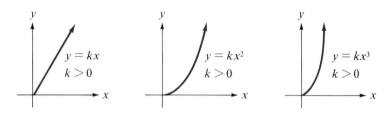

FIGURE 6.11

The value of k determines how rapidly the graph increases; in particular, k is the slope of the linear variation $y = kx$.

Notice that the graph of a direct variation always passes through the origin, so when the independent variable is zero the dependent variable is zero also. Thus, the functions $y = 3x + 2$ and $y = 0.4x^2 - 2.3$ are *not* considered direct variations, although they are increasing functions for positive x.

The constant k in the equation $y = kx^n$ that defines a direct variation is called the **constant of variation.** If we know one pair of associated values for the variables, we can find the constant of variation. We can then use the constant to express one of the variables as a function of the other.

EXAMPLE 5 The speed v at which a particle falls in a certain medium varies directly with the time t it falls. A particle is falling at a speed of 20 feet per second 4 seconds after being dropped. Express v as a function of t.

Solution Since v varies directly with t, there is a positive constant k for which

$$v = kt.$$

Substitute $v = 20$ when $t = 4$ to get

$$20 = k(4),$$

or

$$k = 5.$$

Thus, the functional relationship between v and t is given by

$$v = 5t.$$

Once we have constructed a model, values for one of the variables can readily be obtained from known values of the other variable. In Example 5 above, we can obtain the velocity of the particle at any time t by substituting for t in the equation $v = 5t$. Thus, for t equal to 6, 8, and 10 seconds, respectively,

$$v_1 = 5(6) \qquad v_2 = 5(8) \qquad v_3 = 5(10)$$
$$= 30; \qquad = 40; \qquad = 50.$$

The graph of the function $v = 5t$ is shown in Figure 6.12.

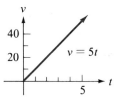

FIGURE 6.12

Inverse Variation

A second type of variation is defined by the equation

$$y = \frac{k}{x^n},$$

where k is a positive constant, n is a natural number, and $x \neq 0$. Here y is said to **vary inversely** with (or be **inversely proportional** to) the nth power of x.

EXAMPLE 6 a. The time T required to travel a given distance D varies inversely with the speed R, since

$$D = RT \quad (D \text{ constant}),$$

or

$$T = \frac{D}{R}.$$

The constant of variation is D.

b. For an ideal gas at constant temperature T, the volume V and pressure P vary inversely, since

$$V = \frac{kT}{P} \quad (k, T \text{ constants}).$$

The constant of variation is kT.

c. For a right circular cylinder with constant volume V, the height h varies inversely with the square of the radius r, since

$$V = \pi r^2 h \quad (V \text{ constant}),$$

or

$$h = \frac{V}{\pi r^2}.$$

The constant of variation is V/π.

In each of the examples above, as the independent variable increases through positive values the dependent variable *decreases*. An inverse variation is an example of a **decreasing** function. The graphs of some typical inverse variations are shown in Figure 6.13.

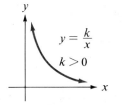
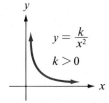

FIGURE 6.13

EXAMPLE 7 The weight of an object varies inversely with the square of its distance from the center of the earth. How much would a 120-pound astronaut weigh on a spaceship 500 miles above the earth? Consider the radius of the earth to be 3963 miles.

Solution If w represents the weight of the astronaut on the earth's surface and d is her distance from the center of the earth,

$$w = \frac{k}{d^2}.$$

Since $w = 120$ when $d = 3963$,

$$120 = \frac{k}{(3963)^2},$$

so $k = 1,884,644,280$ and

$$w = \frac{1,884,644,280}{d^2}.$$

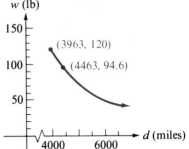

To find the astronaut's weight 500 miles above the earth, we substitute $d = 3963 + 500 = 4463$ into the function to find

$$w = \frac{1,884,644,280}{(4463)^2}$$

$$= 94.6.$$

The astronaut would weigh 94.6 pounds. The graph of the function is shown in the figure.

EXERCISE 6.4

A

■ **a.** *Construct a mathematical model.* **b.** *Solve.*
See Examples 1–3. See the Review of Elementary Topics for appropriate formulas from geometry.

1. The set-up cost for printing a book is $5000, after which it costs $10 for each book printed. Express the printing costs C for a book as a function of the number n of books printed. Find the printing costs for 640 books.

2. The set-up cost to produce a certain kind of computer chip is $12,000, after which it costs $24 to produce each chip. Express the cost C of production as a function of the number n of chips produced. Find the cost of producing 8000 chips.

3. The cost of building a brick wall is $15 per running foot, and the cost of fencing material is $7.50 per running foot. A rancher is building a square corral with three sides brick wall and one side fencing. Express the cost C as a function of the side length s of the corral. Find the cost of a square corral with sides 60 feet in length.

4. The cost of wire mesh fencing is $5.50 per linear foot, and a gate costs $60. Express as a function of its width w the cost C of fencing a rectangular field with two gates if the length of the field is twice its width. Find the cost of fencing a rectangular field with a width of 140 feet.

5. Fran wants to fence a rectangular area of 3200 square feet to grow vegetables for her family of three. Express the perimeter P of the garden as a function of its width w. Find the perimeter of the garden if its width is 40 feet.

6. Denise makes a frame for a circular stained glass window from a strip of aluminum. Express the area A of the window as a function of the length l of the aluminum. Find the area of a window framed by a 72-inch length of aluminum.

7. Express the radius r of a cylindrical water tank as a function of its circumference C. What is the radius of a cylindrical water tank with a circumference of 47 meters?

8. A round bucket just fits inside the square shaft of a well. Express the radius r of the bucket as a function of the side s of the well. How large a bucket will fit in a well measuring 30 inches on a side?

9. The radius of a cylindrical can should be one-half its height. Express the volume V of such a can as a function of its height. What is the volume of a can with a height of 10 centimeters?

10. The Twisty-Freez machine dispenses soft ice cream in a cone-shaped peak with a height three times the radius of its base. Express the volume V of Twisty-Freez that comes in a round dish with diameter d. How much Twisty-Freez comes in a 3-inch dish?

11. Phoebe invests $3000 at 8% interest compounded annually. Express the amount A accumulated in Phoebe's account as a function of the time t in years. How much will Phoebe have after 3 years?

12. Wiley wants to invest $8500 for 5 years in an account that compounds interest annually. Express the amount A accumulated at the end of that time as a function of the interest rate r. How much will Wiley have if he can secure an interest rate of 11%?

13. Ruth received $20,000 in an insurance settlement and plans to invest the money in two accounts: a long-term T-bill paying 12.5% simple interest annually and a savings account that pays 6% simple interest annually. Express her annual income I from the two investments as a function of the amount x she puts into the T-bill. What will her income be if she puts $10,000 into the T-bill?

14. Peyman wants to borrow $25,000 to open a restaurant. He plans to obtain part of the money as a 5-year note from a lending institution at 15% simple annual interest. His father will loan him the rest at 9% simple annual interest. Express the annual interest I on the two loans as a function of the amount x that his father loans him. If Peyman's father loans him $10,000, what will his annual interest payment be?

15. The guy wires used to secure a shortwave radio antenna are attached to a point two-thirds of the way up the antenna and anchored at a distance from the antenna's base equal to one-third of its height. Express the length L of the wire as a function of the height h of the antenna. How long a wire is needed to support an antenna 20 feet high?

16. Cal is building a new garage that is 20 feet wide, and the "pitch," or slope, of the roof is ⅔. Express the area A of the roof as a function of the length l of the garage. If the garage is 25 feet long, what is the area of its roof?

17. A UFO is hovering directly over your head at an altitude of 700 feet. It begins descending at a rate of 10 feet per second. At the same time, you start to run at 25 feet per second. Express the distance D between you and the UFO as a function of time t in seconds after its descent begins. How far are you from the UFO 1 minute later?

18. An owl at an altitude of 100 feet spots a mouse directly below it and stoops at a rate of 40 feet per second. At the same time, the mouse begins to run at 15 feet per second. Express the distance D between the owl and the mouse as a function of time t in seconds after the owl stoops. How far is the owl from the mouse after 2 seconds?

■ *For Problems 19–30, do the following.*
   a. *Find a function relating the variables.*
   b. *Solve for the specified value.*
   c. *Make a rough sketch of the graph of the function. See Examples 4–7.*

19. y varies directly with the square of x, and $y = 24$ when $x = 6$. Find y when $x = 2$.

20. y varies directly with the cube of x, and $y = 120$ when $x = 2$. Find y when $x = 20$.

21. y varies inversely with x, and $y = 56$ when $x = 200$. Find y when $x = 8$.

22. y varies inversely with the square of x, and $y = 3.6$ when $x = 25$. Find y when $x = 0.4$.

23. If we neglect the effect of air resistance, the distance s that an object falls varies directly with the square of the time t it falls. If a pebble falls 400 feet from the southern rim of the Grand Canyon in 5 seconds, how far will it fall in 10 seconds?

24. The length L of a pendulum varies directly with the square of its period T, the time required for the pendulum to make one complete swing back and forth. The pendulum on a grandfather clock is 3¼ feet long and has a period of 2 seconds. How long is the Foucault pendulum in the Panthéon in Paris, with a period of 17 seconds?

25. At constant temperature, the pressure P of a sample of gas varies inversely with its volume V. A cylinder contains 500 cubic centimeters of gas at a pressure of 15 kilograms per square centimeter, and the gas is compressed by a piston. What is the pressure of the gas when the volume inside the cylinder is reduced to 62.5 cubic centimeters?

26. The amount of force F one must exert on a lever to raise a heavy object varies inversely with the length L of the lever. If it takes 100 pounds of force to lift a large stone planter using a 3-foot lever, how much force is necessary to lift the planter using a 4-foot lever?

27. Water pressure p varies directly with the depth d beneath the surface of the water. The pressure at the bottom of a 12-foot swimming pool is 748.8 pounds per square foot. What is the water pressure on a skin diver at a depth of 100 feet?

28. Hubble's law says that distant galaxies are receding from us at a rate V that varies directly with their distance D. (The speeds of these galaxies are measured using a phenomenon called redshifting.) If a galaxy in Ursa Major is 980,000,000 light-years away and is receding at 15,000 kilometers per second, how far away is a galaxy in the constellation Hydra, which is receding at 61,000 kilometers per second?

29. The frequency f of a guitar string at a given tension varies inversely with its length. The fifth string is 65 centimeters long and is tuned to A (with a frequency of 220 vibrations per second). How far from the bridge should the fret for C (256 vibrations per second) be placed?

30. The current I that flows through an electrical wire varies inversely with the resistance R of the wire. If an iron with a resistance of 12 ohms draws 10 amps of current, how much current is drawn by a toaster with a resistance of 9.6 ohms?

B

■ *Construct a mathematical model.*

31. The cost of wire fencing is $7.50 per linear foot. A rancher wants to enclose a rectangular area of 100 square feet with this fencing. Express the cost C of the fence as a function of the width w of the rectangle.

32. The cost of split rail fencing is $10 per linear foot. A rancher wants to enclose a rectangular area of 125 square feet adjacent to his barn. The barn will act as one length of the enclosure, and split rail will be used for the other three sides. Express the cost C of the fence as a function of the width w of the enclosure.

33. A rope 200 feet long is cut into two pieces. The shorter piece is shaped to enclose a square, and the longer piece is shaped to enclose a rectangle whose length is four times its width. Express the total area A enclosed by the two pieces as a function of the length x of the shorter piece.

34. A rope 100 feet long is cut into two pieces. The shorter piece is shaped to enclose a square, and the longer piece is shaped to enclose a circle. Express the total area A enclosed by the two pieces as a function of the length x of the shorter piece.

35. A circular disk of radius 10 inches has a uniform border of width x, which is highly polished, and the area inside the border is painted. The polishing process costs 15 cents per square inch, and paint costs 8 cents per square inch. Express the cost C of finishing the disk as a function of the width of the border.

36. A 50-foot by 50-foot space includes a cement sidewalk border of uniform width x, and the interior is planted with roses. The cost of laying cement is $10 per square foot, and the cost of planting roses is $25 per square foot. Express the cost C of the garden as a function of the width x of the sidewalk.

37. A motorist passes a service station on the highway and 4 miles later turns right onto a county road. After he has traveled for 3 miles, the car's engine throws a rod, and the motorist decides to walk back to the service station. He can walk 4 miles per hour along the road and estimates that he can make 3 miles per hour through the fields. If he heads for a point on the highway x miles before the junction of the county road, express the time t that it will take him to reach the service station as a function of x. (See figure on page 304.)

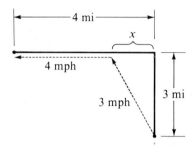

38. The telephone company is laying a cable from an island 12 miles offshore to a relay station 18 miles from the point on the shore directly opposite the island, as shown in the figure. It costs $50 per foot to lay cable underwater and $30 per foot to lay cable underground. Express the cost C of laying the cable as a function of the distance x from the relay station to the point where the cable leaves the water.

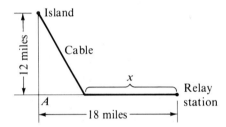

■ *Solve.*

39. The intensity of illumination I from a light source varies inversely with the square of the distance. If you double your distance from a reading lamp, what happens to the illumination?

40. The resistance R of a wire varies inversely with the square of its diameter d. If you replace an old wire with a new one whose diameter is two-thirds of the old one, what happens to the resistance?

41. The wind resistance R experienced by a vehicle on the freeway varies directly with the square of its speed v. If you decrease your speed by 10%, what happens to the wind resistance?

42. The power P generated by a battery varies directly with its electrical potential V. If a battery loses 30% of its potential, what happens to the power it generates?

43. a. If x varies directly with y and y varies directly with z, does x vary directly with z?
    b. If x varies inversely with y and y varies inversely with z, does x vary inversely with z?

44. a. If $xy = k$ (k a constant), how does x vary with y?
    b. If $x/y = k$ (k a constant), how does x vary with y?

6.5

QUADRATIC FUNCTIONS

In Section 6.3 we considered the graphs of some standard functions, including a simple quadratic function, $y = x^2$. In this section we study techniques for graphing more general quadratic functions.

A function of the form

$$f(x) = ax^2 + bx + c \quad (a \neq 0) \tag{1}$$

is called a **quadratic function**. The graphs of quadratic functions are parabolas and are similar in shape to the graph of $y = x^2$. The constants a, b, and c in Equation (1) determine the relative size and position of the graph. Some examples of parabolas are shown in Figure 6.14.

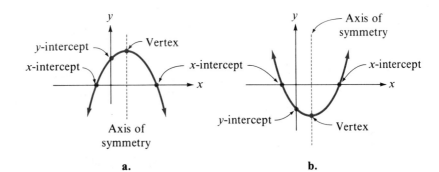

FIGURE 6.14

The highest point on the graph in Figure 6.14a is called the **vertex** of the parabola. It corresponds to the maximum value of the function. If the parabola opens upward, as in Figure 6.14b, then the vertex is the lowest point on the graph and it corresponds to the minimum function value. A parabola is symmetric about a vertical line through its vertex called the **axis of symmetry**. The **y-intercept** is the point where the parabola intersects the y-axis. We shall see that a parabola may intersect the x-axis in zero, one, or two points called the **x-intercepts**. If there are two x-intercepts, they are equidistant from the axis of symmetry.

The intercepts and the vertex of a parabola can be found by analyzing the coefficients a, b, and c. Once these points are found, it is fairly easy to sketch an accurate graph. We begin by considering some special cases.

The Graphs of $y = ax^2$ and $y = x^2 + c$

First, consider the graphs of $f(x) = 2x^2$ and $g(x) = -3x^2$ shown in Figure 6.15.

x	f(x)	g(x)
-2	8	-12
-1	2	-3
0	0	0
1	2	-3
2	8	-12

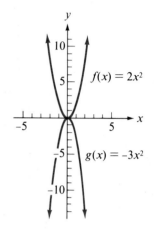

FIGURE 6.15

In general, the graph of $y = ax^2$ opens upward if $a > 0$ and opens downward if $a < 0$. The magnitude of a determines how "wide" or "narrow" the parabola is. For functions of the form $y = ax^2$, the vertex, the x-intercept, and the y-intercept all coincide at the origin.

Now consider the graphs of $f(x) = x^2 + 4$ and $g(x) = x^2 - 4$ shown in Figure 6.16.

x	f(x)	g(x)
-2	8	0
-1	5	-3
0	4	-4
1	5	-3
2	8	0

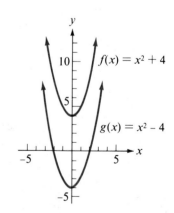

FIGURE 6.16

The graph of $f(x) = x^2 + 4$ is shifted *upward* four units as compared with the graph of $y = x^2$, and the graph of $g(x) = x^2 - 4$ is shifted *downward* four units. Thus, the vertex of the graph of f is the point $(0, 4)$ and the vertex of the graph of g is the point $(0, -4)$. The x-intercepts of the graph of g can be found by setting $g(x)$ equal to zero and solving for x:

$$0 = x^2 - 4$$
$$= (x - 2)(x + 2),$$

so the x-intercepts are 2 and -2. The graph of f has no x-intercepts, since the equation

$$0 = x^2 + 4$$

(obtained by setting $g(x)$ equal to zero) has no real solutions.

EXAMPLE 1 Graph the following functions.

    a. $g(x) = x^2 - 3$ b. $h(x) = -2x^2$

Solutions   a. The graph of $g(x) = x^2 - 3$ is shifted downward by three units. The vertex is the point $(0, -3)$, and the x-intercepts are the solutions of the equation

$$0 = x^2 - 3,$$

or $\sqrt{3}$ and $-\sqrt{3}$.

  b. The graph of $h(x) = -2x^2$ opens downward and is narrower than the graph of $y = x^2$. Its vertex is the point $(0, 0)$.

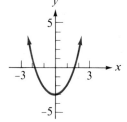

The Graph of $y = ax^2 + bx$

We next consider quadratic functions of the form

$$y = ax^2 + bx.$$

We can find the x-intercepts of the graph by setting y equal to zero and solving for x:

$$0 = ax^2 + bx$$
$$0 = x(ax + b).$$

Thus,

$x = 0$ or $ax + b = 0$

$x = 0$ $\qquad x = -\dfrac{b}{a}.$

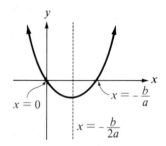

FIGURE 6.17

The x-intercepts are 0 and $-b/a$. The axis of symmetry is located halfway between the intercepts, so, taking their average, we have

$$x = \dfrac{0 + \left(\dfrac{-b}{a}\right)}{2} = \dfrac{-b}{2a}.$$

Thus, the axis of symmetry is the vertical line $x = -b/2a$. (See Figure 6.17.) Since the vertex of the parabola lies on the axis of symmetry, its x-coordinate must be $-b/2a$. We find the y-coordinate of the vertex by evaluating the function at $x = -b/2a$.

EXAMPLE 2 Graph the following functions.

a. $f(x) = 2x^2 + 8x$ \qquad b. $g(x) = -x^2 + 5x$

Solutions a. Since $a = 2$ and $2 > 0$, the parabola opens upward. Determine the x-intercepts by setting $f(x)$ equal to zero:

$$0 = 2x^2 + 8x$$
$$= 2x(x + 4).$$

Thus,

$2x = 0$ or $x + 4 = 0$

$x = 0$ $\qquad\qquad x = -4.$

The x-intercepts are 0 and −4. The vertex has x-coordinate

$$y = \frac{-b}{2a} = \frac{-8}{2(2)}$$
$$= \frac{-8}{4} = -2.$$

The y-coordinate of the vertex is

$$y = f(-2) = 2(-2)^2 + 8(-2)$$
$$= 8 - 16 = -8.$$

Thus, the vertex is the point $(-2, -8)$. The graph of $f(x) = 2x^2 + 8x$ is shown in the figure.

b. Since $a = -1$ and $-1 < 0$, the parabola opens downward. Determine the x-intercepts by setting $g(x)$ equal to zero:

$$0 = -x^2 + 5x$$
$$= -x(x - 5).$$

Thus,

$$-x = 0 \quad \text{or} \quad x - 5 = 0$$
$$x = 0 \qquad\qquad x = 5.$$

The x-intercepts are 0 and 5. The vertex has x-coordinate

$$x = \frac{-b}{2a} = \frac{-5}{2(-1)}$$
$$= \frac{-5}{-2} = \frac{5}{2}.$$

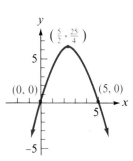

The y-coordinate of the vertex is

$$y = g\left(\frac{5}{2}\right) = -\left(\frac{5}{2}\right)^2 + 5\left(\frac{5}{2}\right)$$
$$= -\frac{25}{4} + \frac{25}{2} = \frac{25}{4}.$$

Thus, the vertex is the point $(5/2, 25/4)$. The graph of $g(x) = -x^2 + 5x$ is shown in the figure.

The General Case:
$y = ax^2 + bx + c$

As an example of the general case $y = ax^2 + bx + c$, consider the function

$$y = 2x^2 + 8x + 6. \qquad (2)$$

The graph of $y = 2x^2 + 8x + 6$ is shifted six units upward from the graph of $y = 2x^2 + 8x$, which was considered in Example 2a. (See Figure 6.18.) Notice that the x-coordinate of the vertex will not be affected by an upward shift, so the formula $x = -b/2a$ for the x-coordinate of the vertex still holds. The y-coordinate of the vertex can still be found by evaluating the function at $x = -b/2a$. We have

$$x = \frac{-b}{2a} = \frac{-8}{2(2)} = -2$$

and

$$y = 2(-2)^2 + 8(-2) + 6$$
$$= 8 - 16 + 6 = -2,$$

so the vertex is the point $(-2, -2)$. (Notice that this point is shifted six units upward from the vertex of $y = 2x^2 + 8x$.)

We find the x-intercepts of the graph by setting y equal to zero:

$$0 = 2x^2 + 8x + 6$$
$$= 2(x + 1)(x + 3).$$

Thus,

$$x + 1 = 0 \quad \text{or} \quad x + 3 = 0$$
$$x = -1 \qquad\qquad x = -3.$$

The x-intercepts are -1 and -3.

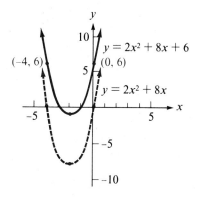

FIGURE 6.18

The y-intercept of the graph is found by setting x equal to zero:

$$y = 2(0)^2 + 8(0) + 6$$
$$= 6.$$

Note that the y-intercept, 6, is just the constant term in Equation (2).

Because of the symmetry of the parabola about its axis, we can locate another point symmetric to the y-intercept. Since the y-intercept lies two units to the right of the axis of symmetry, there will be another point with y-coordinate 6 two units to the left of that axis, with coordinates $(-4, 6)$.

EXAMPLE 3 a. Locate the vertex of $y = -x^2 + 5x - 4$.
  b. Graph the equation.

Solutions a. The x-coordinate of the vertex is

$$x = \frac{-b}{2a} = \frac{-5}{2(-1)} = \frac{5}{2},$$

and the y-coordinate is

$$y = -\left(\frac{5}{2}\right)^2 + 5\left(\frac{5}{2}\right) - 4$$

$$= \frac{-25}{4} + \frac{25}{2} - \frac{16}{4}$$

$$= \frac{-25 + 50 - 16}{4} = \frac{9}{4}.$$

Thus, the vertex is the point $(5/2, 9/4)$.

b. To find the x-intercepts of the graph, set y equal to zero:

$$0 = -x^2 + 5x - 4$$
$$= -(x^2 - 5x + 4)$$
$$= -(x - 4)(x - 1).$$

Thus,

$x - 4 = 0$ or $x - 1 = 0$
$x = 4$ $x = 1$.

The x-intercepts are 4 and 1. Set x equal to zero to find the y-intercept:

$$y = -0^2 + 5(0) - 4 = -4.$$

The y-intercept is -4. Since the y-intercept is $5/2$ units to the left of the axis of symmetry, there will be another point $5/2$ units to the right, at $(5, -4)$.

The graph of $y = -x^2 + 5x - 4$ is shown in the figure.

Number of x-Intercepts

The graph of a quadratic function f may have two, one, or no x-intercepts, according to the number of real solutions of the quadratic equation $f(x) = 0$. Consider the three functions graphed in Figure 6.19. The graph of $f(x) = x^2 - 4x + 3$ has two x-intercepts, since the equation $f(x) = x^2 - 4x + 3 = 0$ has two real solutions. The graph of $g(x) = x^2 - 4x + 4$ has only one x-intercept, since the equation $x^2 - 4x + 4 = 0$ has one repeated real solution. The solutions of the equation $x^2 - 4x + 6 = 0$ are complex and hence do not appear on the graph, so the graph of $h(x) = x^2 - 4x + 6$ has no x-intercepts.

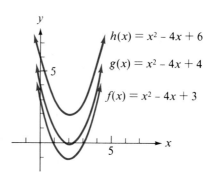

FIGURE 6.19

Graphing Parabolas

Once we have located the vertex of the parabola, the x-intercepts, the y-intercept, and its symmetric point, we can sketch a reasonably accurate graph. If the graph has no x-intercepts, it may be necessary to locate one or two additional points by evaluation in order to sketch the graph. We summarize this procedure as follows.

To Graph $f(x) = ax^2 + bx + c$:

1. Determine whether the parabola opens upward (if $a > 0$) or downward (if $a < 0$).
2. Locate the vertex.
   a. The x-coordinate of the vertex is $x = -b/2a$.
   b. The y-coordinate of the vertex is $y = f(-b/2a)$.
3. Locate the x-intercepts (if any) by setting $f(x)$ equal to zero and solving for x.
4. Locate the y-intercept by evaluating $f(0)$. Locate the point symmetric to the y-intercept with respect to the axis of symmetry.
5. If necessary, locate one or two additional points on the graph by evaluation.

EXAMPLE 4 Graph $f(x) = -2x^2 + x - 1$.

Solution
1. Since $a = -2$ and $-2 < 0$, the parabola opens downward.
2. Compute the coordinates of the vertex:

$$x = \frac{-b}{2a} = \frac{-1}{2(-2)} = \frac{1}{4};$$

$$y = -2\left(\frac{1}{4}\right)^2 + \left(\frac{1}{4}\right) - 1$$

$$= -2\left(\frac{1}{16}\right) + \frac{1}{4} - 1$$

$$= \frac{-1 + 2 - 8}{8} = \frac{-7}{8}.$$

The vertex has coordinates $(\frac{1}{4}, -\frac{7}{8})$.

3. Set y equal to zero to find the x-intercepts:

$$0 = -2x^2 + x - 1;$$

$$x = \frac{-1 \pm \sqrt{(1)^2 - 4(-2)(-1)}}{2(-2)}$$

$$= \frac{-1 \pm \sqrt{-7}}{-4}.$$

The solutions are imaginary, so there are no x-intercepts. (Or, note that there are no x-intercepts because the parabola opens downward from a vertex whose y-coordinate is negative.)

4. Since $f(0) = -1$, the y-intercept is the point $(0, -1)$. The y-intercept lies $\frac{1}{4}$ unit to the left of the axis of symmetry (the line $x = \frac{1}{4}$), so its symmetric point lies $\frac{1}{4}$ unit to the right of the axis, at $(\frac{1}{2}, -1)$.

5. Since the three points located above are not sufficient to provide a good picture of the graph, locate two additional points by evaluation, setting $x = 1$ and $x = -1$:

$$f(1) = -2(1)^2 + (1) - 1 = -2,$$

so $(1, -2)$ lies on the graph;

$$f(-1) = -2(-1)^2 + (-1) - 1 = -4,$$

so $(-1, -4)$ lies on the graph.
Use symmetry to sketch the graph as shown.

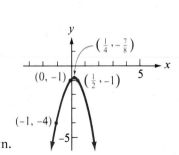

314 CHAPTER 6 ■ FUNCTIONS

EXAMPLE 5 Graph $f(x) = x^2 + 3x + 1$.

Solution
1. Since $a = 1$ and $1 > 0$, the parabola opens upward.
2. Compute the coordinates of the vertex:

$$x = \frac{-b}{2a} = \frac{-3}{2(1)} = \frac{-3}{2};$$

$$y = \left(\frac{-3}{2}\right)^2 + 3\left(\frac{-3}{2}\right) + 1$$

$$= \frac{9}{4} - \frac{9}{2} + 1$$

$$= \frac{9 - 18 + 4}{4} = \frac{-5}{4}.$$

The vertex has coordinates $(-3/2, -5/4)$.

3. Set y equal to zero to find the x-intercepts:

$$0 = x^2 + 3x + 1;$$

$$x = \frac{-3 \pm \sqrt{3^2 - 4(1)(1)}}{2(1)}$$

$$= \frac{-3 \pm \sqrt{5}}{2}.$$

The solutions are irrational, so find decimal approximations for the x-intercepts:

$$\frac{-3 + \sqrt{5}}{2} \approx -0.4;$$

$$\frac{-3 - \sqrt{5}}{2} \approx -2.6.$$

Plot the x-intercepts at $(-2.6, 0)$ and $(-0.4, 0)$.

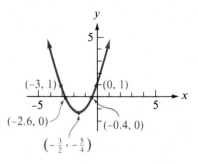

4. Since $f(0) = 1$, the y-intercept is the point $(0, 1)$. The y-intercept lies $3/2$ units to the right of the axis of symmetry (the line $x = -3/2$), so its symmetric point lies $3/2$ units to the left of the axis, at $(-3, 1)$. These points are sufficient to give a reasonable picture of the graph.

Maximum or Minimum Values

Many applied problems involve finding the largest or smallest value for a particular function. For example, the owner of a cannery might want to build a container that holds a certain volume but has the smallest possible surface area and hence the least cost for materials. A theater manager, knowing that attendance is inversely related to ticket price, might want to calculate the ticket price that will maximize his total revenue.

In general, the methods of calculus are needed to solve such problems. However, if the problem can be modeled by a quadratic function, we need only compute the coordinates of the vertex to find the maximum or minimum value of the function.

EXAMPLE 6

a. Find the dimensions of the largest rectangular area that can be enclosed by 36 yards of fence.
b. Sketch the graph of the function used to model the problem.

Solutions

a. The length of the fence is the perimeter of the enclosed rectangle. Which rectangle with perimeter 36 yards has the largest area? First, note that rectangles with the same perimeter can have different areas! Consider the examples below.

Width	Length	Perimeter	Area
2	16	36	32
4	14	36	56
6	12	36	72
8	10	36	80

Let x represent the width of the rectangle so that its length is $18 - x$. The area A of the rectangle is then given by

$$A(x) = x(18 - x)$$
$$= 18x - x^2,$$

a quadratic function of x. Since $a = -1$ and $-1 < 0$, the graph of the function opens downward, and the vertex is the point with the largest value of A. To find the coordinates of the vertex, set

$$x = \frac{-b}{2a} = \frac{-18}{2(-1)} = 9$$

and then evaluate

$$A(9) = 18(9) - (9)^2$$
$$= 162 - 81 = 81.$$

b.

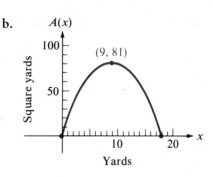

Thus, the largest area that can be enclosed is 81 square yards, by forming a rectangle of width 9 yards and length $18 - 9 = 9$ yards.

EXERCISE 6.5

A

- *Graph. See Example 1.*

1. $y = 3x^2$
2. $y = -2x^2$
3. $y = x^2 - 9$
4. $y = x^2 + 3$
5. $y = \frac{-1}{8}x^2$
6. $y = \frac{2}{3}x^2$
7. $y = x^2 + 6$
8. $y = x^2 - 3$

- *See Example 2.*

9. $y = x^2 - 4x$
10. $y = x^2 + 2x$
11. $y = 3x^2 + 6x$
12. $y = 2x^2 - 6x$
13. $y = -2x^2 + 5x$
14. $y = -3x^2 - 8x$

- *Find the coordinates of the vertex. See Example 3.*

15. $y = 3x^2 - 6x + 4$
16. $y = -2x^2 + 5x - 1$
17. $y = 3 - 5x + x^2$
18. $y = 2 + 3x - x^2$
19. $y = \frac{1}{2}x^2 - \frac{2}{3}x + \frac{1}{3}$
20. $y = \frac{-3}{4}x^2 + \frac{1}{2}x - \frac{1}{4}$
21. $y = 2.3 - 7.2x - 0.8x^2$
22. $y = 5.1 - 0.2x + 4.6x^2$

- *Graph. Specify the coordinates of the vertex and the intercepts. See Examples 3–5.*

23. $y = x^2 - 5x + 4$
24. $y = x^2 + x - 6$
25. $y = -2x^2 + 7x + 4$
26. $y = -3x^2 + 2x + 8$
27. $y = 0.6x^2 + 0.6x - 1.2$
28. $y = 0.5x^2 - 0.25x - 0.75$
29. $y = x^2 + 4x + 7$
30. $y = x^2 - 6x + 10$

31. $y = -2x^2 + x - 3$
32. $y = -3x^2 + x - 2$
33. $y = x^2 + 2x - 1$
34. $y = x^2 - 6x + 2$
35. $y = -2x^2 + 6x - 3$
36. $y = -2x^2 - 8x - 5$

- For Problems 37–44, do the following.
  a. Find the maximum or minimum value.
  b. Sketch the graph. See Example 6.

37. The equation $d = 64t - 16t^2$ gives the distance d in feet above the ground of a toy water rocket t seconds after it is launched. When will the rocket reach its greatest height, and what will that height be?

38. The equation $h = -12 + 32t - 16t^2$ gives the height h in feet of an object tossed into the air from the bottom of a trench 12 feet deep. When will the object reach its greatest height, and what will that height be?

39. What is the area A of the largest rectangle that can be enclosed by 100 inches of twine? (*Hint:* Let x represent the width of the rectangle.)

40. Find the dimensions of the rectangle of greatest area that can be roped off with 80 yards of rope. (*Hint:* Let x represent the width of the rectangle.)

41. A farmer plans to fence a grazing area along a river with 300 yards of fence as shown. What is the largest area he can enclose?

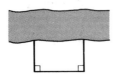

42. A breeder of horses wants to fence two grazing areas along a river with 600 meters of fence as shown. What is the largest area she can enclose?

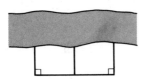

43. An entrepreneur buys a motel with 40 units. The previous owner charged $24 per night for a room and on the average filled 32 rooms per night at that price. The entrepreneur discovers that for every $2 he raises the price another room stands vacant. What price should the entrepreneur charge for a room in order to maximize his revenue? (*Hint:* Let x represent the *number* of $2 price increases.)

44. A travel agent offers a group rate of $2400 per person for a week in London if 16 people sign up for the tour. For each additional person who signs up, the price per person is reduced by $100. How many people must sign up for the tour in order for the travel agent to maximize her revenue?

45. A grocer finds that he will sell x pounds of oranges per day if he charges $80 - 2x$ cents per pound.
  a. At what price will he maximize his revenue?
  b. If his costs are $20x + 120$ cents, what price will maximize his profit?

46. Starbright Skylights sells x skylights per month if it charges $180 - \frac{1}{3}x$ dollars per skylight.
  a. At what price will Starbright maximize its revenue?
  b. If the company's costs are $60x + 500$ dollars, what price will maximize its profit?

B

47. Find the equations of two different parabolas with vertex $(0, 2)$.
48. Find the equations of two different parabolas with vertex $(0, -3)$.
49. Find the equations of two different parabolas with y-intercept $(0, -5)$.
50. Find the equations of two different parabolas with y-intercept $(0, 4)$.
51. Find the equations of two different parabolas with x-intercepts $(2, 0)$ and $(-3, 0)$.
52. Find the equations of two different parabolas with x-intercepts $(-1, 0)$ and $(4, 0)$.
53. Find the equations of two different parabolas with vertex $(2, -3)$.
54. Find the equations of two different parabolas with vertex $(-4, 2)$.

CHAPTER REVIEW

A

[6.1]
■ *Which of the following tables describe functions? Why or why not?*

1.

x	-2	-1	0	1	2	3
y	6	0	1	2	6	8

2.

p	3	-3	2	-2	-2	0
q	2	-1	4	-4	3	0

CHAPTER REVIEW 319

3.

Student	Score on IQ Test	Score on SAT Test
(A)	118	640
(B)	98	450
(C)	110	590
(D)	105	520
(E)	98	490
(F)	122	680

4.

Student	Correct Answers on Math Quiz	Quiz Grade
(A)	13	85
(B)	15	89
(C)	10	79
(D)	12	82
(E)	16	91
(F)	18	95

5. If the function described in Problem 1 is called f, find $f(-2)$ and $f(0)$.
6. If the function described in Problem 4 is called g, find $g(12)$ and $g(15)$.

■ *Evaluate each function for the given values.*

7. $f(x) = 2x^2 - 5x + 1$, $f(2)$ and $f(-2)$
8. $g(x) = \dfrac{x-2}{x+3}$, $g\left(\dfrac{1}{2}\right)$ and $g\left(-\dfrac{3}{4}\right)$
9. $F(t) = \sqrt{1 + 4x^2}$, $F(0)$ and $F(-3)$
10. $H(t) = t^2 + 2t$, $H(2a)$ and $H(a+1)$
11. $f(x) = 2 - 3x$, $f(2) + f(3)$ and $f(2+3)$
12. $f(x) = 2x^2 - 4$, $f(a) + f(b)$ and $f(a+b)$
13. Given the function $P(x) = x^2 - 6x + 5$, compute $P(0)$; find all values of x for which $P(x) = 0$.
14. Given the function $R(x) = \sqrt{4 - x^2}$, compute $R(0)$; find all values of x for which $R(x) = (0)$.

[6.2]

■ *Consider the following graphs for Problems 15 and 16.*

15. a. Find $f(-2)$ and $f(2)$.
    b. For what value(s) of t is $f(t) = 4$?
    c. Find the x- and y-intercepts of the graph.
    d. What is the maximum value of f? For what value(s) of t does f take on its maximum value?

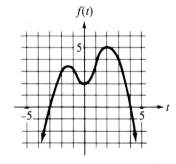

16. a. Find $P(-3)$ and $P(3)$.
    b. For what value(s) of z is $P(z) = 2$?
    c. Find the x- and y-intercepts of the graph.
    d. What is the minimum value of P? For what value(s) of z does P take on its minimum value?

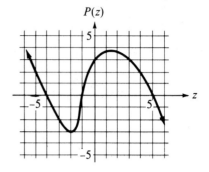

■ Graph each function over the specified domain. Specify the range of the function.

17. $f(t) = -t^2 + 3t; \quad -2 \leq t \leq 4$

18. $g(s) = \sqrt{s - z}; \quad 2 \leq s \leq 6$

19. $F(x) = \dfrac{1}{x + 2}; \quad -4 \leq x \leq 4, \quad x \neq -2$

20. $H(x) = \dfrac{1}{2 - x}; \quad -4 \leq x \leq 4, \quad x \neq 2$

■ Which of the following graphs represent functions?

21.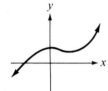

22.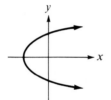

23.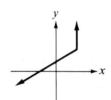

24.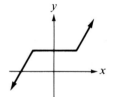

[6.3]
■ Graph each function.

25. $f(t) = -2t + 4$

26. $g(s) = -\dfrac{2}{3}s - 2$

27. $f(x) = \begin{cases} x + 1 & \text{if } x \leq 0 \\ x^2 & \text{if } x > 0 \end{cases}$

28. $g(x) = \begin{cases} x - 1 & \text{if } x \leq 1 \\ x^3 & \text{if } x > 1 \end{cases}$

29. $H(x) = \begin{cases} x^2 & \text{if } x \leq 0 \\ \sqrt{x} & \text{if } x > 0 \end{cases}$

30. $F(x) = \begin{cases} |x| & \text{if } x \leq 0 \\ \dfrac{1}{x} & \text{if } x > 0 \end{cases}$

31. $S(x) = \begin{cases} x^3 & \text{if } x \leq 1 \\ |x| & \text{if } x > 1 \end{cases}$

32. $T(x) = \begin{cases} \dfrac{1}{x^2} & \text{if } x < 0 \\ \sqrt{x} & \text{if } x \geq 0 \end{cases}$

[6.4]

33. A new computer cost the Checks and Balances accounting firm $30,000. It depreciates in value by $5000 every 3 years. Express the value V of the computer as a function of the number of years t since its purchase. How much is the computer worth after 6 years?

34. A large aquarium has a square cross section 2 feet by 2 feet. The glass for the sides of the aquarium costs $8 per square foot, and the bottom of the aquarium costs $5 per square foot. Express the cost C of the aquarium as a function of its length l. How much will a 6-foot-long aquarium cost?

35. Festival Hall holds 3000 people. Season tickets cost $18 per event and single-event seats cost $25. If the hall is sold out for an evening of Vivaldi, express the total revenue R as a function of the number of season tickets x. What is the revenue if 250 season tickets are sold?

36. A recipe for salad dressing calls for 4 ounces (½ cup) of flavored salad oil. Julia mixes peanut oil, which is 35% saturated fat, with safflower oil, which is 10% saturated fat. Express the amount of saturated fat S in the mixture as a function of the amount x of peanut oil used. How many ounces of saturated fat are in the dressing if Julia uses 2 ounces of peanut oil?

37. Express the area A of an equilateral triangle as a function of the length of a side s. Find the area of an equilateral triangle with sides 4 centimeters in length.

38. The hypotenuse c of a right triangle is 12 centimeters long. Express the area A of the triangle as a function of the shortest side x. Find the area of such a right triangle with a short side of 4 centimeters.

39. The distance s a particle falls in a certain medium varies directly with the square of the length of time t it falls. If a particle falls 28 centimeters in 4 seconds, express the distance it will fall as a function of the time it falls. Find the distance a particle will fall in 6 seconds.

40. The volume V of a gas varies directly with the absolute temperature T and inversely with the pressure P of the gas. If $V = 40$ when $T = 300$ and $P = 30$, express the volume of the gas as a function of the absolute temperature and the pressure of the gas. Find the volume when $T = 320$ and $P = 40$.

41. The demand for bottled water is inversely proportional to the price per bottle. If Droplets can sell 600 bottles at $8 each, how many bottles can it sell at $10 each?

42. The intensity of illumination from a light source varies inversely with the distance from the source. If a reading lamp has an intensity of 100 lumens at a distance of 3 feet, what is its intensity 8 feet away?

[6.5]

■ *Graph. Specify the coordinates of the vertex and the intercepts.*

43. $y = \dfrac{1}{2}x^2$

44. $y = x^2 - 4$

45. $y = x^2 - 9x$

46. $y = -2x^2 - 4x$

47. $y = x^2 - x - 12$
48. $y = -2x^2 + x - 4$
49. $y = -x^2 + 2x + 4$
50. $y = x^2 - 3x + 4$

■ *Find the maximum value. Sketch the graph.*

51. A farmer inherits an apple orchard on which 60 trees are planted. Each tree yields 12 bushels of apples. Experimentation has shown that for each additional tree planted per acre, the yield per tree decreases by ½ bushel. How many trees should be planted per acre in order to maximize the total apple harvest?

52. A small company manufactures radios. When it charges $20 for a radio, it sells 500 radios per month. For each dollar the price is increased, 10 fewer radios are sold per month. What should the company charge for a radio in order to maximize its monthly revenue?

B

■ *For each function, calculate* $\dfrac{f(x + h) - f(x)}{h}$.

53. $f(x) = x^2 + 2x$
54. $f(x) = \dfrac{1}{x}$

■ *For Problems 55 and 56, find an equation that describes the function.*

55.

x	g(x)
2	12
3	8
4	6
6	4
8	3
12	2

56.

x	F(x)
−2	8
−1	1
0	0
1	−1
2	−8
3	−27

57. The length L of a pendulum varies directly with the square of its period T (the time required for the pendulum to make one complete swing back and forth). If a certain pendulum is replaced by a new one four-fifths as long as the old one, what happens to the period?

58. The weight of an object varies inversely with the square of its distance from the center of the earth. How far from the center of the earth must an object be in order to weigh one-third of its weight on the surface of the earth? Consider the radius of the earth to be 3963 miles.

7
More About Functions and Graphs

7.1

GRAPHING TECHNIQUES

Translations

A great many useful functions are variations of the standard functions introduced in Section 6.3. Their graphs can be obtained, without constructing tables of values, by modifying the basic graphs in Figures 6.6 through 6.9. We have already seen some examples of this idea in our study of quadratic functions. Recall the graphs of $f(x) = x^2 + 4$ and $g(x) = x^2 - 4$ shown in Figure 6.16. The graphs are reproduced in Figure 7.1.

x	$f(x)$	$g(x)$
-2	8	0
-1	5	-3
0	4	-4
1	5	-3
2	8	0

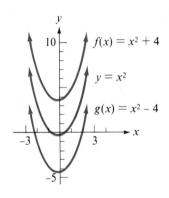

FIGURE 7.1

The graphs of $y = f(x)$ and $y = g(x)$ are said to be **translations** of the graph of $y = x^2$; that is, they are shifted to a different location in the plane but retain the same size and shape as the original graph. In general, we have the following graphing principle.

323

Vertical Translations

Compared with the graph of $y = f(x)$,

1. the graph of $y = f(x) + k$ $(k > 0)$ is shifted *upward* k units;
2. the graph of $y = f(x) - k$ $(k > 0)$ is shifted *downward* k units.

EXAMPLE 1 Graph the following functions.

a. $g(x) = |x| + 3$
b. $h(x) = \dfrac{1}{x} - 2$

Solutions

a. The graph of $g(x) = |x| + 3$ is a translation of the basic graph of $y = |x|$, shifted upward three units.

b. The graph of $h(x) = 1/x - 2$ is a translation of the basic graph of $y = 1/x$, shifted downward two units. We indicate the horizontal asymptote, $y = -2$, by a dotted line.

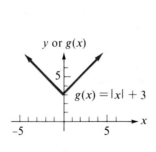
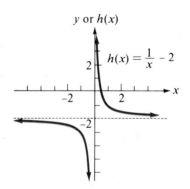

Now consider the graphs of $f(x) = (x + 2)^2$ and $g(x) = (x - 2)^2$ shown in Figure 7.2.

7.1 ■ GRAPHING TECHNIQUES   325

x	f(x)	g(x)
-3	1	25
-2	0	16
-1	1	9
0	4	4
1	9	1
2	16	0
3	25	1

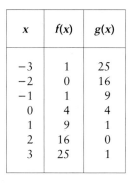

FIGURE 7.2

Compared with the graph of the basic function $y = x^2$, the graph of $f(x) = (x + 2)^2$ is shifted two units to the left and the graph of $g(x) = (x - 2)^2$ is shifted two units to the right. In general, we have the following.

Horizontal Translations

Compared with the graph of $y = f(x)$,

1. the graph of $y = f(x + h)$ $(h > 0)$ is shifted h units to the *left*;
2. the graph of $y = f(x - h)$ $(h > 0)$ is shifted h units to the *right*.

EXAMPLE 2   Graph the following functions.

a. $g(x) = \sqrt{x + 1}$ b. $h(x) = \dfrac{1}{(x - 3)^2}$

Solutions   a. The graph of $g(x) = \sqrt{x + 1}$ is a translation of the basic graph of $y = \sqrt{x}$, shifted one unit to the left.

b. The graph of $h(x) = \dfrac{1}{(x - 3)^2}$ is a translation of the basic graph of $y = 1/x^2$, shifted three units to the right. We indicate the vertical asymptote, $x = 3$, by a dotted line.

(Graphs are on page 326.)

326   CHAPTER 7 ■ MORE ABOUT FUNCTIONS AND GRAPHS

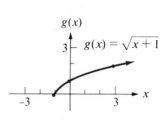 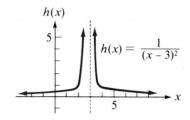

The graphs of some functions involve both horizontal and vertical translations.

EXAMPLE 3   Graph $f(x) = (x + 4)^3 + 2$.

Solution   Analyze the function by performing the translations separately. First, consider the graph of $y = (x + 4)^3$, which is a translation of the basic graph of $y = x^3$, shifted four units to the left. The graph of $f(x)$ is then obtained by shifting this graph upward two units.

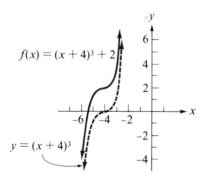

Scale Factors   We have seen that *adding* a constant to the expression defining a function results in a translation of its graph. We now investigate the effect of *multiplying* by a constant. Consider the graphs of the functions

$$f(x) = 2x^2, \quad g(x) = \frac{1}{2}x^2, \quad \text{and} \quad h(x) = -x^2$$

shown in Figure 7.3.

x	$f(x)$	$g(x)$	$h(x)$
-2	8	2	-4
-1	2	$\frac{1}{2}$	-1
0	0	0	0
1	2	$\frac{1}{2}$	-1
2	8	2	-4

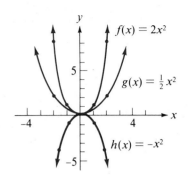

FIGURE 7.3

If we compare each of these graphs with the graph of $y = x^2$, we see that the graph of $f(x) = 2x^2$ is stretched, or expanded, vertically by a factor of 2. The y-coordinate of each point on the graph of $y = x^2$ has been doubled, so each point on the graph of f is twice as far from the x-axis as its counterpart with the same x-coordinate on the basic graph $y = x^2$. The graph of $g(x) = \frac{1}{2}x^2$ is compressed vertically by a factor of $\frac{1}{2}$; each point is half as far from the x-axis as its counterpart on the graph of $y = x^2$. The graph of $h(x) = -x^2$ is reflected about the x-axis.

In general, we have the following graphing principles.

Scale Factors

Compared with the graph of $y = f(x)$, the graph of $y = af(x)$, where $a \neq 0$, is

1. expanded vertically by a factor of a if $|a| > 1$;
2. compressed vertically by a factor of a if $0 < |a| < 1$;
3. reflected about the x-axis if $a < 0$.

The constant a is called the **scale factor** for the graph.

EXAMPLE 4 Graph the following functions.

a. $g(x) = 3\sqrt[3]{x}$

b. $h(x) = -\dfrac{1}{2}|x|$

Solutions a. The graph of $g(x) = 3\sqrt[3]{x}$ is an expansion of the basic graph of $y = \sqrt[3]{x}$ by a factor of 3. Each point on the basic graph has its y-coordinate tripled.

b. The graph of $h(x) = -\frac{1}{2}|x|$ is a compression of the basic graph of $y = |x|$ by a factor of ½, combined with a reflection about the x-axis. It may be helpful to graph the function in two steps, as shown in the figure.

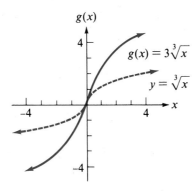
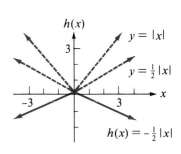

EXERCISE 7.1

A

■ *Graph the following functions. See Examples 1 and 2.*

1. $f(x) = |x| - 4$
2. $g(x) = (x + 1)^3$
3. $g(s) = \sqrt[3]{s - 4}$
4. $f(s) = s^2 + 3$
5. $F(t) = \frac{1}{t^2} + 1$
6. $G(t) = \sqrt{t - 2}$
7. $G(r) = (r + 2)^3$
8. $F(r) = \frac{1}{r - 4}$
9. $H(d) = \sqrt{d} - 3$
10. $h(d) = \sqrt[3]{d} + 5$
11. $h(v) = \frac{1}{v + 6}$
12. $H(v) = \frac{1}{v^2} - 2$

■ *See Example 3.*

13. $f(x) = 2 + (x - 3)^2$
14. $f(x) = (x + 4)^2 - 1$
15. $g(z) = \frac{1}{z + 2} - 3$
16. $g(z) = \frac{1}{z - 1} + 1$
17. $F(u) = \sqrt{u + 4} + 4$
18. $F(u) = \sqrt{u - 3} - 5$
19. $G(t) = |t - 5| - 1$
20. $G(t) = |t + 4| + 2$
21. $h(p) = (p + 2)^3 - 5$
22. $h(p) = (p - 2)^3 + 3$

7.1 ■ GRAPHING TECHNIQUES   329

23. $H(w) = \dfrac{1}{(w-1)^2} + 6$ 24. $H(w) = \dfrac{1}{(w+3)^2} - 3$

25. $f(t) = \sqrt[3]{t-8} - 1$ 26. $f(t) = \sqrt[3]{t+1} + 8$

■ *See Example 4.*

27. $F(t) = 4t^2$ 28. $G(t) = \dfrac{1}{3}t^2$ 29. $f(x) = \dfrac{1}{3}|x|$ 30. $H(x) = 3|x|$

31. $h(z) = \dfrac{2}{z^2}$ 32. $g(z) = \dfrac{2}{z}$ 33. $G(v) = -2\sqrt{v}$ 34. $F(v) = -4\sqrt[3]{v}$

35. $g(s) = \dfrac{-1}{2}s^3$ 36. $f(s) = \dfrac{-1}{8}s^3$ 37. $H(x) = \dfrac{1}{3x}$ 38. $h(x) = \dfrac{1}{4x^2}$

B

■ *Give an equation for the function graphed in each figure.*

39.
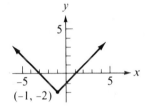

40.
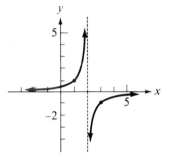

41.
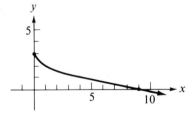

42.
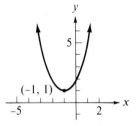

43.
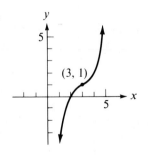

44.
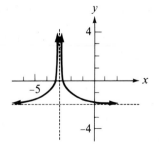

■ a. By completing the square, write each equation in the form
$y = (x - p)^2 + q$.
b. Graph, using horizontal and vertical translations.

45. $y = x^2 - 4x + 7$
46. $y = x^2 - 2x - 1$
47. $y = x^2 + 2x - 3$
48. $y = x^2 + 4x + 5$

7.2

POLYNOMIAL FUNCTIONS

In Section 6.3 we graphed linear functions of the form

$$f(x) = a_1 x + a_0,$$

and in Section 6.5 we graphed quadratic functions defined by

$$f(x) = a_2 x^2 + a_1 x + a_0.$$

Linear and quadratic functions are examples of **polynomial functions,** which are defined in general by

$$f(x) = a_n x^n + a_{n-1} x^{n-1} + \cdots + a_1 x + a_0,$$

where the coefficients $a_n, a_{n-1}, \ldots, a_0$ are real numbers and n, a positive integer or zero, is the degree of the function. Examples of polynomial functions are

$$P(x) = 2x^4 - x^3 + 6x - 5$$

and

$$Q(x) = x^6 - 2.$$

A linear function is a polynomial function of degree one, and a quadratic function is a polynomial function of degree two.

In this section and the next we consider some properties of polynomial functions that will help us to sketch their graphs. To start, we can get some idea of the graph by plotting a sufficient number of ordered pairs (x, y).

EXAMPLE 1 Graph the following polynomial functions.

a. $P(x) = x^3 - 4x$
b. $Q(x) = x^4 - 4x^2$

Solutions For each function, obtain several ordered pairs by evaluating the function. Connect the points with smooth curves to obtain the graphs shown in the figures.

a.

x	$P(x)$
-3	-15
-2	0
-1	3
0	0
1	-3
2	0
3	15

b.

x	$Q(x)$
-3	45
-2	0
-1	-3
0	0
1	-3
2	0
3	45

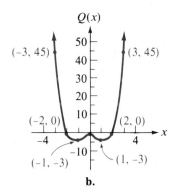

Notice that the graph of P, a polynomial of degree three, starts at the lower left and extends to the upper right, much as the graph of $y = x^3$ does. This is because for large values of $|x|$ the behavior of the polynomial is dominated by its term of highest degree. Thus, a polynomial of *odd* degree (if its lead coefficient is positive) will have negative values for large negative x and positive values for large positive x. A polynomial of *even* degree, such as Q in Example 1, will have positive values for both large positive *and* large negative x. Its graph will start at the upper left and extend to the upper right, as does the graph of $y = x^2$. We can make the following general statement:

For large values of $|x|$, the graph of a polynomial function of degree n (and positive lead coefficient) is similar in shape to

1. *the graph of $y = x^2$ if n is even;*
2. *the graph of $y = x^3$ if n is odd.*

Evaluating Polynomials (Synthetic Substitution)

There is a convenient method for evaluating polynomials that can be used to create tables of values quickly. Consider the polynomial

$$P(x) = 2x^3 - 4x^2 + 2x + 1.$$

By repeatedly isolating the constant term and factoring x (or a power of x) from the remaining terms, we can rewrite the polynomial with nested parentheses as follows:

$$\begin{aligned} P(x) &= (2x^3 - 4x^2 + 2x) + 1 \\ &= (2x^2 - 4x + 2)x + 1 \\ &= [(2x^2 - 4x) + 2]x + 1 \\ &= [(2x - 4)x + 2]x + 1. \end{aligned}$$

(Notice that we factor out x from the *right* of each expression; this will facilitate computation.)

It is now possible, by starting with the innermost parentheses and following the rules for order of operations, to evaluate the polynomial by a sequence of multiplications and additions. For example,

$$\begin{aligned} P(3) &= [(2 \cdot 3 - 4)3 + 2]3 + 1 \quad &(1) \\ &= [(6 - 4)3 + 2]3 + 1 \\ &= [(2)3 + 2]3 + 1 \\ &= (6 + 2)3 + 1 \\ &= (8)3 + 1 \\ &= 24 + 1 = 25. \end{aligned}$$

We can simplify the calculations even further by noting that the numbers in color in Equation (1) are just the coefficients of the polynomial P. To evaluate $P(3)$ we multiply the first coefficient, 2, by 3 and add the second coefficient, -4. Then we multiply the result by 3 and add the third coefficient, 2. Finally, we multiply the result by 3 and add the last coefficient, 1. The entire algorithm can be written in the following form:

$$\begin{array}{r|rrrr} 3 & 2 & -4 & 2 & 1 \\ & & 6 & 6 & 24 \\ \hline & 2 & 2 & 8 & 25. \end{array}$$

The first row gives the value of x, 3, and the coefficients of P. "Bring down" the 2 to start the algorithm. Then *multiply* the first entry on the bottom row by 3 and record the product on the middle row. Finally, *add* down to get the next entry in the bottom row. Repeat these two steps (multiply by 3 and add the result to the next entry of the top row) until you come to the end of the row. The first stage of the algorithm is illustrated by arrows in the example below:

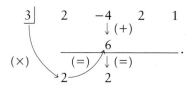

The value of $P(3)$, which is 25, appears as the last entry in the bottom row. This method for evaluating polynomials is often called **synthetic substitution**.

EXAMPLE 2 If $P(x) = 3x^3 - 4x - 1$, evaluate $P(2)$ by synthetic substitution.

Solution Write the number 2 and the coefficients of P as shown on line (1) below. Note that the coefficient of the x^2 term is zero:

$$\begin{array}{c|cccc} 2 & 3 & 0 & -4 & -1 \\ & & & & \\ \hline & 3 & & & \end{array} \begin{array}{c} (1) \\ (2) \\ (3) \end{array}$$

"Bring down" the 3 from line (1) to line (3). Multiply 2 by 3 and write the product on line (2), as shown by the arrows:

$$\begin{array}{c|cccc} 2 & 3 & 0 & -4 & -1 \\ & & 6 & & \\ \hline & 3 & & & \end{array} \begin{array}{c} (1) \\ (2) \\ (3) \end{array}$$

Add the 0 and 6 and write the sum on line (3):

$$\begin{array}{c|cccc} 2 & 3 & 0 & -4 & -1 \\ & & 6 & & \\ \hline & 3 & 6 & & \end{array} \begin{array}{c} (1) \\ (2) \\ (3) \end{array}$$

Repeat the two steps above: multiply the last number on line (3) by 2, record the product on line (2), and add the product to the number directly above on line (1). Write the sum on line (3). Continue until you reach the end of line (1):

$$\begin{array}{c|cccc} 2 & 3 & 0 & -4 & -1 \\ & & 6 & 12 & 16 \\ \hline & 3 & 6 & 8 & 15 \end{array} \begin{array}{c} (1) \\ (2) \\ (3) \end{array}$$

The last entry on line (3) is $P(2)$. Thus, $P(2) = 15$.

With practice, you should be able to eliminate line (2) by performing the two steps "multiply" and "add" in your head. You can then create a table of values by performing several synthetic substitutions in succession.

EXAMPLE 3 Use synthetic substitution to verify the table of values for the polynomial of Example 1a, $P(x) = x^3 - 4x$.

Solution Write the coefficients of P on the top row and the values of x on the left. Note that the coefficient of x^2 and the constant term are both zero. To find $P(-3)$ we "bring down" 1 and calculate mentally to obtain the entries in the first row:

$$-3 \times 1 + 0 = -3; \qquad -3 \times (-3) + (-4) = 5;$$
$$-3 \times 5 + 0 = -15.$$

Entries in the other rows are obtained similarly.

	1	0	−4	0
−3⌋	1	−3	5	−15
−2⌋	1	−2	0	0
−1⌋	1	−1	−3	3
0⌋	1	0	−4	0
1⌋	1	1	−3	−3
2⌋	1	2	0	0
3⌋	1	3	5	15

The values $P(x)$ appear in the last column of the table. Thus, $P(-3) = -15$, $P(-2) = 0$, and so on.

Using Calculators

Of course, when using a calculator to obtain values for $P(x)$ we need not write down each intermediate step in the synthetic substitution algorithm. The entire calculation can be done in one string of operations. On most calculators we must remember to press $\boxed{=}$ after each addition or subtraction. A keying sequence for Equation (1) on page 332,

$$P(3) = [(2 \cdot 3 - 4)3 + 2]3 + 1,$$

appears as follows:

$P(3) = 2\ \boxed{\times}\ 3\ \boxed{-}\ 4\ \boxed{=}\ \boxed{\times}\ 3\ \boxed{+}\ 2\ \boxed{=}\ \boxed{\times}\ 3\ \boxed{+}\ 1\ \boxed{=}.$

EXAMPLE 4 If $P(x) = 3x^4 - 5x^3 + 3x - 2,$ use a calculator to find the following.

    a. $P(-1)$ b. $P(2.6)$

Solutions   a.   First, write the coefficients of P:

$$3 \quad -5 \quad 0 \quad 3 \quad -2.$$

Multiply each coefficient by -1 and add the next coefficient to the result:

$$P(-1) = [([3(-1) - 5](-1) + 0)(-1) + 3](-1) - 2$$
$$= 3.$$

A keying sequence for this calculation is

$3\ \boxed{\times}\ (-1)\ \boxed{-}\ 5\ \boxed{=}\ \boxed{\times}\ (-1)\ \boxed{+}\ 0\ \boxed{=}\ \boxed{\times}\ (-1)\ \boxed{+}\ 3\ \boxed{=}\ \boxed{\times}\ (-1)\ \boxed{-}\ 2\ \boxed{=}.$

    b.   To calculate $P(2.6)$, multiply each coefficient by 2.6 and add the next coefficient to the result:

$$P(2.6) = [([3(2.6) - 5](2.6) + 0)(2.6) + 3](2.6) - 2$$
$$= 55.0128.$$

A keying sequence for this calculation is

$3\ \boxed{\times}\ 2.6\ \boxed{-}\ 5\ \boxed{=}\ \boxed{\times}\ 2.6\ \boxed{+}\ 0\ \boxed{=}\ \boxed{\times}\ (2.6)\ \boxed{+}\ 3\ \boxed{=}\ \boxed{\times}\ 2.6\ \boxed{-}\ 2\ \boxed{=}.$

The Factor Theorem

In Section 6.5 we saw that the x-intercepts of the graph of a quadratic function f occur at values of x for which $f(x) = 0,$ that is, at the real-valued solutions of the equation $ax^2 + bx + c = 0.$ The same relation holds true for polynomials of higher degree.

Solutions of the equation $P(x) = 0$ are called **zeros** of the function P. In Example 1a on page 330,

$$P(x) = x^3 - 4x$$
$$= x(x - 2)(x + 2) = 0$$

when $x = 0$, $x = 2$, or $x = -2$, so the zeros of P are 0, 2, and -2. Note that each zero of P corresponds to a factor of $P(x)$. This result suggests the following theorem, which holds for any polynomial P.

Factor Theorem

Let $P(x)$ be a polynomial with real-number coefficients. Then $(x - a)$ is a factor of $P(x)$ if and only if $P(a) = 0$.

Since a polynomial function P of degree n can have at most n linear factors of the form $(x - a)$, it follows that P can have at most n distinct zeros, or, equivalently, that an equation of the form $P(x) = 0$, where P is of degree n, can have at most n distinct solutions, some of which may be complex.

EXAMPLE 5 Find the zeros of the polynomials.

a. $f(x) = x^3 + 6x^2 + 9x$ b. $g(x) = x^3 - 2x^2 + 5x$

Solutions   a. Factor the polynomial to obtain

$$f(x) = x(x^2 + 6x + 9)$$
$$= x(x + 3)(x + 3).$$

By the factor theorem, the zeros of f are 0, -3, and -3.

b. Factor the polynomial to obtain

$$g(x) = x(x^2 - 2x + 5).$$

By the quadratic formula, the roots of

$$x^2 - 2x + 5 = 0$$

are

$$x = \frac{2 \pm \sqrt{4 - 20}}{2} = 1 \pm 2i.$$

Thus, the zeros of g are 0, $1 + 2i$, and $1 - 2i$. The factored form of g is

$$g(x) = x[x - (1 + 2i)][x - (1 - 2i)].$$

In Example 5a, since -3 is a repeated solution of $f(x) = 0$, we say that -3 is a zero of *multiplicity* two.

EXERCISE 7.2

A

■ *Graph by plotting a sufficient number of ordered pairs. See Example 1.*

1. $y = x^3 - 3x$
2. $y = 9x - x^3$
3. $y = -x^4 + 9x^2 - 8$
4. $y = x^4 - 10x^2 + 9$

■ *Use synthetic substitution to evaluate each polynomial at the given values. See Example 2.*

5. $P(x) = 3x^3 - 2x^2 + 5x - 4$; $x = 3, -2$
6. $Q(x) = x^4 - 10x^3 + 5x^2 - 3x + 6$; $x = 2, -3$
7. $S(x) = 2x^5 - 3x^3 + x^2 - x + 2$; $x = 2, -1$
8. $T(x) = 4x^4 - 2x^3 + 3x^2 - 5$; $x = 3, -4$

■ *Use synthetic substitution to make a table of values for the polynomial function. See Example 3.*

9. $Q(x) = 2x^3 + 9x^2 + 7x - 6$
10. $P(x) = x^3 - 3x^2 - 6x + 8$
11. $D(x) = x^4 - 3x^3 - 10x^2 + 24x$
12. $N(x) = x^4 + 5x^3 - x^2 - 5x$

■ *Use a calculator and synthetic substitution to evaluate each polynomial function at the given values. See Example 4.*

13. $P(x) = 2x^4 - 6x^2 - 4x + 3$; $x = 1.2, -3.1$
14. $Q(x) = -3x^3 + 3x^2 + 2x - 5$; $x = 2.3, -2.3$
15. $P(x) = 1.2x^5 - 27x^3 - 8.3x^2 - 7$; $x = 4, -3$
16. $Q(x) = 0.05x^6 - 1.8x^4 + x^3 + 12x$; $x = 3, -2$

■ *Find the zeros of each polynomial. See Example 5.*

17. $P(x) = x^4 + 4x^2$
18. $P(x) = x^3 + 3x$
19. $P(x) = x^3 - 8x$
20. $P(x) = 4x^4 - 9x^2$
21. $P(x) = 2x^3 + x^2 + 2x + 1$
22. $P(x) = x^3 - 2x^2 + 3x - 6$
23. $P(x) = x^4 + 3x^3 + x^2$
24. $P(x) = x^4 + 4x^3 + 3x^2$

B

■ *Use the factor theorem to determine whether the given binomial is a factor of the given polynomial.*

25. $x - 2$; $x^3 - 3x^2 + 2x + 2$
26. $x - 1$; $2x^3 - 5x^2 + 4x - 1$
27. $x + 3$; $3x^3 + 11x^2 + x - 15$
28. $x + 1$; $2x^2 - 5x^2 + 3x + 3$

■ *Verify that the given value is a solution of the equation and find the other solutions.*

29. $x^3 - 2x^2 + 1 = 0$; $x = 1$
30. $x^3 + 2x^2 - 1 = 0$; $x = -1$
31. $x^4 - 3x^3 - 10x^2 + 24x = 0$; $x = -3$
32. $x^4 + 5x^3 - x^2 - 5x = 0$; $x = -5$

7.3

GRAPHING POLYNOMIAL FUNCTIONS

We can now use the tools developed in Section 7.2, namely the factor theorem and synthetic substitution, to graph polynomial functions. We begin by considering the x-intercepts of the graph.

x-Intercepts of a Polynomial Function

As noted on page 336, a polynomial can have at most n distinct zeros, and therefore its graph can have at most n x-intercepts. However, if some of the zeros of P are complex numbers they will not appear on the graph, so the graph may have *fewer* than n x-intercepts. (Recall that the same is true of quadratic functions.)

EXAMPLE 1 Find the number of x-intercepts of the graph of each polynomial function.

a. $R(x) = x^4 - 10x^2 + 9$
b. $S(x) = 2x^4 + 2x^3 + 4x^2$

Solutions   **a.** To find the zeros of R, factor the polynomial:

$$R(x) = x^4 - 10x^2 + 9$$
$$= (x^2 - 9)(x^2 - 1)$$
$$= (x - 3)(x + 3)(x - 1)(x + 1).$$

By the factor theorem, the zeros of R are 3, -3, 1, and -1. Since R has four real-number zeros, the graph of R has four x-intercepts.

b. To find the zeros of S, factor as far as possible:

$$S(x) = 2x^4 + 2x^3 + 4x^2$$
$$= 2x^2(x^2 + x + 2).$$

Now, the solutions of $x^2 + x + 2 = 0$ are complex, so S has only one (repeated) real zero, namely, 0, which corresponds to the factors $2x^2$. Thus, the graph of S has only one x-intercept.

The appearance of the graph near an x-intercept is determined by the multiplicity of the zero there. All the zeros of the polynomial R in Example 1 are of multiplicity one, and the graph *crosses* the x-axis at each intercept, as shown in Figure 7.4a. However, the polynomial S has a zero of multiplicity two at $x = 0$. The graph of S *touches* the x-axis without crossing, as shown in Figure 7.4b.

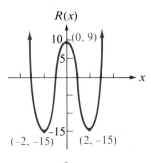
a.

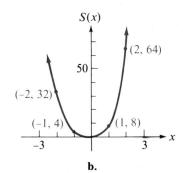
b.

FIGURE 7.4

To see what happens in general, compare the graphs of the three polynomials in Figure 7.5. $L(x) = x - 2$ has a zero of multiplicity one at $x = 2$, and its graph crosses the x-axis there. $Q(x) = (x - 2)^2$ has a zero of multiplicity two at $x = 2$, and its graph touches the x-axis at $x = 2$ but changes direction without crossing; that is, the graph has its vertex there. $C(x) = (x - 2)^3$ has a zero of multiplicity three at $x = 2$, and its graph makes an S-shaped curve at the intercept, like the graph of $y = x^3$.

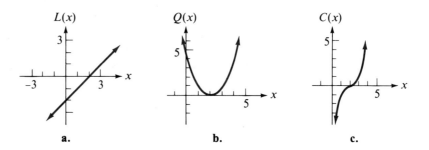

FIGURE 7.5 a. b. c.

Near its x-intercepts the graph of a polynomial takes one of the characteristic shapes illustrated above.

EXAMPLE 2 Graph the polynomial

$$f(x) = (x + 2)^3(x - 1)(x - 3)^2.$$

Solution The polynomial has degree six, an even number, so its graph starts at the upper left and extends to the upper right. Its y-intercept is

$$f(0) = (2)^3(-1)(-3)^2 = -72.$$

f has a zero of multiplicity three at $x = -2$, a zero of multiplicity one at $x = 1$, and a zero of multiplicity two at $x = 3$.

Evaluate the polynomial at several points between the intercepts to fill in the graph:

$$f(-1) = (1)^3(-2)(-4)^2 = -32;$$
$$f(2) = (4)^3(1)(-1)^2 = 64.$$

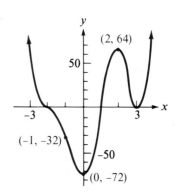

Number of Turning Points

Note that in Example 1 of Section 7.2 on page 000 the graph of the third-degree polynomial P changes direction twice, and the graph of the fourth-degree polynomial Q changes direction three times. In fact, it can be shown that the graph of a polynomial of degree n changes direction at most $n - 1$ times.

EXAMPLE 3 Graph $g(x) = x^3 - 2x^2 - 5x + 6$.

Solution Use synthetic substitution to evaluate the function for selected values of x.

x	$g(x)$
-3	-24
-2	0
-1	8
0	6
1	0
2	-4
3	0
4	18

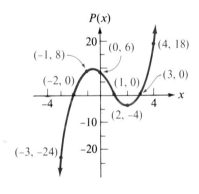

Plot the points and connect them with a smooth curve to obtain the graph.

Note that the graph in Example 3 crosses the x-axis three times and has two "turning points," or points where the graph changes direction, between $x = -3$ and $x = 4$. If we evaluate the polynomial for x-values less than -3 or greater than 4 we will find that the graph continues as indicated in the figure, without any more turning points.

It is possible for the graph of a polynomial of degree n to change direction *fewer* than $n - 1$ times. Consider the graph of $Q(x) = x^4 - 2$. Setting $Q(x) = x^4 - 2 = 0$, we obtain $x^4 = 2$. Thus, the polynomial $x^4 - 2$ has two real zeros, $\pm\sqrt[4]{2}$, and its graph has two x-intercepts, at approximately ± 1.2. We can find a few more points on the graph by evaluation.

x	$Q(x)$
-2	14
-1	-1
0	-2
1	-1
2	14

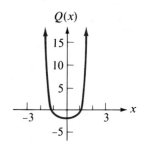

FIGURE 7.6

The graph is shown in Figure 7.6. Although the polynomial is of degree four, its graph changes direction only once. As a second example, note that the polynomial of degree six graphed in Example 2 has only three turning points. We summarize our results as follows.

A polynomial of degree n

1. *can have at most n x-intercepts, which correspond to the real-valued zeros of the polynomial, and*
2. *can change direction at most $n - 1$ times.*

These observations provide some guidelines for graphing polynomials.

EXAMPLE 4 Graph $h(x) = x^3 + 4x^2 + 4x + 16$.

Solution To find the x-intercepts of the graph, factor the polynomial:

$$x^3 + 4x^2 + 4x + 16 = x^2(x + 4) + 4(x + 4)$$
$$= (x^2 + 4)(x + 4).$$

The zeros of the polynomial are $2i$, $-2i$, and -4, so the graph has one x-intercept, at $x = -4$. The y-intercept is $h(0) = 16$. To find some additional points on the graph, use synthetic substitution or the factored form of the polynomial to obtain several additional function values.

x	h(x)
-5	-29
-4	0
-3	13
-2	16
-1	15
0	16
1	25

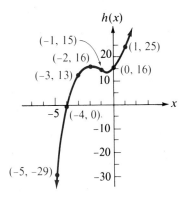

Graph the function over the interval of tabulated values and note that the graph changes direction *twice* between $x = -5$ and $x = 1$. Since h is of degree *three*, the graph cannot change direction again. Plotting additional points with x-values less than -5 or greater than 1 would not change the appearance of the graph. To complete the graph, simply continue the left-hand branch downward and the right-hand branch upward as shown.

Observe that the graph in Example 4 *does not* change direction exactly at the ordered pair $(-1, 15)$. To find the exact coordinates of the turning points for polynomials of degree greater than two requires the methods of calculus. However, we can if necessary obtain approximations for the turning point above by evaluating the function at x-values close to -1.

EXERCISE 7.3

A

■ *Graph. See Example 2.*

1. $f(x) = (x + 3)(x - 1)^2$
2. $g(x) = (x + 4)^2(x - 2)$
3. $G(x) = (x - 2)^2(x + 2)^2$
4. $F(x) = (x - 1)^2(x - 3)^2$
5. $h(x) = x^3(x + 2)(x - 2)$
6. $H(x) = (x + 1)^3(x - 2)^2$
7. $P(x) = (x + 4)^2(x + 1)^2(x - 1)^2$
8. $Q(x) = x^2(x - 5)(x - 1)^2(x + 2)$
9. $q(x) = (x + 3)^2(x + 1)^3$
10. $p(x) = (x + 2)^3(x - 1)^3$

■ *Graph. Specify the x-intercepts. See Examples 1–4.*

11. $S(x) = x^3 + x^2 - 6x$
12. $T(x) = x^3 - 4x^2 + 3x$
13. $g(x) = 16x - x^3$
14. $f(x) = 12x - x^3$
15. $w(x) = x^3 + 3x^2 - x - 3$
16. $v(x) = x^3 + x^2 - 25x - 25$
17. $F(x) = x^4 - 16x^2$
18. $G(x) = x^4 - x^2$
19. $k(x) = x^4 - 10x^2 + 16$
20. $m(x) = x^4 - 15x^2 + 36$
21. $p(x) = x^4 + x^3 - x - 1$
22. $q(x) = x^4 - x^3 + x - 1$
23. $H(x) = x^4 - x^3 - 4x^2 + 4x$
24. $h(x) = x^4 + 3x^3 - x^2 - 3x$
25. $r(x) = (x^2 - 1)(x + 3)^2$
26. $s(x) = (x^2 - 9)(x - 1)^2$

B

■ *Refer to Problems 29–32 of Exercise 7.2 to sketch the graphs of the following polynomials.*

27. $y = x^3 - 2x^2 + 1$
28. $y = x^3 + 2x^2 - 1$
29. $y = x^4 - 3x^3 - 10x^2 + 24x$
30. $y = x^4 + 5x^3 - x^2 - 5x$

7.4

RATIONAL FUNCTIONS

A function of the form

$$f(x) = \frac{P(x)}{Q(x)},$$

where $P(x)$ and $Q(x)$ are polynomials (and $Q(x)$ is not the zero polynomial) is called a **rational function**. Thus,

$$f(x) = \frac{3}{2x + 1}, \quad g(x) = \frac{x}{x^2 - 1},$$

$$h(x) = \frac{x^2 - 4x}{x^2 + 6x + 9}, \quad \text{and} \quad r(x) = \frac{x + 2}{x^4 + x^3}$$

are examples of rational functions. Unlike polynomial functions, which are defined for all values of x, a rational function $P(x)/Q(x)$ is undefined at $x = a$ if $Q(a) = 0$. These values are not in the domain of the function.

EXAMPLE 1 Find the domains of the functions f and g defined above.

Solution The domain of f is the set of all real numbers except $-\frac{1}{2}$, since the denominator, $2x + 1$, equals zero when $x = -\frac{1}{2}$.

The domain of g is the set of all real numbers except 1 and -1, since the denominator, $x^2 - 1$, equals zero when $x = 1$ or $x = -1$.

Vertical Asymptotes

A polynomial function is defined for all values of x, and its graph is a smooth curve without any breaks or holes. A rational function, on the other hand, is undefined at x-values for which the denominator equals zero, and its graph will exhibit breaks or holes at those values.

For example, consider the graph of $f(x) = \dfrac{1}{x - 3}$ shown in Figure 7.7. From our study of translations in Section 7.1 we know that the graph looks like the graph of $y = \dfrac{1}{x}$ but shifted three units to the right. The graph has a vertical asymptote at $x = 3$. It cannot be drawn in one smooth curve as can a polynomial; it is broken into two pieces separated at $x = 3$.

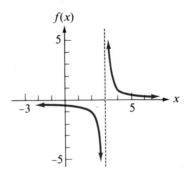

FIGURE 7.7

In general, we have the following result.

> **Finding Vertical Asymptotes**
>
> If $Q(a) = 0$ but $P(a) \neq 0$, then the graph of the rational function $R(x) = P(x)/Q(x)$ has a vertical asymptote at $x = a$.

346 CHAPTER 7 ■ MORE ABOUT FUNCTIONS AND GRAPHS

If $P(a)$ and $Q(a)$ are both zero, then the graph of R may have a "hole" at $x = a$ rather than an asymptote. This possibility is considered in Problems 31–34 at the end of this section.

Near a vertical asymptote the graph of a rational function has one of four characteristic shapes, as illustrated in Figure 7.8.

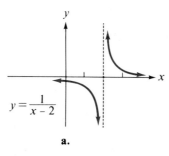
a.

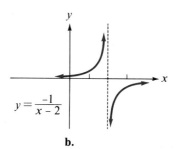
b.

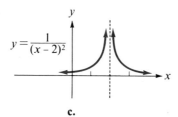
c.

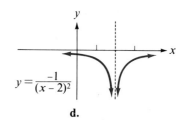
d.

FIGURE 7.8

To determine the appearance of a graph near a vertical asymptote it is usually sufficient to find the function values at one or two points on each side of the asymptote.

EXAMPLE 2 Find the vertical asymptotes of the graph of

$$f(x) = \frac{2}{1 - x}$$

and sketch the graph.

Solution Since the denominator, $1 - x$, equals zero when $x = 1$, and the numerator, 2, does *not* equal zero when $x = 1$, the line $x = 1$ is a vertical asymptote for the graph of f.

To sketch the graph, evaluate the function at points on each side of the asymptote:

$$f(0) = \frac{2}{1-0} = 2;$$

$$f(-1) = \frac{2}{1-(-1)} = 1;$$

$$f(2) = \frac{2}{1-2} = -2;$$

$$f(3) = \frac{2}{1-3} = -1.$$

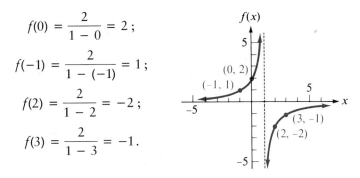

The graph, shown in the figure, has a shape similar to Figure 7.8a.

Horizontal Asymptotes

In the graph of $f(x) = \dfrac{2}{1-x}$ shown in Example 2, note that as $|x|$ gets large, that is, as we move away from the origin along the x-axis in either direction, the corresponding values of y get close to zero but never *equal* zero for any value of x. The graph approaches, but never coincides with, the x-axis (the line $y = 0$). Thus, the graph has a horizontal asymptote at $y = 0$.

To find the horizontal asymptotes of more general rational functions we can take advantage of the fundamental principle of fractions. As an example, consider the behavior of the function

$$f(x) = \frac{2x^2}{x^2 + 1}$$

for very large and very small values of x.

x	$f(x)$
4	1.882
5	1.923
10	1.980
100	1.999

x	$f(x)$
-4	1.882
-5	1.923
-10	1.980
-100	1.999

We see that as $|x|$ gets very large the function values approach 2. This is because for large values of x the term of highest degree dominates the behavior of a polynomial; the contribution of the other terms is negligible by comparison. The fraction $\dfrac{2x^2}{x^2+1}$ "looks like" $\dfrac{2x^2}{x^2}$, or 2. This becomes apparent if we divide the numerator and denominator of the rational fraction by x^2, the highest power of x that appears in the fraction:

$$\frac{2x^2}{x^2+1} \div \frac{x^2}{x^2} = \frac{2}{1+\dfrac{1}{x^2}}.$$

For large x-values, the term $1/x^2$ in the denominator is very small and the rational function is approximately equal to $2/1$, or 2, as we found with more effort in the table above. Hence, the graph of f, shown in Figure 7.9, has a horizontal asymptote $y = 2$.

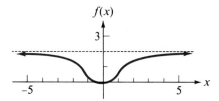

FIGURE 7.9

The discussion above suggests the following method for finding horizontal asymptotes of the graphs of rational functions.

Finding Horizontal Asymptotes

1. Divide the numerator and denominator of the rational function by the highest power of x that appears in the expression.
2. For large values of $|x|$, any terms of the form k/x^n, where k is a constant, are approximately zero and can be ignored.
3. Simplify the resulting expression. If the expression is a constant c, then the rational function has a horizontal asymptote at $y = c$.

EXAMPLE 3 Find any horizontal asymptotes of the graph of each function.

a. $f(x) = \dfrac{3x}{x^2 - 5x + 4}$

b. $g(x) = \dfrac{4x^2 + 1}{2x^2 - x}$

c. $h(x) = \dfrac{x^4 + 1}{x^2 + 2}$

Solutions a. Divide the numerator and denominator of $f(x)$ by x^2:

$$\frac{3x}{x^2 - 5x + 4} \div \frac{x^2}{x^2} = \frac{\dfrac{3}{x}}{1 - \dfrac{5}{x} + \dfrac{4}{x^2}}.$$

For large values of $|x|$,

$$f(x) \approx \frac{0}{1} = 0,$$

so the graph of f has a horizontal asymptote at $y = 0$.

b. Divide the numerator and denominator of $g(x)$ by x^2:

$$\frac{4x^2 + 1}{2x^2 - x} \div \frac{x^2}{x^2} = \frac{4 + \dfrac{1}{x^2}}{2 - \dfrac{1}{x}}.$$

For large values of $|x|$,

$$g(x) \approx \frac{4}{2} = 2,$$

so the graph of g has a horizontal asymptote at $y = 2$.

c. Divide the numerator and denominator of $h(x)$ by x^4:

$$\frac{x^4 + 1}{x^2 + 2} \div \frac{x^4}{x^4} = \frac{1 + \dfrac{1}{x^4}}{\dfrac{1}{x^2} + \dfrac{2}{x^4}}.$$

For large values of $|x|$,

$$h(x) \approx \frac{x^2}{0}.$$

Since this expression is undefined, the graph does not have a horizontal asymptote. The function values continue to increase as $|x|$ increases.

Graphing Rational Functions

The vertical and horizontal asymptotes can be thought of as a framework on which to build a sketch of the graph. After the x- and y-intercepts have been located, plotting a few points on each side of each vertical asymptote is usually sufficient to complete the picture.

As an example, we graph the function

$$g(x) = \frac{6}{x^2 - x - 6}.$$

First, find the asymptotes. The denominator

$$x^2 - x - 6 = (x - 3)(x + 2)$$

equals zero when $x = 3$ or $x = -2$. Since the numerator, 6, does not equal zero at either of these values, the lines $x = 3$ and $x = -2$ are vertical asymptotes.

To find any horizontal asymptotes, divide the numerator and denominator by x^2:

$$\frac{6}{x^2 - x - 6} \div \frac{x^2}{x^2} = \frac{\frac{6}{x^2}}{1 - \frac{1}{x} - \frac{6}{x^2}}.$$

So, for large values of $|x|$,

$$g(x) \approx \frac{0}{1},$$

and the horizontal asymptote is $y = 0$.

Since the numerator is constant, the graph has no x-intercepts. The y-intercept is $g(0) = -1$.

Note that the vertical asymptotes divide the plane into three regions: region I includes all points with $x < -2$, region II includes $-2 < x < 3$, and region III includes $x > 3$. To fill in the graph, evaluate the function at a few points in each region and use the asymptotes to direct the shape of the graph. In region I we evaluate $g(-3)$:

$$g(-3) = \frac{6}{(-3)^2 - (-3) - 6} = \frac{6}{6} = 1.$$

Since the graph never crosses the x-axis, as x approaches -2 from the left the corresponding y-values must *increase* without bound. (If the opposite were true, the graph would have to cross the x-axis somewhere between -3 and -2.) Similarly, in region III, since

$$g(4) = \frac{6}{4^2 - 4 - 6} = \frac{6}{6} = 1,$$

the graph must *increase* as x approaches 3 from the right.

In region II the graph must "rise" from the vertical asymptote $x = -2$, reach a maximum value without crossing the x-axis, and then "sink" toward the vertical asymptote $x = 3$.

As with the graphs of polynomial functions, we cannot determine the precise turning point without more advanced methods. However, we can sharpen the picture somewhat by evaluating the function at $x = 1$ and $x = 2$ to obtain the points $(1, -1)$ and $(2, -\frac{3}{2})$. The completed graph is shown in Figure 7.10.

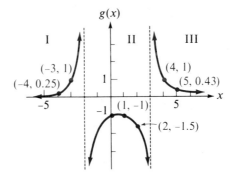

FIGURE 7.10

EXAMPLE 4 Graph $f(x) = \dfrac{2x - 4}{x^2 - 9}$.

Solution The denominator of the function, $x^2 - 9 = (x - 3)(x + 3)$, equals zero when $x = 3$ or $x = -3$. Since the numerator does not equal zero at either of those values, the vertical asymptotes are $x = 3$ and $x = -3$.

To find the horizontal asymptote, divide the numerator and denominator by x^2.

$$\frac{2x - 4}{x^2 - 9} \div \frac{x^2}{x^2} = \frac{\dfrac{2}{x} - \dfrac{4}{x^2}}{1 - \dfrac{9}{x^2}}.$$

So, for large values of $|x|$,

$$f(x) \approx \frac{0}{1} = 0,$$

and the horizontal asymptote is $y = 0$. $f(0) = \frac{4}{9} \approx 0.44$ is the y-intercept, and since $f(x) = 0$ when $2x - 4 = 0$, the x-intercept is 2. Since the graph cannot cross the x-axis except at $x = 2$, it is necessary to plot only a few additional points to complete the graph:

$f(-4) = \frac{-12}{7} \approx -1.71;$

$f(-2) = \frac{-8}{-5} = 1.60;$

$f\left(\frac{5}{2}\right) = \frac{-4}{11} \approx -0.36;$

$f(4) = \frac{4}{7} \approx 0.57.$

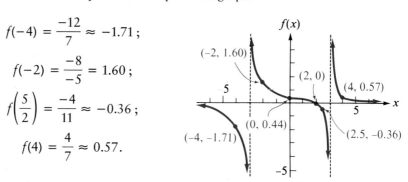

Use the asymptotes to direct the branches of the curve through these points.

EXERCISE 7.4

A

■ a. Determine the vertical asymptotes of the graph of each function.
b. Give the domain of each function. See Examples 1 and 2.

1. $y = \dfrac{2}{x + 3}$
2. $y = \dfrac{1}{x - 4}$
3. $y = \dfrac{3}{(x - 2)(x + 3)}$
4. $y = \dfrac{4}{(x + 1)(x - 4)}$
5. $y = \dfrac{2x}{x^2 - x - 6}$
6. $y = \dfrac{2x + 1}{x^2 - 3x + 2}$

■ Determine any vertical or horizontal asymptotes of the graph of each function. See Example 3.

7. $y = \dfrac{x}{x^2 - 9}$
8. $y = \dfrac{2x - 4}{x^2 + 5x + 4}$
9. $y = \dfrac{x - 4}{2x - 1}$
10. $y = \dfrac{2x + 1}{x - 3}$
11. $y = \dfrac{2x^2}{x^2 - 3x - 4}$
12. $y = \dfrac{x^2}{x^2 - x - 12}$

■ *Graph each function after first identifying all asymptotes and intercepts. See Example 4.*

13. $y = \dfrac{1}{x + 3}$

14. $y = \dfrac{1}{x - 3}$

15. $y = \dfrac{2}{(x - 4)(x + 1)}$

16. $y = \dfrac{4}{(x + 2)(x - 1)}$

17. $y = \dfrac{2}{x^2 - 5x + 4}$

18. $y = \dfrac{4}{x^2 - x - 6}$

19. $y = \dfrac{x}{x + 3}$

20. $y = \dfrac{x}{x - 2}$

21. $y = \dfrac{x + 1}{x + 2}$

22. $y = \dfrac{x - 1}{x - 3}$

23. $y = \dfrac{2x}{x^2 - 4}$

24. $y = \dfrac{x}{x^2 - 9}$

25. $y = \dfrac{x - 2}{x^2 + 5x + 4}$

26. $y = \dfrac{x + 1}{x^2 - x - 6}$

27. $y = \dfrac{x^2 - 1}{x^2 - 4}$

28. $y = \dfrac{2x^2}{x^2 - 1}$

29. $y = \dfrac{x}{x^2 + 3}$

30. $y = \dfrac{x^2 + 2}{x^2 + 4}$

B

■ a. Find the domain of the function.
b. Reduce the fraction to lowest terms.
c. Graph the function. [*Hint:* The graph of the original function is identical to the graph of the function in (b) except that certain points are excluded from the domain.] Indicate a "hole" in the graph by an open circle.

31. $y = \dfrac{x^2 - 4}{x - 2}$

32. $y = \dfrac{x^2 - 1}{x + 1}$

33. $y = \dfrac{x + 1}{x^2 - 1}$

34. $y = \dfrac{x - 3}{x^2 - 9}$

7.5

INVERSE FUNCTIONS

It is often convenient to think of a function as a machine or process that acts on the elements of the domain (the "input" values) and produces the elements of the range (the "output" values). (See Figure 7.11.) For instance, from the input values 1, 2, 3, and 4 the function $f(x) = 2x$ produces the output values 2, 4, 6, and 8, respectively. This is just another way of saying that $f(1) = 2$, $f(2) = 4$, $f(3) = 6$, and $f(4) = 8$.

The **inverse** of a function f is obtained by running the function machine in reverse: the elements of the *range* of f are used as the input values, and the output values produced by the machine are the corresponding elements of the

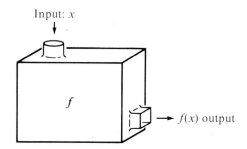

FIGURE 7.11

domain of f. If g is the inverse of the function f above, then g turns 2 into 1, 4 into 2, 6 into 3, and 8 into 4. (See Figure 7.12.) In other words, $g(2) = 1$, $g(4) = 2$, $g(6) = 3$, and $g(8) = 4$. Thus, the domain of g is the range of f, and the range of g is the domain of f.

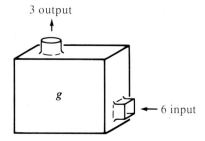

FIGURE 7.12

Notice that if (a, b) is an ordered pair obtained from the function f, then (b, a) will be an ordered pair from the inverse function g. The function $f(x) = 2x$ produces the ordered pairs

$$(1, 2), \quad (2, 4), \quad (3, 6), \quad \text{and} \quad (4, 8),$$

and its inverse function g produces the ordered pairs

$$(2, 1), \quad (4, 2), \quad (6, 3), \quad \text{and} \quad (8, 4).$$

To obtain the ordered pairs of the inverse function g, we need only interchange the components of the ordered pairs of f.

This observation suggests that the inverse of a function defined by an equation can be obtained by interchanging the variables in the equation. Thus, if the function f is defined by the equation $y = 2x$, then an equation for the inverse function g is found by interchanging the variables x and y to obtain $x = 2y$. Solving this new equation for y, we find $y = x/2$ or $g(x) = x/2$.

EXAMPLE 1 Find the inverse of the function $f(x) = 4x - 3$.

Solution Write the equation for f in the form

$$y = 4x - 3.$$

Interchange the variables to obtain

$$x = 4y - 3$$

and solve for y:

$$y = \frac{x}{4} + \frac{3}{4}.$$

The inverse function is $g(x) = \frac{x}{4} + \frac{3}{4}.$

The inverse function g "undoes" the effect of the function f. For example, the function $f(x) = 2x$ doubles the elements of its domain, and its inverse function, $g(x) = x/2$, halves them. If we apply the function f to a given input value and then apply the function g to the output from f, the end result will be the original input value. It is like feeding a number into the function machine and then feeding the result back in with the machine running in reverse. For the function $f(x) = 2x$ and its inverse, $g(x) = x/2$, if we choose $x = 5$ as an input value, we find that $f(5) = 2(5) = 10$, and $g(10) = {}^{10}\!/_2 = 5$. (See Figure 7.13 on page 356.)

EXAMPLE 2
a. Find the inverse of the function $f(x) = x^3 + 2$.
b. Show that the inverse function "undoes" the effect of f on $x = 2$.
c. Show that f "undoes" the effect of the inverse function on $x = -25$.

Solutions a. Write the equation for f in the form

$$y = x^3 + 2.$$

Interchange the variables to obtain

$$x = y^3 + 2$$

and solve for y:

$$y^3 = x - 2$$
$$y = \sqrt[3]{x - 2}.$$

The inverse function is $g(x) = \sqrt[3]{x - 2}$.

b. $f(2) = 2^3 + 2 = 10$ and $g(10) = \sqrt[3]{10 - 2} = 2$.
c. $g(-25) = \sqrt[3]{-25 - 2} = -3$ and $f(-3) = (-3)^3 + 2 = -25$.

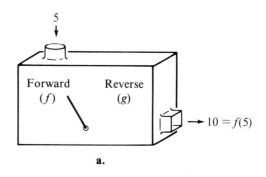

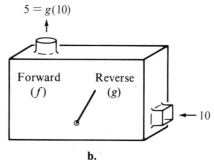

FIGURE 7.13

Graph of the Inverse

The graphs of a function and its inverse are related in an interesting way. To see this, we first observe in Figure 7.14 that the graphs of the ordered pairs (a, b) and (b, a) are always located symmetrically with respect to the graph of $y = x$. Now, for every ordered pair (a, b) in f, the ordered pair (b, a) is in the inverse of f. Thus, the graphs of $y = f(x)$ and its inverse, $y = g(x)$, are reflections of each other about the line $y = x$. Figure 7.15 shows the graphs of $y = 4x - 3$ and its inverse, $y = x/4 + 3/4$, from Example 1, along with the graph of $y = x$.

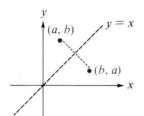

FIGURE 7.14

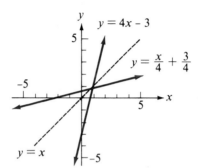

FIGURE 7.15

EXAMPLE 3 Graph the function $f(x) = x^3 + 2$ and its inverse, $g(x) = \sqrt[3]{x - 2}$, which we found in Example 2, on the same set of axes.

Solution The graph of f is the graph of $y = x^3$ translated two units upward. The graph of g is the graph of $y = \sqrt[3]{x}$ translated two units to the right. The two graphs are symmetric about the line $y = x$.

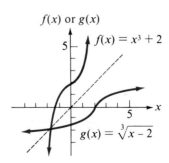

The Horizontal Line Test

It is not always true that the inverse of a function is itself a function. For example, consider the function $f(x) = x^2$, or $y = x^2$. To find its inverse we first interchange x and y to obtain $x = y^2$ and then solve for y to get $y = \pm\sqrt{x}$. The graphs of f and its inverse are shown in Figure 7.16. Since the graph of the inverse does not pass the vertical line test, it is *not* a function.

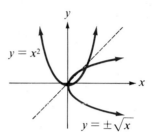

FIGURE 7.16

For many applications it is important to know whether or not the inverse of f is a function. This can be determined from the graph of f. Note that when we interchange x and y to find a formula for the inverse, horizontal lines of the form $y = k$ become vertical lines of the form $x = k$. Thus, if the graph of the *inverse* is to pass the vertical line test, then the graph of the *original function* must pass the horizontal line test, namely, that no horizontal line should intersect the graph in more than one point. Notice that the graph of $f(x) = x^2$ does *not* pass the horizontal line test, so we would not expect its inverse to be a function.

The Horizontal Line Test

If no horizontal line intersects the graph of a function more than once, then the inverse is also a function.

EXAMPLE 4 Which of the following functions have inverses that are also functions?

Solutions In each case, to determine whether the inverse is a function, apply the horizontal line test. Since no horizontal line intersects their graphs more than once, the functions pictured in (a) and (c) have inverses that are also functions.

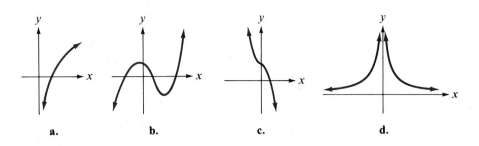

Inverse Notation

If the inverse of a function f is also a function, then the inverse is often denoted by the symbol f^{-1}, read "f inverse." For example, since the function $f(x) = x^3 + 2$ passes the horizontal line test (see Example 3), its inverse is a function and can be denoted by $f^{-1}(x) = \sqrt[3]{x - 2}$.

EXAMPLE 5 If $h(x) = 2x - 6$, find $h^{-1}(10)$.

Solution First, find the inverse function for $y = 2x - 6$. Interchange x and y to get

$$x = 2y - 6$$

and solve for y:

$$2y = x + 6$$
$$y = \frac{x}{2} + 3.$$

The inverse function is

$$h^{-1}(x) = \frac{x}{2} + 3.$$

Now evaluate the inverse function at $x = 10$:

$$h^{-1}(10) = \frac{10}{2} + 3 = 8.$$

Common Error Although the same symbol, $^{-1}$, is used for both reciprocals and inverse functions, the two notions are not equivalent. In Example 5 above, note that $h^{-1}(x)$ does *not* indicate $\dfrac{1}{2x-6}$. To avoid confusion, we use the notation $\dfrac{1}{h}$ to refer to the reciprocal of the function h.

EXERCISE 7.5

A

■ *Find the inverse of each function. See Example 1.*

1. $f(x) = x + 2$
2. $f(x) = x - 3$
3. $f(x) = 2x$
4. $f(x) = \dfrac{x}{5}$
5. $f(x) = 2x - 6$
6. $f(x) = 3x + 1$
7. $f(x) = \dfrac{3-x}{2}$
8. $f(x) = \dfrac{5-x}{3}$
9. $f(x) = x^3 + 1$
10. $f(x) = x^3 - 8$
11. $f(x) = \sqrt[3]{x}$
12. $f(x) = \dfrac{1}{x}$
13. $f(x) = \dfrac{1}{x-1}$
14. $f(x) = \sqrt[3]{x+1}$
15. $f(x) = \sqrt[3]{x} + 4$
16. $f(x) = \dfrac{1}{x} - 3$

■ *For Problems 17 and 18, see Example 2.*

17. a. Find the inverse g of the function $f(x) = (x - 2)^3$.
    b. Show that g "undoes" the effect of f on $x = 4$.
    c. Show that f "undoes" the effect of g on $x = -8$.

18. a. Find the inverse g of the function $f(x) = \dfrac{2}{x+1}$.
    b. Show that g "undoes" the effect of f on $x = 3$.
    c. Show that f "undoes" the effect of g on $x = -1$.

■ *For Problems 19–34, for each of Problems 1–16, graph the function and its inverse on the same set of axes, along with the graph of $y = x$. See Example 3.*

■ *Which of the following functions have inverses that are also functions? See Example 4.*

35. a. 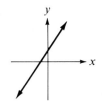   b. 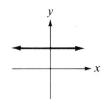   c. 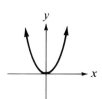   d.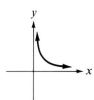

36. a. 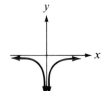   b. 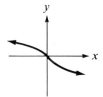   c. 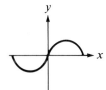   d.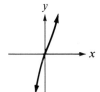

37. a. $f(x) = x$ b. $f(x) = x^2$
38. a. $f(x) = x^3$ b. $f(x) = x^4$
39. a. $f(x) = \dfrac{1}{x}$ b. $f(x) = \dfrac{1}{x^2}$
40. a. $f(x) = \sqrt{x}$ b. $f(x) = \sqrt[3]{x}$

■ *For Problems 41–44, see Example 5.*

41. If $F(t) = \dfrac{2}{3}t + 1$, find $F^{-1}(5)$.

42. If $G(s) = \dfrac{s - 3}{4}$, find $G^{-1}(-2)$.

43. If $m(v) = 6 - \dfrac{2}{v}$, find $m^{-1}(-3)$.

44. If $p(z) = 1 - 2z^3$, find $p^{-1}(7)$.

B

45. $f(x) = \dfrac{x + 2}{x - 1}$. Find $f^{-1}(x)$.

46. $f(x) = \dfrac{3x + 1}{x - 3}$. Find $f^{-1}(x)$.

47. $f(x) = x^3 + x + 1$. Find the following.
   a. $f^{-1}(1)$ b. $f^{-1}(3)$

48. $f(x) = x^5 + x^3 + 7$. Find the following.
   a. $f^{-1}(7)$ b. $f^{-1}(5)$

49. If $f(-1) = 0$, $f(0) = 1$, $f(1) = -2$, and $f(2) = -1$, find the following.
   a. $f^{-1}(1)$
   b. $f^{-1}(-1)$
50. If $f^{-1}(-2) = 1$, $f^{-1}(-1) = -2$, $f^{-1}(0) = 0$, and $f^{-1}(1) = -1$, find the following.
   a. $f(-1)$
   b. $f(1)$
51. Let $f(x) = 2^x$.
   a. Fill in the following table for f.
   b. Graph f.
   c. By interchanging the components of the ordered pairs in (a), make a table for f^{-1}.
   d. Graph f^{-1} on the same set of axes with f.

x	f(x)
-2	
-1	
0	
1	
2	

52. Repeat Problem 51 with $f(x) = \left(\dfrac{1}{2}\right)^x$.

CHAPTER REVIEW

A

[7.1]

■ Graph each function.

1. $g(x) = |x| + 2$
2. $F(t) = \dfrac{1}{t} - 2$
3. $f(s) = \sqrt{s} + 3$
4. $h(r) = (r - 2)^3$
5. $f(x) = (x - 2)^2 - 4$
6. $g(u) = \sqrt{u + 2} - 3$
7. $G(t) = |t + 2| - 3$
8. $H(t) = \dfrac{1}{(t - 2)^2} + 3$
9. $h(s) = -2\sqrt{s}$
10. $g(s) = \dfrac{1}{2}|s|$

[7.2]

■ Use synthetic substitution to evaluate the polynomial at the given value.

11. $4x^3 - 2x^2 + x - 1$; $x = 2$
12. $2x^5 - x^3 + 3x^2 - 1$; $x = -3$

■ Use a calculator and synthetic substitution to evaluate the polynomial function at the given value.

13. $R(x) = 2x^3 - 5x^2 - 2x + 2$; $x = 1.3$
14. $P(x) = 1.4x^4 - x^2 + 2.3x + 1$; $x = -2.1$

■ *Find the zeros of each polynomial function.*

15. $Q(x) = x^5 - 4x^3$
16. $R(x) = 2x^3 + 3x^2 - 2x$

[7.3]
■ *Graph. Specify the x-intercepts.*

17. $f(x) = (x - 2)(x + 1)^2$
18. $g(x) = (x - 3)^2(x + 2)$
19. $G(x) = x^2(x - 1)(x + 3)$
20. $F(x) = (x + 1)^2(x - 2)^2$
21. $V(x) = 4x^3 - x^5$
22. $H(x) = x^4 - 9x^2$
23. $P(x) = x^3 + x^2 - x - 1$
24. $q(x) = x^3 - x^2 - 4x + 4$
25. $y = x^3 + x^2 - 2x$
26. $y = x^3 - 2x^2 + x - 2$
27. $y = x^4 - 7x^2 + 6$
28. $y = x^4 + x^3 - 3x^2 - 3x$

[7.4]
■ *Graph each function after first identifying all asymptotes and intercepts.*

29. $y = \dfrac{1}{x - 4}$
30. $y = \dfrac{2}{x^2 - 3x - 10}$
31. $y = \dfrac{x - 2}{x + 3}$
32. $y = \dfrac{x - 1}{x^2 - 2x - 3}$
33. $y = \dfrac{3x^2}{x^2 - 4}$
34. $y = \dfrac{2x^2 - 2}{x^2 - 9}$

[7.5]
■ *Find the inverse g(x) of each function.*

35. $f(x) = x + 4$
36. $f(x) = \dfrac{x - 2}{4}$
37. $f(x) = x^3 - 1$
38. $f(x) = \dfrac{1}{x + 2}$
39. $f(x) = \dfrac{1}{x} + 2$
40. $f(x) = \sqrt[3]{x} - 2$

41. Graph the function in Problem 35. Graph its inverse on the same set of axes.
42. Graph the function in Problem 37. Graph its inverse on the same set of axes.
43. If $F(t) = \dfrac{3}{4}t + 2$, find $F^{-1}(2)$.
44. If $G(x) = \dfrac{1}{x} - 4$, find $F^{-1}(3)$.

B

■ *By completing the square, write each equation in the form $y = (x - p)^2 + q$ and then graph the equation.*

45. $y = x^2 - 6x + 6$
46. $y = x^2 + 2x - 4$

47. Verify that -1 is a solution of $y = x^4 - 2x^3 - x^2 + 2x$ and find the other solutions. Graph the equation.
48. Verify that -3 is a solution of $y = x^4 + x^3 - 9x^2 - 9x$ and find the other solutions. Graph the equation.
49. If $f(x) = \dfrac{x - 1}{x + 3}$, find $f^{-1}(x)$.
50. If $f(x) = \dfrac{2x + 4}{x - 2}$, find $f^{-1}(x)$.

8
Exponential and Logarithmic Functions

In previous chapters we have considered several kinds of functions, including linear and quadratic functions, polynomial functions of higher degree, rational functions, and some simple examples of radical functions. We now turn our attention to another class of functions, the exponential functions, which are used to model relationships in many fields, from biology to business.

8.1

EXPONENTIAL GROWTH AND DECAY

We first consider some examples of population growth.

EXAMPLE 1 In a laboratory experiment a colony of 100 bacteria is established and the growth of the colony is monitored. The experimenters discover that the colony triples in population every day.

a. Write a function that gives the population of the colony at any time t in days.
b. How many bacteria are present after 5 days? After 36 hours?
c. Graph the function found in (a).

Solutions a. The population of the colony is 100 when the experiment starts (at $t = 0$) and triples every day thereafter. If P represents the population of the colony after t days, then

when $t = 0$, $P = 100$;
when $t = 1$, $P = 100 \cdot 3$;
when $t = 2$, $P = (100 \cdot 3) \cdot 3 = 100 \cdot 3^2$;
when $t = 3$, $P = (100 \cdot 3^2) \cdot 3 = 100 \cdot 3^3$.

In general, on the tth day the original population of 100 has been multiplied by 3 t times, so the population on day t is $100 \cdot 3^t$, or

$$P(t) = 100 \cdot 3^t.$$

b. To find the population at any particular time, evaluate the function for the appropriate value of t. Thus, after 5 days the population is

$$P(5) = 100 \cdot 3^5$$
$$= 100(243) = 24{,}300.$$

After 36 hours, or 1.5 days, the population is

$$P(1.5) = 100 \cdot 3^{1.5}$$
$$= 100(5.196) = 519.6^*.$$

Since there will never be a fraction of a bacterium, the population of the colony after 36 hours is 519 bacteria. (The 520th bacterium is not formed yet.)

c. To graph the function $P(t) = 100 \cdot 3^t$, tabulate several function values. Then connect the points with a smooth curve to obtain the graph shown in the figure.

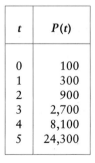
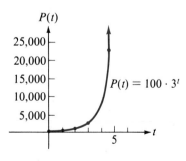

*We used a calculator to obtain 5.196 as an approximate value for $3^{1.5}$. Although many of the calculations in this chapter involve approximations, we will continue to use an "$=$" sign instead of "$\approx$."

EXAMPLE 2 Under ideal conditions the number of rabbits in a certain area can double every 3 months. A rancher estimates that there are 60 rabbits living on his land.

a. Write a function that gives the rabbit population after t months.
b. How many rabbits are present after 2 years? After 8 months?
c. Graph the function found in (a).

Solutions a. If P represents the population of rabbits after t months, then

$$\text{when} \quad t = 0, \quad P = 60;$$
$$\text{when} \quad t = 3, \quad P = 60 \cdot 2;$$
$$\text{when} \quad t = 6, \quad P = (60 \cdot 2) \cdot 2 = 60 \cdot 2^2;$$
$$\text{when} \quad t = 9, \quad P = (60 \cdot 2^2) \cdot 2 = 60 \cdot 2^3.$$

In general, the original population of 60 is multiplied by 2 every 3 months, so the population in month t is $60 \cdot 2^{t/3}$, or $P = 60 \cdot 2^{t/3}$.

b. To find the population at any particular time, evaluate the function for the appropriate value of t. Thus, after 2 years, or 24 months, the population is

$$P(24) = 60 \cdot 2^{24/3}$$
$$= 60 \cdot 2^8 = 15{,}360.$$

After 8 months the population is

$$P(8) = 60 \cdot 2^{8/3}$$
$$= 60(6.350) = 381.$$

c. To graph the function $P(t) = 60 \cdot 2^{t/3}$, tabulate several function values and plot points. Connect them with a smooth curve to obtain the graph shown in the figure.

t	$P(t)$
0	60
1	75
2	95
3	120
4	151
5	190
6	240

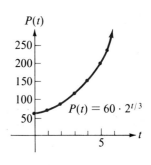

8.1 ■ EXPONENTIAL GROWTH AND DECAY

The functions found above describe **exponential growth.** During each time interval of a fixed length the population is *multiplied* by a certain constant amount. In Example 1 the bacteria population grew by a factor of 3 every day. In Example 2 the rabbit population grew by a factor of 2 every 3 months.

Functions that describe exponential growth can be put into the standard form

$$P(t) = P_0 a^t,$$

where $P_0 = P(0)$ is the **initial value** of the function and a is the **growth factor.** In Example 1 we have

$$P(t) = 100 \cdot 3^t,$$

so $P_0 = 100$ and $a = 3$. In Example 2 we have

$$P(t) = 60 \cdot 2^{t/3}$$
$$= 60 \cdot (2^{1/3})^t,$$

so $P_0 = 60$ and $a = 2^{1/3}$.

If the units are the same, a population with a larger growth factor grows faster than one with a smaller growth factor.

EXAMPLE 3 A lab technician compares the growth of two species of bacteria. She starts two colonies of 50 bacteria each. Species A doubles in population every 2 days, and species B triples every 3 days. Find the growth factor for each species.

Solution A function describing the growth of species A is

$$P(t) = 50 \cdot 2^{t/2} = 50 \cdot (2^{1/2})^t,$$

so the growth factor for species A is $2^{1/2}$, or approximately 1.41. For species B,

$$P(t) = 50 \cdot 3^{t/3} = 50 \cdot (3^{1/3})^t,$$

so the growth factor for species B is $3^{1/3}$, or approximately 1.44. Thus, species B grows faster than species A.

Percent Increase

Other phenomena besides populations can exhibit exponential growth. For example, if the interest on a savings account is compounded annually, the amount of money in the account grows exponentially.

Consider a principal of $100 invested at 5% interest compounded annually. At the end of 1 year the amount is

$$A = P + Prt$$
$$= P(1 + rt)$$
$$= 100[1 + 0.05(1)]$$
$$= 100(1.05) = 105.$$

This amount becomes the new principal, so at the end of the second year

$$A = P(1 + rt)$$
$$= 105[1 + 0.05(1)]$$
$$= 105(1.05) = 110.25.$$

Note that to find the amount at the end of each year we *multiply* the principal by a factor of $1 + r = 1.05$. Thus, we can express the amount at the end of the second year as

$$A = [100(1.05)](1.05)$$
$$= 100(1.05)^2.$$

We organize our results into a table as follows.

	Principal	Amount
First year	100	$100(1.05)$
Second year	$100(1.05)$	$[100(1.05)](1.05) = 100(1.05)^2$
Third year	$100(1.05)^2$	$[100(1.05)^2](1.05) = 100(1.05)^3$

Continuing in this way we find that the amount of money accumulated after t years is $100(1.05)^t$. In general, for an initial investment of P dollars at an interest rate $100r$ compounded annually, the amount accumulated after t years is

$$A(t) = P(1 + r)^t.$$

This function describes exponential growth with an initial value of P and a growth factor of $1 + r$. Note that the interest rate $100r$, which indicates the *percent increase* in the account each year, corresponds to a *growth factor* of $1 + r$. (See Example 6 on page 145 for another example of percent increase.)

8.1 ■ EXPONENTIAL GROWTH AND DECAY

The notion of percent increase is often used to describe other quantities that grow exponentially.

EXAMPLE 4 During a period of rapid inflation prices rose by 12% every 6 months. At the beginning of the inflationary period a pound of butter cost $2.

a. Write a function that gives the price of a pound of butter t years after inflation began.
b. How much did a pound of butter cost after 3 years? After 15 months?
c. Graph the function found in (a).

Solutions a. The *percent increase* in the cost of butter is 12% every 6 months. Therefore, the *growth factor* for the cost of butter is $1 + 0.12 = 1.12$ every half year. If P represents the price of the butter after t years, then

$$\text{when} \quad t = 0, \quad P = 2;$$
$$\text{when} \quad t = \frac{1}{2}, \quad P = 2(1.12);$$
$$\text{when} \quad t = 1, \quad P = 2(1.12)^2;$$
$$\text{when} \quad t = \frac{3}{2}, \quad P = 2(1.12)^3;$$
$$\text{when} \quad t = 2, \quad P = 2(1.12)^4.$$

In general, after t years of inflation the original price of $2 has been multiplied $2t$ times by a factor of 1.12. Thus, $P(t) = 2(1.12)^{2t}$.

b. To find the price of a pound of butter at any time after inflation began, evaluate the function at the appropriate value of t. After 3 years the price was

$$P(3) = 2(1.12)^{2(3)}$$
$$= 2(1.12)^6 = 3.95.$$

After 15 months, or 1.25 years, the price was

$$P(1.25) = 2(1.12)^{2(1.25)}$$
$$= 2(1.12)^{2.5} = 2.66.$$

c. To graph the function $P(t) = 2(1.12)^{2t}$, evaluate the function for several values. Then connect the points with a smooth curve to obtain the graph shown in the figure on page 370.

t	$P(t)$
0	2.00
1	2.51
2	3.15
3	3.95
4	4.95

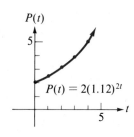

Exponential Decay

In the examples above exponential growth was modeled by increasing functions of the form

$$P(t) = P_0 a^t,$$

where $a > 1$. The function $P(t) = P_0 a^t$ is a *decreasing* function if we have $0 < a < 1$. In this case the function is said to describe **exponential decay**, and the constant a is called the **decay factor**.

EXAMPLE 5 A small coal-mining town has been losing population since 1930, when there were 5000 inhabitants. At each census thereafter (taken at 10-year intervals) the population has been approximately 9/10 of its earlier figure.

a. Write a function that gives the population of the town t years after 1930.
b. What was the population of the town in 1980? In 1985?
c. Graph the function found in (a).

Solutions a. If P represents the population t years after 1930, then

when $t = 0$, $P = 5000$;
when $t = 10$, $P = 5000(0.9)$;
when $t = 20$, $P = 5000(0.9)^2$;
when $t = 30$, $P = 5000(0.9)^3$.

In general, we find that after t years the original population of 5000 has been multiplied $t/10$ times by 0.9, so the population t years after 1930 is $5000 \cdot (0.9)^{t/10}$, or $P(t) = 5000(0.9)^{t/10}$.

b. In 1980, 50 years had elapsed since 1930, so the population was

$$P(50) = 5000(0.9)^{50/10}$$
$$= 5000(0.9)^5 = 2952.45,$$

or 2952. In 1985 (55 years after 1930) the population was

$$P(55) = 5000(0.9)^{55/10}$$
$$= 5000(0.9)^{5.5} = 2800.94,$$

or 2801.

c. To graph the function $P(t) = 5000(0.9)^{t/10}$, tabulate several function values. Then connect the points to obtain the graph shown in the figure.

t	P(t)
0	5000
10	4500
20	4050
30	3645
40	3280

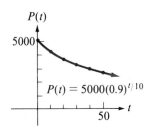

On page 368 we noted that a percent *increase* of 100r corresponds to a growth factor of $1 + r$. A percent *decrease* of 100r in a quantity corresponds to a decay factor of $1 - r$.

EXAMPLE 6 A plastic window coating 1 millimeter thick decreases the light coming through the window by 25%.

a. Write a function that gives the amount of light that will come through a layer of the window coating x millimeters thick.

b. How much light will come through 5 millimeters of the coating? Through ½ millimeter?

c. Graph the function found in (a).

Solutions a. Each millimeter of coating reduces the fraction of light coming through the window by 25%, so the decay factor for the light is $1 - 0.25$, or 0.75. If P represents the fraction of light that comes through the window and x represents the thickness of the coating in millimeters, then

$$\begin{aligned} \text{when} \quad x &= 0, \quad P = 1.00\,; \\ \text{when} \quad x &= 1, \quad P = 1.00(0.75)\,; \\ \text{when} \quad x &= 2, \quad P = 1.00(0.75)^2; \\ \text{when} \quad x &= 3, \quad P = 1.00(0.75)^3. \end{aligned}$$

In general, the amount of light that comes through x millimeters of coating is $P(x) = 1.00(0.75)^x$, or $P(x) = (0.75)^x$.

b. For 5 millimeters of coating,

$$P(5) = (0.75)^5 = 0.2373,$$

so approximately 24% of the original light shines through. For a layer of coating ½ millimeter thick,

$$P(0.5) = (0.75)^{0.5} = 0.8660,$$

so approximately 87% of the light shines through.

c. To graph the function $P(x) = (0.75)^x$, evaluate the function for several values. Then connect the points with a smooth curve to obtain the graph shown in the figure.

x	$P(x)$
0	1.00
1	0.75
2	0.56
3	0.42
4	0.32

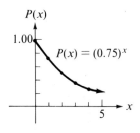

The functions defined by the models above are examples of **exponential functions**. We shall discuss these functions in more detail in the following sections.

Linear Growth and Exponential Growth

It may be helpful to compare the notions of linear growth and exponential growth. Consider the two functions

$$L(t) = 5 + 2t \quad \text{and} \quad E(t) = 5 \cdot 2^t \quad (t \geq 0),$$

whose graphs are shown in Figure 8.1.

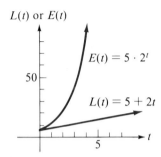

FIGURE 8.1

L is a linear function with y-intercept 5 and slope 2; E is an exponential function with initial value 5 and growth factor 2. The growth factor of an exponential function is in a sense analogous to the slope of a linear function: each measures how quickly the function is increasing (or decreasing). However, for each unit that t increases, 2 units are *added* to the value of $L(t)$, whereas the value of $E(t)$ is *multiplied* by 2. An exponential function with growth factor 2 grows much more rapidly than a linear function with slope 2.

EXAMPLE 7 A solar energy company sold $80,000 worth of solar collectors last year, its first year of operation. This year its sales rose to $88,000, an increase of 10%. The marketing department must estimate its projected sales for the next 3 years.

  a. If the marketing department predicts that sales will grow linearly, what should it expect the sales total to be next year? Graph the projected sales figures over the next 3 years, assuming that sales will grow linearly.
  b. If the marketing department predicts that sales will grow exponentially, what should it expect the sales total to be next year? Graph the projected sales figures over the next 3 years, assuming that sales will grow exponentially.

Solutions a. Let $L(t)$ represent the company's total sales t years after starting business, where $t = 0$ is considered the first year of operation, and assume that sales grow linearly. Then L is a linear function of the form $L(t) = mt + b$. Since $L(0) = 80{,}000$, the intercept b is 80,000. The slope m of the function is

$$\frac{\Delta S}{\Delta t} = \frac{8000 \text{ dollars}}{1 \text{ year}} = 8000,$$

where $\Delta S = 8000$ is the increase in sales during the first year. Thus, $L(t) = 8000t + 80{,}000$, and the expected sales total for next year is

$$L(2) = 8000(2) + 80{,}000 = 96{,}000.$$

b. Let $E(t)$ represent the company's sales under the assumption that sales will grow exponentially. Then E is a function of the form $E(t) = E_0 \cdot a^t$. The percent increase in sales over the first year was $r = 0.10$, so the growth rate is $a = 1 + r = 1.10$. The initial value E_0 of the function is 80,000. Thus, $E(t) = 80{,}000(1.10)^t$, and the expected sales total for next year is

$$E(2) = 80{,}000(1.10)^2 = 96{,}800.$$

Evaluate each function at several points to obtain the graphs shown in the figure.

t	$L(t)$	$E(t)$
0	80,000	80,000
1	88,000	88,000
2	96,000	96,800
3	104,000	106,480
4	112,000	117,128

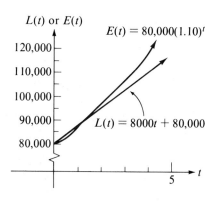

EXERCISE 8.1

A

- For Problems 1–16, do the following.
  a. Write a function that describes exponential growth or decay.
  b. Evaluate the function at the given values.
  c. Graph the function.

See Examples 1 and 2.

1. A colony of bacteria starts with 300 organisms and doubles every week. How many bacteria will there be after 8 weeks? After 5 days?

2. A population of 24 fruit flies triples every month. How many fruit flies will there be after 6 months? After 3 weeks? (Assume that a month equals 4 weeks.)

3. A typical beehive contains 20,000 insects. The population can increase in size by a factor of 2.5 every 6 weeks. How many bees will there be after 4 weeks? After 20 weeks?

4. A rancher who started with 800 head of cattle finds that his herd increases by a factor of 1.8 every 3 years. How many head of cattle will he have after 1 year? After 10 years?

- See Example 4.

5. A sum of $4000 is invested in an account that pays 8% interest compounded annually. How much is in the account after 2 years? After 10 years?

6. Otto invests $600 in an account that pays 7.3% interest compounded annually. How much is in Otto's account after 3 years? After 6 years?

7. Since 1963 housing prices have risen an average of 5% per year. Paul bought a house for $20,000 in 1963. How much was the house worth in 1975? In 1990?

8. Housing prices in Los Angeles have risen an average of 10% per year since 1982. If Marlene bought a house for $135,000 in 1982, how much was the house worth in 1985? In 1989?

9. Sales of Windsurfers have increased 12% per year since 1980. If Sunsails sold 1500 Windsurfers in 1980, how many did it sell in 1986? How many should it expect to sell in 1992?

10. Sales of personal computers have increased 23% per year since 1980. If Compucalc sold 500 personal computers in 1980, how many did it sell in 1985? How many should it expect to sell in 1995?

- See Examples 5 and 6.

11. During a vigorous spraying program the mosquito population was reduced to three-fourths of its previous size every 2 weeks. If the mosquito population was originally estimated at 250,000, how many mosquitos remained after 3 weeks of spraying? After 8 weeks?

12. The number of perch in Hidden Lake has declined to half of its previous value every 5 years since 1960, when the perch population was estimated at 8000. How many perch were there in 1970? In 1988?

13. Scuba divers find that the water in Emerald Lake filters out 15% of the sunlight for each 4 feet that they descend. How much sunlight penetrates to a depth of 20 feet? To a depth of 45 feet?

14. Arch's motorboat cost $15,000 in 1980 and has depreciated by 10% every 3 years. How much was the boat worth in 1989? In 1990?

15. Plutonium 238 is a radioactive element that decays over time into a less harmful element at a rate of 0.8% per year. A power plant has 50 pounds of plutonium 238 to dispose of. How much plutonium 238 will be left after 10 years? After 100 years?

16. Iodine 131 is a radioactive element that decays at a rate of 8.3% per day. How much of a 12-gram sample will be left after 1 week? After 15 days?

■ *Determine which population grows faster. See Example 3.*

17. A researcher starts two populations of fruit flies of different species, each with 30 flies. Species A increases by 30% in 6 days and species B increases by 20% in 4 days. Which species multiplies more rapidly?

18. A biologist isolates two strains of a particular virus and monitors the growth of each, starting with samples of 0.01 gram. Strain A increases by 10% in 8 hours and strain B increases by 12% in 9 hours. Which strain grows more rapidly?

■ *Compare linear and exponential growth. See Example 7.*

19. At a large university six students start a rumor that final exams have been canceled. After 2 hours nine students (including the first six) have heard the rumor.
    a. Assuming that the rumor grows linearly, write a function that gives the number of students who have heard the rumor at time t. Graph the function.
    b. Repeat (a) assuming that the rumor grows exponentially.

20. Over the weekend the Midland Infirmary identifies five cases of Asian flu. Three days later it has treated a total of nine cases.
    a. Assuming that the number of flu cases grows linearly, write a function that gives the number of people infected at time t. Graph the function.
    b. Repeat (a) assuming that the flu spreads exponentially.

B

21. An eccentric millionaire offers you a summer job for the month of June. She will pay you 1 cent for your first day of work and will double your wages every day thereafter. How much will you make on June 15? On June 30?

22. If you place one grain of wheat on the first square of a chessboard, two grains on the second square, four grains on the third square, and so on, how many grains of wheat should be placed on the last (64th) square?

23. You receive a chain letter with a list of six names. You are instructed to send $10 to the name at the top of the list, then cross it out and add your name to the bottom of the list. You should then send copies of the letter to six friends.
    a. If the chain is not broken, how much money should you receive?
    b. If you are one of the six people who receive the original letter, how many people will have received the letter before you start receiving money?
    c. Why are chain letters illegal?

24. A friend asks you to take part in a pyramid scheme selling pet supplies. Each salesperson must pay 30% of his monthly earnings, or at least $25, to his manager. If a salesperson can recruit six more salespersons, he becomes their manager and does not have to sell any more pet supplies himself. However, he must still pay 30% of his monthly earnings to *his* manager.
    a. What is the minimum amount that a level 1 manager, with six salespersons under him, will make each month? What is the minimum amount that a level 2 manager, with six level 1 managers under him, will make each month? Find the minimum monthly earnings of managers up to level 6.
    b. How many salespersons (including managers at lower levels) does a level 6 manager have working under him?
    c. Why are pyramid schemes a risky venture?

25. The population of the state of Texas was 9,579,700 in 1960. In 1970 the population was 11,196,700. What was the annual rate of growth to the nearest hundredth of a percent?

26. The population of the state of Florida was 4,951,600 in 1960. In 1970 the population was 6,789,400. What was the annual rate of growth to the nearest hundredth of a percent?

27. a. The population of Rainville was 10,000 in 1950 and doubled in 20 years. What was the annual rate of growth to the nearest hundredth of a percent?
    b. The population of Elmira was 350,000 in 1950 and doubled in 20 years. What was the annual rate of growth to the nearest hundredth of a percent?
    c. If a population doubles in 20 years, does the rate of growth depend on the size of the original population?
    d. The population of Grayling doubled in 20 years. What was the annual rate of growth to the nearest hundredth of a percent?

28. a. The population of Boomtown was 300 in 1908 and tripled in 7 years. What was the annual rate of growth to the nearest hundredth of a percent?
    b. The population of Fairview was 15,000 in 1962 and tripled in 7 years. What was the annual rate of growth to the nearest hundredth of a percent?
    c. If a population triples in 7 years, does the rate of growth depend on the size of the original population?
    d. The population of Pleasant Lake tripled in 7 years. What was the annual rate of growth to the nearest hundredth of a percent?

8.2 EXPONENTIAL FUNCTIONS

In Section 8.1 we studied a number of functions that described exponential growth or decay. More formally, we define an **exponential function** to be one of the form

$$f(x) = b^x, \text{ where } b > 0 \text{ and } b \neq 1.$$

The positive constant b is called the **base** of the exponential function. We do not allow $b < 0$ as a base because if b is negative, then b^x is not a real number for some values of x. (For example, if $b = -4$, then $f(\frac{1}{2}) = (-4)^{1/2}$ is an imaginary number.) We also exclude $b = 1$ as a base because $1^x = 1$ for all values of x, and hence the function $f(x) = 1^x$ is not exponential but is actually the constant function $f(x) = 1$.

Powers with Irrational Exponents

If b is a positive number not equal to 1, then b^x is defined for all real values of x. In Chapter 3 we defined b^x for rational values of x by $b^{m/n} = \sqrt[n]{b^m}$. We now assume that powers such as $2^{\sqrt{3}}$ and $(\frac{1}{2})^\pi$, in which the exponent is an *irrational number,* are also defined. Many calculators give decimal approximations for such powers directly. Alternatively, approximate values can be found by using decimal approximations for the irrational exponent.

EXAMPLE 1 Find decimal approximations for the following.

a. $2^{\sqrt{3}}$ b. $\left(\dfrac{1}{2}\right)^\pi$

Solutions a. A calculator keying sequence for evaluating $2^{\sqrt{3}}$ is

2 $\boxed{y^x}$ 3 $\boxed{\sqrt{}}$ $\boxed{=}$ 3.321997085.

Or, approximate $\sqrt{3}$ by 1.732 and calculate $2^{1.732} = 3.321880096$. Thus, to three decimal places, $2^{\sqrt{3}} \approx 3.322$.

b. A calculator keying sequence for evaluating $(\frac{1}{2})^\pi$ is

0.5 $\boxed{y^x}$ $\boxed{\pi}$ $\boxed{=}$ 0.113314732.

Or, approximate π by 3.142 and calculate $0.5^{3.142} = 0.113282742$. Thus, to three decimal places, $(\frac{1}{2})^\pi \approx 0.113$.

Graphs of Exponential Functions

The graphs of exponential functions have two characteristic shapes, depending on whether the base b is greater than 1 or less than 1. As typical examples, consider the graphs of $f(x) = 2^x$ and $g(x) = (½)^x$ shown in Figure 8.2.

x	$f(x)$
-3	$\frac{1}{8}$
-2	$\frac{1}{4}$
-1	$\frac{1}{2}$
0	1
1	2
2	4
3	8

x	$g(x)$
-3	8
-2	4
-1	2
0	1
1	$\frac{1}{2}$
2	$\frac{1}{4}$
3	$\frac{1}{8}$

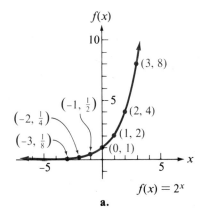
a. $f(x) = 2^x$

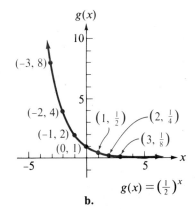
b. $g(x) = \left(\frac{1}{2}\right)^x$

FIGURE 8.2

Notice that $f(x) = 2^x$ is an increasing function and $g(x) = (½)^x$ is a decreasing function. In general, exponential functions have the following properties.

Properties of Exponential Functions, $f(x) = b^x$

1. Domain: all real numbers.
2. Range: all positive real numbers.
3. If $b > 1$, the function is increasing; if $0 < b < 1$, the function is decreasing.

Notice also that the negative x-axis is an asymptote for exponential functions with $b > 1$, and for exponential functions with $0 < b < 1$ the positive x-axis is an asymptote.

EXAMPLE 2 Compare the graphs of $f(x) = 3^x$ and $g(x) = 4^x$.

Solution Evaluate each function for several convenient values.

x	f(x)	g(x)
−2	$\frac{1}{9}$	$\frac{1}{16}$
−1	$\frac{1}{3}$	$\frac{1}{4}$
0	1	1
1	3	4
2	9	16

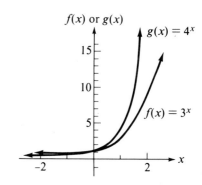

Plot the points for each function and connect them with smooth curves. Notice that $g(x) = 4^x$ grows more rapidly than $f(x) = 3^x$. Both graphs cross the y-axis at $(0, 1)$.

EXAMPLE 3 Graph $h(x) = 2^{-x}$.

Solution Note that $2^{-x} = 1/2^x = (\frac{1}{2})^x$, so the function $h(x) = 2^{-x}$ is the same as the function $g(x) = (\frac{1}{2})^x$, whose graph is shown in Figure 8.2b.

The graphing techniques of Section 7.1 involving horizontal and vertical translations can be used to graph many exponential functions. Compared to the graph of $f(x) = b^x$, the graph of $f(x) + k = b^x + k$ is shifted k units upward if k is positive or k units downward if k is negative. The graph of $f(x + k) = b^{x+k}$ is shifted k units to the left if k is positive or k units to the right if k is negative.

EXAMPLE 4 Graph the following functions.

a. $y = 2^x + 3$
b. $y = 2^{x+3}$

Solutions
a. Translate the graph of $y = 2^x$ three units upward to obtain the graph in the figure.
b. Translate the graph of $y = 2^x$ three units to the left to obtain the graph in the figure.

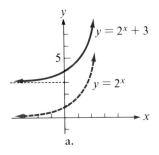
a.

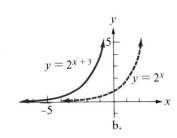
b.

EXAMPLE 5 Graph the following functions.

a. $y = 3 - 2^x$
b. $y = 3 + 2^{-x}$

Solutions
a. Reflect the graph of $y = 2^x$ about the x-axis to obtain the graph of $y = -2^x$. Then translate this graph three units upward.
b. Translate the graph of $y = 2^{-x} = (\frac{1}{2})^x$ three units upward to obtain the graph in the figure.

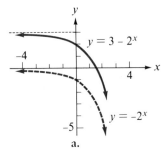
a.

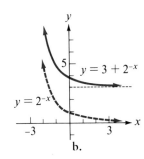
b.

Exponential Equations

An **exponential equation** is one in which the variable is part of an exponent. For example, the equation

$$3^x = 81 \tag{1}$$

is exponential. Many exponential equations can be solved by writing both sides of the equation as powers with the same base. To solve Equation (1) we would write

$$3^x = 3^4,$$

which is true if and only if $x = 4$. In general, if two equivalent powers have the same base, then their exponents must be equal also.

Sometimes the laws of exponents can be used to express both sides of an equation as single powers of a common base.

EXAMPLE 6 Solve the following equations.

a. $3^{x-2} = 9^3$
b. $27 \cdot 3^{-2x} = 9^{x+1}$

Solutions a. Using the fact that $9 = 3^2$, write each side of the equation as a power of 3:

$$3^{x-2} = (3^2)^3$$
$$3^{x-2} = 3^6.$$

Now equate the exponents to obtain

$$x - 2 = 6$$
$$x = 8.$$

b. Write each factor as a power of 3:

$$3^3 \cdot 3^{-2x} = (3^2)^{x+1}.$$

Use the laws of exponents to simplify each side:

$$3^{3-2x} = 3^{2x+2}.$$

Equate the exponents to obtain

$$3 - 2x = 2x + 2$$
$$-4x = -1$$
$$x = \frac{1}{4}.$$

Exponential equations arise frequently in the study of exponential growth.

EXAMPLE 7 During the summer a population of fleas doubles in number every 5 days. If a population starts with 10 fleas, how long will it be before there are 10,240 fleas?

Solution Let P represent the number of fleas present after t days. The original population of 10 is multiplied by a factor of 2 every 5 days, or

$$P(t) = 10 \cdot 2^{t/5}.$$

Set $P = 10{,}240$ and solve for t:

$$10{,}240 = 10 \cdot 2^{t/5}.$$

Divide both sides by 10 to obtain

$$1024 = 2^{t/5}$$
$$2^{10} = 2^{t/5}.$$

Equate the exponents to get

$$10 = \frac{t}{5},$$

or $t = 50$. The population will grow to 10,240 fleas in 50 days.

It is not always so easy to express both sides of the equation as powers of the same base. In the following sections we will develop more general methods for solving exponential equations.

EXERCISE 8.2

A

■ *Find decimal approximations for the following expressions. See Example 1.*

1. $3^{\sqrt{2}}$
2. $8^{\sqrt{5}}$
3. $4^{\pi-1}$
4. $5^{\pi+2}$
5. $-0.6^{2\sqrt{3}}$
6. $-1.2^{\sqrt{13}/2}$
7. $6^{-\sqrt{5}}$
8. $10^{-\sqrt{2}}$
9. $2.8(9)^{\sqrt{7}}$
10. $0.3(11)^{\sqrt{3}}$
11. $8 - 4^{\sqrt{13}}$
12. $9 - 7^{\sqrt{8}}$

■ *Graph the following functions. Choose appropriate scales for the axes. See Examples 2 and 3.*

13. $f(x) = 5^x$
14. $g(x) = 10^x$
15. $h(t) = 3^{-t}$
16. $q(t) = 5^{-t}$
17. $G(z) = -4^z$
18. $F(z) = -3^z$
19. $P(x) = \left(\dfrac{1}{10}\right)^x$
20. $R(x) = \left(\dfrac{1}{4}\right)^x$
21. $y = \left(\dfrac{1}{2}\right)^{-x}$
22. $y = \left(\dfrac{1}{3}\right)^{-x}$
23. $g(t) = 1.3^t$
24. $h(t) = 2.4^t$
25. $N = 0.8^x$
26. $P = 0.7^x$

■ *See Examples 4 and 5.*

27. $y = 3^x - 5$
28. $y = 4^x + 2$
29. $y = 4^{t-3}$
30. $y = 3^{t+2}$
31. $f(h) = 1 - 2^{-h}$
32. $g(p) = 5^{-p} + 10$
33. $N(t) = 20 + 10^{t+2}$
34. $N(t) = 10^{t-3} - 50$

■ *Solve each equation. See Example 6.*

35. $2^x = 32$
36. $5^x = 125$
37. $5^{x+2} = 25^{4/3}$
38. $3^{x-1} = 27^{1/2}$
39. $3^{2x-1} = \dfrac{\sqrt{3}}{9}$
40. $2^{3x-1} = \dfrac{\sqrt{2}}{16}$
41. $4 \cdot 2^{x-3} = 8^{-2x}$
42. $9 \cdot 3^{x+2} = 81^{-x}$
43. $27^{4x+2} = 81^{x-1}$
44. $16^{2-3x} = 64^{x+5}$
45. $10^{x^2-1} = 1000$
46. $5^{x^2-x-4} = 25$

■ *Solve. See Example 7.*

47. During an introductory advertising campaign in a large city, the makers of Chip-O's corn chips estimated that after t days of advertising the number of people who had heard of Chip-O's was given by $N(t) = 100 \cdot 8^{t/4}$. How many days should they run the campaign in order for Chip-O's to be familiar to 51,200 people?

48. A nationwide association of cosmetologists finds that news of a new product will spread among its members according to the formula $N(t) = 20 \cdot 9^{t/5}$, where N is the number of cosmetologists who have tried the product after t weeks. How long will it be before 14,580 cosmetologists have tried a new product?

49. Before the advent of antibiotics an outbreak of cholera might spread through a city so that the number of cases doubled every 6 days. If 26 cases had been discovered on July 5, when should hospitals have expected to be treating 106,496 cases?

50. An outbreak of ungulate fever can sweep through the livestock in a region so that the number of animals affected triples every 4 days. If a rancher does not act quickly after discovering four cases of ungulate fever, how long will it be until 324 head are affected?

51. A color television set loses 30% of its value every 2 years. How long will it be before a $700 television set depreciates to $343?

52. A mobile home loses 20% of its value every 3 years. How long will it be before a $20,000 mobile home depreciates to $12,800?

B

■ *Which of the following tables could describe exponential functions?*

53.

a.
x	y
0	3
1	6
2	12
3	24
4	48

b.
t	P
0	6
1	7
2	10
3	15
4	22

c.
x	N
0	2
1	6
2	34
3	110
4	258

d.
p	R
0	405
1	135
2	45
3	15
4	5

54.

a.
t	y
1	100
2	50
3	$33\frac{1}{3}$
4	25
5	20

b.
x	P
1	$\frac{1}{2}$
2	1
3	2
4	4
5	8

c.
h	a
0	70
1	7
2	0.7
3	0.07
4	0.007

d.
t	Q
0	0
1	$\frac{1}{4}$
2	1
3	$\frac{9}{4}$
4	4

■ *Fill in the given tables. Graph each pair of functions on the same set of axes.*

55.
x	$f(x) = x^2$	$g(x) = 2^x$
-2		
-1		
0		
1		
2		
3		
4		
5		
6		

56.
x	$f(x) = x^3$	$g(x) = 3^x$
-2		
-1		
0		
1		
2		
3		
4		
5		
6		

8.3

LOGARITHMS

Suppose that a colony of bacteria doubles in size every day. If the colony starts with 50 bacteria, how long will it be before there are 800 bacteria?

We solved problems of this type in Section 8.2 by writing and solving an appropriate exponential equation. The function

$$P(t) = 50 \cdot 2^t$$

gives the number of bacteria present on day t, so we must solve the equation

$$800 = 50 \cdot 2^t.$$

Dividing both sides by 50 yields

$$16 = 2^t.$$

The solution to this equation is the answer to the question "To what power must we raise 2 in order to get 16?" The value of t that solves the equation is called the base 2 **logarithm** of 16. Since $2^4 = 16$, the base 2 logarithm of 16 is 4. This can be written more succinctly as

$$\log_2 16 = 4.$$

In general, we make the following definition.

The **base b logarithm of x**, written $\log_b x$, is the exponent to which b must be raised in order to yield x.

EXAMPLE 1
a. $\log_3 9 = 2$ because $3^2 = 9$.
b. $\log_5 125 = 3$ because $5^3 = 125$.
c. $\log_4 \dfrac{1}{16} = -2$ because $4^{-2} = \dfrac{1}{16}$.
d. $\log_5 \sqrt{5} = \dfrac{1}{2}$ because $5^{1/2} = \sqrt{5}$.

Note in particular that

$$\log_b b = 1 \quad \text{since} \quad b^1 = b,$$
$$\log_b 1 = 0 \quad \text{since} \quad b^0 = 1,$$

and

$$\log_b b^x = x \quad \text{since} \quad b^x = b^x.$$

EXAMPLE 2 a. $\log_2 2 = 1$ b. $\log_5 1 = 0$ c. $\log_3 3^4 = 4$

From the definition of a logarithm and the examples above we see that the statements

$$y = \log_b x \quad \text{and} \quad x = b^y \tag{1}$$

are equivalent. From Equation (1) we see that the logarithm, y, is the same as the *exponent* in $x = b^y$. Thus, a logarithm is an exponent; it is the exponent to which b must be raised to yield x.

Every equation concerning logarithms can be rewritten in exponential form, and vice versa, by using the equations in (1). Thus,

$$3 = \log_2 8 \quad \text{and} \quad 8 = 2^3$$

are equivalent statements, just as

$$5 = \sqrt{25} \quad \text{and} \quad 25 = 5^2$$

are equivalent statements. The operation of taking a base b logarithm is the *inverse* of raising the base b to a power, just as extracting square roots is the inverse of squaring a number.

EXAMPLE 3 Rewrite each equation in exponential form.

a. $\log_{10} 0.001 = z$
b. $\log_3 20 = t$
c. $\log_b(3x + 1) = 3$
d. $\log_q p = w$

Solutions In each expression identify the base b and the exponent or logarithm y. Rewrite the expression in the form $x = b^y$.

    a. $10^z = 0.001$ b. $3^t = 20$

    c. $b^3 = 3x + 1$ d. $q^w = p$

EXAMPLE 4   Rewrite each equation in logarithmic form.

    a. $2^{-1} = \dfrac{1}{2}$ b. $a^{1/5} = 2.8$

    c. $6^{1.5} = T$ d. $M^v = 3K$

Solutions In each expression identify the base b and the exponent or logarithm y. Rewrite the expression in the form $y = \log_b x$.

    a. $\log_2 \dfrac{1}{2} = -1$ b. $\log_a 2.8 = \dfrac{1}{5}$

    c. $\log_6 T = 1.5$ d. $\log_M 3K = v$

Simple equations involving logarithms can sometimes be solved by rewriting them in exponential form.

EXAMPLE 5   Solve for the unknown value in each equation.

    a. $\log_2 x = 3$ b. $\log_b 2 = \dfrac{1}{2}$

    c. $\log_3 (2x - 1) = 4$ d. $2(\log_3 x) - 1 = 4$

Solutions Write each equation in exponential form and solve for the variable.

    a. $2^3 = x$ b. $b^{1/2} = 2$

        $x = 8$ $(b^{1/2})^2 = 2^2$

                                                  $b = 4$

    c. $2x - 1 = 3^4$ d. $2(\log_3 x) = 5$

        $2x = 82$ $\log_3 x = \dfrac{5}{2}$

         $x = 41$ $x = 3^{5/2}$

Estimating Logarithms

Although we can easily evaluate $\log_3 81$ ($\log_3 81 = 4$ because $3^4 = 81$) or $\log_{10} 100$ ($\log_{10} 100 = 2$ because $10^2 = 100$), it is not so easy to evaluate an expression like $\log_2 11$. We know that if $\log_2 11 = y$, then $2^y = 11$. Since 11 is not an integral power of 2, we cannot find $\log_2 11$ without the aid of a calculator or a table of logarithms. However, we can estimate the value of $\log_2 11$ between two integers. Note that $2^3 = 8$ and $2^4 = 16$. Since 11 is between 8 and 16, if $2^y = 11$, then y is between 3 and 4. Thus, we can say that $\log_2 11$ is a number between 3 and 4, or $3 < \log_2 11 < 4$.

EXAMPLE 6 Estimate the following logarithms between two integers.

a. $\log_3 7$ b. $\log_{10} 286$

Solutions

a. If $\log_3 7 = y$, then $3^y = 7$. Now, $3^1 = 3$ and $3^2 = 9$, and 7 is between 3 and 9, so y is between 1 and 2. Thus, $1 < \log_3 7 < 2$.

b. If $\log_{10} 286 = y$, then $10^y = 286$. Now, $10^2 = 100$ and $10^3 = 1000$, and 286 is between 100 and 1000, so y is between 2 and 3. Thus, $2 < \log_{10} 286 < 3$.

Base 10 Logarithms with a Calculator*

Logarithms are used to solve exponential equations. For instance, to solve the equation

$$16 \cdot 10^t = 360$$

we first divide both sides by 16 to obtain

$$10^t = 22.5,$$

then use Equation (1) on page 387 to rewrite the equation in the form

$$t = \log_{10} 22.5.$$

(Recall that $\log_{10} 22.5$ means "To what power must we raise 10 in order to get 22.5?") Because 22.5 is not an integral power of 10, the value of $\log_{10} 22.5$ is not immediately apparent. However, approximate values of $\log_{10} x$ can be found with a calculator by using the key labeled $\boxed{\log}$. To find $\log_{10} 22.5$ we press

$$22.5 \;\; \boxed{\log}$$

*Logarithms are treated using tables in the Appendix.

and the calculator displays 1.352182518. The logarithms of many numbers are irrational, so their decimal representations are nonrepeating and nonterminating. The calculator gives a decimal approximation with as many digits as its display will allow. We can then round off the answer to whatever accuracy we need.

Notice that the calculator key is labeled $\boxed{\log}$ rather than $\boxed{\log_{10}}$. Base 10 logarithms are used frequently in applications and consequently are called **common logarithms**. The subscript 10 is often omitted, so "log x" is understood to mean "$\log_{10} x$."

EXAMPLE 7 Approximate the following logarithms to two decimal places.

    a. $\log_{10} 6.5$ **b.** $\log_{10} 256$

Solutions   **a.** The keying sequence 6.5 $\boxed{\log}$ yields 0.812913356, so to two decimal places $\log_{10} 6.5 = 0.81$.

  b. The keying sequence 256 $\boxed{\log}$ yields 2.408239965, so to two decimal places $\log_{10} 256 = 2.41$.

To solve exponential equations involving powers of 10 we can use the following steps.

1. Isolate the power on one side of the equation.
2. Rewrite the equation in logarithmic form.
3. Use a calculator, if necessary, to evaluate the logarithm.
4. Solve for the variable.

EXAMPLE 8 Solve the equation

$$38 = 95 - 15 \cdot 10^{0.4x}.$$

Solution First, isolate the power of 10: subtract 95 from both sides of the equation and divide by -15 to obtain

$$-57 = -15 \cdot 10^{0.4x}$$
$$3.8 = 10^{0.4x}.$$

Rewrite the equation in logarithmic form as

$$\log_{10} 3.8 = 0.4x.$$

Use a calculator to approximate the logarithm to two decimal places and solve.

$$0.58 = 0.4x$$
$$x = 1.45.$$

EXAMPLE 9 The value of a large tractor originally worth \$30,000 depreciates exponentially according to the formula $V(t) = 30{,}000(10)^{-0.04t}$, where t is in years. When will the tractor be worth half its original value?

Solution We want to find the value of t for which $V(t) = 15{,}000$. That is, we want to solve the equation

$$15{,}000 = 30{,}000(10)^{-0.04t}.$$

Divide both sides by 30,000 to obtain

$$0.5 = 10^{-0.04t}.$$

Rewrite the equation in logarithmic form as

$$\log_{10} 0.5 = -0.04t$$

and use a calculator to approximate the logarithm, yielding

$$-0.3010 = -0.04t,$$

from which

$$t = 7.525.$$

The tractor will be worth \$15,000 in approximately 7½ years.

At this stage it seems we will be able to solve only exponential equations in which the base is 10. However, in Section 8.5 we will see how the properties of logarithms enable us to solve exponential equations with any base, knowing only the values for base 10 logarithms.

EXERCISE 8.3

A

- *Find each logarithm without using a calculator. See Examples 1 and 2.*

1. $\log_7 49$
2. $\log_2 32$
3. $\log_4 64$
4. $\log_3 27$
5. $\log_3 \sqrt{3}$
6. $\log_5 \sqrt{5}$
7. $\log_5 \dfrac{1}{5}$
8. $\log_3 \dfrac{1}{3}$
9. $\log_4 4$
10. $\log_{10} 10$
11. $\log_{10} 1$
12. $\log_6 1$
13. $\log_8 8^5$
14. $\log_7 7^6$
15. $\log_{10} 10^{-4}$
16. $\log_{10} 10^{-6}$
17. $\log_{10} 10{,}000$
18. $\log_{10} 1000$
19. $\log_{10} 0.1$
20. $\log_{10} 0.001$

- *Rewrite each equation in exponential form. See Example 3.*

21. $\log_{16} 256 = w$
22. $\log_9 729 = y$
23. $\log_b 9 = -2$
24. $\log_b 8 = -3$
25. $\log_{10} A = -2.3$
26. $\log_{10} C = -4.5$
27. $\log_4 36 = 2q - 1$
28. $\log_5 3 = 6 - 2p$
29. $\log_u v = w$
30. $\log_m n = p$

- *Rewrite each equation in logarithmic form. See Example 4.*

31. $8^{-1/3} = \dfrac{1}{2}$
32. $64^{-1/6} = \dfrac{1}{2}$
33. $t^{3/2} = 16$
34. $v^{5/3} = 12$
35. $0.8^{1.2} = M$
36. $3.7^{2.5} = Q$
37. $x^{5t} = W - 3$
38. $z^{-3t} = 2P + 5$
39. $3^{-0.2t} = 2N_0$
40. $10^{1.3t} = 3M_0$

- *Solve for the unknown value. See Example 5.*

41. $\log_b 8 = 3$
42. $\log_b 625 = 4$
43. $\log_4 x = 3$
44. $\log_{1/2} x = 5$
45. $\log_2 \dfrac{1}{2} = y$
46. $\log_5 \dfrac{1}{5} = y$
47. $\log_b 10 = \dfrac{1}{2}$
48. $\log_b 0.1 = -1$
49. $\log_2(3x - 1) = 5$
50. $\log_5(9 - 4x) = 3$
51. $3(\log_7 x) + 5 = 7$
52. $5(\log_2 x) + 6 = -14$

- *Estimate each logarithm between two integers. See Example 6.*

53. $\log_2 25$
54. $\log_3 100$
55. $\log_{10} 50$
56. $\log_{10} 7$
57. $\log_8 5$
58. $\log_6 24$
59. $\log_3 67.9$
60. $\log_5 86.3$

- *Use a calculator to approximate each logarithm to four decimal places. See Example 7.*

61. $\log_{10} 54.3$
62. $\log_{10} 27.9$
63. $\log_{10} 2344$
64. $\log_{10} 1476$
65. $\log_{10} 0.073$
66. $\log_{10} 0.00614$
67. $\log_{10} 0.6942$
68. $\log_{10} 0.0104$

■ *Solve for x. See Example 8.*

69. $10^x = 200$
70. $10^x = 6$
71. $10^{-3x} = 5$
72. $10^{-5x} = 76$
73. $25 \cdot 10^{0.2x} = 80$
74. $8 \cdot 10^{1.6x} = 312$
75. $12.2 = 2(10^{1.4x}) - 11.6$
76. $163 = 3(10^{0.7x}) - 49.3$
77. $3(10^{-1.5x}) - 14.7 = 17.1$
78. $4(10^{-0.6x}) + 16.1 = 28.2$
79. $80(1 - 10^{-0.2x}) = 65$
80. $250(1 - 10^{-0.3x}) = 100$

■ *Solve. For Problems 81–86, use the relationship* $P(a) = 30(10)^{-0.09a}$ *between altitude a in miles and atmospheric pressure P in inches of mercury. See Example 9.*

81. The elevation of Mount Everest, the highest mountain in the world, is 29,028 feet. What is the atmospheric pressure at the top?
82. The elevation of Mount McKinley, the highest mountain in the United States, is 20,320 feet. What is the atmospheric pressure at the top?
83. How high above sea level is the atmospheric pressure 20.2 inches of mercury?
84. How high above sea level is the atmospheric pressure 16.1 inches of mercury?
85. Find the height above sea level at which the atmospheric pressure is equal to one-half the pressure at sea level.
86. Find the height above sea level at which the atmospheric pressure is equal to one-fourth the pressure at sea level.
87. The population of the state of California increased during the years 1960 to 1970 according to the formula $P(t) = 15{,}717{,}000(10)^{0.0104t}$, where t is measured in years since 1960.
    a. What was the population in 1970?
    b. Assuming the same rate of growth, estimate the population of California in the years 1980, 1990, and 2000.
88. The population of the state of New York increased during the years 1960 to 1970 according to the formula $P(t) = 16{,}782{,}000(10)^{0.0036t}$, where t is measured in years since 1960.
    a. What was the population in 1970?
    b. Assuming the same rate of growth, estimate the population of New York in the years 1980, 1990, and 2000.
89. Using the formula in Problem 87, estimate when the population of California will reach 20,000,000. When will the population reach 30,000,000?
90. Using the formula in Problem 88, estimate when the population of New York will reach 20,000,000. When will the population reach 30,000,000?

B

■ *Simplify each expression.*

91. $\log_2(\log_4 16)$
92. $\log_5(\log_5 5)$
93. $\log_{10}[\log_3(\log_5 125)]$
94. $\log_{10}[\log_2(\log_3 9)]$
95. $\log_2[\log_2(\log_2 16)]$
96. $\log_4[\log_2(\log_3 81)]$
97. $\log_b(\log_b b)$
98. $\log_b(\log_a a^b)$

8.4

LOGARITHMIC FUNCTIONS

Inverse of the Exponential Function

Recall that the graph of every exponential function $f(x) = b^x$, where $b > 0$ and $b \neq 1$, looks like one of the graphs in Figure 8.3, depending on whether the base b is greater than 1 or less than 1. Both of the graphs in Figure 8.3 pass the horizontal line test, so the exponential function $f(x) = b^x$ has an inverse that is also a function.

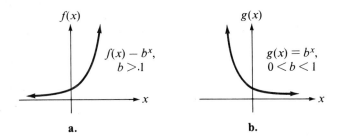

FIGURE 8.3   a.   b.

To find the inverse of $f(x) = b^x$ we write the function in the form $y = b^x$ and interchange the variables to obtain $x = b^y$. Using Equation (1) on page 387, we then write this equation in logarithmic form as $y = \log_b x$. Thus, the inverse of the exponential function $f(x) = b^x$ is the **logarithmic function** $y = \log_b x$.

EXAMPLE 1 Graph the function $f(x) = 10^x$ and its inverse, $f^{-1}(x) = \log_{10} x$ on the same axes.

Solution Make a table of values for the function $f(x) = 10^x$. A table of values for the inverse function, $f^{-1}(x) = \log_{10} x$, can be obtained by interchanging the components of each ordered pair in the table for f.

x	$f(x)$
-2	0.01
-1	0.1
0	1
1	10
2	100

x	$f^{-1}(x)$
0.01	-2
0.1	-1
1	0
10	1
100	2

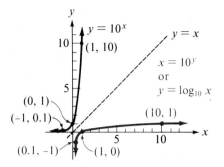

Plot each set of points and connect them with smooth curves to obtain the graphs shown in the figure.

Note that the graphs of $y = \log_b x$ and $y = b^x$ are symmetric about the line $y = x$. Recall that the domain of f^{-1} is the range of f and the range of f^{-1} is the domain of f. Thus, the logarithmic function $y = \log_b x$ has the following properties.

Properties of Logarithmic Functions $y = \log_b x$

1. Domain: all positive real numbers.
2. Range: all real numbers.
3. The graphs of $y = \log_b x$ and $y = b^x$ are symmetric about the line $y = x$.

Notice that because the domain of a logarithmic function includes only the *positive* real numbers, *the logarithm of a negative number or zero is undefined.*

While the exponential function increases very rapidly for positive domain values, its inverse, the logarithmic function, grows extremely slowly. Observe that for the common logarithmic function $y = \log_{10} x$,

$$\text{if } 0 < x \leq 1, \text{ then } \log_{10} x \leq 0;$$
$$\text{if } 1 < x \leq 10, \text{ then } 0 < \log_{10} x \leq 1.$$

(See Figure 8.4 on page 396.) In fact,

$$\text{if } 10^m < x \leq 10^n, \text{ then } m < \log_{10} x \leq n.$$

A calculator can be used to approximate the values of expressions involving $\log_{10} x$.

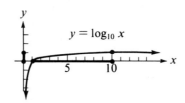

FIGURE 8.4

EXAMPLE 2 Let $f(x) = \log_{10} x$. Evaluate the following.

    a. $f(35)$ b. $f(-8)$ c. $2f(16) + 1$

Solutions
a. $f(35) = \log_{10} 35 = 1.544$
b. Since -8 is not in the domain of f, $f(-8)$, or $\log_{10}(-8)$, is undefined.
c. $2f(16) + 1 = 2(\log_{10} 16) + 1$
                    $= 2(1.204) + 1 = 3.408$

EXAMPLE 3 Evaluate the expression $T = \dfrac{\log_{10}\left(\dfrac{M_f}{M_0} + 1\right)}{k}$ for $k = 0.028$, $M_f = 1832$, and $M_0 = 15.3$.

Solution Follow the order of operations and calculate as shown:

$$T = \frac{\log_{10}\left(\dfrac{1832}{15.3} + 1\right)}{0.028} = \frac{\log_{10}(120.739)}{0.028}$$

$$= \frac{2.082}{0.028} = 74.35.$$

A calculator keying sequence for this computation is

    1832 $\boxed{\div}$ 15.3 $\boxed{+}$ 1 $\boxed{=}$ $\boxed{\log}$ $\boxed{\div}$ 0.028 .

EXAMPLE 4 If $f(x) = \log_{10} x$, find x so that $f(x) = -3.2$.

Solution Solve the equation $\log_{10} x = -3.2$. Rewriting in exponential form yields

$$x = 10^{-3.2} = 0.00063.$$

In Example 4 the expression $10^{-3.2}$ can be evaluated in two different ways with a calculator. One way is to use the $\boxed{y^x}$ key and press

$$10 \;\boxed{y^x}\; 3.2 \;\boxed{+/-}\; \boxed{=},$$

which gives the answer as 0.00063. Alternatively, we can use the key labeled $\boxed{10^x}$ and press

$$3.2 \;\boxed{+/-}\; \boxed{10^x},$$

which gives the same answer as before.

On many calculators the $\boxed{10^x}$ function is operated by pressing $\boxed{\text{INV}}\;\boxed{\log}$ or $\boxed{\text{2nd}}\;\boxed{\log}$. This is because the function $y = 10^x$ is the inverse of the function $y = \log_{10} x$. In this case the keying sequence would be

$$3.2 \;\boxed{+/-}\; \boxed{\text{INV}}\; \boxed{\log}.$$

Logarithmic Models

We have seen that while exponential functions grow more rapidly than polynomial functions, logarithmic functions grow very slowly.

There are a number of phenomena that can be modeled by logarithmic functions.

EXAMPLE 5 The acidity of a substance is determined by the concentration of hydrogen ions, denoted by [H$^+$], in the substance. Chemists use a logarithmic scale called pH to measure acidity, where

$$\text{pH} = -\log_{10}[\text{H}^+]. \tag{1}$$

The lower the pH value, the more acidic the substance.

a. Calculate the pH of a solution with a hydrogen ion concentration of 3.98×10^{-5}.
b. The water in a swimming pool should be maintained at a pH of 7.5. What is the hydrogen ion concentration of the water?

Solutions a. Use a calculator to evaluate Equation (1) with $[\text{H}^+] = 3.98 \times 10^{-5}$:

$$\text{pH} = -\log_{10}(3.98 \times 10^{-5}) = 4.4.$$

b. Solve the equation

$$7.5 = -\log_{10}[\text{H}^+]$$

for [H$^+$]. First, write

$$-7.5 = \log_{10}[\text{H}^+],$$

then rewrite the equation in exponential form to get

$$[\text{H}^+] = 10^{-7.5} = 3.2 \times 10^{-8}.$$

The hydrogen ion concentration of the water is 3.2×10^{-8}.

EXAMPLE 6 The "loudness" of a sound is measured in decibels D by

$$D = 10 \log_{10}\left(\frac{I}{10^{-12}}\right), \tag{2}$$

where I is the intensity of its sound waves (in watts per square meter).

a. A whisper generates about 10^{-10} watts per square meter at a distance of 3 feet. Find the number of decibels for a whisper 3 feet away.
b. Normal conversation registers at about 40 decibels. How many times more intense than a whisper is normal conversation?

Solutions a. Using Equation (2) with $I = 10^{-10}$:

$$D = 10 \log_{10}\left(\frac{10^{-10}}{10^{-12}}\right) = 10 \log_{10} 10^2$$
$$= 10(2) = 20 \text{ decibels.}$$

b. Let I_w stand for the intensity of a whisper and let I_c stand for the intensity of normal conversation. We are looking for the ratio I_c/I_w. From (a),

$$I_w = 10^{-10},$$

and from Equation (2) we have

$$40 = 10 \log_{10}\left(\frac{I_c}{10^{-12}}\right).$$

Dividing both members of the equation by 10 and rewriting in exponential form, we have

$$\frac{I_c}{10^{-12}} = 10^4,$$

or

$$I_c = 10^4(10^{-12}) = 10^{-8},$$

so

$$\frac{I_c}{I_w} = \frac{10^{-8}}{10^{-10}} = 10^2.$$

Normal conversation is 100 times more intense than a whisper.

EXAMPLE 7 The magnitude of an earthquake is measured by comparing the amplitude A of its seismographic trace with the amplitude A_0 of the smallest detectable earthquake. The log of their ratio is the Richter magnitude M. Thus,

$$M = \log_{10}\left(\frac{A}{A_0}\right). \tag{3}$$

a. A moderate earthquake registering 3.5 on the Richter scale occurred in Los Angeles in 1986. What would be the magnitude of an earthquake 100 times as powerful as the Los Angeles quake?

b. How many times more powerful than the Los Angeles quake was the San Francisco earthquake of 1989, which registered 7.1 on the Richter scale?

Solutions a. Let A_L represent the amplitude of the Los Angeles quake and let A_H represent the amplitude of a quake 100 times more powerful. From Equation (3) we have

$$3.5 = \log_{10}\left(\frac{A_L}{A_0}\right),$$

or, rewriting in exponential form,

$$\frac{A_L}{A_0} = 10^{3.5}.$$

Now, $A_H = 100 A_L$, so

$$\frac{A_H}{A_0} = \frac{100 A_L}{A_0}$$
$$= 100\left(\frac{A_L}{A_0}\right) = 10^2(10^{3.5})$$
$$= 10^{5.5}.$$

Thus, from Equation (3) the magnitude of the more powerful quake is

$$\log_{10}\left(\frac{A_H}{A_0}\right) = \log_{10} 10^{5.5}$$
$$= 5.5.$$

b. Let A_S stand for the amplitude of the San Francisco earthquake. Then we are looking for the ratio A_S/A_L. From Equation (3) we have

$$3.5 = \log_{10}\left(\frac{A_L}{A_0}\right) \quad \text{and} \quad 7.1 = \log_{10}\left(\frac{A_S}{A_0}\right).$$

Rewriting each equation in exponential form, we have

$$\frac{A_L}{A_0} = 10^{3.5} \quad \text{and} \quad \frac{A_S}{A_0} = 10^{7.1},$$

or

$$A_L = 10^{3.5} A_0 \quad \text{and} \quad A_S = 10^{7.1} A_0.$$

Thus,

$$\frac{A_S}{A_L} = \frac{10^{7.1} A_0}{10^{3.5} A_0} = 10^{3.6}.$$

The San Francisco earthquake was $10^{3.6}$, or approximately 3981, times as powerful as the Los Angeles quake.

Notice that in Example 7a an earthquake 100, or 10^2, times as strong is only two units greater in magnitude on the Richter scale. In general, a *difference* of K units on the Richter scale corresponds to a *factor* of 10^K in the intensity of the quake.

EXERCISE 8.4

A

■ Graph each function and its inverse on the same set of axes. See Example 1.

1. $f(x) = 2^x$
2. $f(x) = 3^x$
3. $f(x) = \left(\frac{1}{3}\right)^x$
4. $f(x) = \left(\frac{1}{2}\right)^x$

- In the following exercises, $f(x) = \log_{10} x$. Evaluate. See Example 2.

5. $f(487)$
6. $f(93)$
7. $f(2.16)$
8. $f(6.95)$
9. $f(-5)$
10. $f(0)$
11. $6f(28)$
12. $3f(41)$
13. $18 - 5f(3)$
14. $15 - 4f(7)$
15. $\dfrac{2}{5 + f(0.6)}$
16. $\dfrac{3}{2 + f(0.2)}$

- Evaluate each expression for the given values. See Example 3.

17. $t = \dfrac{1}{k} \log_{10} \dfrac{C_H}{C_L}$; for $k = 0.05$, $C_H = 2$, and $C_L = 0.5$

18. $k = \dfrac{\log_{10} P - \log_{10} P_0}{t}$; for $P = 35{,}000$, $P_0 = 18{,}000$, and $t = 26$

19. $R = \dfrac{1}{L} \log_{10}\left(\dfrac{P}{L - P}\right)$; for $L = 8500$ and $P = 3600$

20. $T = \dfrac{H \log_{10} \dfrac{N}{N_0}}{\log_{10} \dfrac{1}{2}}$; for $H = 5730$, $N = 180$, and $N_0 = 920$

21. $M = \sqrt{\dfrac{\log_{10} H}{k \log_{10} H_0}}$; for $H = 0.93$, $H_0 = 0.02$, and $k = 0.006$

22. $h = a - \sqrt{\dfrac{\log_{10} B}{t}}$; for $a = 56.2$, $B = 78$, and $t = 0.3$

- In Problems 23–28 $f(x) = \log_{10} x$. Solve for x. See Example 4.

23. $f(x) = 1.41$
24. $f(x) = 2.3$
25. $f(x) = 0.52$
26. $f(x) = 0.8$
27. $f(x) = -1.3$
28. $f(x) = -1.69$

- For Problems 29–32, see Example 5.

29. The hydrogen ion concentration of vinegar is about 6.3×10^{-4}. Calculate the pH of vinegar.
30. The hydrogen ion concentration of spinach is about 3.2×10^{-6}. Calculate the pH of spinach.
31. The pH of lime juice is 1.9. Calculate its hydrogen ion concentration.
32. The pH of ammonia is 9.8. Calculate its hydrogen ion concentration.

- For Problems 33–38, see Example 6.

33. A lawn mower generates a noise of intensity 10^{-2} watts per square meter. Find the decibel level of the sound of a lawn mower.
34. A jet airplane generates 100 watts per square meter at a distance of 100 feet. Find the decibel level for a jet airplane.
35. The loudest sound emitted by any living source is made by the blue whale. Its whistles have been measured at 188 decibels and are detectable 500 miles away. Find the intensity of the blue whale's whistle in watts per square meter.

36. The loudest sound created in a laboratory registered at 210 decibels. The energy from such a sound is sufficient to bore holes in solid material. Find the intensity of a 210-decibel sound.

37. At a concert by The Who in 1976, the sound level 50 meters from the stage registered 120 decibels. How many times more intense was this than a 90-decibel sound (the threshold of pain for the human ear)?

38. The loudest scientifically measured shouting by a human being registered 123.2 decibels. How many times more intense was this than normal conversation at 40 decibels?

■ *For Problems 39–42, see Example 7.*

39. In 1964 an earthquake in Alaska measured 8.4 on the Richter scale. An earthquake measuring 4.0 is considered small and causes little damage. How many times stronger was the Alaska quake than one measuring 4.0?

40. On April 30, 1986, an earthquake in Mexico City measured 7.0 on the Richter scale. On September 21 a second earthquake, measuring 8.1, hit Mexico City. How many times stronger was the September quake than the one in April?

41. A small earthquake measured 4.2 on the Richter scale. What would be the magnitude of an earthquake three times as strong?

42. Earthquakes measuring 3.0 on the Richter scale often go unnoticed. What would be the magnitude of a quake 200 times as strong as a 3.0 quake?

B

43. Let $f(x) = 3^x$ and $g(x) = \log_3 x$.
    a. Compute $f(4)$.
    b. Compute $g[f(4)]$.
    c. Explain why $\log_3 3^x = x$ for any x.
    d. Compute $\log_3 3^{1.8}$. Simplify $\log_3 3^a$.

44. Let $f(x) = \log_2 x$ and $g(x) = 2^2$.
    a. Compute $f(32)$.
    b. Compute $g[f(32)]$.
    c. Explain why $2^{\log_2 x} = x$ for $x > 0$.
    d. Compute $2^{\log_2 6}$. Simplify $2^{\log_2 Q}$.

45. What is the inverse operation for "raise 6 to the power x"?

46. What is the inverse operation for "take the log base 5 of x"?

47. The log base 3 of a number is 5. What is the number?

48. Four raised to a certain power is 32. What is the exponent?

49. How large must x be before the graph of $y = \log_{10} x$ reaches a height of 5?

50. How large must x be before the graph of $y = \log_2 x$ reaches a height of 10?

51. For what values of x is $\log_b(x - 9)$ defined?

52. For what values of x is $\log_b(16 - 3x)$ defined?

53. a. Complete the following chart.

x	x^2	$\log_{10} x$	$\log_{10} x^2$
1			
2			
3			
4			
5			
6			

b. Do you notice a relationship between $\log_{10} x$ and $\log_{10} x^2$? State the relationship as an equation.

54. a. Complete the following chart.

x	$\dfrac{1}{x}$	$\log_{10} x$	$\log_{10} \dfrac{1}{x}$
1			
2			
3			
4			
5			
6			

b. Do you notice a relationship between $\log_{10} x$ and $\log_{10} 1/x$? State the relationship as an equation.

■ *Assuming that the relationships you found in Exercise 53 and 54 hold for logarithms to any base, complete the following charts and use them to graph the given functions.*

55.

x	$y = \log_e x$
1	0
2	0.693
4	
16	
$\dfrac{1}{2}$	
$\dfrac{1}{4}$	
$\dfrac{1}{16}$	

56.

x	$y = \log_f x$
1	0
2	0.431
4	
16	
$\dfrac{1}{2}$	
$\dfrac{1}{4}$	
$\dfrac{1}{16}$	

8.5

PROPERTIES OF LOGARITHMS

Since logarithms are actually exponents, they have several properties that can be derived from the laws of exponents.

Properties of Logarithms

If $x, y > 0$, then

1. $\log_b(xy) = \log_b x + \log_b y$
2. $\log_b \dfrac{x}{y} = \log_b x - \log_b y$
3. $\log_b x^m = m \log_b x$

As an example of Property (1), note that

$$\log_2 32 = \log_2(4 \cdot 8) = \log_2 4 + \log_2 8$$
$$5 = 2 + 3$$

As an example of Property (2),

$$\log_2 8 = \log_2 \frac{16}{2} = \log_2 16 - \log_2 2$$
$$3 = 4 - 1$$

As an example of Property (3),

$$\log_2 64 = \log_2(4)^3 = 3 \log_2 4$$
$$6 = 3 \cdot 2$$

The three properties above can be used to simplify expressions involving logarithms or to write them in more convenient forms for solving exponential and logarithmic equations.

EXAMPLE 1 Simplify $\log_b \sqrt{\dfrac{xy}{z}}$.

Solution First, express $\sqrt{xy/z}$ using a fractional exponent:

$$\log_b \sqrt{\dfrac{xy}{z}} = \log_b \left(\dfrac{xy}{z}\right)^{1/2}.$$

By Property (3),

$$\log_b \left(\dfrac{xy}{z}\right)^{1/2} = \dfrac{1}{2} \log_b \left(\dfrac{xy}{z}\right).$$

By Property (2),

$$\dfrac{1}{2} \log_b \left(\dfrac{xy}{z}\right) = \dfrac{1}{2} (\log_b xy - \log_b z),$$

and by Property (1),

$$\dfrac{1}{2} (\log_b xy - \log_b z) = \dfrac{1}{2} (\log_b x + \log_b y - \log_b z).$$

Thus,

$$\log_b \sqrt{\dfrac{xy}{z}} = \dfrac{1}{2} (\log_b x + \log_b y - \log_b z).$$

EXAMPLE 2 Given that $\log_b 2 = 0.6931$ and $\log_b 3 = 1.0986$, find the value of $\log_b 12$.

Solution Using the properties of logarithms, express $\log_b 12$ in terms of $\log_b 2$ and $\log_b 3$. Since $12 = 2^2 \cdot 3$,

$$\begin{aligned}
\log_b 12 &= \log_b (2^2 \cdot 3) \\
&= \log_b 2^2 + \log_b 3 \\
&= 2 \log_b 2 + \log_b 3 \\
&= 2(0.6931) + 1.0986 \\
&= 2.4848.
\end{aligned}$$

We can also use the three properties of logarithms to write sums and differences of logarithms as a single logarithm.

EXAMPLE 3 Express $\frac{1}{2}(\log_b x - \log_b y)$ as a single logarithm with a coefficient of 1.

Solution By Property (2),

$$\frac{1}{2}(\log_b x - \log_b y) = \frac{1}{2}\log_b\left(\frac{x}{y}\right).$$

By Property (3),

$$\frac{1}{2}\log_b\left(\frac{x}{y}\right) = \log_b\left(\frac{x}{y}\right)^{1/2}.$$

Therefore,

$$\frac{1}{2}(\log_b x - \log_b y) = \log_b\left(\frac{x}{y}\right)^{1/2}.$$

Common Errors

Note that

$$\log_b(x + y) \neq \log_b x + \log_b y$$

and

$$\log_b \frac{x}{y} \neq \frac{\log_b x}{\log_b y}.$$

Solving Logarithmic Equations

The properties of logarithms are useful in solving equations in which the variable is part of a logarithmic expression.

EXAMPLE 4 Solve $\log_{10}(x + 1) + \log_{10}(x - 2) = 1$.

Solution Use Property (1) of logarithms to rewrite the left-hand side as a single logarithm:

$$\log_{10}(x + 1)(x - 2) = 1.$$

Once the left-hand side is expressed as a *single* logarithm we can rewrite the equation in exponential form to get

$$(x + 1)(x - 2) = 10^1,$$

from which

$$x^2 - x - 2 = 10$$
$$x^2 - x - 12 = 0$$
$$(x - 4)(x + 3) = 0.$$

Thus,

$$x = 4 \quad \text{or} \quad x = -3.$$

The number -3 is not a solution of the original equation because $\log_{10}(x + 1)$ and $\log_{10}(x - 2)$ are not defined for $x = -3$. The solution of the original equation is 4.

Solving Exponential Equations

The properties of logarithms also enable us to solve exponential equations in which the base is not 10. For example, to solve the equation

$$5^x = 7 \tag{1}$$

we could rewrite the equation in logarithmic form to obtain the solution

$$x = \log_5 7.$$

However, if we want a decimal approximation for the solution we begin by taking the base 10 logarithm of both sides of Equation (1) to get

$$\log_{10}(5^x) = \log_{10} 7.$$

Using Property (3), we rewrite the left-hand side as

$$x \log_{10} 5 = \log_{10} 7$$

and divide both sides by $\log_{10} 5$ to get

$$x = \frac{\log_{10} 7}{\log_{10} 5}.$$

Using a calculator, we find

$$x = \frac{0.8451}{0.6990} = 1.2091.$$

Common Error

Note that

$$\frac{\log_{10} 7}{\log_{10} 5} \neq \log_{10} 7 - \log_{10} 5.$$

EXAMPLE 5 Solve $1640 = 80 \cdot 6^{0.03x}$.

Solution Divide both sides by 80 to obtain

$$20.5 = 6^{0.03x}.$$

Take the base 10 logarithm of both sides of the equation and use Property (3) of logarithms to get

$$\log_{10} 20.5 = \log_{10} 6^{0.03x}$$
$$= 0.03x \log_{10} 6.$$

Solve for x and use a calculator to evaluate the answer:

$$x = \frac{\log_{10} 20.5}{0.03 \log_{10} 6}$$

$$= \frac{1.3118}{0.03(0.7782)} = 56.19.$$

EXAMPLE 6 The population of Silicon City was 6500 in 1970 and has been tripling every 12 years. When will the population reach 75,000?

Solution The population of Silicon City grows according to the formula $P(t) = 6500 \cdot 3^{t/12}$, where t is the number of years after 1970. We want to find the value of t for which $P(t) = 75,000$; that is, we want to solve the equation

$$75,000 = 6500 \cdot 3^{t/12}.$$

Divide both sides by 6500 to get

$$11.5385 = 3^{t/12},$$

then take the base 10 logarithm of both sides and solve:

$$\log_{10} 11.5385 = \log_{10} 3^{t/12}$$

$$\log_{10} 11.5385 = \frac{t}{12} \log_{10} 3$$

$$t = \frac{12(\log_{10} 11.5385)}{\log_{10} 3}.$$

Use a calculator to evaluate the answer:

$$t = \frac{12(1.0621)}{0.4771} = 26.71.$$

The population of Silicon City will reach 75,000 about 27 years after 1970, or in 1997.

Solving Formulas

The techniques discussed above can also be used to solve formulas involving exponential or logarithmic expressions for one variable in terms of the others.

EXAMPLE 7 Solve $P = Cb^{kt}$ for t $(C, k \neq 0)$.

Solution First, express the power b^{kt} in terms of the other variables:

$$b^{kt} = \frac{P}{C}.$$

Write the exponential equation in logarithmic form:

$$kt = \log_b \frac{P}{C}.$$

Multiply each member by $1/k$:

$$t = \frac{1}{k} \log_b \frac{P}{C}.$$

EXAMPLE 8 Solve $N = N_0 \log_b(ks)$ for s.

Solution First, express $\log_b(ks)$ in terms of the other variables:

$$\log_b(ks) = \frac{N}{N_0} \qquad (N_0 \neq 0).$$

Write the logarithmic equation in exponential form:

$$ks = b^{N/N_0},$$

or

$$s = \frac{1}{k} b^{N/N_0} \qquad (k \neq 0).$$

EXERCISE 8.5

A

■ Use Properties (1), (2), and (3) on page 404 to write each expression in terms of simpler logarithms. Assume that all variables denote positive real numbers. See Example 1.

1. $\log_b 2x$
2. $\log_b xy$
3. $\log_b \dfrac{x}{y}$
4. $\log_b \dfrac{y}{x}$
5. $\log_b \dfrac{xy}{z}$
6. $\log_b \dfrac{x}{yz}$
7. $\log_b x^3$
8. $\log_b x^{1/3}$
9. $\log_b \sqrt{x}$
10. $\log_b \sqrt[5]{y}$
11. $\log_b \sqrt[3]{x^2}$
12. $\log_b \sqrt{x^3}$
13. $\log_b x^2 y^3$
14. $\log_b x^{1/3} y^2$
15. $\log_b \dfrac{x^{1/2} y}{z^2}$
16. $\log_b \dfrac{xy^2}{z^{1/2}}$
17. $\log_{10} \sqrt[3]{\dfrac{xy^2}{z}}$
18. $\log_{10} \sqrt{\dfrac{2L}{R^2}}$
19. $\log_{10} 2\pi \sqrt{\dfrac{l}{g}}$
20. $\log_{10} 2y \sqrt[3]{\dfrac{x}{y}}$
21. $\log_{10} \sqrt{(s-a)(s-b)}$
22. $\log_{10} \sqrt{s^2(s-a)^3}$

■ Given that $\log_b 2 = 0.6931$, $\log_b 3 = 1.0986$, and $\log_b 5 = 1.6094$, find the value of each expression. See Example 2.

23. $\log_b 6$
24. $\log_b 10$
25. $\log_b \dfrac{2}{5}$
26. $\log_b \dfrac{3}{2}$
27. $\log_b 9$
28. $\log_b 25$
29. $\log_b \dfrac{15}{2}$
30. $\log_b \dfrac{6}{5}$
31. $\log_b (0.002)^3$
32. $\log_b \sqrt{50}$
33. $\log_b 75$
34. $\log_b \dfrac{0.08}{15}$

■ Express as a single logarithm with a coefficient of 1. See Example 3.

35. $\log_b 8 - \log_b 2$
36. $\log_b 5 + \log_b 2$
37. $2 \log_b x + 3 \log_b y$
38. $\dfrac{1}{4} \log_b x - \dfrac{3}{4} \log_b y$
39. $-2 \log_b x$
40. $-\log_b x$
41. $\dfrac{1}{2}(\log_{10} y + \log_{10} x - 3 \log_{10} z)$
42. $\dfrac{1}{3}(\log_{10} x - 2 \log_{10} y - \log_{10} z)$
43. $\dfrac{1}{2} \log_b 16 + 2(\log_b 2 - \log_b 8)$
44. $\dfrac{1}{2}(\log_b 6 + 2 \log_b 4) - \log_b 2$

■ Solve each logarithmic equation. See Example 4.

45. $\log_{10} x + \log_{10} 2 = 3$
46. $\log_6 3 + \log_6 x = 2$
47. $\log_{10} x + \log_{10}(x + 21) = 2$
48. $\log_{10}(x + 3) + \log_{10} x = 1$
49. $\log_8(x + 5) - \log_8 2 = 1$
50. $\log_{10}(x - 1) - \log_{10} 4 = 2$

51. $\log_{10}(x+2) + \log_{10}(x-1) = 1$
52. $\log_4(x+8) + \log_4(x+2) = 2$
53. $\log_3(x-2) - \log_3(x+1) = 3$
54. $\log_{10}(x+3) - \log_{10}(x-1) = 1$

■ Solve each equation by using logarithms base 10. See Example 5.

55. $2^x = 7$
56. $3^x = 4$
57. $3^{x+1} = 8$
58. $2^{x-1} = 9$
59. $4^{x^2} = 15$
60. $3^{x^2} = 21$
61. $4.26^{-x} = 10.3$
62. $2.13^{-x} = 8.1$
63. $25 \cdot 3^{2.1x} = 47$
64. $12 \cdot 5^{1.5x} = 85$
65. $3600 = 20 \cdot 8^{-0.2x}$
66. $0.06 = 50 \cdot 4^{-0.6x}$

■ Solve. See Example 6.

67. A culture of *Salmonella* bacteria is started with 0.01 gram and triples in weight every 16 hours. How long will it take for the culture to weigh 0.5 gram?

68. In 1975 Summit City used 4.2×10^6 kilowatt-hours of electricity, and the demand for electricity has increased by a factor of 1.5 every 10 years. When will Summit City need 10 million kilowatt-hours annually?

69. If the annual rate of growth of Hickory Corners is 3.7%, how long will it take for the population to double?

70. In 1986 the inflation rate in Bolivia was 8000% annually. How long did it take for prices to double?

71. The concentration of a certain drug injected into the bloodstream decreases by 20% each hour as the drug is eliminated from the body. If the initial dose creates a concentration of 0.7 milligrams per milliliter and the minimum safe concentration is 0.4 milligrams per milliliter, when should the second dose be administered?

72. A small pond is tested for pollution and the concentration of toxic chemicals is found to be 80 parts per million. Clean water enters the pond from a stream, mixes with the polluted water, then leaves the pond so that the pollution level is reduced by 10% each month. How long will it be before the concentration of toxic chemicals reaches a safe level of 25 parts per million?

73. Radioactive potassium 42, which is used by cardiologists as a tracer, decays at a rate of 5.4% per hour.
    a. Find the half-life of potassium 42, that is, the time it takes for one-half of a sample of the isotope to decay.
    b. How long will it take for three-fourths of the sample to decay? For seven-eighths of the sample?

74. Radium 226 decays at a rate of 0.4% per year.
    a. Find the half-life of radium 226.
    b. How long will it take for three-fourths of a sample of radium 226 to decay? For seven-eighths of the sample?

■ Solve each formula for the specified variable. See Examples 7 and 8.

75. $A = A_0(10^{kt} - 1)$, for t
76. $B = B_0(1 - 10^{-kt})$, for t
77. $w = pv^q$, for q
78. $L = p^a q^b$, for b
79. $t = T \log_{10}\left(1 + \dfrac{A}{k}\right)$, for A
80. $\log_{10} R = \log_{10} R_0 + kt$, for R

B

■ *Verify that each statement is true.*

81. $\log_b 4 + \log_b 8 = \log_b 64 - \log_b 2$
82. $\log_b 24 - \log_b 2 = \log_b 3 + \log_b 4$
83. $2 \log_b 6 - \log_b 9 = 2 \log_b 2$
84. $4 \log_b 3 - 2 \log_b 3 = \log_b 9$
85. $\dfrac{1}{2} \log_b 12 - \dfrac{1}{2} \log_b 3 = \dfrac{1}{3} \log_b 8$
86. $\dfrac{1}{4} \log_b 8 + \dfrac{1}{4} \log_b 2 = \log_b 2$
87. Show by a numerical example that $\log_{10}(x + y)$ is not equivalent to the expression $\log_{10} x + \log_{10} y$.
88. Show by a numerical example that $\log_{10} \dfrac{x}{y}$ is not equivalent to $\dfrac{\log_{10} x}{\log_{10} y}$.

■ *For Problems 89–96, use the following information. If P dollars is invested at an annual interest rate r (expressed as a decimal) compounded n times yearly, the amount A after t years is given by*

$$A = P\left(1 + \dfrac{r}{n}\right)^{nt}.$$

89. Find the compounded amount of $5000 invested at 12% for 10 years when compounded annually and when compounded quarterly.
90. Find the compounded amount of $800 invested at 7% for 15 years when compounded semiannually and when compounded monthly.
91. How long would it take $1000 to grow to $2000 if compounded semiannually at 6%? If compounded quarterly at 6%?
92. How long would it take $500 to grow to $1000 if compounded monthly at 8%? If compounded daily at 8%? (Use 365 days for a year.)
93. What rate of interest is required so that $1000 will yield $1900 after 5 years if the interest is compounded monthly?
94. What rate of interest is required so that $400 will yield $600 after 3 years if the interest is compounded quarterly?
95. How long will it take a sum of money to triple if it is invested at 10% compounded daily?
96. How long will it take a sum of money to increase fivefold if it is invested at 10% compounded quarterly?

8.6

THE NATURAL BASE

There is another base for logarithmic and exponential functions that is often used in applications. This base is an irrational number called *e*, where

$$e = 2.71828182845\ldots.$$

8.6 ■ THE NATURAL BASE

The number e is essential for the study of numerous advanced topics and is often referred to as the **"natural base."**

The Natural Logarithmic Function

The base e logarithm of a number x, or $\log_e x$, is called the **natural logarithm** of x and is denoted by **ln** x. Thus,

$$\ln x = \log_e x.$$

The function $y = \ln x$ is called the **natural logarithmic function**. Recall that ln x, or $\log_e x$, means "To what power must we raise e to get x?" Because e is irrational, a calculator or a table of values is necessary for any calculations involving the natural base. To evaluate the natural log of a number, use the key labeled [ln].

EXAMPLE 1 Evaluate the following natural logarithms.

    a. ln 6.6   b. ln 0.7

Solutions
a. Use the keying sequence 6.6 [ln] to find ln 6.6 = 1.8871.
b. Use the keying sequence 0.7 [ln] to find ln 0.7 = −0.3567.

The Natural Exponential Function

The **natural exponential function** is the function $y = e^x$. Values for e^x can be obtained with a calculator by using the key labeled [e^x]. On some calculators, e^x is evaluated by pressing [INV] [ln] or [2nd] [ln]. This is because the function $y = e^x$ is the inverse of the function $y = \ln x$.

EXAMPLE 2 Evaluate the following powers.

    a. $e^{2.4}$ b. $e^{-4.7}$

Solutions
a. Use the keying sequence 2.4 [e^x] or 2.4 [INV] [ln] to find $e^{2.4} = 11.023$.
b. Use the keying sequence 4.7 [+/−] [e^x] or 4.7 [+/−] [INV] [ln] to find $e^{-4.7} = 0.0090953$.

The graphs of the functions $y = e^x$ and $y = \ln x$ are shown in Figure 8.5.

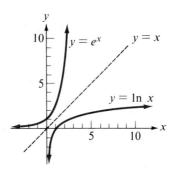

FIGURE 8.5

Solving Equations

The techniques for solving logarithmic and exponential equations with other bases also apply to equations with base e.

EXAMPLE 3 Solve each equation for x.

    a. $e^x = 0.24$ b. $\ln x = 3.5$

Solutions a. Rewrite the equation in logarithmic form and evaluate using a calculator:

$$x = \ln 0.24 = -1.427.$$

  b. Rewrite the equation in exponential form and evaluate:

$$x = e^{3.5} = 33.115.$$

In more complicated exponential equations we isolate the power on one side of the equation before rewriting in logarithmic form.

EXAMPLE 4 Solve $140 = 20\, e^{0.4x}$.

Solution First, divide each side by 20 to obtain

$$7 = e^{0.4x}$$

and then rewrite the equation in logarithmic form as

$$0.4x = \ln 7.$$

Thus,

$$x = \frac{\ln 7}{0.4},$$

which can be evaluated on some calculators as

$$7 \; \boxed{\ln} \; \boxed{\div} \; 0.4 \; \boxed{=} \; 4.8648.$$

EXAMPLE 5 Solve $P = \dfrac{a}{1 + be^{-kt}}$ for t.

Solution Multiply both sides of the equation by the denominator, $1 + be^{-kt}$, to get

$$P(1 + be^{-kt}) = a,$$

then isolate the power, e^{-kt}:

$$1 + be^{-kt} = \frac{a}{P}$$

$$be^{-kt} = \frac{a}{P} - 1 = \frac{a - P}{P}$$

$$e^{-kt} = \frac{a - P}{bP}.$$

Take the natural logarithm of both sides to get

$$\ln e^{-kt} = \ln \frac{a - P}{bP},$$

or

$$-kt = \ln \frac{a - P}{bP},$$

then solve for t:

$$t = \frac{-1}{k} \ln \frac{a - P}{bP}.$$

Exponential Growth

In Section 8.1 we considered functions of the form

$$P(t) = P_0 \cdot a^t, \tag{1}$$

which describe exponential growth when $a > 1$ and exponential decay when $0 < a < 1$. Exponential growth and decay can also be modeled by functions of the form

$$P(t) = P_0 \cdot e^{kt}, \tag{2}$$

where we have substituted e^k for the growth factor a in Equation (1) so that

$$P(t) = P_0 \cdot a^t$$
$$= P_0 \cdot (e^k)^t = P_0 \cdot e^{kt}.$$

The value of k can be found by solving the equation $a = e^k$ for k, which yields $k = \ln a$.

For instance, in Example 1 on page 364 we found that a colony of bacteria grew according to the formula

$$P(t) = 100 \cdot 3^t.$$

We can express this function in the form $P(t) = 100 \cdot e^{kt}$ if we set

$$3 = e^k \quad \text{or} \quad k = \ln 3 = 1.0986.$$

Thus, the growth law for the colony of bacteria can be written

$$P(t) = 100 \cdot e^{1.0986t}.$$

EXAMPLE 6 Express the decay law $N(t) = 60(0.8)^t$ in the form $N(t) = N_0 e^{kt}$.

Solution Since $0.8 = e^k$, $k = \ln 0.8 = -0.2231$. $N_0 = 60$, so the decay law is

$$N(t) = 60e^{-0.2231t}.$$

Growth and decay laws given in terms of the natural base are convenient to work with because values for e^x and $\ln x$ are easily found with the aid of a calculator.

EXAMPLE 7 Many savings institutions offer accounts on which the interest is compounded continuously. The amount accumulated in such an account after t years at interest rate r is given by the function

$$A(t) = P e^{rt},$$

where P is the principal invested.

    a. If $500 is invested in an account offering 8% interest compounded continuously, how much will the account be worth after 5 years?

    b. How long will it be before the account is worth $1000?

Solutions   a. Evaluate $A(t)$ for $t = 5$, with $P = 500$ and $r = 0.08$:

$$A(5) = 500 \, e^{0.08(5)}$$
$$= 500 \, e^{0.4}$$
$$= 500(1.4918) = 745.91.$$

The account will be worth $745.91 after 5 years.

    b. Substitute 1000 for A and solve the equation:

$$1000 = 500 \, e^{0.08t}.$$

Divide both sides by 500 to get

$$2 = e^{0.08t},$$

then rewrite the equation in logarithmic form as

$$0.08t = \ln 2 = 0.6931.$$

Solve for t to find $t = 8.6643$. The account will be worth $1000 after approximately 8.7 years.

EXAMPLE 8 Radioactive elements decay into more stable isotopes in such a way that the amount of radioactive material left at time t is given by an exponential decay law. A scientist starts with 25 grams of krypton 91, which decays according to the formula $N(t) = N_0 \, e^{-0.07t}$, where t is in seconds.

    a. How much krypton 91 (to the nearest hundredth of a gram) is left after 15 seconds?

    b. How long does it take for 60% of the krypton 91 to decay?

Solutions a. Evaluate $N(t)$ for $t = 15$, with $N_0 = 25$:

$$N(15) = 25\,e^{-0.07(15)}$$
$$= 25\,e^{-1.05}$$
$$= 25(0.3499) = 8.7484.$$

There are approximately 8.75 grams of krypton 91 left after 15 seconds.

b. If 60% of the krypton 91 has decayed, then 40% of the original 25 grams, or 10 grams, is left. Substitute 10 for N and solve the equation:

$$10 = 25\,e^{-0.07t}.$$

Divide both sides by 25 to get

$$0.4 = e^{-0.07t}$$

and rewrite the equation in logarithmic form as

$$-0.07t = \ln 0.4 = -0.9163.$$

Solving for t yields $t = 13.0899$. It takes approximately 13 seconds for 60% of the krypton 91 to decay.

EXERCISE 8.6

A

■ *Evaluate each logarithm. See Example 1.*

1. $\ln 3.9$
2. $\ln 6.3$
3. $\ln 16$
4. $\ln 55$
5. $\ln 0.3$
6. $\ln 0.7$
7. $\ln 1$
8. $\ln e$

■ *Find each power. See Example 2.*

9. $e^{0.4}$
10. $e^{0.73}$
11. $e^{2.34}$
12. $e^{3.16}$
13. $e^{-1.2}$
14. $e^{-2.3}$
15. $e^{-0.4}$
16. $e^{-0.62}$

■ *Solve for x. See Example 3.*

17. $e^x = 1.9$
18. $e^x = 2.1$
19. $e^x = 45$
20. $e^x = 60$
21. $e^x = 0.3$
22. $e^x = 0.9$
23. $\ln x = 1.42$
24. $\ln x = 2.03$
25. $\ln x = 0.63$
26. $\ln x = 0.59$
27. $\ln x = -2.6$
28. $\ln x = -3.4$

■ See Example 4.

29. $6.21 = 2.3\, e^{1.2x}$
30. $22.26 = 5.3\, e^{0.4x}$
31. $7.74 = 1.72\, e^{0.2x}$
32. $14.15 = 4.03\, e^{1.4x}$
33. $6.4 = 20\, e^{0.3x} - 1.8$
34. $4.5 = 4\, e^{2.1x} + 3.3$
35. $46.52 = 3.1\, e^{1.2x} + 24.2$
36. $1.23 = 1.3\, e^{2.1x} - 17.1$
37. $16.24 = 0.7\, e^{-1.3x} - 21.7$
38. $55.68 = 0.6\, e^{-0.7x} + 23.1$

■ Solve each equation for the specified variable. See Example 5.

39. $y = e^{kt}$, for t
40. $\dfrac{T}{R} = e^{t/2}$, for t
41. $y = k(1 - e^{-t})$, for t
42. $B - 2 = (A + 3)e^{-t/3}$, for t
43. $T = T_0 \ln(k + 10)$, for k
44. $P = P_0 + \ln 10k$, for k

■ Express each growth or decay law in the form $N(t) = N_0\, e^{kt}$. See Example 6.

45. $N(t) = 100 \cdot 2^t$
46. $N(t) = 50 \cdot 3^t$
47. $N(t) = 1200(0.6)^t$
48. $N(t) = 300(0.8)^t$
49. $N(t) = 10(1.15)^t$
50. $N(t) = 1000(1.04)^t$

■ Solve. See Examples 7 and 8.

51. The number of bacteria present in a culture is given by $N = N_0\, e^{0.04t}$, where N_0 is the number of bacteria present at time $t = 0$ and t is the time in hours.
   a. If 6000 bacteria were present at $t = 0$, how many were present 10 hours later?
   b. How much time must elapse (to the nearest tenth of an hour) for 6000 bacteria to increase to 10,000?

52. The amount of a radioactive element present at any time t is given by $y = y_0\, e^{-0.4t}$, where t is measured in seconds and y_0 is the amount present initially.
   a. How much of the element (to the nearest hundredth of a gram) would remain after 3 seconds if 40 grams were present initially?
   b. How much time must elapse (to the nearest hundredth of a second) for 40 grams to be reduced to 12 grams?

53. The intensity I (in lumens) of a light beam after passing through t centimeters of a filter having an absorption coefficient of 0.1 is given by $I = 1000\, e^{-0.1t}$.
   a. What is the intensity (to the nearest tenth of a lumen) of a light beam that has passed through 0.6 centimeter of the filter?
   b. How many centimeters (to the nearest tenth) of the filter will reduce the illumination to 800 lumens?

54. The voltage V across a capacitor in a certain circuit is given by the formula $V = 100(1 - e^{-0.5t})$, where t is the time in seconds.
   a. What is the voltage (to the nearest tenth of a volt) after 10 seconds?
   b. How much time must elapse (to the nearest hundredth of a second) for the voltage to reach 75 volts?

55. Hope invests $2000 at 9½% interest compounded continuously.
   a. How much will Hope's account be worth after 7 years?
   b. How long will it take for the account to grow to $5000?
56. D. G.'s savings account pays 6.25% interest compounded continuously.
   a. If D. G. deposits $500 today, how much will be in his account 6 months from now?
   b. How much should D. G. deposit now in order to have $1200 in 2 years?
57. $600 compounded continuously for 4 years yields $809.92. What is the interest rate to the nearest tenth of a percent?
58. How long will it take $300 to yield $650 at 9% interest compounded continuously?
59. All living things contain a certain amount of the isotope carbon 14. When an organism dies the carbon 14 decays according to the formula $N = N_0 e^{-0.000124t}$. Scientists can estimate the age of an organic object by measuring the amount of carbon 14 remaining.
   a. When the Dead Sea scrolls were discovered in 1947 they had 78.8% of their original carbon 14. How old were the Dead Sea scrolls then?
   b. What is the half-life of carbon 14; that is, how long does it take for half of an object's carbon 14 to decay?
60. The absorption of X rays by a lead plate of thickness t inches is given by the formula $I = I_0 e^{-1.88t}$, where I_0 is the X-ray count at the source and I is the X-ray count behind the lead plate.
   a. What percent of an X-ray beam will penetrate a lead plate ½ inch thick?
   b. How thick should the lead plate be in order to screen out 70% of the X rays?

B

61. The population of Citrus Valley was 20,000 in 1970. In 1980 it was 35,000.
   a. Find a growth law of the form $P(t) = P_0 e^{kt}$ for the population of Citrus Valley.
   b. If it continues at the same rate of growth, what will the population of Citrus Valley be in 2000?
62. In 1981 a copy of *Time* magazine cost $1.50. In 1988 the cover price was $2.00.
   a. Find a growth law of the form $P(t) = P_0 e^{kt}$ for the price of *Time*.
   b. If the price continues to increase at the same rate, what will *Time* cost in 2000?
63. The half-life (the time it takes for half of a sample of radioactive material to decay) of iodine 131 is approximately 8 days. Find a decay law of the form $N(t) = N_0 e^{kt}$, where $k < 0$, for iodine 131.
64. The half-life of hydrogen 3 is 12.5 years. Find a decay law of the form $N(t) = N_0 e^{kt}$, where $k < 0$, for hydrogen 3.

65. Follow the steps below to calculate $\log_8 20$.
    a. Let $x = \log_8 20$. Write the equation in exponential form.
    b. Take the logarithm base 10 of both sides of the equation.
    c. Simplify and solve for x.
66. Follow the steps in Problem 65 to calculate $\log_8 5$.
67. Use Problem 65 to find a formula for calculating $\log_8 Q$, where Q is any positive number.
68. Find a formula for calculating $\log_b Q$, where b is any positive number and $b \neq 1$.
69. Find a formula for calculating $\ln Q$ in terms of $\log_{10} Q$.
70. Find a formula for calculating $\log_{10} Q$ in terms of $\ln Q$.

CHAPTER REVIEW

A

[8.1]

- *For Problems 1–4, do the following.*
    a. *Write a function that describes exponential growth or decay.*
    b. *Evaluate the function at the given values.*

1. The number of computer science degrees awarded by Monroe College has increased by a factor of ³⁄₂ every 5 years since 1969. If the college granted 8 degrees in 1969, how many did it award in 1979? In 1990?
2. The price of public transportation has been rising by 10% per year since 1975. If it cost $0.25 to ride the bus in 1975, how much did it cost in 1985? How much will it cost in the year 2000 if the current trend continues?
3. A certain medication is eliminated from the body at a rate of 15% per hour. If an initial dose of 100 milligrams is taken at 8 A.M., how much is left at 12 noon? At 6 P.M.?
4. After the World Series sales of T-shirts and other memorabilia declines 30% per week. If $200,000 worth of souvenirs were sold during the Series, how much will be sold 4 weeks later? Six weeks after the Series?

[8.2]

- *Find a decimal approximation for each expression.*

5. $7^{\sqrt{2}}$
6. $3^{\pi+1}$
7. $0.2(5)^{\sqrt{3}}$
8. $6 - 2^{\sqrt{7}}$

- *Graph each function.*

9. $f(t) = 1.2^t$
10. $g(t) = 0.6^{-t}$
11. $P(x) = 2^x - 3$
12. $R(x) = 2^{x+3}$

■ Solve each equation.

13. $3^{x+2} = 9^{1/3}$
14. $2^{x-1} = 8^{-2x}$
15. $4^{2x+1} = 8^{x-3}$
16. $3^{x^2-4} = 27$

[8.3]
■ Find each logarithm.

17. $\log_2 16$
18. $\log_4 2$
19. $\log_3 \dfrac{1}{3}$
20. $\log_7 7$
21. $\log_{10} 10^{-3}$
22. $\log_{10} 0.0001$

■ Write each equation in exponential form.

23. $\log_2 3 = x - 2$
24. $\log_n q = p - 1$

■ Write each equation in logarithmic form.

25. $0.3^{-2} = x + 1$
26. $4^{0.3t} = 3N_0$

■ Solve for the unknown value.

27. $\log_3 \dfrac{1}{3} = y$
28. $\log_3 x = 4$
29. $\log_b 16 = 2$
30. $\log_2 (3x - 1) = 3$

■ Solve for x.

31. $4 \cdot 10^{1.3x} = 20.4$
32. $127 = 2(10^{0.5x}) - 17.3$
33. $3(10^{-0.7x}) + 6.1 = 9$
34. $40(1 - 10^{-1.2x}) = 30$

[8.4]
■ Evaluate each expression.

35. $k = \dfrac{1}{t}(\log_{10} N - \log_{10} N_0)$; for $t = 2.3$, $N = 12{,}000$, and $N_0 = 9{,}000$

36. $P = \dfrac{1}{k}\sqrt{\dfrac{\log_{10} N}{t}}$; for $k = 0.4$, $N = 48$, and $t = 1.2$

37. $h = k \log_{10}\left(\dfrac{N}{N - N_0}\right)$; for $k = 1.2$, $N = 6400$, and $N_0 = 2000$

38. $Q = \dfrac{1}{t}\left(\dfrac{\log_{10} M}{\log_{10} N}\right)$; for $t = 0.3$, $M = 180$, and $N = 460$

[8.5]
■ Write each expression in terms of simpler logarithms. (Assume that all variables denote positive real numbers.)

39. $\log_b\left(\dfrac{xy^{1/3}}{z^2}\right)$
40. $\log_b \sqrt{\dfrac{L^2}{2R}}$
41. $\log_{10}\left(x\sqrt[3]{\dfrac{x}{y}}\right)$
42. $\log_{10} \sqrt{(s - a)(s - g)^2}$

■ *Write each expression as a single logarithm with a coefficient of 1.*

43. $\dfrac{1}{3}(\log_{10} x - 2\log_{10} y)$

44. $\dfrac{1}{2}\log_{10}(3x) - \dfrac{2}{3}\log_{10} y$

45. $\dfrac{1}{3}\log_{10} 8 - 2(\log_{10} 8 - \log_{10} 2)$

46. $\dfrac{1}{2}(\log_{10} 9 + 2\log_{10} 4) + 2\log_{10} 5$

■ *Solve each logarithmic equation.*

47. $\log_3 x + \log_3 4 = 2$
48. $\log_2(x + 2) - \log_2 3 = 6$
49. $\log_{10}(x - 1) + \log_{10}(x + 2) = 3$
50. $\log_{10}(x + 2) - \log_{10}(x - 3) = 1$

■ *Solve each equation by using base 10 logarithms.*

51. $3^{x-2} = 7$
52. $4 \cdot 2^{1.2x} = 64$
53. $1200 = 24 \cdot 6^{-0.3x}$
54. $0.08 = 12 \cdot 3^{-1.5x}$

55. Solve $N = N_0(10^{kt})$ for t.
56. Solve $Q = R_0 + R\log_{10} kt$ for t.

[8.6]

■ *Solve each equation.*

57. $e^x = 4.7$
58. $e^x = 0.5$
59. $\ln x = 6.02$
60. $\ln x = -1.4$
61. $4.73 = 1.2\, e^{0.6x}$
62. $1.75 = 0.3\, e^{-1.2x}$

63. Solve $y = 12\, e^{-kt} + 6$ for t.
64. Solve $N = N_0 + 4\ln(k + 10)$ for k.
65. Express $N(t) = 600(0.4)^t$ in the form $N(t) = N_0\, e^{kt}$.
66. Express $N(t) = 100(1.06)^t$ in the form $N(t) = N_0\, e^{kt}$.

Summary for Part II

The section in which the symbol or property is first used is shown in parentheses.

Symbols

Δx:	change in x	(5.3)
m:	slope of a line	(5.3)
f, g, etc.:	names of functions	(6.1)
$f(x)$, etc.:	f of x, or the value of f at x	(6.1)
f^{-1}:	f inverse, or the inverse of f	(7.5)
$\log_b x$:	base b logarithm of x	(8.3)
$\log x$:	base 10 logarithm of x	(8.3)
$\ln x$:	base e logarithm of x	(8.6)

Properties

Slope of a line

$$m = \frac{\Delta y}{\Delta x} = \frac{y_2 - y_1}{x_2 - x_1} \quad (x_1 \neq x_2) \quad (5.3)$$

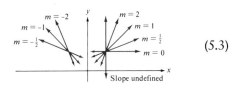
(5.3)

Two line segments with slopes m_1 and m_2 are
  parallel if $m_1 = m_2$,
  perpendicular if $m_1 m_2 = -1$ (5.4)

Forms of linear equations
  $ax + by = c$ standard form (5.2)
  $y - y_1 = m(x - x_1)$ point-slope form (5.3)
  $y = mx + b$ slope-intercept form (5.4)

Vertical line test
A graph represents a function if every vertical line intersects the graph in at most one point.

Basic graphs

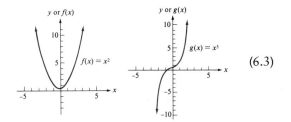
(6.3)

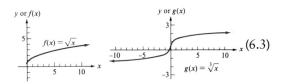
(6.3)

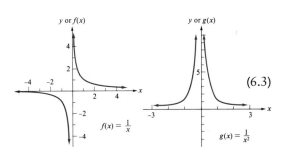
(6.3)

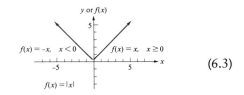
(6.3)

Vertical translations (6.4)

Compared with the graph of $y = f(x)$,

1. the graph of $y = f(x) + k$ ($k > 0$) is shifted upward k units;
2. the graph of $y = f(x) - k$ ($k > 0$) is shifted downward k units.

Horizontal translations (6.4)

Compared with the graph of $y = f(x)$,

1. the graph of $y = f(x + h)$ ($h > 0$) is shifted h units to the left;
2. the graph of $y = f(x - h)$ ($h > 0$) is shifted h units to the right.

Scale factor effect (6.4)

Compared with the graph of $y = f(x)$, the graph of $y = af(x)$ ($a \neq 0$) is

1. expanded vertically by a factor of a if $|a| > 1$;
2. compressed vertically by a factor of a if $0 < |a| < 1$;
3. reflected about the x-axis if $a < 0$.

Factor theorem (7.2)

Let $P(x)$ be a polynomial with real-number coefficients. Then

$(x - a)$ is a factor of $P(x)$
if and only if $P(a) = 0$.

Vertical asymptotes (7.4)

If $Q(a) = 0$ but $P(a) \neq 0$, the graph of the rational function $R(x) = P(x)/Q(x)$ has a vertical asymptote at $x = a$.

Horizontal asymptotes (7.4)

1. Divide the numerator and denominator of $P(x)/Q(x)$ by the highest power of x in the expression. For large values of $|x|$, any terms of the form k/x^n (k is a constant) are approximately zero and can be ignored.
2. If the resulting expression is a constant c, the graph of the function has a horizontal asymptote $y = c$.

Exponential and logarithmic functions

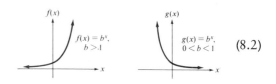 (8.2)

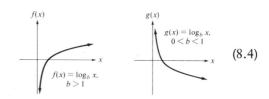 (8.4)

$y = \log_b x$ if and only if $x = b^y$. (8.3)

Properties of logarithms (8.5)

If $x, y > 0$, then

I. $\log_b(xy) = \log_b x + \log_b y$;

II. $\log_b \dfrac{x}{y} = \log_b x - \log_b y$;

III. $\log_b x^p = p \log_b x$.

Cumulative Review Exercises for Part II

■ *The numbers in brackets refer to the sections in which such problems are first considered.*

1. Simplify $\log_{10}(\log_5 5)$. [8.3]
2. If $f(x) = x^2 - 2x - 1$, find $2f(3) - 3$. [6.1]
3. If $f(x) = -\sqrt{x^2 + 1}$, find an equation defining its inverse. [7.5]
4. Find an approximation to three decimal places for $\pi^{\sqrt{3}}$. [8.2]
5. Find an equation for the line that includes the points $(-2, 5)$ and $(1, 3)$. [5.3]
6. Evaluate $f(x) = \dfrac{x+1}{x-3}$ for $x = -\dfrac{1}{2}$. [6.1]
7. Simplify $\log_{10} 2 + \log_{10} 5$. [8.5]
8. Find the slope of the graph of $\dfrac{2}{3}x - \dfrac{1}{2}y = \dfrac{1}{6}$. [5.4]
9. If $P(x) = x^4 - 2x^2 + 5x - 1$, evaluate $P(3)$ by synthetic substitution. [7.2]
10. Write an equation for the line that is perpendicular to the graph of $3x - 2y = 4$ and passes through the point $(2, -1)$. [5.4]
11. Find an equation for the line with x-intercept $\frac{1}{3}$ and y-intercept $-\frac{3}{2}$. [5.2]
12. Find all horizontal and vertical asymptotes of $y = \dfrac{3x^3 - 1}{x^3 - x}$. [7.4]
13. Find the zeros of $P(x) = x^5 - 7x^3 + 12x$. [7.2]
14. Show that $(-1, -1)$, $(9, 7)$, $(5, 12)$, and $(-5, 4)$ are the vertices of a rectangle. [5.4]
15. z varies inversely with the square of y. If $z = 6$ when $y = 5$, find z when $y = 4$. [6.4]
16. If $f(x) = 7x^4 - 2x^3 + x + 5$, find $f^{-1}(5)$. [7.5]

■ *Solve each equation.*

17. $38.6 = 12.2\, e^{-1.6x} + 13.2$ [8.6]

18. $\log_{10} x + \log_{10}(x - 3) = 1$ [8.5]
19. $\log_{64} x = \dfrac{1}{3}$ [8.3]
20. $5^{x+2} = 7$ [8.3]
21. $\ln x = -0.26$ [8.6]
22. Solve for y explicitly in terms of x: $2x^2y + 5y = xy - 2$.   [5.1]
23. Solve for k: $P = P_0 e^{-kt} + C$.   [8.6]
24. Graph the line whose slope is $-\frac{3}{2}$ and whose y-intercept is -1.   [5.4]

■ **Graph.**

25. $3x + 2y = 9$ [5.2]
26. $f(x) = 2x - 6$ [6.3]
27. $f(x) = x^2 - x - 6$ [6.5]
28. $P(t) = \sqrt{t} + 2$ [7.1]
29. $h(x) = \dfrac{-2}{x^2}$ [6.3]
30. $Q(x) = (x + 2)(x - 1)(x - 3)$ [7.3]
31. $y = \log_3 x$ [8.4]
32. $g(s) = \dfrac{1}{s} - 2$ [7.1]
33. $h(t) = 2^{t-3} + 1$ [8.2]
34. $F(t) = 1.4^t$ [8.2]
35. $P(x) = x^3 + x^2 - 4x - 4$ [7.3]
36. $R(x) = \dfrac{x - 1}{x + 3}$ [7.4]
37. $G(x) = \dfrac{2x}{x - 1}$ [7.4]
38. $f(t) = |t - 1| + 2$ [7.1]
39. $f(s) = \sqrt[3]{s - 1} + 1$ [7.1]
40. $f(x) = -2x^2 + x + 3$ [6.5]
41. $R(x) = \dfrac{x^2 - 4}{x^2 + 1}$ [7.4]
42. $P(x) = x^3 - 4x^2 + 4x$ [7.3]
43. $g(t) = \begin{cases} t + 2 & \text{if } t \leq 0 \\ t^2 & \text{if } t > 0 \end{cases}$ [6.3]
44. $F(x) = \begin{cases} x^3 & \text{if } x \leq 0 \\ \sqrt{x} & \text{if } x > 0 \end{cases}$ [6.3]

■ **Graph** $y = f^{-1}(x)$ **for the following functions.**

45. $f(x) = \sqrt[3]{x}$ [7.4]
46. $f(x) = 4x - 2$ [7.4]
47. Give the domain and range of the graph shown below.   [6.2]

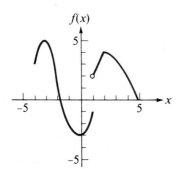

48. For the graph shown in problem 47, do the following. [6.2]
    a. Find all x for which $f(x) = 2$.
    b. Find $f(2) - f(-1)$.
49. The population of Abingdon doubled in 25 years. What was its annual growth factor? [8.1]
50. Nicola spends Saturday afternoon reading Tolstoy's *War and Peace*. At 2 P.M. she is on page 378, and at 3:30 she is on page 438.
    a. Write a linear equation that expresses Nicola's page number in terms of time.
    b. What is the slope of the line?
    c. What is the significance of the slope in terms of the problem? [5.3]
51. What rate of interest is required for an investment to double in 8 years if the interest is compounded monthly? [8.5]
52. Find the maximum possible area for a rectangle with a perimeter of 80 inches. [6.5]
53. Housing prices are increasing in Sunnyvale at 23% annually. How much will a $90,000 house cost in 5 years? [8.1]
54. A rectangular tool chest has a square base, and its height is half its length. Express its surface area S as a function of its length x. Find the surface area of a tool chest with a base that measures 20 inches on a side. [6.4]
55. The amount of brass needed (and hence the cost) for a souvenir replica of the Statue of Liberty is proportional to the cube of its height. The large-sized souvenir is 60% taller than the smaller one. How much more does the large one cost? [6.4]
56. A certain radioactive isotope decays so that the amount present after t seconds is $N = N_0 e^{-0.2t}$, where N_0 is the initial amount. How long will it be until only 25% of the isotope is left? [8.6]
57. Define function, domain, and range. [6.2]
58. Compare linear and exponential growth. [8.1]
59. Describe a method of determining from the graph of a function whether its inverse is also a function. [7.5]
60. Sketch graphs that illustrate the following. [6.4]
    a. s varies directly with the square of t.
    b. v varies inversely with n.

PART III

Additional Algebraic Topics

Systems of Linear Equations

9.1

SYSTEMS IN TWO VARIABLES

In this chapter we consider problems that can be modeled by a system of two or more linear equations and study methods for solving such systems. Systems of equations in several variables are very useful for applied problems that involve more than one unknown quantity.

As an example, suppose that the owner of a sporting goods store stocks two kinds of sleeping bags, a standard model and a down-filled model for colder temperatures. From past experience she estimates that she will sell twice as many of the standard variety as of the down filled. She has room to stock 60 sleeping bags at a time. How many of each variety should she order?

To solve the problem we write an equation to represent each of the conditions described. If x stands for the number of standard sleeping bags that the owner should order and y stands for the number of down filled, then

$$x = 2y$$

and (1)

$$x + y = 60.$$

A pair of linear equations in two variables such as (1) above is called a 2×2 **system** of linear equations. A **solution** to the system is an ordered pair (x, y) that satisfies each equation in the system. To get a better idea of what this means, consider the graphs of the two equations in Figure 9.1 on page 432.

Recall that every point on the graph of an equation represents a solution to the equation. A solution to *both* equations thus corresponds to a point on *both* graphs. In other words, a solution to the system is a point where the two

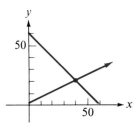

FIGURE 9.1

graphs intersect. From Figure 9.1 it appears that the point (40, 20) is the intersection point, so we would expect that the values $x = 40$ and $y = 20$ are a solution to the problem. We can check the solution by substituting the values into *each* equation of System (1).

EXAMPLE 1 Solve the system

$$x + y = 5$$
$$x - y = 1$$

by graphing.

Solution Use the intercept method of graphing to obtain the graphs in the figure. It appears that the two lines intersect at the point (3, 2). Verify that (3, 2) is indeed a solution to the system by substituting (3, 2) into each equation and observing that a true statement results in each case.

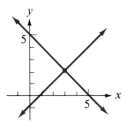

Solution of Systems by Substitution

Graphing equations is a time-consuming process, and it is not always possible to determine precisely the intersection point of two equations. Therefore, the solutions to systems of equations are usually sought by algebraic methods. You are probably already familiar with the method of substitution, which involves solving one of the equations for one of the variables in terms of the other, then substituting this expression into the second equation.

To solve System (1) on page 431 by substitution note that the first equation is already solved for x in terms of y: $x = 2y$. Substitute $2y$ for x in the second equation to obtain

$$2y + y = 60,$$

or

$$3y = 60.$$

Solving for y we find $y = 20$, and substituting this value into either original equation yields $x = 40$.

Linear Combinations

The method of substitution does not generalize easily to larger linear systems, so we will concentrate our efforts on a second method that depends on the following properties.

---

1. Any ordered pair that satisfies the equation

$$ax + by = c$$

also satisfies the equation

$$Aax + Aby = Ac,$$

where A is any real number.

2. Any ordered pair that satisfies the equations

$$a_1 x + b_1 y = c_1$$
$$a_2 x + b_2 y = c_2$$

also satisfies the equation

$$(a_1 + a_2)x + (b_1 + b_2)y = c_1 + c_2.$$

Property (1) states that we can multiply each term of a linear equation by a constant (except zero) without changing its solution. Property (2) states that if we add the like terms in two linear equations, we obtain an equation whose solutions include the solutions of the original system. A new equation obtained by performing the operations described in Property (2) above is called a **linear combination** of the original equations.

We can use linear combinations to solve a system of equations by *eliminating* one of the variables, resulting in an equation in one variable. The technique is illustrated in the following examples.

EXAMPLE 2 Solve the system

$$2x + 3y = 8 \qquad (1)$$
$$3x - 4y = -5 \qquad (2)$$

by the method of linear combinations.

Solution Multiply each equation by an appropriate constant so that the coefficients of one of the variables (choose either x or y) will have the same value but opposite signs. If we choose to eliminate x we multiply Equation (1) by 3 and Equation (2) by -2 to obtain

$$6x + 9y = 24 \qquad (1a)$$
$$-6x + 8y = 10. \qquad (2a)$$

Add the corresponding terms of (1a) and (2a). The x-terms are "eliminated," yielding an equation in one variable,

$$17y = 34,$$

or

$$y = 2.$$

Thus, the y-component of the solution is 2. To find the x-component substitute 2 for y in either (1) or (2) and solve for x. (If the y-value is correct, both equations will yield the same x-value.) If we use (1), then

$$2x + 3(2) = 8,$$

or

$$x = 1.$$

Verify the solution by substituting $x = 1$ and $y = 2$ into Equation (2). The ordered pair (1, 2) is the solution of the system.

If either equation in a system has fractional coefficients, it is helpful to clear the fractions before attempting the method of linear combinations.

EXAMPLE 3 Solve the system by linear combinations:

$$\frac{2}{3}x - y = 2 \qquad (1)$$

$$x + \frac{1}{2}y = 7. \qquad (2)$$

Solution Multiply each side of Equation (1) by 3 and each side of Equation (2) by 2 to clear fractions:

$$2x - 3y = 6 \qquad (1a)$$
$$2x + y = 14. \qquad (2a)$$

To eliminate the variable x multiply Equation (2a) by -1 and add the result to Equation (1a) to get

$$-4y = -8$$
$$y = 2.$$

Substitute 2 for y in (1), (2), (1a), or (2a) and solve for x. Using Equation (2) we find

$$x + \frac{1}{2}(2) = 7$$
$$x = 6.$$

Now verify the solution by substituting $x = 6$ and $y = 2$ into Equation (1). The solution to the system is the ordered pair $(6, 2)$.

Inconsistent and Dependent Systems

It is not always the case that a system of linear equations has a unique solution as in the examples above. This is because two straight lines do not always intersect at a single point. In fact, there are three possibilities, as illustrated in Figure 9.2 on page 436.

    a. The graphs may be the same line.
    b. The graphs may be parallel but distinct lines.
    c. The graphs may intersect at one and only one point.

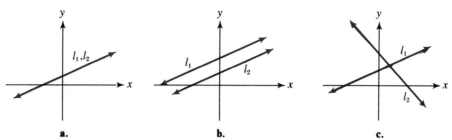

FIGURE 9.2   a.   b.   c.

These geometric possibilities lead, respectively, to the following cases for the solution of a linear system.

a. All the solutions of one equation are also solutions of the second equation and hence are solutions of the system. In this case the equations are said to be **dependent** and the system has infinitely many solutions.
b. The graphs of the equations do not intersect, so there is no solution to the system. The system is said to be **inconsistent**.
c. The system has one and only one solution. The system is said to be **consistent** and the equations are **independent**.

It may not be immediately apparent from the equations in a system whether there is one solution, no solution, or infinitely many solutions. However, the method of linear combinations will reveal which of the three cases applies. For example, consider the system

$$x + y = 5 \qquad (1)$$
$$2x + 2y = 3. \qquad (2)$$

If we multiply Equation (1) by -2 and add the result to Equation (2) we obtain

$$0x + 0y = -7. \qquad (3)$$

Since there are no values of x and y that satisfy Equation (3), it has no solution, and therefore the system has no solution. We conclude that the system is *inconsistent*. In fact, we can see from the graph of the system in Figure 9.3 that the lines are parallel and hence have no intersection point.

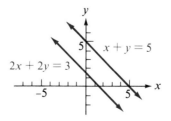

FIGURE 9.3

Now consider the system

$$2x - 3y = 1 \qquad (4)$$
$$-4x + 6y = -2. \qquad (5)$$

If we multiply Equation (4) by 2 and add the result to Equation (5) we obtain

$$0x + 0y = 0. \qquad (6)$$

Since *all* values of x and y satisfy Equation (6), the system has infinitely many solutions—in fact, the two equations of the system have the same graph, as shown in Figure 9.4. The equations are *dependent;* every point on the line represents a solution to the system.

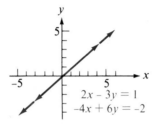

FIGURE 9.4

We summarize these results as follows.

---

1. If an equation of the form

$$0x + 0y = k \qquad (k \neq 0)$$

is obtained as a linear combination of the equations in a system, then the system is inconsistent.

2. If an equation of the form

$$0x + 0y = 0$$

is obtained as a linear combination of the equations in a system, then the equations are dependent.

EXAMPLE 4 Solve each system.

a. $2x = 2 - 3y$
   $6y = 7 - 4x$

b. $3x - 4 = y$
   $2y + 8 = 6x$

Solutions a. First, rewrite the system in standard form as

$$2x + 3y = 2 \qquad (1)$$
$$4x + 6y = 7. \qquad (2)$$

Multiply Equation (1) by -2 and add the result to Equation (2) to obtain

$$0x + 0y = 3.$$

The system is inconsistent and has no solution.

b. Rewrite the system in standard form as

$$3x - y = 4 \qquad (3)$$
$$-6x + 2y = -8. \qquad (4)$$

Multiply Equation (3) by 2 and add the result to Equation (4) to obtain

$$0x + 0y = 0.$$

The system is dependent and has infinitely many solutions.

Applications   In Chapter 4 we solved a variety of problems by writing equations in one variable to model the problem. This usually required us to express several quantities in terms of a *single* variable. In most situations it is easier to assign *different* variables to represent different quantities, but we must then write two or more equations describing the conditions of the problem. This results in a system of equations.

EXAMPLE 5   A lumber company offers split rail fencing for sale in two options. One option consists of four posts and six rails for \$31; the other includes three posts and four rails for \$22. What are the costs of one post and one rail?

Solution   Represent each unknown quantity by a separate variable:

cost of one post: p;
cost of one rail: r.

Represent the two conditions stated in the problem by two equations:

$$4p + 6r = 31 \qquad (1)$$
$$3p + 4r = 22. \qquad (2)$$

Solve the system by the method of linear combinations; multiply Equation (1) by 3 and Equation (2) by -4:

$$12p + 18r = 93 \qquad (3)$$
$$-12p - 16r = -88. \qquad (4)$$

Add the corresponding terms of the two equations to eliminate p, yielding

$$2r = 5$$
$$r = 2.5.$$

Substitute 2.5 for r in (1) or (2); using Equation (2),

$$3p + 4(2.5) = 22$$
$$p = 4.$$

Now verify the solution by substituting $r = 2.5$ and $p = 4$ into Equation (1). Thus, one post costs $4.00 and one rail costs $2.50.

In economics the number of an item that the public will buy usually decreases as the price increases. The **demand function** for a product gives the number of units of the product that the public will buy in terms of the price per unit. On the other hand, producers will be willing to supply more units of a product if the price increases. The **supply function** for a product gives the number of units that the manufacturer will supply in terms of the price per unit. The price at which the supply and demand are equal is called the **equilibrium price**. This is the price at which the consumer and the manufacturer agree to do business.

EXAMPLE 6 A woolens mill can produce $400x$ yards of fine suit fabric if it charges x dollars per yard. Its clients in the garment industry will buy $6000 - 100x$ yards of wool fabric at a price of x dollars per yard.

a. Graph the supply function, $y = S(x) = 400x$, and the demand function, $y = D(x) = 6000 - 100x$, on the same set of axes.
b. Find the equilibrium price. How many yards of fabric will change hands at that price?

Solutions a. The graphs of $S(x) = 400x$ and $D(x) = 6000 - 100x$ are shown in the figure.

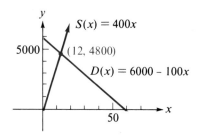

b. To find the equilibrium price, solve the system

$$y = 400x$$
$$y = 6000 - 100x.$$

The most efficient way to solve this system is to use substitution to write

$$400x = 6000 - 100x,$$

or

$$500x = 6000,$$

from which we find $x = 12$. Thus, the equilibrium price is $12 per yard. The mill will sell 400(12), or 4800, yards of fabric at that price.

EXERCISE 9.1

A

■ Solve each system by linear combinations. Sketch the graphs of the equations. See Examples 1 and 2.

1. $2x - 3y = 6$
   $x + 3y = 3$

2. $x + 2y = -6$
   $2x - 3y = 16$

3. $3x + y = 7$
   $2x = 5y - 1$

4. $2x = y + 7$
   $2y = 14 - 3x$

5. $2x - 3y = -4$
   $5x + 2y = 9$

6. $3x + 5y = 1$
   $2x - 3y = 7$

■ *Solve by linear combinations. See Examples 2 and 3.*

7. $3y = 2x - 8$
    $4y + 11 = 3x$

8. $4x - 3 = 3y$
    $25 + 5x = -2y$

9. $\frac{2}{3}x - y = 4$
    $x - \frac{3}{4}y = 6$

10. $\frac{1}{7}x - \frac{3}{7}y = 1$
    $\frac{1}{2}x - \frac{1}{4}y = -1$

11. $\frac{x}{4} = \frac{y}{3} - \frac{5}{12}$
    $\frac{y}{5} = \frac{1}{2} - \frac{x}{10}$

12. $\frac{x}{3} = \frac{2y}{3} + 2$
    $\frac{y}{3} = \frac{x}{6} - 1$

13. $\frac{x}{2} = \frac{7}{6} - \frac{y}{3}$
    $\frac{x}{4} = \frac{3}{4} - \frac{y}{4}$

14. $\frac{2x}{3} + \frac{8y}{9} = \frac{4}{3}$
    $\frac{x}{3} = 2 + \frac{y}{2}$

15. $4.8x - 3.5y = 5.44$
    $2.7x + 1.3y = 8.29$

16. $6.4x + 2.3y = -14.09$
    $-5.2x - 3.7y = -25.37$

17. $0.9x = 25.78 + 1.03y$
    $0.25x + 0.3y = 85.7$

18. $0.02x = 0.6y - 78.72$
    $1.1y = 0.4x + 108.3$

■ *Solve for x and y.*

19. $x - y = a$
    $x + y = b$

20. $ax + y = b$
    $x - cy = d$

21. $ax + by = c$
    $dx + ey = f$

22. $ax + by = c$
    $ax - by = d$

■ *Identify each system as dependent, inconsistent, or consistent and independent. See Example 4.*

23. $x + 3y = 6$
    $2x + 6y = 12$

24. $3x - 2y = 6$
    $6x - 4y = 8$

25. $2x = y + 1$
    $8x - 4y = 3$

26. $6x = 1 - 2y$
    $12x + 4y = 2$

27. $x - 3y = 4$
    $2x + y = 6$

28. $2x + y = 4$
    $x - 3y = 2$

29. $2x - 5y = 6$
    $\frac{15y}{2} + 9 = 3x$

30. $-3x = 4y + 8$
    $\frac{1}{2}x + \frac{4}{3} = \frac{-2}{3}y$

■ *Solve each problem using a system of equations. See Example 5.*

31. The admission at a baseball game was $1.50 for adults and $0.85 for children. The ticket office took in $93.10 for 82 paid admissions. How many adults and how many children attended the game?

32. There were 42 passengers on an airplane flight for which first-class fare was $80 and tourist fare was $64. If receipts for the flight totaled $2880, how many first-class and how many tourist passengers were on the flight?

33. In a recent election 7179 votes were cast for two candidates. If 6 votes had been switched from the winner to the loser, the loser would have won by one vote. How many votes were cast for each candidate?

34. Delbert answered 13 true-false and 9 fill-in questions correctly on his last test and got a score of 71. If he had answered 9 true-false and 13 fill-ins correctly, he would have made a 93. How many points was each type of problem worth?

35. Dan invested $2000, part in bonds paying 10% and the rest in a certificate account at 8%. Find the amount Dan invested at each rate if his yearly income from the two investments is $184.

36. Carmella has $1200 invested in two stocks; one returns 8% per year and the other returns 12% per year. How much did she invest in each stock if the income from the 8% stock is $3 more than the income from the 12% stock?

37. A silversmith needs 40 pounds of a 48% silver alloy to finish a collection of jewelry. How many pounds of 45% silver alloy should he melt with a 60% silver alloy to obtain the alloy he needs?

38. A chemist plans to make 10 liters of a 17% acid solution by mixing a 20% acid solution with a 15% acid solution. How much of each should she use?

39. An airplane travels 1260 miles in the same time that an automobile travels 420 miles. If the rate of the airplane is 120 miles per hour faster than the rate of the automobile, find the rate of each.

40. Two cars start together and travel in the same direction, one going twice as fast as the other. At the end of 3 hours they are 96 miles apart. How fast is each traveling?

41. Because of prevailing easterly winds a flight from Detroit to Denver, a distance of 1120 miles, takes 4 hours on Econoflite, while the return trip takes 3.5 hours. Find the speed of the airplane and the speed of the wind.

42. On a breezy day a speed skater covered the 100-meter straightaway heading into the wind in 14.5 seconds and made the 100-meter backstretch, with the wind at his back, in 10.5 seconds. Find the speed of the wind and the skater's speed in still air in meters per second and then in miles per hour.

43. A cup of rolled oats provides 11 grams of protein and 310 calories. A cup of rolled wheat flakes provides 8.5 grams of protein and 290 calories. A new breakfast cereal combines wheat and oats to provide 10 grams of protein and 302 calories per cup. How much of each grain does 1 cup of the cereal include?

44. Acme Motor Company is opening a new plant to produce chassis for two of its models, a sports coupe and a wagon. Each sports coupe requires a riveter for 3 hours and a welder for 4 hours; each wagon requires a riveter for 4 hours and a welder for 5 hours. If the plant has available 120 hours of riveting and 155 hours of welding per day, how many of each model of chassis can it produce in a day?

45. The manager of Books for Cooks plans to spend $400 stocking a new diet cookbook. The paperback version costs her $5, and the hardback costs $10. She finds that she will sell three times as many paperbacks as hardbacks. How many of each should she buy?

46. The mathematics department has $19,920 to set up a new computer lab. It will need one printer for every four terminals it purchases. If a printer costs $280 and a terminal costs $760, how many of each should be bought?

■ *Solve each problem. See Example 6.*

47. A record company determines that each production run to manufacture a record involves an initial set-up cost of $20 and $0.40 for each record produced. The records sell for $1.20 each.
    a. Express the cost C of production in terms of the number x of records produced.
    b. Express the revenue R in terms of the number x of records sold.
    c. Graph the revenue and cost functions on the same set of axes.
    d. How many records must be sold for the record company to break even on a particular production run?

48. The Bread Alone Bakery has a daily overhead of $80. It costs $0.60 to produce each loaf of bread, and the bread sells for $1.50 per loaf.
    a. Express the cost C in terms of the number x of loaves produced.
    b. Express the revenue R in terms of the number x of loaves sold.
    c. Graph the cost and revenue functions on the same set of axes.
    d. How many loaves must the bakery sell to break even on a given day?

49. A farmer can afford to produce $50x$ bushels of wheat if he can sell them at x cents per bushel, and the market will buy $2100 - 20x$ bushels at x cents per bushel.
    a. Graph the supply function, $S(x) = 50x$.
    b. Graph the demand function, $D(x) = 2100 - 20x$, on the same set of axes.
    c. Find the equilibrium price, that is, the price at which supply equals demand. How many bushels of wheat will be sold at that price?

50. Mel's Pool Service can clean $1.5x$ pools per week if it charges x dollars per pool, and the public will book $120 - 2.5x$ pool cleanings at x dollars per pool.
    a. Graph the supply function, $S(x) = 1.5x$.
    b. Graph the demand function, $D(x) = 120 - 2.5x$, on the same set of axes.
    c. Find the equilibrium price, that is, the price at which supply equals demand. How many pools will be cleaned at that price?

51. Dash Phone Company charges a monthly fee of $10 plus $0.09 per minute for long-distance calls. Friendly Phone Company charges $15 per month plus $0.05 per minute.
    a. On the same set of axes graph the monthly charges for each company as a function of minutes of long-distance calls.
    b. How many minutes of long-distance calls would result in equal bills from the two companies?

52. The Olympus Health Club charges an initial fee of $230 and $13 monthly dues. The Valhalla Health Club charges $140 initially and $16 per month.
    a. On the same set of axes graph the cost of belonging to each club as a function of time.
    b. After how many months of membership would the costs of belonging to the two clubs be equal?

B

53. Find a and b so that the solution of the system below is $(1, 2)$:
$$ax + by = 4$$
$$bx - ay = -3.$$

54. Find a and b so that the solution of the system below is $(-3, -1)$:
$$2ax - by = -7$$
$$bx + ay = -17.$$

55. Use a system of equations to find the equation of the line that passes through the points $(-1, 2)$ and $(-3, 0)$. (*Hint:* Find m and b so that both points satisfy $y = mx + b$.)

56. Use a system of equations to find the equation of the line that passes through the points $(-2, 4)$ and $(5, -3)$. (See the hint for Problem 55.)

■ Solve each system by using the substitutions $u = 1/x$, $v = 1/y$. Solve for u and v, then for x and y.

57. $\dfrac{1}{x} + \dfrac{1}{y} = 7$

$\dfrac{2}{x} + \dfrac{3}{y} = 16$

58. $\dfrac{1}{x} + \dfrac{2}{y} = -\dfrac{11}{12}$

$\dfrac{1}{x} + \dfrac{1}{y} = -\dfrac{7}{12}$

59. $\dfrac{5}{x} - \dfrac{6}{y} = -3$

$\dfrac{10}{x} + \dfrac{9}{y} = 1$

60. $\dfrac{1}{x} + \dfrac{2}{y} = 11$

$\dfrac{1}{x} + \dfrac{2}{y} = -1$

61. $\dfrac{1}{x} - \dfrac{1}{y} = 4$

$\dfrac{2}{x} - \dfrac{1}{2y} = 11$

62. $\dfrac{2}{3x} + \dfrac{3}{4y} = \dfrac{7}{12}$

$\dfrac{4}{x} - \dfrac{3}{4y} = \dfrac{7}{4}$

63. Show that the equations
$$a_1 x + b_1 y = c_1 \quad (a_1, b_1, c_1 \neq 0)$$
$$a_2 x + b_2 y = c_2 \quad (a_2, b_2, c_2 \neq 0)$$

are dependent if and only if
$$\frac{a_1}{a_2} = \frac{b_1}{b_2} = \frac{c_1}{c_2}.$$

(*Hint:* Write the equations in slope-intercept form.)

64. Show that the system
$$a_1 x + b_1 y = c_1 \quad (a_1, b_1, c_1 \neq 0)$$
$$a_2 x + b_2 y = c_2 \quad (a_2, b_2, c_2 \neq 0)$$

is inconsistent if and only if
$$\frac{a_1}{a_2} = \frac{b_1}{b_2} \neq \frac{c_1}{c_2}.$$

(*Hint:* Write the equations in slope-intercept form.)

9.2

SYSTEMS IN THREE VARIABLES

Some problems involve three (or more) unknown quantities. The notion of a system of equations can be extended to any number of variables, and there are efficient techniques for solving linear systems in several variables. In this section we introduce some of these techniques for systems of three linear equations in three variables.

A solution to an equation in three variables, such as

$$x + 2y - 3z = -4,$$

is an **ordered triple** of numbers (x, y, z). For example, $(0, -2, 0)$ and $(-1, 0, 1)$ are solutions of the equation above but $(1, 1, 1)$ is not. As with an equation in two variables, there are infinitely many solutions to an equation in three variables.

A solution to a *system* of three linear equations in three variables is an ordered triple that satisfies each equation in the system. For example, the system

$$3x - y - z = 9$$
$$-x + 2y - z = -2$$
$$x + 3z = -4$$

has as a solution the ordered triple $(2, -1, -2)$. You can verify that each of the equations of the system is satisfied by substituting 2 for x, -1 for y, and -2 for z.

The equations in the system above are said to be in *standard form;* that is,

$$ax + by + cz = d.$$

Back-Substitution

To investigate the solution of 3×3 systems, we first consider a special case in which the second equation has at most two variables and the third equation has only one variable. Such systems are easy to solve with a technique called **back-substitution.**

EXAMPLE 1 Solve the system

$$x + 2y + 3z = 2$$
$$-2y - 4z = -2$$
$$3z = -3$$

by back-substitution.

Solution Solve the third equation to find $z = -1$. Then substitute -1 for z in the second equation and solve for y to find

$$-2y - 4(-1) = -2$$
$$y = 3.$$

Finally, substitute -1 for z and 3 for y into the first equation to find

$$x + 2(3) + 3(-1) = 2$$
$$x = -1.$$

The solution is the ordered triple $(-1, 3, -1)$.

Linear Combinations

We can solve a 3×3 linear system with the method of linear combinations, although the procedure is longer for a larger system. Consider the system

$$x + 2y - 3z = -4 \quad (1)$$
$$2x - y + z = 3 \quad (2)$$
$$3x + 2y + z = 10. \quad (3)$$

The idea is to eliminate one of the variables from each of the three equations by considering them in pairs. This results in a 2×2 system that can be solved by the technique of Section 9.1.

We can choose any one of the three variables to eliminate first. For the example above we will eliminate the xs. We then choose two of the equations, say, (1) and (2), and use a linear combination: we multiply Equation (1) by -2 and add the result to Equation (2) to produce

$$-2x - 4y + 6z = 8 \quad (1a)$$
$$\underline{2x - y + z = 3} \quad (2a)$$
$$-5y + 7z = 11. \quad (4)$$

Next, we choose a different pair of equations, say (1) and (3), and *eliminate the variable x again:* we multiply Equation (1) by -3 and add the result to Equation (3) to obtain

$$-3x - 6y + 9z = 12 \quad (1b)$$
$$\underline{3x + 2y + z = 10} \quad (3)$$
$$-4y + 10z = 22. \quad (5)$$

We now form a 2 × 2 system from Equations (4) and (5),

$$-5y + 7z = 11 \qquad (4)$$
$$-4y + 10z = 22, \qquad (5)$$

and eliminate one of the variables. If we choose to eliminate y, we add 4 times Equation (4) to -5 times Equation (5) to obtain

$$-20y + 28z = 44 \qquad (4a)$$
$$\underline{20y - 50z = -110} \qquad (5a)$$
$$-22z = -66. \qquad (6)$$

Finally, we form a new system with Equation (6), one of the equations from the 2 × 2 system [we choose Equation (4)] and one of the original equations [we choose Equation (1)]:

$$x + 2y - 3z = -4 \qquad (1)$$
$$-5y + 7z = 11 \qquad (4)$$
$$-22z = -66. \qquad (6)$$

This new system has the same solutions as the original system, and it can be solved by back-substitution. We first solve Equation (6) to find $z = 3$. Substituting 3 for z in Equation (4), we find

$$-5y + 7(3) = 11$$
$$y = 2. \qquad (4)$$

Substituting 3 for z and 2 for y into Equation (1), we find

$$x + 2(2) - 3(3) = -4$$
$$x = 1. \qquad (1)$$

The solution to the system is the ordered triple $(1, 2, 3)$. To verify the solution we substitute these values into the other two original equations, (2) and (3).

We summarize the method of linear combinations for solving a 3 × 3 linear system as follows.

To Solve a 3 × 3 Linear System:

1. Clear each equation of fractions and put the equations in standard form.
2. Choose two of the equations and eliminate one of the variables by forming a linear combination.
3. Choose a different pair of equations and eliminate the *same* variable.
4. Form a 2 × 2 system with the equations found in Steps (2) and (3). Eliminate one of the variables from this 2 × 2 system by using a linear combination.
5. Form a new 3 × 3 system with the result of Step (4), one of the equations from the 2 × 2 system, and one of the original equations. Use back-substitution to obtain the components of the solution.

EXAMPLE 2 Solve the system

$$x + 2y + z = -3 \tag{1}$$
$$\frac{1}{3}x - y + \frac{1}{3}z = 2 \tag{2}$$
$$x + \frac{1}{2}y + z = \frac{5}{2}. \tag{3}$$

Solution Follow the steps outlined above.

1. Multiply each side of Equation (2) by 3 and each side of Equation (3) by 2 to obtain the equivalent system

$$x + 2y - z = -3 \tag{1}$$
$$x - 3y + z = 6 \tag{2a}$$
$$2x + y + 2z = 5. \tag{3a}$$

2. Eliminate x from Equations (1) and (2a): multiply Equation (1) by -1 and add the result to Equation (2a) to get

$$\begin{array}{r} -x - 2y + z = 3 \\ x - 3y + z = 6 \\ \hline -5y + 2z = 9. \end{array} \tag{4}$$

3. Eliminate x from Equations (1) and (3a): multiply Equation (1) by -2 and add the result to Equation (3a) to get

$$-2x - 4y + 2z = 6$$
$$\underline{2x + y + 2z = 5}$$
$$-3y + 4z = 11. \quad (5)$$

4. Form the 2 × 2 system consisting of Equations (4) and (5):

$$-5y + 2z = 9 \quad (4)$$
$$-3y + 4z = 11. \quad (5)$$

Eliminate the variable z by adding -2 times Equation (4) to Equation (5) to obtain

$$10y - 4z = -18$$
$$\underline{-3y + 4z = 11}$$
$$7y = -7 \quad (6)$$

5. Form a new 3 × 3 system using Equations (1), (4), and (6):

$$x + 2y - z = -3 \quad (1)$$
$$-5y + 2z = 9 \quad (4)$$
$$7y = -7. \quad (6)$$

Use back-substitution to find the solution. Solving Equation (6) yields $y = -1$. Substitute -1 for y in Equation (4) and solve for z to find

$$-5(-1) + 2z = 9$$
$$z = 2. \quad (4)$$

Finally, substitute -1 for y and 2 for z into Equation (1) and solve for x:

$$x + 2(-1) - 2 = -3$$
$$x = 1. \quad (1)$$

The solution is the ordered triple $(1, -1, 2)$.

Inconsistent and Dependent Systems

Ordered triples of the form (x, y, z) can be represented geometrically as points in a three-dimensional Cartesian coordinate system. In this coordinate system the graph of a linear equation in three variables such as

$$ax + by + cz = d$$

is a plane, and a solution of a system of three such linear equations is a point that lies on the intersection of three planes.

In Section 9.1 we saw that a solution to a 2×2 system represented the intersection of two lines in the plane and that there were three possibilities, depending on the relative positions of the two lines. Figure 9.5 shows the different ways in which three planes may intersect in space.

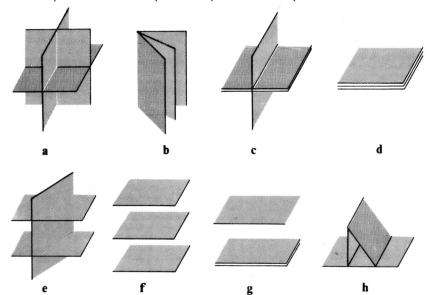

FIGURE 9.5

In Case (a) the three planes intersect in a single point, so the corresponding system of three equations has a unique solution. In Cases (b), (c), and (d) the intersection is a line or a plane, so the corresponding system has infinitely many solutions. We say that such a system is **dependent**. In Cases (e), (f), (g), and (h) the three planes have no common intersection, so the corresponding system has no solution. In this case the system is said to be **inconsistent**.

The results on page 437 for identifying dependent and inconsistent systems can be extended to 3×3 linear systems. That is, if at any step in forming linear combinations we obtain an equation of the form

$$0x + 0y + 0z = k \qquad (k \neq 0),$$

or
$$0x + 0y + 0z = 0,$$

then the system does not have a unique solution. In the first case the system is inconsistent and has no solution, and in the second case the system is dependent and has an infinite number of solutions.

EXAMPLE 3 Solve the system

$$3x + y - 2z = 1 \quad (1)$$
$$6x + 2y - 4z = 5 \quad (2)$$
$$-2x - y + 3z = -1. \quad (3)$$

Solution To eliminate y from Equations (1) and (2), multiply Equation (1) by -2 and add the result to Equation (2) to obtain

$$\begin{array}{r} -6x - 2y + 4z = -2 \\ 6x + 2y - 4z = 5 \\ \hline 0x + 0y + 0z = 3. \end{array}$$

Since the resulting linear combination has no solution, the system is *inconsistent*.

EXAMPLE 4 Solve the system

$$-x + 3y - z = -2 \quad (1)$$
$$2x + y - 4z = 6 \quad (2)$$
$$2x - 6y + 2z = 4. \quad (3)$$

Solution To eliminate x from Equations (1) and (3), multiply Equation (1) by 2 and add Equation (3) to obtain

$$\begin{array}{r} -2x + 6y - 2z = -4 \\ 2x - 6y + 2z = 4 \\ \hline 0x + 0y + 0z = 0. \end{array}$$

Since the resulting linear combination vanishes, the system is *dependent*.

Applications The following examples illustrate problems that can be modeled by a system of three linear equations. In writing such systems we must be careful that the conditions represented by each equation are independent of the conditions represented by the other equations.

EXAMPLE 5 One angle of a triangle measures 4° less than twice the second angle, and the third angle is 20° greater than the sum of the first two. Find the measure of each angle.

Solution Represent the measure of each angle by a separate variable:

$$\text{first angle: } x;$$
$$\text{second angle: } y;$$
$$\text{third angle: } z.$$

Write the three conditions stated in the problem as three equations:

$$x = 2y - 4$$
$$z = x + y + 20$$
$$x + y + z = 180.$$

Rewrite each equation in standard form:

$$x - 2y = -4 \qquad (1)$$
$$x + y - z = -20 \qquad (2)$$
$$x + y + z = 180. \qquad (3)$$

Since Equation (1) has no z-term, it will be most efficient to eliminate the variable z from Equations (2) and (3). Add the two equations to obtain

$$x + y - z = -20$$
$$\underline{x + y + z = 180}$$
$$2x + 2y = 160. \qquad (4)$$

Form a 2 × 2 system from Equations (1) and (4). Add the two equations to eliminate the variable y, yielding

$$x - 2y = -4 \qquad (1)$$
$$\underline{2x + 2y = 160} \qquad (4)$$
$$3x = 156,$$

or

$$x = 52.$$

Substitute 52 for x in Equation (4) to find

$$2(52) + 2y = 160$$
$$y = 28.$$

Substitute 52 for x and 28 for y in Equation (3) to find

$$52 + 28 + z = 180$$
$$z = 100.$$

Thus, the angles measure 52°, 28°, and 100°.

EXAMPLE 6 A manufacturer of office supplies makes three types of file cabinet: two drawer, four drawer, and horizontal. The manufacturing process is divided into three phases: assembly, painting, and finishing. A two-drawer cabinet requires 3 hours to assemble, 1 hour to paint, and 1 hour to finish. The four-drawer model takes 5 hours to assemble, 90 minutes to paint, and 2 hours to finish. The horizontal cabinet takes 4 hours to assemble, 1 hour to paint, and 3 hours to finish. The manufacturer employs enough workers for 500 hours of assembly time, 150 hours of painting, and 230 hours of finishing per week. How many of each type of file cabinet should he make in order to use all the hours available?

Solution Represent the number of each model of file cabinet by a different variable:

number of two-drawer cabinets: x;
number of four-drawer cabinets: y;
number of horizontal cabinets: z.

Write three equations describing the time constraints in each of the three manufacturing phases (for example, the assembly phase requires $3x$ hours for the two-drawer cabinets, $5y$ hours for the four-drawer cabinets, and $4z$ hours for the horizontal cabinets; the sum of these times should be the time available, 500 hours):

$$3x + 5y + 4z = 500 \quad \text{(Assembly time)}$$
$$x + \frac{3}{2}y + z = 150 \quad \text{(Painting time)}$$
$$x + 2y + 3z = 230. \quad \text{(Finishing time)}$$

Clear the fractions from the second equation to obtain the system

$$3x + 5y + 4z = 500 \tag{1}$$
$$2x + 3y + 2z = 300 \tag{2}$$
$$x + 2y + 3z = 230. \tag{3}$$

Subtract Equation (1) from three times Equation (3) to obtain

$$y + 5z = 190, \tag{4}$$

and subtract Equation (2) from twice Equation (3) to obtain

$$y + 4z = 160. \tag{5}$$

Now subtract Equation (5) from Equation (4) to find

$$z = 30 \tag{6}$$

and form the system

$$x + 2y + 3z = 230 \tag{3}$$
$$y + 4z = 160 \tag{5}$$
$$z = 30. \tag{6}$$

Solving this system by back-substitution yields $z = 30$, $y = 40$, and $x = 60$. The manufacturer should make 60 two-drawer cabinets, 40 four-drawer cabinets, and 30 horizontal cabinets.

EXAMPLE 7 Find values for a, b, and c so that the points $(1, 3)$, $(3, 5)$, and $(4, 9)$ lie on the graph of the parabola $y = ax^2 + bx + c$.

Solution Substitute the coordinates of each point into the equation for the parabola to obtain

$$3 = a(1)^2 + b(1) + c$$
$$5 = a(3)^2 + b(3) + c$$
$$9 = a(4)^2 + b(4) + c,$$

or, equivalently,

$$a + b + c = 3 \tag{1}$$
$$9a + 3b + c = 5 \tag{2}$$
$$16a + 4b + c = 9. \tag{3}$$

To solve the system, first eliminate the variable c. Subtract Equation (1) from Equation (2) to obtain

$$8a + 2b = 2, \tag{4}$$

and subtract Equation (1) from Equation (3) to get

$$15a + 3b = 6. \tag{5}$$

Now eliminate b from Equations (4) and (5): add -3 times Equation (4) to 2 times Equation (5) to get

$$\begin{aligned} -24a - 6b &= -6 \\ \underline{30a + 6b} &= \underline{12} \\ 6a &= 6, \end{aligned}$$

or $a = 1$. Substitute 1 for a in Equation (4) to find

$$\begin{aligned} 8(1) + 2b &= 2 \\ b &= -3. \end{aligned}$$

Finally, substitute -3 for b and 1 for a in Equation (1) to find

$$\begin{aligned} 1 + (-3) + c &= 3 \\ c &= 5. \end{aligned}$$

Thus, the equation of the parabola is

$$y = x^2 - 3x + 5.$$

EXERCISE 9.2

A

■ *Solve. See Example 1.*

1. $\begin{aligned} x + y + z &= 2 \\ 3y + z &= 5 \\ -4y &= -8 \end{aligned}$

2. $\begin{aligned} 2x + 3y - z &= -7 \\ y - 2z &= -6 \\ 5z &= 15 \end{aligned}$

3. $\begin{aligned} 2x - y - z &= 6 \\ 5y + 3z &= -8 \\ 13y &= -13 \end{aligned}$

4. $x + y + z = 1$
   $x + 4y = 1$
   $3x = 3$

5. $2x + z = 5$
   $3y + 2z = 6$
   $5x = 20$

6. $3x - y = 6$
   $x - 2z = -7$
   $13x = 13$

■ Solve. See Examples 1 and 2.

7. $x + y + 2z = 0$
   $2x - 2y + z = 8$
   $3x + 2y + z = 2$

8. $x - 2y + 4z = -3$
   $3x + y - 2z = 12$
   $2x + y - 3z = 11$

9. $x - 2y + z = -1$
   $2x + y - 3z = 3$
   $3x + 3y - 2z = 10$

10. $x - 5y - z = 2$
    $3x - 9y + 3z = 6$
    $x - 3y - z = -6$

11. $4x + z = 3$
    $2x - y = 2$
    $3y + 2z = 0$

12. $3y + z = 3$
    $-2x + 3y = 7$
    $3x + 2z = -6$

13. $2x + 3y - 2z = 5$
    $3x - 2y - 5z = 5$
    $5x + 2y + 3z = 9$

14. $3x - 4y + 2z = 20$
    $4x + 3y - 3z = -4$
    $2x - 5y + 5z = 24$

15. $4x + 6y + 3z = -3$
    $2x - 3y - 2z = 5$
    $-6x + 6y + 2z = -5$

16. $3x + 4y + 6z = 2$
    $-2x + 2y - 3z = 1$
    $4x - 10y + 9z = 0$

17. $x - \frac{1}{2}y - \frac{1}{2}z = 4$
    $x - \frac{3}{2}y - 2z = 3$
    $\frac{1}{4}x + \frac{1}{4}y - \frac{1}{4}z = 0$

18. $x + 2y + \frac{1}{2}z = 0$
    $x + \frac{3}{5}y - \frac{2}{5}z = \frac{1}{5}$
    $4x - 7y - 7z = 6$

19. $x + y - z = 2$
    $\frac{1}{2}x - y + \frac{1}{2}z = -\frac{1}{2}$
    $x + \frac{1}{3}y - \frac{2}{3}z = \frac{4}{3}$

20. $x + y - 2z = 3$
    $x - \frac{1}{3}y + \frac{1}{3}z = \frac{5}{3}$
    $\frac{1}{2}x - \frac{1}{2}y - z = \frac{3}{2}$

21. $x = -y$
    $x + z = \frac{5}{6}$
    $y - 2z = -\frac{7}{6}$

22. $x = y + \frac{1}{2}$
    $y = z + \frac{5}{4}$
    $2z = x - \frac{7}{4}$

■ Solve. If there is no unique solution, so state. See Examples 1–4.

23. $3x - 2y + z = 6$
    $2x + y - z = 2$
    $4x + 2y - 2z = 3$

24. $x + 3y - z = 4$
    $-2x - 6y + 2z = 1$
    $x + 2y - z = 3$

25. $2x + 3y - z = -2$
    $x - y + \frac{1}{2}z = 2$
    $4x - \frac{1}{3}y + 2z = 8$

26. $3x + 6y + 2z = -2$
    $\frac{1}{2}x - 3y - z = 1$
    $4x + y + \frac{1}{3}z = -\frac{1}{3}$

27. $2x + y = 6$
    $x - z = 4$
    $3x + y - z = 10$

28. $x - 2y + z = 5$
    $-x + y = -2$
    $y - z = -3$

29. $x = 2y - 7$
    $y = 4z + 3$
    $z = 3x + y$

30. $x = y + z$
    $y = 2x - z$
    $z = 3x - y$

31. $\frac{1}{2}x + y = \frac{1}{2}z$
    $x - y = -z - 2$
    $-x - 2y = -z + \frac{4}{3}$

32. $x = \frac{1}{2}y - \frac{1}{2}z + 1$
    $x = 2y + z - 1$
    $x = \frac{1}{2}y - \frac{1}{2}z + \frac{1}{4}$

33. $x - y = 0$
    $2x + 2y + z = 5$
    $2x + y - \frac{1}{2}z = 0$

34. $x + y = 1$
    $2x - y + z = -1$
    $x - 3y - z = -\frac{2}{3}$

■ *Solve each problem using a system of equations. See Examples 5–7.*

35. A box contains $6.25 in nickels, dimes, and quarters. There are 85 coins in all, with 3 times as many nickels as dimes. How many coins of each kind are there?

36. A man has $446 in 10-dollar, 5-dollar, and 1-dollar bills. There were 94 bills in all and 10 more 5-dollar bills than 10-dollar bills. How many bills of each kind did he have?

37. The perimeter of a triangle is 155 inches. Side x is 20 inches shorter than side y, and side y is 5 inches longer than side z. Find the lengths of the sides of the triangle.

38. One angle of a triangle measures 10° more than a second angle, and the third angle is 10° more than six times the measure of the smallest angle. Find the measure of each angle.

39. Vegetable Medley is made of carrots, green beans, and cauliflower. The package says that 1 cup of Vegetable Medley provides 29.4 milligrams of vitamin C and 47.4 milligrams of calcium. One cup of carrots contains 9 milligrams of vitamin C and 48 milligrams of calcium. One cup of green beans contains 15 milligrams of vitamin C and 63 milligrams of calcium. One cup of cauliflower contains 69 milligrams of vitamin C and 26 milligrams of calcium. How much of each vegetable is in 1 cup of Vegetable Medley?

40. The Java Shoppe sells a house brand of coffee that is only 2.25% caffeine for $6.60 per pound. The house brand is a mixture of Colombian coffee that sells for $6 per pound and is 2% caffeine, French roast that sells for $7.60 per pound and is 4% caffeine, and Sumatran at $6.80 per pound and 1% caffeine. How much of each variety is in a pound of house brand?

41. The ABC Psychological Testing Service offers three types of reports on test results: score only, evaluation, and narrative report. Each score-only test takes 3 minutes to score using an optical scanner and 1 minute to print the interpretation on the computer. Each evaluation takes 3 minutes to score, 4 minutes to analyze, and 2 minutes to print. Each narrative report takes 3 minutes to score, 5 minutes to analyze, and 8 minutes to print. If ABC Services uses its optical scanner 7 hours per day, has 8 hours in which to analyze results, and has two printers that run for a total of 12 hours per day, how many of each type of report can it complete each day?

42. Reliable Auto Company wants to ship 1700 Status Sedans to three major dealers in Los Angeles, Chicago, and Miami. From past experience Reliable figures that it will sell twice as many sedans in Los Angeles as in Chicago. It costs $230 to ship

a sedan to Los Angeles, $70 to Chicago, and $160 to Miami. If Reliable Auto has $292,000 to pay for shipping costs, how many sedans should it ship to each city?

43. Ace, Inc. produces three kinds of wooden rackets: tennis rackets, Ping-Pong paddles, and squash rackets. After the pieces are cut each racket goes through three phases of production: gluing, sanding, and finishing. A tennis racket takes 3 hours to glue, 2 hours to sand, and 3 hours to finish. A Ping-Pong paddle takes 1 hour to glue, 1 hour to sand, and 1 hour to finish. A squash racket takes 2 hours to glue, 2 hours to sand, and 2½ hours to finish. Ace has available 95 man-hours in its gluing department, 75 man-hours in sanding, and 100 man-hours in finishing per day. How many of each racket should it make in order to use all the available manpower?

44. A farmer has 1300 acres on which to plant wheat, corn, and soybeans. The seed costs $6 for an acre of wheat, $4 for an acre of corn, and $5 for an acre of soybeans. An acre of wheat requires 5 acre-feet of water during the growing season, while an acre of corn requires 2 acre-feet and an acre of soybeans requires 3 acre-feet. If the farmer has $6150 to spend on seed and can count on 3800 acre-feet of water, how many acres of each crop should he plant in order to use all his resources?

B

■ Solve each system. (Hint: Use substitutions $u = 1/x$, $v = 1/y$, and $w = 1/z$; solve for u, v, and w, then solve for x, y, and z.)

45. $\dfrac{1}{x} + \dfrac{1}{y} - \dfrac{1}{z} = 1$

$\dfrac{2}{x} - \dfrac{2}{y} + \dfrac{1}{z} = 1$

$\dfrac{-3}{x} + \dfrac{1}{y} - \dfrac{1}{z} = -3$

46. $\dfrac{4}{x} - \dfrac{2}{y} + \dfrac{1}{z} = 4$

$\dfrac{3}{x} - \dfrac{1}{y} + \dfrac{2}{z} = 0$

$\dfrac{-1}{x} + \dfrac{3}{y} - \dfrac{2}{z} = 0$

47. $\dfrac{1}{x} + \dfrac{2}{y} - \dfrac{2}{z} = 3$

$\dfrac{2}{x} - \dfrac{4}{y} + \dfrac{2}{z} = -2$

$\dfrac{4}{x} - \dfrac{2}{y} - \dfrac{4}{z} = 5$

48. $\dfrac{2}{x} - \dfrac{1}{y} - \dfrac{1}{z} = -1$

$\dfrac{4}{x} - \dfrac{2}{y} + \dfrac{1}{z} = -5$

$\dfrac{2}{x} + \dfrac{1}{y} - \dfrac{4}{z} = 4$

49. Three solutions of the equation $ax + by + cz = 1$ are $(-2, 0, 4)$, $(6, -1, 0)$, and $(0, 3, 0)$. Find the coefficients a, b, and c.

50. Three solutions of the equation $ax + by + cz = 1$ are $(0, 4, 2)$, $(-1, 3, 0)$, and $(-1, 0, 2)$. Find the coefficients a, b, and c.

51. Find values for a, b, and c so that the graph of the parabola $y = ax^2 + bx + c$ includes the points $(-1, 0)$, $(2, 12)$, and $(-2, 8)$.

52. Find values for a, b, and c so that the graph of the parabola $y = ax^2 + bx + c$ includes the points $(-1, 2)$, $(1, 6)$, and $(2, 11)$.

9.3 DETERMINANTS AND CRAMER'S RULE

Second-Order Determinants

We now introduce a mathematical tool called a *determinant* that can be used to solve linear systems. An expression of the form

$$\begin{vmatrix} a_1 & b_1 \\ a_2 & b_2 \end{vmatrix}$$

is called a **determinant**. The numbers a_1, b_1, a_2, and b_2 are called **elements** of the determinant. Because this determinant has two rows and two columns of elements, it is called a **2 × 2 determinant**, or a determinant of order two. We define this determinant as follows:

$$\begin{vmatrix} a_1 & b_1 \\ a_2 & b_2 \end{vmatrix} = a_1 b_2 - a_2 b_1.$$

In other words, the value of the determinant is obtained by multiplying the elements on the diagonals, first from upper left to lower right and then from lower left to upper right, then subtracting the second product from the first. The process can be shown schematically as

$$\begin{vmatrix} a_1 & b_1 \\ a_2 & b_2 \end{vmatrix} = a_1 b_2 - a_2 b_1.$$

Note that a determinant represents a single number.

EXAMPLE 1

a. $\begin{vmatrix} 1 & 2 \\ -1 & 3 \end{vmatrix} = (1)(3) - (-1)(2) = 5$

b. $\begin{vmatrix} 0 & -1 \\ -1 & 7 \end{vmatrix} = (0)(7) - (-1)(-1) = -1$

Solution of 2 × 2 Linear Systems

Determinants can be used to solve systems of linear equations. We first consider 2 × 2 systems of the form

$$a_1 x + b_1 y = c_1 \tag{1}$$
$$a_2 x + b_2 y = c_2. \tag{2}$$

To solve this system by linear combinations we multiply Equation (1) by $-a_2$ and Equation (2) by a_1, then add to get

$$-a_1 a_2 x - a_2 b_1 y = -a_2 c_1$$
$$\underline{a_1 a_2 x + a_1 b_2 y = a_1 c_2}$$
$$a_1 b_2 y - a_2 b_1 y = a_1 c_2 - a_2 c_1.$$

To solve this last equation for y we factor the left side to get

$$(a_1 b_2 - a_2 b_1) y = a_1 c_2 - a_2 c_1,$$

from which

$$y = \frac{a_1 c_2 - a_2 c_1}{a_1 b_2 - a_2 b_1} \qquad (a_1 b_2 - a_2 b_1 \neq 0). \tag{3}$$

Now, the numerator of (3) is just the value of the determinant

$$\begin{vmatrix} a_1 & c_1 \\ a_2 & c_2 \end{vmatrix},$$

which we designate as D_y. The denominator is the value of the determinant

$$\begin{vmatrix} a_1 & b_1 \\ a_2 & b_2 \end{vmatrix},$$

which we designate as D; so Equation (3) can be written as follows:

$$y = \frac{D_y}{D} = \frac{\begin{vmatrix} a_1 & c_1 \\ a_2 & c_2 \end{vmatrix}}{\begin{vmatrix} a_1 & b_1 \\ a_2 & b_2 \end{vmatrix}} \qquad (D \neq 0). \tag{4}$$

The elements of the determinant in the denominator of (4) are the coefficients of the variables in (1) and (2). The elements of the determinant in the numerator of (4) are identical to those in the denominator, except that *the elements in the column containing the coefficients of y have been replaced by c_1 and c_2*, the constant terms of (1) and (2).

By an analogous procedure we can show the following:

$$x = \frac{D_x}{D} = \frac{\begin{vmatrix} c_1 & b_1 \\ c_2 & b_2 \end{vmatrix}}{\begin{vmatrix} a_1 & b_1 \\ a_2 & b_2 \end{vmatrix}} \quad (D \neq 0). \tag{5}$$

Note that the elements of the determinant in the numerator of (5) are identical to those in the denominator, except that *the elements in the column containing the coefficients of x have been replaced by c_1 and c_2*.

Equations (4) and (5) together give the components of the solution of the system. The use of determinants in this way is known as **Cramer's rule** for the solution of a system of linear equations.

EXAMPLE 2 Solve the following system using Cramer's rule:

$$2x - 3y = 6$$
$$2x + y = 14.$$

Solution First, compute the determinant D:

$$D = \begin{vmatrix} 2 & -3 \\ 2 & 1 \end{vmatrix} = (2)(1) - (2)(-3) = 8.$$

The elements of D_x are obtained from the elements of D by replacing the coefficients of x with the corresponding constants, 6 and 14:

$$D_x = \begin{vmatrix} 6 & -3 \\ 14 & 1 \end{vmatrix} = (6)(1) - (14)(-3) = 48.$$

The elements of D_y are obtained from the elements of D by replacing the coefficients of y with the constants 6 and 14:

$$D_y = \begin{vmatrix} 2 & 6 \\ 2 & 14 \end{vmatrix} = (2)(14) - (2)(6) = 16.$$

Values for x and y can now be determined by Cramer's rule:

$$x = \frac{D_x}{D} = \frac{48}{8} = 6; \qquad y = \frac{D_y}{D} = \frac{16}{8} = 2.$$

The solution is the ordered pair $(6, 2)$.

If the value of the determinant D is zero, then the system is either dependent or inconsistent, depending on the values of the constants c_1 and c_2. In either case the system does not have a unique solution.

Third-Order Determinants

We define a 3×3, or **third-order, determinant** as follows:

$$\begin{vmatrix} a_1 & b_1 & c_1 \\ a_2 & b_2 & c_2 \\ a_3 & b_3 & c_3 \end{vmatrix} = a_1 b_2 c_3 - a_1 b_3 c_2 + a_3 b_1 c_2 \\ - a_2 b_1 c_3 + a_2 b_3 c_1 - a_3 b_2 c_1. \qquad (6)$$

We can rewrite (6) in a simpler form by using 2×2 determinants called minors. The **minor** of an element in a determinant is the smaller determinant that remains after deleting the row and column in which that element appears. In the determinant (6), for example, the minor of the element a_1 is obtained by deleting the first row and the first column of the determinant as shown below:

$$\begin{vmatrix} \cancel{a_1} & \cancel{b_1} & \cancel{c_1} \\ \cancel{a_2} & b_2 & c_2 \\ \cancel{a_3} & b_3 & c_3 \end{vmatrix} \rightarrow \begin{vmatrix} b_2 & c_2 \\ b_3 & c_3 \end{vmatrix}.$$

Similarly, the minor of element b_1 is

$$\begin{vmatrix} \cancel{a_1} & \cancel{b_1} & \cancel{c_1} \\ a_2 & \cancel{b_2} & c_2 \\ a_3 & \cancel{b_3} & c_3 \end{vmatrix} \rightarrow \begin{vmatrix} a_2 & c_2 \\ a_3 & c_3 \end{vmatrix},$$

and the minor of the element c_1 is

$$\begin{vmatrix} \cancel{a_1} & \cancel{b_1} & \cancel{c_1} \\ a_2 & b_2 & \cancel{c_2} \\ a_3 & b_3 & \cancel{c_3} \end{vmatrix} \rightarrow \begin{vmatrix} a_2 & b_2 \\ a_3 & b_3 \end{vmatrix}.$$

A 3 × 3 determinant can be expressed in terms of the minors on its first row. Note that by suitably factoring pairs of terms on the right side (6) can be rewritten in the form

$$\begin{vmatrix} a_1 & b_1 & c_1 \\ a_2 & b_2 & c_2 \\ a_3 & b_3 & c_3 \end{vmatrix} = a_1(b_2c_3 - b_3c_2) - b_1(a_2c_3 - a_3c_2) + c_1(a_2b_3 - a_3b_2). \quad (7)$$

We observe that the sums enclosed in parentheses on the right side of (7) are the respective minors of the elements a_1, b_1, and c_1. Therefore, (7) can be written

$$\begin{vmatrix} a_1 & b_1 & c_1 \\ a_2 & b_2 & c_2 \\ a_3 & b_3 & c_3 \end{vmatrix} = a_1 \begin{vmatrix} b_2 & c_2 \\ b_3 & c_3 \end{vmatrix} - b_1 \begin{vmatrix} a_2 & c_2 \\ a_3 & c_3 \end{vmatrix} + c_1 \begin{vmatrix} a_2 & b_2 \\ a_3 & b_3 \end{vmatrix}. \quad (8)$$

The right side of (8) is called the **expansion** of the determinant by minors about the first row.

EXAMPLE 3 Evaluate the determinant

$$\begin{vmatrix} 1 & 2 & -3 \\ 0 & 2 & -1 \\ 1 & 1 & 0 \end{vmatrix}.$$

Solution Expand in minors about the first row:

$$\begin{vmatrix} 1 & 2 & -3 \\ 0 & 2 & -1 \\ 1 & 1 & 0 \end{vmatrix} = 1 \begin{vmatrix} 2 & -1 \\ 1 & 0 \end{vmatrix} - 2 \begin{vmatrix} 0 & -1 \\ 1 & 0 \end{vmatrix} + (-3) \begin{vmatrix} 0 & 2 \\ 1 & 1 \end{vmatrix}$$

$$= 1(0 + 1) - 2(0 + 1) - 3(0 - 2)$$

$$= 1 - 2 + 6 = 5.$$

Suppose that instead of factoring the right side of (6) as in (7) above we factor it as

$$\begin{vmatrix} a_1 & b_1 & c_1 \\ a_2 & b_2 & c_2 \\ a_3 & b_3 & c_3 \end{vmatrix} = a_1(b_2c_3 - b_3c_2) - a_2(b_1c_3 - b_3c_1) + a_3(b_1c_2 - b_2c_1). \quad (9)$$

Then we have the expansion of the determinant by minors *about the first column:*

$$\begin{vmatrix} a_1 & b_1 & c_1 \\ a_2 & b_2 & c_2 \\ a_3 & b_3 & c_3 \end{vmatrix} = a_1 \begin{vmatrix} b_2 & c_2 \\ b_3 & c_3 \end{vmatrix} - a_2 \begin{vmatrix} b_1 & c_1 \\ b_3 & c_3 \end{vmatrix} + a_3 \begin{vmatrix} b_1 & c_1 \\ b_2 & c_2 \end{vmatrix}.$$

With the proper use of signs it is possible to expand a determinant by minors about *any* row or column. A helpful device for determining the signs of the terms in an expansion of a third-order determinant by minors is the array of alternating signs

$$\begin{matrix} + & - & + \\ - & + & - \\ + & - & + \end{matrix},$$

which we call the **sign array** for the determinant. To obtain an expansion of (6) about a given row or column we prefix the appropriate sign from the sign array to each term in the expansion.

EXAMPLE 4 Evaluate the determinant in Example 3 by expanding in minors about the second row.

Solution The second row of the sign array is $- + -$. Multiply each element of the second row by its minor and prefix each product by the appropriate sign to write

$$\begin{vmatrix} 1 & 2 & -3 \\ 0 & 2 & -1 \\ 1 & 1 & 0 \end{vmatrix} = -0 \begin{vmatrix} 2 & -3 \\ 1 & 0 \end{vmatrix} + 2 \begin{vmatrix} 1 & -3 \\ 1 & 0 \end{vmatrix} - (-1) \begin{vmatrix} 1 & 2 \\ 1 & 1 \end{vmatrix}$$
$$= 0 + 2(0 + 3) + 1(1 - 2)$$
$$= 6 - 1 = 5,$$

the same answer we found in Example 3.

You can verify that the result is the same when the determinant is expanded about any row or column.

Cramer's Rule for 3 × 3 Systems

A system of three linear equations in three variables can be solved by Cramer's rule in the same way that we solved 2 × 2 systems. If the system has the form

$$a_1 x + b_1 y + c_1 z = d_1$$
$$a_2 x + b_2 y + c_2 z = d_2$$
$$a_3 x + b_3 y + c_3 z = d_3,$$

then:

$$x = \frac{D_x}{D}, \quad y = \frac{D_y}{D}, \quad z = \frac{D_z}{D} \quad (D \neq 0),$$

where

$$D = \begin{vmatrix} a_1 & b_1 & c_1 \\ a_2 & b_2 & c_2 \\ a_3 & b_3 & c_3 \end{vmatrix}, \quad D_x = \begin{vmatrix} d_1 & b_1 & c_1 \\ d_2 & b_2 & c_2 \\ d_3 & b_3 & c_3 \end{vmatrix},$$

$$D_y = \begin{vmatrix} a_1 & d_1 & c_1 \\ a_2 & d_2 & c_2 \\ a_3 & d_3 & c_3 \end{vmatrix}, \quad D_z = \begin{vmatrix} a_1 & b_1 & d_1 \\ a_2 & b_2 & d_2 \\ a_3 & b_3 & d_3 \end{vmatrix}.$$

Note that the elements of the determinant D in each denominator are the coefficients of the variables of the system. The numerators are formed from D by replacing the elements in the first, second, and third columns, respectively, by d_1, d_2, and d_3.

EXAMPLE 5 Solve the following system using Cramer's rule:

$$x + 2y - 3z = -4$$
$$2x - y + z = 3$$
$$3x + 2y + z = 10.$$

Solution The determinant D, whose elements are the coefficients of the variables, is given by

$$D = \begin{vmatrix} 1 & 2 & -3 \\ 2 & -1 & 1 \\ 3 & 2 & 1 \end{vmatrix}.$$

Expand the determinant about the first column to obtain

$$D = \begin{vmatrix} 1 & 2 & -3 \\ 2 & -1 & 1 \\ 3 & 2 & 1 \end{vmatrix} = 1\begin{vmatrix} -1 & 1 \\ 2 & 1 \end{vmatrix} - 2\begin{vmatrix} 2 & -3 \\ 2 & 1 \end{vmatrix} + 3\begin{vmatrix} 2 & -3 \\ -1 & 1 \end{vmatrix}$$
$$= -3 - 16 - 3 = -22.$$

Replace the first column in D by $-4, 3,$ and 10 to find

$$D_x = \begin{vmatrix} -4 & 2 & -3 \\ 3 & -1 & 1 \\ 10 & 2 & 1 \end{vmatrix}.$$

Expand D_x about the third column. This gives

$$D_x = \begin{vmatrix} -4 & 2 & -3 \\ 3 & -1 & 1 \\ 10 & 2 & 1 \end{vmatrix} = -3\begin{vmatrix} 3 & -1 \\ 10 & 2 \end{vmatrix} - 1\begin{vmatrix} -4 & 2 \\ 10 & 2 \end{vmatrix} + 1\begin{vmatrix} -4 & 2 \\ 3 & -1 \end{vmatrix}$$
$$= -48 + 28 - 2 = -22.$$

In similar fashion we can compute D_y and D_z:

$$D_y = \begin{vmatrix} 1 & -4 & -3 \\ 2 & 3 & 1 \\ 3 & 10 & 1 \end{vmatrix} = -44, \quad D_z = \begin{vmatrix} 1 & 2 & -4 \\ 2 & -1 & 3 \\ 3 & 2 & 10 \end{vmatrix} = -66.$$

We then have

$$x = \frac{D_x}{D} = \frac{-22}{-22} = 1, \quad y = \frac{D_y}{D} = \frac{-44}{-22} = 2, \text{ and } z = \frac{D_z}{D} = \frac{-66}{-22} = 3.$$

The solution of the system is the ordered triple $(1, 2, 3)$.

As was the case for a linear system in two variables, if $D = 0$ for a linear system in three variables, the system does not have a unique solution.

EXERCISE 9.3

A

■ *Evaluate. See Example 1.*

1. $\begin{vmatrix} 1 & 0 \\ 2 & 1 \end{vmatrix}$
2. $\begin{vmatrix} 3 & -2 \\ 4 & 1 \end{vmatrix}$
3. $\begin{vmatrix} -5 & -1 \\ 3 & 3 \end{vmatrix}$
4. $\begin{vmatrix} 1 & -2 \\ -1 & 2 \end{vmatrix}$
5. $\begin{vmatrix} -1 & 6 \\ 0 & -2 \end{vmatrix}$
6. $\begin{vmatrix} 20 & 3 \\ -20 & -2 \end{vmatrix}$
7. $\begin{vmatrix} -2 & -1 \\ -3 & -4 \end{vmatrix}$
8. $\begin{vmatrix} -1 & -5 \\ -2 & -6 \end{vmatrix}$

■ *Solve each system using Cramer's rule. See Example 2.*

9. $2x - 3y = -1$
   $x + 4y = 5$
10. $3x - 4y = -2$
    $x - 2y = 0$
11. $3x - 4y = -2$
    $6x + 12y = 36$
12. $2x - 4y = 7$
    $x - 2y = 1$
13. $\frac{1}{3}x - \frac{1}{2}y = 0$
    $\frac{1}{2}x + \frac{1}{4}y = 4$
14. $\frac{2}{3}x + y = 1$
    $x - \frac{4}{3}y = 0$
15. $x - 2y = 5$
    $\frac{2}{3}x - \frac{4}{3}y = 6$
16. $\frac{1}{2}x + y = 3$
    $-\frac{1}{4}x - y = -3$
17. $x - 3y = 1$
    $y = 1$
18. $2x - 3y = 12$
    $x = 4$
19. $ax + by = 1$
    $bx + ay = 1$
20. $x + y = a$
    $x - y = b$

■ *Evaluate each determinant. See Examples 3 and 4.*

21. $\begin{vmatrix} 2 & 0 & 1 \\ 1 & 1 & 2 \\ -1 & 0 & 1 \end{vmatrix}$
22. $\begin{vmatrix} 1 & 3 & 1 \\ -1 & 2 & 1 \\ 0 & 2 & 0 \end{vmatrix}$
23. $\begin{vmatrix} 2 & -1 & 0 \\ -3 & 1 & 2 \\ 1 & -3 & 1 \end{vmatrix}$
24. $\begin{vmatrix} 2 & 4 & -1 \\ -1 & 3 & 2 \\ 4 & 0 & 2 \end{vmatrix}$
25. $\begin{vmatrix} 1 & 2 & 3 \\ 3 & -1 & 2 \\ 2 & 0 & 2 \end{vmatrix}$
26. $\begin{vmatrix} 1 & 0 & 0 \\ 0 & 1 & 2 \\ 0 & 3 & 4 \end{vmatrix}$
27. $\begin{vmatrix} -1 & 0 & 2 \\ -2 & 1 & 0 \\ 0 & 1 & -3 \end{vmatrix}$
28. $\begin{vmatrix} 2 & 1 & 4 \\ 3 & 2 & 6 \\ 5 & -3 & 10 \end{vmatrix}$
29. $\begin{vmatrix} 2 & 5 & -1 \\ 1 & 0 & 2 \\ 0 & 0 & 1 \end{vmatrix}$
30. $\begin{vmatrix} 2 & 3 & 1 \\ 0 & 1 & 0 \\ -4 & 2 & 1 \end{vmatrix}$
31. $\begin{vmatrix} a & b & 1 \\ a & b & 1 \\ 1 & 1 & 1 \end{vmatrix}$
32. $\begin{vmatrix} a & a & a \\ 1 & 2 & 3 \\ 4 & 5 & 6 \end{vmatrix}$
33. $\begin{vmatrix} x & 0 & 0 \\ 0 & x & 0 \\ 0 & 0 & x \end{vmatrix}$
34. $\begin{vmatrix} 0 & 0 & x \\ 0 & x & 0 \\ x & 0 & 0 \end{vmatrix}$
35. $\begin{vmatrix} x & y & 0 \\ x & y & 0 \\ 0 & 0 & 1 \end{vmatrix}$
36. $\begin{vmatrix} 0 & a & b \\ a & 0 & a \\ b & a & 0 \end{vmatrix}$
37. $\begin{vmatrix} a & b & 0 \\ b & 0 & b \\ 0 & b & a \end{vmatrix}$
38. $\begin{vmatrix} 0 & b & 0 \\ b & a & b \\ 0 & b & 0 \end{vmatrix}$

■ *Solve each system using Cramer's rule. If a unique solution does not exist* $(D = 0),$ *say so. See Example 5.*

39. $x + y = 2$
    $2x - z = 1$
    $2y - 3z = -1$

40. $2x - 6y + 3z = -12$
    $3x - 2y + 5z = -4$
    $4x + 5y - 2z = 10$

41. $x - 2y + z = -1$
    $3x + y - 2z = 4$
    $y - z = 1$

42. $2x + 5z = 9$
    $4x + 3y = -1$
    $3y - 4z = -13$

43. $2x + 2y + z = 1$
    $x - y + 6z = 21$
    $3x + 2y - z = -4$

44. $4x + 8y + z = -6$
    $2x - 3y + 2z = 0$
    $x + 7y - 3z = -8$

45. $x + y + z = 0$
    $2x - y - 4z = 15$
    $x - 2y - z = 7$

46. $x + y - 2z = 3$
    $3x - y + z = 5$
    $3x + 3y - 6z = 9$

47. $x - 2y + 2z = 3$
    $2x - 4y + 4z = 1$
    $3x - 3y - 3z = 4$

48. $3x - 2y + 5z = 6$
    $4x - 4y + 3z = 0$
    $5x - 4y + z = -5$

49. $\frac{1}{4}x - z = -\frac{1}{4}$
    $x + y = \frac{2}{3}$
    $3x + 4z = 5$

50. $2x - \frac{2}{3}y + z = 2$
    $\frac{1}{2}x - \frac{1}{3}y - \frac{1}{4}z = 0$
    $4x + 5y - 3z = -1$

51. $x + 4z = 3$
    $y + 3z = 9$
    $2x + 5y - 5z = -5$

52. $2x + y = 18$
    $y + z = -1$
    $3x - 2y - 5z = 38$

B

■ *Show that each statement is true for every real value of each variable.*

53. $\begin{vmatrix} a & a \\ b & b \end{vmatrix} = 0$

54. $\begin{vmatrix} a_1 & b_1 \\ a_2 & b_2 \end{vmatrix} = -\begin{vmatrix} a_2 & b_2 \\ a_1 & b_1 \end{vmatrix}$

55. $\begin{vmatrix} a_1 & b_1 \\ a_2 & b_2 \end{vmatrix} = -\begin{vmatrix} b_1 & a_1 \\ b_2 & a_2 \end{vmatrix}$

56. $\begin{vmatrix} ka_1 & b_1 \\ ka_2 & b_2 \end{vmatrix} = k\begin{vmatrix} a_1 & b_1 \\ a_2 & b_2 \end{vmatrix}$

57. $\begin{vmatrix} ka & a \\ kb & b \end{vmatrix} = 0$

58. $\begin{vmatrix} a_1 & b_1 \\ ka_2 & kb_2 \end{vmatrix} = k\begin{vmatrix} a_1 & b_1 \\ a_2 & b_2 \end{vmatrix}$

59. Show that if both $D_y = 0$ and $D_x = 0,$ it follows that $D = 0$ when c_1 and c_2 are not both 0, and the equations in the system

$$a_1x + b_1y = c_1$$
$$a_2x + b_2y = c_2$$

are dependent. (*Hint:* Show that the first two determinant equations imply that $a_1c_2 = a_2c_1$ and $b_1c_2 = b_2c_1$ and that the rest follows from the formation of a proportion with these equations.)

60. Show that for the system given in Problem 59, if $D = 0$ and $D_x = 0,$ then $D_y = 0.$

61. Show that for all values of x, y, and z,

$$\begin{vmatrix} x & x & a \\ y & y & b \\ z & z & c \end{vmatrix} = 0.$$

(*Hint:* Expand about the elements of the third column.) Make a conjecture about determinants that contain two identical columns.

62. Show that

$$\begin{vmatrix} 0 & 0 & 0 \\ a & b & c \\ d & e & f \end{vmatrix} = 0$$

for all values of a, b, c, d, e, and f. Make a conjecture about determinants that contain a row of zero elements.

63. Show that

$$\begin{vmatrix} 1 & 2 & 3 \\ 4 & 5 & 6 \\ 0 & 0 & 1 \end{vmatrix} = -\begin{vmatrix} 4 & 5 & 6 \\ 1 & 2 & 3 \\ 0 & 0 & 1 \end{vmatrix}.$$

Make a conjecture about the result of interchanging any two rows of a determinant.

64. Show that

$$\begin{vmatrix} 2 & 0 & 1 \\ 4 & 1 & -2 \\ 6 & 1 & 1 \end{vmatrix} = 2\begin{vmatrix} 1 & 0 & 1 \\ 2 & 1 & -2 \\ 3 & 1 & 1 \end{vmatrix}.$$

Make a conjecture about the result of factoring a common factor from each element of a column in a determinant.

65. Show that the graph of

$$\begin{vmatrix} x & y & 1 \\ 4 & -1 & 1 \\ 2 & 3 & 1 \end{vmatrix} = 0$$

is a line containing the points $(4, -1)$ and $(2, 3)$.

66. Show that the slope-intercept form of the equation of a line can be written

$$\begin{vmatrix} x & y & 1 \\ 0 & b & 1 \\ 1 & m & 0 \end{vmatrix} = 0.$$

9.4 SOLUTION OF SYSTEMS USING MATRICES

In previous sections we solved linear systems by using linear combinations and determinants. In this section we consider another mathematical tool called a *matrix* (plural: matrices) that has wide application in mathematics. In particular, we shall see how matrices can be used to solve linear systems.

A **matrix** is a rectangular array of elements or **entries.** (In this book the entries will be real numbers.) These entries are ordinarily displayed in rows and columns, and the entire matrix is enclosed in brackets or parentheses. Thus,

$$\begin{bmatrix} 1 & 2 & 3 \\ 4 & 5 & 6 \\ 7 & 8 & 9 \end{bmatrix}, \quad \begin{bmatrix} 2 & -1 & 3 \\ 4 & 0 & 2 \end{bmatrix}, \quad \text{and} \quad \begin{bmatrix} 4 \\ 5 \\ 6 \end{bmatrix}$$

are matrices. A matrix of **order,** or **dimension,** $n \times m$ (read "n by m") has n (horizontal) rows and m (vertical) columns. The matrices above are 3×3, 2×3, and 3×1, respectively. The first matrix—in which the number of rows is equal to the number of columns—is an example of a **square matrix.**

Coefficient Matrix and Augmented Matrix of a System

For a system of linear equations of the form

$$a_1 x + b_1 y + c_1 z = d_1$$
$$a_2 x + b_2 y + c_2 z = d_2$$
$$a_3 x + b_3 y + c_3 z = d_3$$

the matrices

$$\begin{bmatrix} a_1 & b_1 & c_1 \\ a_2 & b_2 & c_2 \\ a_3 & b_3 & c_3 \end{bmatrix} \quad \text{and} \quad \left[\begin{array}{ccc|c} a_1 & b_1 & c_1 & d_1 \\ a_2 & b_2 & c_2 & d_2 \\ a_3 & b_3 & c_3 & d_3 \end{array} \right]$$

are called the **coefficient matrix** and the **augmented matrix,** respectively. Notice that each *row* of the augmented matrix represents one of the equations of the system. For example, the augmented matrix of the system

$$\begin{array}{r} 3x - 4y + z = 2 \\ -x + 2y = -1 \\ 2x - y - 3z = 4 \end{array} \quad \text{is} \quad \left[\begin{array}{ccc|c} 3 & -4 & 1 & 2 \\ -1 & 2 & 0 & -1 \\ 2 & -1 & -3 & 4 \end{array} \right],$$

and the augmented matrix of the system

$$\begin{array}{r} x - 3y + 2z = 5 \\ 2y - z = 4 \\ 4z = 8 \end{array} \quad \text{is} \quad \left[\begin{array}{ccc|c} 1 & -3 & 2 & 5 \\ 0 & 2 & -1 & 4 \\ 0 & 0 & 4 & 8 \end{array} \right].$$

9.4 ■ SOLUTION OF SYSTEMS USING MATRICES

The augmented matrix of this last system, which has all zero entries in the lower left corner, is said to be in **upper triangular form**. As we saw in Section 9.2, it is easy to find the solution of such a system by back-substitution. If we can change a system of linear equations into an equivalent system in upper triangular form, we can then use back-substitution to find the solution of the system.

Elementary Transformations

In Section 9.1 we used the two properties on page 433 to change a given linear system into an equivalent one, that is, one that has the same solution as the original system. The properties allowed us to perform the following operations on the equations of a system.

1. Multiply both sides of an equation by a nonzero real number.
2. Add a constant multiple of one equation to another equation.

It is clear that we can also perform the following operation without changing the solution of the system.

3. Interchange two equations.

Since each equation of the system corresponds to a row in the augmented matrix for the system, the three operations above correspond to the following **elementary row operations** on the augmented matrix.

Elementary Row Operations

1. Multiply the entries of any row by a nonzero number.
2. Add a constant multiple of one row to another.
3. Interchange two rows.

EXAMPLE 1 a.

$$A = \begin{bmatrix} 1 & 3 & -1 & | & -1 \\ 2 & 1 & 4 & | & 5 \\ 6 & 2 & -1 & | & -12 \end{bmatrix} \quad \text{and} \quad B = \begin{bmatrix} 1 & 3 & -1 & | & -1 \\ 6 & 3 & 12 & | & 15 \\ 6 & 2 & -1 & | & -12 \end{bmatrix}$$

represent equivalent systems because we can multiply each entry in the second row of A by 3 to obtain B.

b.
$$A = \begin{bmatrix} 3 & -1 & 2 & | & 7 \\ 2 & 1 & 4 & | & -5 \\ 3 & 1 & 9 & | & -16 \end{bmatrix} \quad \text{and} \quad B = \begin{bmatrix} 3 & 1 & 9 & | & -16 \\ 2 & 1 & 4 & | & -5 \\ 3 & -1 & 2 & | & 7 \end{bmatrix}$$

represent equivalent systems because we can interchange the first and third rows of A to obtain B.

c. $A = \begin{bmatrix} 1 & 2 & 1 & | & -3 \\ 2 & 0 & -1 & | & 7 \\ 3 & 1 & 2 & | & 10 \end{bmatrix}$ and $B = \begin{bmatrix} 1 & 2 & 1 & | & -3 \\ 0 & -4 & -3 & | & 13 \\ 3 & 1 & 2 & | & 10 \end{bmatrix}$

represent equivalent systems because we can add -2 times each entry of the first row of A to the corresponding entry of the second row of A to obtain B.

EXAMPLE 2 Use row operations to form an equivalent matrix with the given elements:

$$\begin{bmatrix} 1 & -4 & | & -5 \\ 3 & 6 & | & 3 \end{bmatrix} \rightarrow \begin{bmatrix} 1 & -4 & | & -5 \\ 0 & ? & | & ? \end{bmatrix}.$$

Solution To obtain 0 as the first entry in the second row, add -3(row 1) to row 2:

$$-3(\text{row 1}) + \text{row 2} \begin{bmatrix} 1 & -4 & | & -5 \\ 3 & 6 & | & 3 \end{bmatrix} \rightarrow \begin{bmatrix} 1 & -4 & | & -5 \\ 0 & 18 & | & 18 \end{bmatrix}.$$

It is often convenient to perform more than one row operation at a time on a given matrix.

EXAMPLE 3 Use row operations on the first matrix to form an equivalent matrix with the given elements:

$$\begin{bmatrix} 1 & -3 & 1 & | & -4 \\ 3 & -1 & -1 & | & 8 \\ 2 & -2 & 3 & | & -1 \end{bmatrix} \rightarrow \begin{bmatrix} 1 & -3 & 1 & | & -4 \\ 0 & ? & ? & | & ? \\ 0 & 0 & ? & | & ? \end{bmatrix}.$$

9.4 ■ SOLUTION OF SYSTEMS USING MATRICES

Solution Perform the transformation in two steps: first, obtain zeros in the second and third entries of the first column by adding suitable multiples of the first row to the second and third rows:

$$\begin{matrix} \\ -3(\text{row 1}) + \text{row 2} \\ -2(\text{row 1}) + \text{row 3} \end{matrix} \begin{bmatrix} 1 & -3 & 1 & | & -4 \\ 3 & -1 & -1 & | & 8 \\ 2 & -2 & 3 & | & -1 \end{bmatrix} \rightarrow \begin{bmatrix} 1 & -3 & 1 & | & -4 \\ 0 & 8 & -4 & | & 20 \\ 0 & 4 & 1 & | & 7 \end{bmatrix}.$$

Now obtain a zero as the second entry of the third row by adding $-\frac{1}{2}(\text{row 2})$ to row 3:

$$\begin{matrix} \\ \\ -\frac{1}{2}(\text{row 2}) + \text{row 3} \end{matrix} \begin{bmatrix} 1 & -3 & 1 & | & -4 \\ 0 & 8 & -4 & | & 20 \\ 0 & 4 & 1 & | & 7 \end{bmatrix} \rightarrow \begin{bmatrix} 1 & -3 & 1 & | & -4 \\ 0 & 8 & -4 & | & 20 \\ 0 & 0 & 3 & | & -3 \end{bmatrix}.$$

Note that the last matrix is in upper triangular form.

Gaussian Elimination

We are now ready to discuss a general method for solving linear systems called Gaussian elimination. This method can be used to solve linear systems of any size and is well suited for implementation as a computer program.

There are three steps to the method.

Solving a Linear System with Gaussian Elimination

1. Write the augmented matrix for the system.
2. Using row operations, transform the matrix into an equivalent one in upper triangular form.
3. Use back-substitution to find the solution to the system.

EXAMPLE 4 Use Gaussian elimination to solve the system

$$x - 2y = -5$$
$$2x + 3y = 11.$$

Solution The augmented matrix is

$$\begin{bmatrix} 1 & -2 & | & -5 \\ 2 & 3 & | & 11 \end{bmatrix}.$$

Use row operations to obtain 0 in the first entry of the second row:

$$-2(\text{row 1}) + \text{row 2} \begin{bmatrix} 1 & -2 & | & -5 \\ 2 & 3 & | & 11 \end{bmatrix} \to \begin{bmatrix} 1 & -2 & | & -5 \\ 0 & 7 & | & 21 \end{bmatrix}.$$

The last matrix corresponds to the system

$$x - 2y = -5 \qquad (1)$$
$$7y = 21. \qquad (2)$$

Use back-substitution to solve the system. From Equation (2), $y = 3$. Substitute 3 for y in (1) to find

$$x - 2(3) = -5$$
$$x = 1.$$

The solution is the ordered pair $(1, 3)$.

To transform the augmented matrix of a 3 × 3 system into upper triangular form, we can use the following sequence of row operations.

1. Obtain zeros in the *first* entries of the second and third rows by adding suitable multiples of the *first* row to the second and third rows.
2. Obtain a zero in the *second* entry of the third row by adding a suitable multiple of the *second* row to the third row.

EXAMPLE 5 Use Gaussian elimination to solve the system

$$2x - 4y = 6$$
$$3x - 4y + z = 8$$
$$2x - 3z = -11.$$

Solution The augmented matrix for the system is

$$\begin{bmatrix} 2 & -4 & 0 & | & 6 \\ 3 & -4 & 1 & | & 8 \\ 2 & 0 & -3 & | & -11 \end{bmatrix}.$$

For each transformation of the augmented matrix we show the corresponding system of equations. At each step the new system is equivalent to the original one. Begin by obtaining a 1 in the first entry of the first row by multiplying each entry of the first row by ½ (this will make it easier to obtain zeros in the first entries of the second and third rows):

$$1/2(\text{row 1}) \begin{bmatrix} 2 & -4 & 0 & | & 6 \\ 3 & -4 & 1 & | & 8 \\ 2 & 0 & -3 & | & -11 \end{bmatrix} \quad \begin{array}{l} 2x - 4y + 0z = 6 \\ 3x - 4y + z = 8 \\ 2x + 0y - 3z = -11 \end{array}$$

↓

$$\begin{bmatrix} 1 & -2 & 0 & | & 3 \\ 3 & -4 & 1 & | & 8 \\ 2 & 0 & -3 & | & -11 \end{bmatrix} \quad \begin{array}{l} x - 2y + 0z = 3 \\ 3x - 4y + z = 8 \\ 2x + 0y - 3z = -11. \end{array}$$

Next, obtain zeros in the first entries of the second and third rows by adding suitable multiples of the first row to the second and third rows:

$$\begin{array}{l} -3(\text{row 1}) + \text{row 2} \\ -2(\text{row 1}) + \text{row 3} \end{array} \begin{bmatrix} 1 & -2 & 0 & | & 3 \\ 3 & -4 & 1 & | & 8 \\ 2 & 0 & -3 & | & -11 \end{bmatrix} \quad \begin{array}{l} x - 2y + 0z = 3 \\ 3x - 4y + z = 8 \\ 2x + 0y - 3z = -11 \end{array}$$

↓

$$\begin{bmatrix} 1 & -2 & 0 & | & 3 \\ 0 & 2 & 1 & | & -1 \\ 0 & 4 & -3 & | & -17 \end{bmatrix} \quad \begin{array}{l} x - 2y + 0z = 3 \\ 0x + 2y + z = -1 \\ 0x + 4y - 3z = -17. \end{array}$$

Finally, obtain a zero in the second entry of the third row by adding $-2(\text{row 2})$ to row 3:

$$-2(\text{row 2}) + \text{row 3} \begin{bmatrix} 1 & -2 & 0 & | & 3 \\ 0 & 2 & 1 & | & -1 \\ 0 & 4 & -3 & | & -17 \end{bmatrix} \quad \begin{array}{l} x - 2y + 0z = 3 \\ 0x + 2y + z = -1 \\ 0x + 4y - 3z = -17 \end{array}$$

↓

$$\begin{bmatrix} 1 & -2 & 0 & | & 3 \\ 0 & 2 & 1 & | & -1 \\ 0 & 0 & -5 & | & -15 \end{bmatrix} \quad \begin{array}{l} x - 2y + 0z = 3 \\ 0x + 2y + z = -1 \\ 0x + 0y - 5z = -15. \end{array}$$

The system is now in upper triangular form; use back-substitution to find the solution. Solve the last equation to get $z = 3$ and substitute 3 for z in the second equation to find $y = -2$. Finally, substitute 3 for z and -2 for y in the first equation to find $x = -1$. The solution is the ordered triple $(-1, -2, 3)$.

If any step in this procedure results in the equation $0x + 0y + 0z = 0$, or in a contradiction such as $0x + 0y + 0z = d$, $d \neq 0$, then the system is either dependent or inconsistent, respectively. In either case the system does not have a unique solution.

EXERCISE 9.4

A

■ *Perform the given elementary row operation on the given matrix. See Example 1.*

1. Multiply row 2 by -3:
$$\begin{bmatrix} -2 & 1 & | & 0 \\ 3 & -1 & | & 2 \end{bmatrix}.$$

2. Multiply row 1 by $\dfrac{1}{4}$:
$$\begin{bmatrix} 2 & 0 & | & 3 \\ -1 & 5 & | & 4 \end{bmatrix}.$$

3. Add 2(row 1) to row 2:
$$\begin{bmatrix} 1 & -3 & | & 6 \\ -2 & 4 & | & -1 \end{bmatrix}.$$

4. Add -3(row 1) to row 2:
$$\begin{bmatrix} 1 & -4 & | & 8 \\ 3 & -2 & | & 10 \end{bmatrix}.$$

5. Interchange row 1 and row 3:
$$\begin{bmatrix} 0 & -3 & 2 & | & -3 \\ 2 & 6 & -1 & | & 4 \\ 1 & 0 & -2 & | & 5 \end{bmatrix}.$$

6. Interchange row 2 and row 3:
$$\begin{bmatrix} 1 & 6 & 0 & | & -2 \\ 0 & 0 & 5 & | & -10 \\ 0 & 3 & -2 & | & 8 \end{bmatrix}.$$

7. Add -4(row 1) to row 3:
$$\begin{bmatrix} 1 & 2 & 1 & | & -5 \\ 0 & 4 & -2 & | & 3 \\ 4 & -1 & 6 & | & -8 \end{bmatrix}.$$

8. Add 2(row 2) to row 3:
$$\begin{bmatrix} 1 & -7 & 5 & | & 2 \\ 0 & 1 & -3 & | & -1 \\ 0 & -2 & -3 & | & 4 \end{bmatrix}.$$

■ *Use row operations on the first matrix to form an equivalent matrix with the given elements. See Examples 2 and 3.*

9. $\begin{bmatrix} 1 & -3 & | & 2 \\ 2 & 1 & | & 4 \end{bmatrix} \to \begin{bmatrix} 1 & -3 & | & 2 \\ 0 & ? & | & ? \end{bmatrix}$

10. $\begin{bmatrix} -2 & 3 & | & 0 \\ 4 & 1 & | & 6 \end{bmatrix} \to \begin{bmatrix} -2 & 3 & | & 0 \\ 0 & ? & | & ? \end{bmatrix}$

11. $\begin{bmatrix} 2 & 6 & | & -4 \\ 5 & 3 & | & 1 \end{bmatrix} \to \begin{bmatrix} 2 & 6 & | & -4 \\ ? & 0 & | & ? \end{bmatrix}$

12. $\begin{bmatrix} 6 & 4 & | & -2 \\ -1 & -2 & | & -3 \end{bmatrix} \to \begin{bmatrix} 6 & 4 & | & -2 \\ ? & 0 & | & ? \end{bmatrix}$

13. $\begin{bmatrix} 1 & -2 & 2 & | & 1 \\ 2 & 3 & -1 & | & 6 \\ 4 & 1 & -3 & | & 3 \end{bmatrix} \to \begin{bmatrix} 1 & -2 & 2 & | & 1 \\ 0 & ? & ? & | & ? \\ 0 & ? & ? & | & ? \end{bmatrix}$

14. $\begin{bmatrix} 2 & -1 & 3 & | & -1 \\ -4 & 0 & 4 & | & 5 \\ 6 & 2 & -1 & | & -2 \end{bmatrix} \to \begin{bmatrix} 2 & -1 & 3 & | & -1 \\ 0 & ? & ? & | & ? \\ 0 & ? & ? & | & ? \end{bmatrix}$

15. $\begin{bmatrix} -1 & 4 & 3 & | & 2 \\ 2 & -2 & -4 & | & 6 \\ 1 & 2 & 3 & | & -3 \end{bmatrix} \to \begin{bmatrix} -1 & 4 & 3 & | & 2 \\ ? & 0 & ? & | & ? \\ ? & 0 & ? & | & ? \end{bmatrix}$

16. $\begin{bmatrix} 3 & -2 & 4 & | & -4 \\ 2 & 2 & 1 & | & 2 \\ -1 & 1 & 5 & | & -1 \end{bmatrix} \to \begin{bmatrix} 3 & -2 & 4 & | & -4 \\ ? & ? & 0 & | & ? \\ ? & ? & 0 & | & ? \end{bmatrix}$

17. $\begin{bmatrix} -2 & 1 & -3 & | & -2 \\ 4 & 2 & 0 & | & 2 \\ 6 & -1 & 2 & | & 0 \end{bmatrix} \to \begin{bmatrix} -2 & 1 & -3 & | & -2 \\ 0 & ? & ? & | & ? \\ 0 & 0 & ? & | & ? \end{bmatrix}$

18. $\begin{bmatrix} -1 & 2 & 3 & | & 3 \\ 4 & 0 & 1 & | & -6 \\ 2 & 2 & -3 & | & -2 \end{bmatrix} \to \begin{bmatrix} -1 & 2 & 3 & | & 3 \\ 0 & ? & ? & | & ? \\ 0 & 0 & ? & | & ? \end{bmatrix}$

■ Use row operations on the augmented matrix to solve each system. See Example 4 for Problems 19–26.

19. $x + 3y = 11$
    $2x - y = 1$

20. $x - 5y = 11$
    $2x + 3y = -4$

21. $x - 4y = -6$
    $3x + y = -5$

22. $x + 6y = -14$
    $5x - 3y = -4$

23. $2x + y = 5$
    $3x - 5y = 14$

24. $3x - 2y = 16$
    $4x + 2y = 12$

25. $x - y = -8$
    $x + 2y = 9$

26. $4x - 3y = 16$
    $2x + y = 8$

■ See Example 5 for Problems 27–34.

27. $x + 3y - z = 5$
    $3x - y + 2z = 5$
    $x + y + 2z = 7$

28. $x - 2y + 3z = -11$
    $2x + 3y - z = 6$
    $3x - y - z = 3$

29. $2x - y + z = 5$
    $x - 2y - 2z = 2$
    $3x + 3y - z = 4$

30. $x - 2y - 2z = 4$
    $2x + y - 3z = 7$
    $x - y - z = 3$

31. $2x - y - z = -4$
    $x + y + z = -5$
    $x + 3y - 4z = 12$

32. $x - 2y - 5z = 2$
    $2x + 3y + z = 11$
    $3x - y - z = 11$

33. $2x - y = 0$
    $3y + z = 7$
    $2x + 3z = 1$

34. $3x - z = 7$
    $2x + y = 6$
    $3y - z = 7$

B

■ Use elementary row operations to transform each matrix to an equivalent one in upper triangular form:

$$\begin{bmatrix} * & * & * & * & | & * \\ 0 & * & * & * & | & * \\ 0 & 0 & * & * & | & * \\ 0 & 0 & 0 & * & | & * \end{bmatrix}.$$

(Many different answers are possible.)

35. $\begin{bmatrix} 1 & 2 & 1 & 0 & | & -1 \\ 2 & 0 & -1 & 1 & | & 4 \\ -1 & 1 & 0 & 0 & | & -2 \\ 0 & 2 & 1 & 2 & | & 2 \end{bmatrix}$

36. $\begin{bmatrix} 1 & 1 & -1 & 1 & | & 2 \\ 0 & 3 & 2 & -1 & | & 0 \\ 2 & 0 & -2 & 1 & | & 6 \\ -1 & 1 & 3 & -2 & | & -2 \end{bmatrix}$

37. $\begin{bmatrix} 4 & 0 & 2 & 1 & | & 1 \\ 2 & -1 & 1 & -2 & | & -6 \\ 0 & 1 & 3 & 1 & | & 2 \\ -2 & 0 & -1 & -3 & | & -3 \end{bmatrix}$

38. $\begin{bmatrix} 3 & -2 & 1 & 0 & | & 4 \\ 1 & 4 & -2 & 1 & | & -5 \\ 1 & 0 & -1 & 0 & | & -3 \\ -2 & 2 & 1 & -1 & | & 1 \end{bmatrix}$

■ Use row operations on the augmented matrix to solve each system.

39. $\begin{aligned} x + y + z + w &= -1 \\ 2x \phantom{{}+y} + z - w &= 0 \\ -x - y + 2z + w &= -2 \\ 2y - 3z - w &= 6 \end{aligned}$

40. $\begin{aligned} x - y + z - w &= 8 \\ x + 3y \phantom{{}+z} + 2w &= -1 \\ -2x - 2y + z - w &= -3 \\ 3x + y \phantom{{}+z} + w &= 10 \end{aligned}$

41. $\begin{aligned} 3x + y - z - w &= 1 \\ 2x + y + 2z \phantom{{}-w} &= 1 \\ y + z - w &= -3 \\ 2x \phantom{{}+y} + 2z + w &= 5 \end{aligned}$

42. $\begin{aligned} 2x + y + 2z \phantom{{}+w} &= 3 \\ -2x - y \phantom{{}+z} + 3w &= -1 \\ 3x - 2y + z + w &= 7 \\ 3x + y - 3z + 2w &= 20 \end{aligned}$

CHAPTER REVIEW

A

[9.1]

■ Solve each system by linear combinations.

1. $\begin{aligned} x + 5y &= 18 \\ x - y &= -3 \end{aligned}$

2. $\begin{aligned} x + 5y &= 11 \\ 2x + 3y &= 8 \end{aligned}$

3. $\frac{2}{3}x - 3y = 8$

   $x + \frac{3}{4}y = 12$

4. $3x = 5y - 6$

   $3y = 10 - 11x$

■ *State whether each system is dependent, inconsistent, or consistent and independent.*

5. $2x - 3y = 4$
   $x + 2y = 7$

6. $2x - 3y = 4$
   $6x - 9y = 4$

7. $2x - 3y = 4$
   $6x - 9y = 12$

8. $x - y = 6$
   $x + y = 6$

[9.2]

■ *Solve each system using linear combinations.*

9. $x + 3y - z = 3$
   $2x - y + 3z = 1$
   $3x + 2y + z = 5$

10. $x + y + z = 2$
    $3x - y + z = 4$
    $2x + y + 2z = 3$

11. $x + z = 5$
    $y - z = -8$
    $2x + z = 7$

12. $x + 4y + 4z = -20$
    $3x - 2y + z = -4$
    $2x - 4y + z = -4$

13. $\frac{1}{2}x + y + z = 3$

    $x - 2y - \frac{1}{3}z = -5$

    $\frac{1}{2}x - 3y - \frac{2}{3}z = -6$

14. $\frac{3}{4}x - \frac{1}{2}y + 6z = 2$

    $\frac{1}{2}x + y - \frac{3}{4}z = 0$

    $\frac{1}{4}x + \frac{1}{2}y - \frac{1}{2}z = 0$

[9.3]

15. Evaluate $\begin{vmatrix} 3 & -2 \\ 1 & -5 \end{vmatrix}$.

16. Evaluate $\begin{vmatrix} -4 & 0 \\ 2 & -6 \end{vmatrix}$.

■ *Solve each system using Cramer's rule.*

17. $x - 2y = 6$
    $3x + y = 25$

18. $2x + 3y = -2$
    $x - 8y = -39$

19. $\frac{1}{4}x - \frac{1}{3}y = -\frac{5}{12}$

    $\frac{1}{10}x + \frac{1}{5}y = \frac{1}{2}$

20. $\frac{2}{3}x - y = 4$

    $x - \frac{3}{4}y = 6$

- Evaluate each determinant.

21. $\begin{vmatrix} 2 & 1 & 3 \\ 0 & 4 & -1 \\ 2 & 0 & 3 \end{vmatrix}$

22. $\begin{vmatrix} 3 & -1 & 2 \\ -2 & 1 & 0 \\ 2 & 4 & 1 \end{vmatrix}$

- Solve each system using Cramer's rule.

23. $\begin{aligned} x + y &= 3 \\ y + z &= 5 \\ x - y + 2z &= 5 \end{aligned}$

24. $\begin{aligned} 2x + 3y - z &= -2 \\ x - y + z &= 6 \\ 3x - y + z &= 10 \end{aligned}$

25. $\begin{aligned} x + y + z &= 2 \\ 2x - y + z &= -1 \\ x - y - z &= 0 \end{aligned}$

26. $\begin{aligned} x - 2y &= -3 \\ y + 3z &= -1 \\ x - z &= 2 \end{aligned}$

[9.4]

- Use row operations on a matrix to solve each system.

27. $\begin{aligned} x - 2y &= 5 \\ 2x + y &= 5 \end{aligned}$

28. $\begin{aligned} 4x - 3y &= 16 \\ 2x + y &= 8 \end{aligned}$

29. $\begin{aligned} 2x - y &= 7 \\ 3x + 2y &= 14 \end{aligned}$

30. $\begin{aligned} 2x - y + 3z &= -6 \\ x + 2y - z &= 7 \\ 3x + y + z &= 2 \end{aligned}$

31. $\begin{aligned} x + 2y - z &= -3 \\ 2x - 3y + 2z &= 2 \\ x - y + 4z &= 7 \end{aligned}$

32. $\begin{aligned} x + y + z &= 1 \\ 2x - y - z &= 2 \\ 2x - y + 3z &= 2 \end{aligned}$

- Solve each problem using two or three variables.

33. A collection of coins consisting of dimes and quarters has a value of $4.95. How many dimes are in the collection if there are 25 more dimes than quarters?

34. The first-class fare on an airplane flight is $280 and the tourist fare is $160. If 64 passengers paid a total of $12,160 for the flight, how many of each ticket were sold?

35. A woman has invested $8000, part in a bank at 10% and part in a savings and loan association at 12%. If her annual return is $844, how much has she invested at each rate?

36. A sum of $2400 is split between an investment in a mutual fund paying 14% and one in corporate bonds paying 11%. If the return on the 14% investment exceeds that on the 11% investment by $111 per year, how much is invested at each rate?

37. An airplane travels 840 miles in the same time that an automobile travels 210 miles. If the rate of the airplane is 180 miles per hour greater than the rate of the automobile, find the rate of each.

38. One woman drives 180 miles in the same time that a second woman drives 200 miles. Find the speed of each woman if the second woman drives 5 miles per hour faster than the first woman.

39. The perimeter of a triangle is 30 centimeters. The length of one side is 7 centimeters shorter than the second side, and the third side is 1 centimeter longer than the second side. Find the length of each side.

40. One angle of a triangle measures 20° more than the second angle, and the third angle is three times the measure of the first angle. Find the measure of each angle.

B

41. Use a substitution of variables to solve:

$$\frac{4}{x} - \frac{3}{y} = \frac{15}{2}$$
$$\frac{1}{x} - \frac{3}{y} = 0.$$

42. Use a substitution of variables to solve:

$$\frac{3}{x} - \frac{1}{y} = \frac{7}{2}$$
$$\frac{2}{x} + \frac{3}{y} = 7.$$

43. Find a and b so that the graph of $y = ax + b$ passes through the points $(0, 3)$ and $(-2, 2)$.

44. Find a and b so that the graph of $ax + by - 4 = 0$ passes through the points $(2, -2)$ and $(1, 3)$.

45. Three solutions of the equation $ax + by + cz = 1$ are $(0, 2, 1)$, $(6, -1, 2)$, and $(0, 2, 0)$. Find the coefficients a, b, and c.

46. Three solutions of the equation $ax + by + cz = 1$ are $(2, 1, 0)$, $(-1, 3, 2)$, and $(3, 0, 0)$. Find the coefficients a, b, and c.

10 Inequalities

10.1

LINEAR INEQUALITIES

An **inequality** is a statement such as

$$x + 3 \geq 10 \quad \text{or} \quad 2t - 6 < 8t^2 + 1$$

in which two expressions are related by one of the symbols > ("greater than"), < ("less than"), ≥ ("greater than or equal to"), or ≤ ("less than or equal to"). A **solution** of an inequality is a value of the variable that makes the inequality true. We often represent the set of all solutions, or **solution set,** of an inequality as an interval on the real-number line, as shown in the table below.

	Solution Set	Graph
$x > a$	(a, ∞)	----○——→ at a
$x < a$	$(-\infty, a)$	←——○---- at a
$x \geq a$	$[a, \infty)$	----●——→ at a
$x \leq a$	$(-\infty, a]$	←——●---- at a

If each side of an inequality has the form $ax + b$, the inequality is called linear. We solve linear inequalities much as we solved linear equations: by generating a sequence of equivalent inequalities until we arrive at one whose solution is obvious. To do this we first recall some fundamental properties of inequalities. Notice that if we add 5 to both sides of

$$2 < 3,$$

we obtain

$$2 + 5 < 3 + 5,$$

or $7 < 8$, a true statement. Similarly,

$$2 - 5 < 3 - 5.$$

The addition of 5 or -5 to each side of the inequality $2 < 3$ shifts the respective graphs to the right or left on the number line, as shown in Figure 10.1, but leaves their order unchanged.

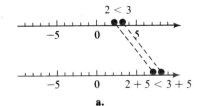
a.
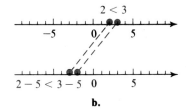
b.

FIGURE 10.1

In general, we have the following property.

---

1. *For all real numbers a, b, and c,*

   if $a < b$, *then* $a + c < b + c$.

Thus, we can add the same expression to both sides of an inequality without changing its solution set.

If we multiply each side of the inequality

$$2 < 3$$

by 2, we have

$$4 < 6,$$

which is a true statement. However, if we multiply each side of

$$2 < 3$$

by -2, we have

$$-4 > -6,$$

where the direction of the inequality must be reversed to make a true statement. These relationships are illustrated in Figure 10.2.

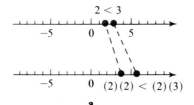
a.

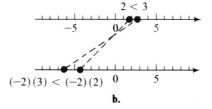
b.

FIGURE 10.2

In general, we have the following properties.

---

2. For all real numbers a, b, and c with $c > 0$,

$$\text{if } a < b, \text{ then } a \cdot c < b \cdot c.$$

3. For all real numbers a, b, and c with $c < 0$,

$$\text{if } a < b, \text{ then } a \cdot c > b \cdot c.$$

Thus, we can multiply both sides of an inequality by a *positive* number without changing its solution set. If we multiply both sides of an inequality by a *negative* number, we must reverse the direction of the inequality.

Note that the three properties stated above also hold if $>$ is replaced by $<$, $\geq$, or $\leq$. We can apply these properties to solve linear inequalities in the same way that we applied the properties of equality to solve linear equations.

EXAMPLE 1 To solve the inequality

$$x + 4 \leq 18 + 3x,$$

first add -4 and $-3x$ to each side to obtain

$$-2x \leq 14.$$

Then multiply each side by $-\frac{1}{2}$ and *reverse the direction* of the inequality to get

$$x \geq -7.$$

The solution set is the interval $[-7, \infty)$. The endpoint of the interval is graphed as a solid dot to show that -7 is included in the set.

We can also use the properties of inequalities to solve **compound inequalities** of the form shown in the next example.

EXAMPLE 2 To solve

$$4 < x + 4 \leq 6,$$

add -4 to each side to obtain

$$4 + (-4) < x + 4 + (-4) \leq 6 + (-4)$$

from which

$$0 < x \leq 2.$$

The solution set is the interval $(0, 2]$. The left endpoint of its graph is an open dot and the right endpoint is a closed dot.

Compound Inequalities

The inequality in Example 2, $4 < x + 4 \leq 6$, is actually a shorter way of writing the pair of inequalities

$$4 < x + 4 \quad \text{and} \quad x + 4 \leq 6.$$

The solution set of this compound inequality consists of all real numbers that are solutions of both $4 < x + 4$ and $x + 4 \leq 6$. Solving each inequality, we find

$$0 < x \quad \text{and} \quad x \leq 2.$$

If we graph each of these solution sets as shown in Figure 10.3, we see that the points in the interval (0, 2] lie in *both* of the solution sets.

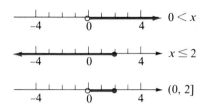

FIGURE 10.3

We call this resulting set the **intersection** of the two solution sets. In general, we make the following definition.

> *The* **intersection** *of two sets A and B consists of all those numbers that are in set A and also in set B.*

We designate the intersection by $A \cap B$, read "the intersection of A and B." Thus, for the compound inequality above, the solution set is $(0, \infty) \cap (-\infty, 2] = (0, 2]$.

The easiest way to find the intersection of two intervals is to graph each interval and see where the graphs "overlap."

EXAMPLE 3 To find $(-2, 5] \cap (1, 8]$, graph each set as shown in the figure. The "overlap" is the set $(1, 5]$. Note that 5 is included in both sets and hence is in the intersection, but 1 is not in the intersection because it is not an element of $(1, 8]$.

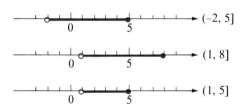

Sets that do not have any members in common are called **disjoint sets.** That is, if

$$A \cap B = \emptyset,$$

where ∅ stands for the empty set, then A and B are disjoint. For example, $(-\infty, -2)$ and $(3, \infty)$ are disjoint, as are $[-7, 2]$ and $[6, 10]$.

To solve a compound inequality joined by the word "and" we solve each simple inequality and then find the intersection of the solution sets.

EXAMPLE 4 Solve the compound inequality

$$2x - 3 > 1 \quad \text{and} \quad 10 - 3x \leq -2.$$

Solution Solve each inequality separately:

$$\begin{aligned} 2x &> 4 & -3x &\leq -12 \\ x &> 2 & x &\geq 4. \end{aligned}$$

The solution consists of all values of x for which $x > 2$ *and* $x \geq 4$. Graph each solution set as shown in the figure. The intersection of the two sets is the interval $[4, \infty)$.

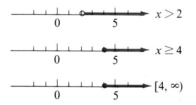

A second type of compound inequality has the form

$$x > 2 \quad \text{or} \quad x < -5.$$

The solution set, shown in Figure 10.4 on page 488, consists of all values for x that satisfy *either* $x > 2$ *or* $x < -5$.

The resulting set is called the **union** of the two solution sets. In general, we make the following definition.

> The **union** of two sets A and B consists of all those numbers that are in either A or B or both.

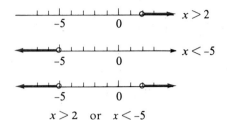

FIGURE 10.4 $x > 2$ or $x < -5$

We denote the union by $A \cup B$, read "the union of A and B." Thus, for the compound inequality above, the solution set is written $(-\infty, -5) \cup (2, \infty)$.

Note that while the compound inequality

$$0 < x \quad \text{and} \quad x \leq 2$$

can be combined into the "double inequality"

$$0 < x \leq 2,$$

it is *incorrect* to write the inequality

$$x > 2 \quad \text{or} \quad x < -5$$

as

$$-5 > x > 2,$$

since this statement implies that $-5 > 2$. In general, a compound inequality involving "or" *cannot* be rewritten as a double inequality.

EXAMPLE 5   To find the union of $[-6, 1)$ and $(-3, 4)$, graph each set as shown in the figure. The union includes all points that appear in either graph. Thus, $[-6, 1) \cup (-3, 4) = [-6, 4)$.

EXAMPLE 6 Solve the compound inequality

$$2x + 5 > 1 \quad \text{or} \quad 8 - 2x < x - 4.$$

Solution Solve each inequality separately:

$$2x > -4 \quad\quad -3x < -12$$
$$x > -2 \quad\quad x > 4.$$

The solution consists of all values of x for which $x > -2$ or $x > 4$. Graph each solution set as shown in the figure. The union of the two sets is the interval $(-2, \infty)$.

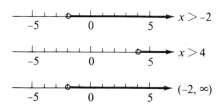

Common Error

It is *incorrect* to write the solution set to Example 6,

$$x > -2 \quad \text{or} \quad x > 4,$$

as $-2 < x > 4$.

Applications

For some applied problems it is appropriate to use an inequality for a model rather than an equation.

EXAMPLE 7 Willard will receive a B in anthropology if his average on five tests is at least 80% but less than 90%. His grades on the first four tests were 98%, 76%, 86%, and 92%. What grade on the fifth test will give him a B in the course?

Solution Let x represent Willard's grade on the fifth test. Then write an inequality expressing the fact that the average of the five tests is greater than or equal to 80 and less than 90:

$$80 \leq \frac{98 + 76 + 86 + 92 + x}{5} < 90.$$

Solve the inequality to obtain

$$80 \leq \frac{352 + x}{5} < 90$$
$$400 \leq 352 + x < 450$$
$$48 \leq x < 98.$$

The solution set is the interval $[48, 98)$. Therefore, any grade equal to or greater than 48 but less than 98 will give Willard a B.

EXERCISE 10.1

A

Solve each inequality and graph its solution set. See Example 1.

1. $3x - 2 > 1 + 2x$
2. $2x + 3 \leq x - 1$
3. $8 - 3x \leq 4(1 - 3x)$
4. $-3(-1 - 2x) \geq 5x - 6$
5. $5x + 10 > 5 - 3(4 - x)$
6. $-6 - 2(x - 4) < 5x + 2$
7. $\dfrac{-2x - 6}{-3} > 2$
8. $\dfrac{-2x - 3}{2} \leq -5$
9. $\dfrac{2x - 3}{3} \leq \dfrac{3x}{-2}$
10. $\dfrac{3x - 4}{-2} > \dfrac{-2x}{5}$
11. $-\dfrac{3}{4}(1 - 3x) > \dfrac{1}{2}(2x - 3)$
12. $\dfrac{1}{3}(2x + 5) \leq -\dfrac{5}{6}(x - 2)$
13. $\dfrac{2}{3}(x - 4) - \dfrac{1}{4}(3 - x) \geq -2$
14. $-\dfrac{3}{2}(2x + 1) - \dfrac{1}{4}(x + 1) < 3$

See Example 2.

15. $-3 < 2x + 1 \leq 7$
16. $-3 \leq 3 - 2x < 9$
17. $-2 > \dfrac{3x + 2}{5} > -4$
18. $0 \geq \dfrac{x + 5}{2} \geq -2$
19. $1.5 \leq \dfrac{x - 2.5}{3} < 1.8$
20. $0.2 < \dfrac{2x - 1.4}{4} \leq 2.6$

Write each nonempty intersection as a single interval. See Example 3.

21. $[-8, -4] \cap [-6, 2)$
22. $(-3, 0] \cap [-1, 4]$
23. $[-5, -3) \cap [-2, 0)$
24. $[-6, 4] \cap (6, 7]$
25. $(0, 3) \cap \emptyset$
26. $(-4, 0] \cap \emptyset$
27. $(-3, 2) \cap (-6, 4)$
28. $(-5, -3) \cap (-7, 4)$

■ *Solve each compound inequality and graph the solution set. See Example 4.*

29. $x < 2$ and $x > -2$
30. $x \leq 5$ and $x \geq 1$
31. $x + 1 \leq 3$ and $-x - 1 \leq 3$
32. $2x - 3 < 5$ and $-2x + 3 < 5$
33. $x - 7 < 2x$ and $-3x \geq 4x + 28$
34. $9 - 3x < 21$ and $-10 < -5x + 15$

■ *Simplify each union (if possible) and graph. See Example 5.*

35. $[-5, -2] \cup [-3, 4]$
36. $[-6, -1] \cup [-5, 1]$
37. $(-4, 2] \cup [2, 4)$
38. $(-4, 2) \cup (2, 4)$
39. $(3, \infty) \cup [5, 8)$
40. $(-6, -2] \cup (-5, -2)$
41. $(-3, -1] \cup (2, 3]$
42. $(-\infty, 2) \cup [3, 4]$

■ *Solve each compound inequality and graph the solution set. See Example 6.*

43. $x + 1 \leq 6$ or $x - 2 > 1$
44. $x - 3 \geq 6$ or $x + 4 < 16$
45. $-4 - 3x < 2$ or $4x - 1 \geq 11$
46. $3x - 1 \leq 8$ or $3 - 2x > 7$
47. $4x - 3 < 1$ or $2x - 3 > 5$
48. $6x + 2 \geq -16$ or $3x + 8 < -10$

■ *Solve each problem. See Example 7.*

49. The Fahrenheit and Celsius temperature scales are related by the formula

$$C = \frac{5F - 160}{9}.$$

The travel brochure for an Austrian ski resort lists the average temperatures in February as between $-23°C$ and $-2°C$. What is the average temperature range in degrees Fahrenheit?

50. The Fahrenheit and Celsius temperature scales are related by the formula

$$F = \frac{9}{5}C + 32.$$

A ski parka has a "comfort range" of $-15°F$ to $25°F$. What is the comfort range in degrees Celsius?

51. Lewis would like to average at least 16 points per game for the remainder of the basketball season. In previous games he scored 12, 15, 19, 17, and 11 points. How many points must he score in the last game to achieve an average of at least 16 points per game?

52. Owen's Market needs to make at least $20,000 per month to stay in business. Over the past 5 months it has made $16,000, $17,600, $19,500, $18,800, and $22,000. How much must it make next month to meet its goal?

53. Barker's Employment Agency charges a commission of $50 plus 15% of the first month's salary for its services. Carter's Career Search charges a commission of $80 plus 12% of the first month's salary. When is it cheaper to use Carter's?

54. Mavis has a choice of two dental insurance plans. Plan A pays 70% of the cost of a regular visit to the dentist after a deductible of $10. Plan B pays 80% of the cost after a deductible of $15. When is plan A cheaper than plan B?

55. In Professor Bunsen's chemistry class, lab projects count as 30% of the grade, tests count as 40%, and the final exam counts as 30%. If Linus has an 86% average on tests and a 73% average on lab projects, what range of scores on the final exam will guarantee Linus a B (greater than or equal to 80% but less than 90%) in the course?

56. A long-distance runner wants to put in 50 training miles per week at an average pace of 6.5 to 7 minutes per mile. If she runs one 12-mile course twice each week at 8 minutes per mile and six 1-mile "sprints" at 5 minutes per mile, at what pace must she run the remaining 20 training miles?

57. Central Bank wants to keep its average interest payments under 8%. If it pays 11% on the $32 million invested in CDs and 12.5% on the $18 million in T-bills, what interest rate can it offer its clients on the $126 million invested in savings accounts?

58. Lacy's Department Stores wants to keep the average salary of its employees at each branch under $19,000 per year. If the downtown store pays its four managers $28,000 per year and its 12 department heads $22,000 per year, how much can it pay its 30 clerks?

10.2

ABSOLUTE-VALUE INEQUALITIES

Before we study absolute-value inequalities it will be helpful to consider some simple equations involving absolute values. First, note that the equation

$$|x| = 5$$

has two solutions, $x = 5$ and $x = -5$. This follows from the definition of absolute value,

$$|x| = \begin{cases} x & \text{if } x \geq 0 \\ -x & \text{if } x < 0 \end{cases}.$$

Thus, in particular, the expression $|ax + b|$ is equal to

$$ax + b \quad \text{if} \quad ax + b \geq 0$$

or to

$$-(ax + b) \quad \text{if} \quad ax + b < 0.$$

Consequently, every absolute-value equation of the form $|ax + b| = c$, where $c > 0$, is equivalent to two linear equations.

The absolute-value equation

$$|ax + b| = c \quad (c > 0)$$

is equivalent to the compound statement

$$ax + b = c \quad \text{or} \quad -(ax + b) = c. \tag{1}$$

EXAMPLE 1 Solve $|2x - 3| = 5$.

Solution The equation is equivalent to

$$
\begin{array}{ll}
2x - 3 = 5 & \text{or} \quad -(2x - 3) = 5 \\
2x = 8 & \quad\quad 2x - 3 = -5 \\
x = 4 & \quad\quad 2x = -2 \\
& \quad\quad x = -1.
\end{array}
$$

The solutions are 4 and -1.

Absolute Value and Distance

An important application of absolute value involves its interpretation as distance. Geometrically, the absolute value of a number gives the distance from the origin to the number. For example, the equation $|x| = 5$ can be interpreted as "x is five units from the origin." Both 5 and -5 are solutions to the equation, since distance is always considered to be positive.

More generally, the distance between two points x and a can be denoted by $|x - a|$. The absolute value ensures that the distance will be computed as positive regardless of whether x or a lies farther to the right on the number line. For example, the distance between 8 and 2 is the same as the distance between 2 and 8, and

$$|8 - 2| = |2 - 8| = 6.$$

EXAMPLE 2 Write each statement using absolute-value notation.

a. x is three units from the origin.
b. p is two units from the point 5.
c. a is within four units of the point -2.

Solutions Restate each sentence in terms of distance.

a. The distance between x and the origin is three units, so $|x| = 3$.
b. The distance between p and 5 is two units, so $|p - 5| = 2$.
c. The distance between a and -2 is less than four units, so $|a - (-2)| < 4$, or $|a + 2| < 4$.

Inequalities Involving Absolute Value

Using the notion of distance for absolute value we can interpret the statement

$$|x| < 4$$

as "The distance from x to the origin is less than four units." The set of solutions for the inequality is shown in Figure 10.5. Note that the solution set can also be described by the compound inequality $-4 < x < 4$.

FIGURE 10.5

Similarly, the absolute-value inequality $|x - 2| < 5$ is equivalent to the compound inequality $-5 < x - 2 < 5$, whose solution set is $-3 < x < 7$. (See Figure 10.6.) We can read the inequality as "The distance between x and 2 is less than five units."

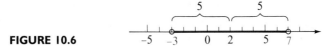

FIGURE 10.6

In general, we have the following property.

The absolute-value inequality

$$|ax + b| < c \quad (c > 0)$$

is equivalent to the compound inequality

$$-c < ax + b < c. \qquad (2)$$

EXAMPLE 3 Solve $|2x + 1| < 7$ and graph the solution set.

Solution From Property (2) above, the inequality is equivalent to

$$-7 < 2x + 1 < 7,$$

from which we have

$$-8 < 2x < 6$$
$$-4 < x < 3.$$

The solution set is $(-4, 3)$, with the graph as shown.

Now consider the statement

$$|x| > 4,$$

which can be interpreted in terms of distance as "The distance from x to the origin is greater than four units." The set of solutions for the inequality is shown in Figure 10.7. Note that the solution set can also be described by the compound inequality $x < -4$ or $x > 4$.

FIGURE 10.7

Similarly, the absolute-value inequality $|x - 2| > 5$ is equivalent to the compound inequality $x - 2 < -5$ or $x - 2 > 5$, whose solution set is $(-\infty, 3) \cup (7, \infty)$. (See Figure 10.8.) We can read the inequality as "The distance between x and 2 is greater than five units."

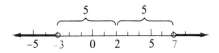

FIGURE 10.8

In general, we have the following property.

The absolute-value inequality

$$|ax + b| > c \quad (c > 0)$$

is equivalent to the compound inequality

$$ax + b < -c \quad \text{or} \quad ax + b > c. \tag{3}$$

Note that the graph in Figure 10.8 is composed of *two* intervals and the compound inequality that describes it *cannot* be written as a single inequality. Thus, $7 < x < -3$ would be an *incorrect* description of the set graphed in Figure 10.8 because 7 is not less than -3.

EXAMPLE 4 Solve $|x + 1| \geq 3$ and graph the solution set.

Solution From Property (3) above, the inequality is equivalent to the compound inequality

$$x + 1 \leq -3 \quad \text{or} \quad x + 1 \geq 3,$$

from which we have

$$x \leq -4 \quad \text{or} \quad x \geq 2.$$

The solution set is $(-\infty, -4] \cup [2, \infty)$, with the graph as shown.

Accuracy

It usually is impossible to make measurements of length, weight, or other physical quantities with complete accuracy. Measuring instruments can never be made perfectly accurate, and readings taken by human investigators will never be absolutely precise. Consequently, scientists and engineers usually include an estimate of the accuracy of any measured quantity.

For instance, the length of a metal bolt might be given as 8 centimeters ± 0.1 centimeter, which means that the actual length of the bolt may be anywhere between 8 − 0.1, or 7.9, centimeters and 8 + 0.1, or 8.1, centimeters. If x represents the actual length of the bolt, we can describe this range of values by the inequality

$$|x - 8| < 0.1.$$

Inaccuracy, or error, in a measurement introduces error into any calculated values based on that measurement. The estimated error in the calculated value can be computed using inequalities.

EXAMPLE 5 A biologist needs a square plastic plate whose perimeter is 12 centimeters, accurate to within 0.2 centimeter. With what accuracy must the sides of the plate be measured?

Solution If x represents the side of the square, then its perimeter is given by $4x$, and we want

$$|4x - 12| < 0.2.$$

Solve the inequality for x to find

$$-0.2 < 4x - 12 < 0.2$$
$$11.8 < 4x < 12.2$$
$$2.95 < x < 3.05.$$

The sides of the plate must be between 2.95 centimeters and 3.05 centimeters; that is, they must be 3 centimeters ± 0.05 centimeter. Thus, the sides must be measured with an accuracy of 0.05 centimeter.

EXERCISE 10.2

A

■ *Solve. See Example 1.*

1. $|x - 4| = 3$
2. $|x - 3| = 2$
3. $|x + 2| = 5$
4. $|x + 6| = 1$
5. $|2x - 1| = 4$
6. $|3x - 1| = 5$
7. $|4 - 3x| = 1$
8. $|6 - 5x| = 4$
9. $|x^2 + 5x + 3| = 3$
10. $|x^2 - 2x - 4| = 4$

■ *Express using absolute value. See Example 2.*

11. x is six units from the origin.
12. a is seven units from the origin.
13. The distance from p to -3 is five units.
14. The distance from q to -7 is two units.
15. t is within three units of 6.
16. w is no more than one unit from -5.
17. b is at least 0.5 unit from -1.
18. m is more than 0.1 unit from 8.

■ *Solve each inequality and graph its solution set. See Example 3.*

19. $|x| < 2$
20. $|x| < 5$
21. $|x + 3| \leq 4$
22. $|x + 1| \leq 8$
23. $|2x - 6| < 3$
24. $|2x + 4| < 6$
25. $|4 - x| \leq 8$
26. $|5 - 3x| \leq 15$
27. $|2x + 0.5| < 0.1$
28. $|4x + 1| < 0.02$

■ *See Example 4.*

29. $|x| > 3$
30. $|x| \geq 5$
31. $|x - 2| > 5$
32. $|x + 5| > 2$
33. $|3 - 2x| \geq 7$
34. $|4 - 3x| > 10$
35. $|3x + 1.5| > 0.5$
36. $|2x - 3.2| \geq 1.4$
37. $|2 + 3x| > 0$
38. $|1 - 7x| > 0$

■ *See Examples 3 and 4.*

39. $|3x + 4| \leq \dfrac{1}{2}$
40. $|2x - 5| > \dfrac{3}{2}$
41. $\left|x - \dfrac{1}{4}\right| < \dfrac{1}{3}$
42. $\left|x + \dfrac{1}{2}\right| \geq \dfrac{1}{3}$
43. $|9.6 + 2.3x| \geq 3.4$
44. $|27 - 5x| > 12.5$
45. $|4x - 5.6| < 0.08$
46. $|3x + 2.1| < 0.01$

■ *Solve. See Example 5.*

47. The mass of a meteorite is measured as 57 grams ± 2 grams. Write an absolute-value inequality that describes the mass of the meteorite.

48. The length of a pendulum is measured as 78 centimeters ± 0.5 centimeter. Write an absolute-value inequality that describes the length of the pendulum.

49. The length of the equator is determined to be 24,900 miles ± 2 miles.
    a. Write an inequality that describes the length of the equator.
    b. Assuming that the earth is a perfect sphere, find the radius of the earth, including an error estimate. (Round to two decimal places.)

50. The circumference of a large cylindrical tank is measured as 56.2 meters ± 0.4 meter.
    a. Write an inequality that describes the circumference of the tank.
    b. Find the radius of the tank, including an error estimate. (Round to two decimal places.)

51. A decorator plans to tile a 12-foot-long room with 8-inch marble tiles. The contractor guarantees that the new floor will fit the room to within ½ inch. Write an inequality that gives the acceptable values for the length of each tile.

52. Ace Trucking Company will haul a load of 5000 pounds, with a 5-pound margin of error, for $300. Five Star Feed Company wants to ship a load of 40-pound feed sacks for $300. Write an inequality that gives the acceptable values for the weight of each sack.

53. In 17 seconds a cheetah was observed to run a distance later measured as 510 yards ± 2 yards. Write an inequality describing the cheetah's speed in yards per second. What is the cheetah's speed in miles per hour?

54. A cyclist sprints for 15 seconds, then measures the distance traveled as 366 yards ± 1 yard. Write an inequality describing the cyclist's speed in yards per second. What is the cyclist's speed in miles per hour?

B

■ *Express using absolute value.*

55. The distance from a to b is twice the distance from b to c.
56. The sum of the distances from P to Q and from P to N is 6.
57. y is closer to x than it is to z.
58. x is farther from d than it is from 5.

■ *Write an absolute-value inequality with the given solution. (Hint: Find the center of the interval a and its "radius" r. The interval can then be expressed as* $|x - a| < r$.)

59. $-3 \leq x \leq 3$ 60. $-4 \leq x \leq 4$ 61. $0 < x < 8$ 62. $-10 < x < 0$
63. $-2 < x < 6$ 64. $-8 < x < -2$ 65. $-7 \leq x \leq -2$ 66. $-5 \leq x \leq 2$

10.3

NONLINEAR INEQUALITIES

Quadratic Inequalities

To develop a method for solving inequalities of the form

$$ax^2 + bx + c < 0$$

we will use a tool called a *sign graph*. For example, consider the inequality

$$x^2 + 4x < 5.$$

We first rewrite the inequality equivalently as

$$x^2 + 4x - 5 < 0,$$

then factor the left side to obtain

$$(x + 5)(x - 1) < 0.$$

Now, the product on the left side will be negative (less than zero) only if the factors $x + 5$ and $x - 1$ have opposite signs, that is, for x-values that satisfy either

$$x + 5 < 0 \quad \text{and} \quad x - 1 > 0 \tag{1}$$

or

$$x + 5 > 0 \quad \text{and} \quad x - 1 < 0. \tag{2}$$

The solutions to compound inequalities (1) and (2) can be found by examining the sign graph shown in Figure 10.9. The sign graph is made of three

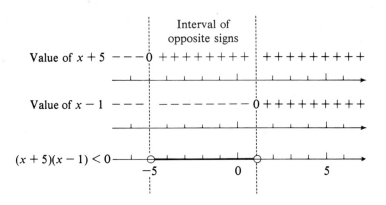

FIGURE 10.9

number lines, one for each of the factors $x + 5$ and $x - 1$ and one for their product. The top row shows the sign of the expression $x + 5$ for different values of x. Thus, $x + 5$ is positive for $x > -5$ and negative for $x < -5$. The second row shows that the factor $x - 1$ is positive for $x > 1$ and negative for $x < 1$.

The bottom row of the sign graph is used to record the sign of the product $(x + 5)(x - 1)$. Note that $(x + 5)(x - 1)$ is *equal* to zero for $x = -5$ and $x = 1$. These values are called **critical numbers** for the inequality. They divide the number line into the three intervals

$$(-\infty, -5), \quad (-5, 1), \quad \text{and} \quad (1, \infty).$$

On each interval the sign of the product $(x + 5)(x - 1)$ is determined by the signs of its factors on that interval. On the interval $(-\infty, -5)$ the two factors $x + 5$ and $x - 1$ are both negative, so their product is positive. On the interval $(-5, 1)$ the two factors have opposite signs, so their product is negative. On the interval $(1, \infty)$ both factors are positive, so their product is also positive. Thus, the solution set for the incquality $x^2 + 4x - 5 < 0$ is the interval $(-5, 1)$.

We can improve our method of solution by making use of the following observation. Note that on each of the intervals determined by the critical numbers the expression $x^2 + 4x - 5$ is either always positive or always negative. To determine which, we need only evaluate the expression at a single number in each interval. Thus, instead of the sign graph we can use a single number line on which the critical numbers are plotted, as shown in Figure 10.10.

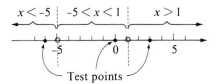

FIGURE 10.10

We next choose a convenient "test point" in each interval; for this example we choose -6, 0, and 2. Evaluate the expression $x^2 + 4x - 5$ at each test point:

$$\text{for} \quad x = -6, \quad x^2 + 4x - 5 = (-6)^2 + 4(-6) - 5$$
$$= 36 - 24 - 5 = 7 > 0;$$
$$\text{for} \quad x = 0, \quad x^2 + 4x - 5 = 0^2 + 4(0) - 5$$
$$= -5 < 0;$$
$$\text{for} \quad x = 2, \quad x^2 + 4x - 5 = 2^2 + 4(2) - 5$$
$$= 4 + 8 - 5 = 7 > 0.$$

The only interval on which $x^2 + 4x - 5 < 0$ is $(-5, 1)$, so this interval is the solution set, as we found above. The graph is shown in Figure 10.11. If the inequality in this example were $x^2 + 4x - 5 \leq 0$, the endpoints -5 and 1 would be included in the solution set and would appear on the graph as closed dots and the solution set would be $[-5, 1]$.

FIGURE 10.11

We summarize the method as follows.

To Solve a Nonlinear Inequality:

1. Rewrite the inequality as an equivalent inequality with zero on one side of the inequality symbol.
2. Plot the critical numbers for the inequality on a number line. This divides the number line into intervals.
3. Choose a test point in each interval and evaluate the inequality at each point to determine which intervals are in the solution set.

EXAMPLE 1 Solve and graph the solution set of

$$x^2 - 3x \geq 4.$$

Solution Write the left side in standard form and factor to obtain

$$x^2 - 3x - 4 \geq 0 \qquad (3)$$
$$(x + 1)(x - 4) \geq 0.$$

a.

The critical numbers are -1 and 4 because $(x + 1)(x - 4) = 0$ for these values. Plot the critical numbers on a number line to find the intervals shown in Figure (a). Choose a test point in each interval, say, -2, 0, and 5, and substitute each into the left side of (3):

$$\text{for } x = -2, \quad x^2 - 3x - 4 = (-2)^2 - 3(-2) - 4$$
$$= 6 > 0;$$
$$\text{for } x = 0, \quad x^2 - 3x - 4 = 0^2 - 3(0) - 4$$
$$= -4 < 0;$$
$$\text{for } x = 5, \quad x^2 - 3x - 4 = 5^2 - 3(5) - 4$$
$$= 6 > 0.$$

The expression $x^2 - 3x - 4$ is positive for $x < -1$ or $x > 4$. Hence, the solution set is $(-\infty, -1] \cup [4, \infty)$. The graph is shown in Figure (b).

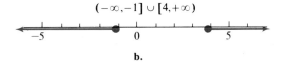

b.

Other Nonlinear Inequalities

The method of critical points used above to solve quadratic inequalities can be applied to polynomial inequalities of any degree and to inequalities involving rational fractions. To do so we extend the definition of critical number as follows.

> A **critical number** for the inequality $P(x)/Q(x) > 0$ or $P(x)/Q(x) < 0$ is a value of x for which either $P(x) = 0$ or $Q(x) = 0$.

We can now proceed with the three steps listed on page 502.

EXAMPLE 2 Solve and graph the solution set of

$$\frac{x}{x-2} \geq 5. \tag{4}$$

Solution Write (4) equivalently as

$$\frac{x}{x-2} - 5 \geq 0,$$

from which

$$\frac{x - 5(x - 2)}{x - 2} \geq 0,$$

or

$$\frac{-4x + 10}{x - 2} \geq 0.$$

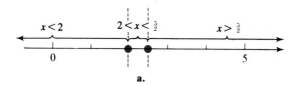

a.

The critical numbers are $5/2$ and 2, because $-4x + 10$ equals zero for $x = 5/2$ and $x - 2$ equals zero for $x = 2$. Plot the critical points on a number line to determine the intervals shown in Figure (a). Choose a test point in each interval, say, 0, $9/4$, and 3, and evaluate $\frac{-4x + 10}{x - 2}$ for each test point. Since

$$\frac{-4(0) + 10}{0 - 2} < 0, \qquad \frac{-4\left(\frac{9}{4}\right) + 10}{\frac{9}{4} - 2} > 0, \quad \text{and} \quad \frac{-4(3) + 10}{3 - 2} < 0,$$

the solution is the interval

$$\left(2, \frac{5}{2}\right].$$

The graph is shown in Figure (b). Note that 2 is *not* a member of the solution set because the left side of (4) is undefined for $x = 2$.

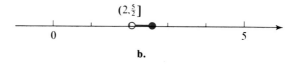

b.

Common Error In Example 2 it would *not* be correct to multiply both sides of Inequality (4) by $x - 2$ to obtain $x \geq 5(x - 2)$. This is because $x - 2$ has different signs for $x > 2$ and $x < 2$.

EXAMPLE 3 Solve and graph the solution set of

$$\frac{x^2 + x - 2}{x - 3} \leq 0. \tag{5}$$

Solution First, find the critical numbers. The numerator,

$$x^2 + x - 2 = (x - 1)(x + 2),$$

equals zero when $x = 1$ or $x = -2$, and the denominator equals zero when $x = 3$. Plot the critical numbers on a number line to determine the four intervals shown in Figure (a).

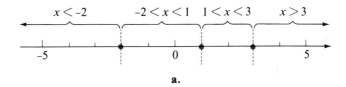

a.

Check the intervals by substituting one value chosen from each interval, say, $-3, 0, 2,$ and 4. Since

$$\frac{(-3 - 1)(-3 + 2)}{-3 - 3} < 0, \qquad \frac{(0 - 1)(0 + 2)}{0 - 3} > 0,$$

$$\frac{(2 - 1)(2 + 2)}{2 - 3} < 0, \quad \text{and} \quad \frac{(4 - 1)(4 + 2)}{4 - 3} > 0,$$

the solution set is the union

$$(-\infty, -2] \cup [1, 3)$$

shown in Figure (b). Note that the critical point 3 is not a solution of the inequality because the left side of (5) is not defined for $x = 3$.

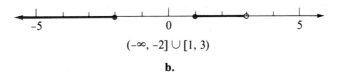

$(-\infty, -2] \cup [1, 3)$

b.

Some inequalities can be solved simply by inspection. For example, the solution set of

$$4x^2 + 6 > 0$$

is $(-\infty, \infty)$ because $4x^2 + 6$ is greater than zero for any real number. For the same reason the solution set of

$$4x^2 + 6 < 0$$

is $\emptyset$.

Applications   Quadratic inequalities are useful as models in certain applied settings.

EXAMPLE 4   The Sub Station sells $120 - \frac{1}{4}p$ submarine sandwiches at lunchtime if it sells them at p cents each. What range of prices can it charge if it wishes to keep its daily revenue from subs over $140?

Solution   The total revenue is given by $R = xp$, where x is the number of subs sold at a price of p cents each. Thus,

$$R = p\left(120 - \frac{1}{4}p\right)$$
$$= 120p - \frac{1}{4}p^2,$$

and the desired price range is the solution of the inequality

$$120p - \frac{1}{4}p^2 > 14{,}000,$$

or

$$p^2 - 480p + 56{,}000 < 0.$$

Factor the left side to obtain

$$(p - 200)(p - 280) < 0.$$

The critical numbers are 200 and 280. Choose a test point in each interval determined by the critical numbers, say, 100, 240, and 300. Evaluate the left side of the inequality at each test point to find

$$(100 - 200)(100 - 280) = 18{,}000 > 0;$$
$$(240 - 200)(240 - 280) = -1600 < 0;$$
$$(300 - 200)(300 - 280) = 2000 > 0.$$

The solution to the inequality is the interval (200, 280), so the Sub Station should charge between $2.00 and $2.80 for a sub in order to achieve the desired revenue of $140.

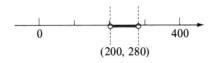

EXERCISE 10.3

A

■ *Solve, and graph the solution set. See Example 1.*

1. $(x + 1)(x - 2) > 0$
2. $(x - 3)(x + 2) > 0$
3. $(t + 3)(t - 4) < 0$
4. $(z + 2)(z + 5) \le 0$
5. $k(k - 2) \le 0$
6. $m(m + 4) > 0$
7. $w^2 - 3w - 4 \ge 0$
8. $p^2 - 5p - 6 \ge 0$
9. $3q^2 + q - 2 > 0$
10. $2r^2 + r - 10 < 0$
11. $y^2 + 2y \le 15$
12. $y^2 - 3y \ge 10$
13. $2z^2 - 7z > 4$
14. $6h^2 + 13h < 15$
15. $9p^2 - 30p + 25 \ge 0$
16. $16k^2 + 24k + 9 > 0$
17. $v^2 < 5$
18. $t^2 \le 7$
19. $4x^2 + 1 < 0$
20. $2y^2 + 5 \le 0$

■ *See Examples 2 and 3.*

21. $\dfrac{y + 3}{y - 2} < 0$
22. $\dfrac{x - 1}{x + 4} > 0$
23. $\dfrac{-3}{x + 5} > 2$
24. $\dfrac{2}{z - 3} < -3$
25. $\dfrac{t}{t + 2} \ge 4$
26. $\dfrac{w + 2}{w - 2} \le 6$
27. $(m - 1)(m + 1)(m - 3) < 0$
28. $(n + 1)(n - 4)n > 0$
29. $(4x + 3)(4x^2 - 1) \ge 0$
30. $(9p^2 - 4)(2p + 5) \le 0$
31. $\dfrac{z + 1}{z^2 - 4z + 3} \le 0$
32. $\dfrac{q^2 - 4}{q + 5} \ge 0$
33. $\dfrac{(2k - 7)^2}{3k + 2} > 0$
34. $\dfrac{(y + 3)^2}{4y - 1} < 0$
35. $\dfrac{r^2 - 6}{5r} \le 1$
36. $\dfrac{b^2 - 16}{3b} \ge -2$
37. $\dfrac{3z}{5z - 3} < -3z$
38. $\dfrac{8t}{6t + 1} > 4t$

■ *Solve. See Example 4.*

39. A fireworks rocket fired from ground level is at a height of $320t - 16t^2$ feet after t seconds. During what interval of time is the rocket higher than 1024 feet?

40. A baseball thrown vertically reaches a height h in feet given by $h = 56t - 16t^2$, where t is measured in seconds. During what period(s) of time is the ball between 40 and 48 feet high?

41. The cost in dollars of manufacturing x pairs of gardening shears is given by $C(x) = -0.02x^2 + 14x + 1600$ for $0 \leq x \leq 700$. How many pairs of shears can be produced if the total cost must be kept under $2800?

42. The cost in dollars of producing $100x$ cashmere sweaters is given by $C(x) = 100x^2 + 4000x + 9000$. How many sweaters can be produced if the total cost must be kept under $185,000?

43. The Locker Room finds that it sells $1200 - 30p$ sweatshirts each month when it charges p dollars per sweatshirt. If it would like its revenue from sweatshirts to be over $9000 per month, in what range should it keep the price of a sweatshirt? (Recall that $R = xp$, where x is the number of items sold at price p.)

44. Green Valley Nursery sells $120 - 10p$ boxes of rose food per month at a price of p dollars per box. If it would like to keep its monthly revenue from rose food over $350, in what range should it price a box of rose food?

B

■ *The inequality $ax^2 + bx + c > 0$ can be solved by graphing the equation $y = ax^2 + bx + c$ and noting for which x-values the graph lies above the x-axis. Use this method to solve the following inequalities.*

45. $x^2 - x - 12 > 0$
46. $x^2 + x + 20 \geq 0$
47. $x^2 + 3x \leq 28$
48. $x^2 - 2x < 3$
49. $x^2 + 4x + 4 \leq 0$
50. $x^2 - 6x + 9 > 0$

■ *The **discriminant** of the quadratic equation $ax^2 + bx + c = 0$ is the expression $D = b^2 - 4ac$. The number of real-valued solutions to the equation are given by the discriminant as follows.*

1. *If $D > 0$, there are two distinct real roots.*
2. *If $D = 0$, there is one real root of multiplicity two.*
3. *If $D < 0$, there are two complex roots.*

Using the discriminant, determine the values of k so that each equation has two real roots.

51. $x^2 + kx + 1 = 0$
52. $x^2 - 2kx + 5 = 0$
53. $kx^2 + 2kx + 1 = 0$
54. $kx^2 - 4kx + 6 = 0$
55. $kx^2 + 4kx + k = 6$
56. $kx^2 - 2kx + 2k = 4$

10.4

LINEAR INEQUALITIES IN TWO VARIABLES; SYSTEMS

The solutions of linear inequalities in two variables, such as

$$ax + by + c > 0 \quad \text{or} \quad ax + by + c < 0,$$

where a, b, and c are real numbers, are ordered pairs. The graph of the solution set forms a region in the plane whose boundary is a straight line. As an example, consider the inequality

$$2x + y - 3 < 0. \tag{1}$$

Rewritten in the form

$$y < -2x + 3 \tag{2}$$

the inequality says that y is less than $-2x + 3$ for each x. Now, the graph of the *equation*

$$y = -2x + 3 \tag{3}$$

is simply a straight line, as illustrated in Figure 10.12. Therefore, to graph (2) we need only observe that any point *below* this line has a y-coordinate *less* than $-2x + 3$ and hence satisfies (2). For example, if $x = 2$, then all ordered pairs of the form $(2, y)$, where

$$y < -2(2) + 3$$

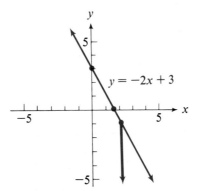

FIGURE 10.12

or

$$y < -1,$$

are in the solution set. The graphs of these points lie below the point $(2, -1)$ on the line, as shown in Figure 10.12.

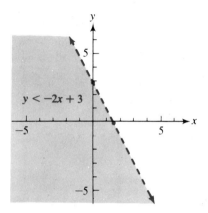

FIGURE 10.13

The solution set of (2) is the entire region below the line. The region is indicated on the graph in Figure 10.13 by shading. The dashed line in Figure 10.13 indicates that the points on the line are *not* elements of the solution set. If the original inequality were

$$2x + y - 3 \leq 0,$$

then the line would be a part of the solution set, and its graph would be shown as a solid line.

In general, the graph of the solution set of the inequalities

$$ax + by + c < 0 \quad \text{or} \quad ax + by + c > 0$$

is a **half-plane** on one side of the line

$$ax + by + c = 0.$$

To determine which of the half-planes is the solution set we can use any point not on the line itself as a test point. Substitute the coordinates of the point into the inequality and note whether a true statement results. If the test point is a solution of the inequality, then the half-plane containing the test point should be shaded; if the test point is not a solution, then the other half-plane should be shaded. The origin, $(0, 0)$, is a convenient test point if it does not lie on the line.

10.4 ■ LINEAR INEQUALITIES IN TWO VARIABLES; SYSTEMS

EXAMPLE 1 Graph $3x - 2y < 6$.

Solution First, graph the line $3x - 2y = 6$ using the intercept method. Since $(0, 0)$ does not lie on the line, choose $(0, 0)$ as the test point. Substitute 0 for x and 0 for y into the inequality to obtain

$$3(0) - 2(0) < 6.$$

Since this is a true statement, shade the half-plane that contains the origin. In this example the edge of the half-plane is a dashed line because the original inequality does not include the "equal to" symbol.

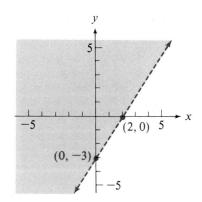

In some situations the inequality $x > 2$ may represent the inequality in two variables $x + 0y > 2$. Its graph is then a region in the plane.

EXAMPLE 2 Graph $x \geq 2$ in the xy-plane.

Solution First, graph the equation $x = 2$; its graph is a vertical line. The solution set of the inequality consists of all ordered pairs for which x is greater than 2. These points lie in the half-plane to the right of the line $x = 2$. In this case the line $x = 2$ is included in the solution set, so the edge of the half-plane is shown as a solid line.

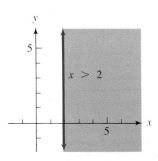

Systems of Inequalities

For some applications we want to find the solutions that are common to two or more inequalities; that is, we want to find the *intersection* of their solution sets. Usually we can find good approximations to such solutions by graphical methods.

EXAMPLE 3 Graph the solution set of the system

$$y > x \quad \text{and} \quad y > 2.$$

Solution Graph the equation $y = x$ and use a test point, say, $(0, 1)$, to see that the upper half-plane should be shaded. Then graph the line $y = 2$ and shade the half-plane above the line to represent the solution set for $y > 2$. Use a dashed line for the boundary of each half-plane. Use heavier shading to indicate the region common to the two half-planes; this is the solution set of the system.

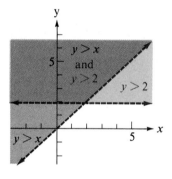

In order to describe the solution set of a system of inequalities it is useful to locate the vertices of its boundary.

EXAMPLE 4 Graph the solution set of the system below and find the coordinates of its vertices:

$$x \geq 0$$
$$y \geq 0$$
$$x - y - 2 \leq 0$$
$$x + 2y - 6 \leq 0.$$

Solution Graph the associated equations,

$$x = 0$$
$$y = 0$$
$$x - y - 2 = 0$$
$$x + 2y - 6 = 0,$$

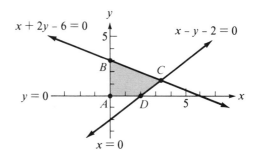

as shown. Use a test point for each inequality to determine which side of the line to shade. With heavier shading or another color indicate the intersection of all four solution sets, as shown in the figure.

To determine the coordinates of the vertices A, B, C, and D, solve simultaneously the equations of the two lines that intersect at the vertex. Thus,

for A, solve the system $\quad x = 0 \quad$ to find $(0, 0)$;
$\qquad\qquad\qquad\qquad\qquad\quad y = 0$

for B, solve the system $\quad x = 0 \quad$ to find $(0, 3)$;
$\qquad\qquad\qquad\qquad\qquad\quad x + 2y = 6$

for C, solve the system $\quad x + 2y = 6 \quad$ to find $(^{10}/_3, ^{4}/_3)$;
$\qquad\qquad\qquad\qquad\qquad\quad x - y = 2$

for D, solve the system $\quad y = 0 \quad$ to find $(2, 0)$.
$\qquad\qquad\qquad\qquad\qquad\quad x - y = 2$

The vertices of the solution set are the points $(0, 0)$, $(0, 3)$, $(^{10}/_3, ^{4}/_3)$, and $(2, 0)$.

EXERCISE 10.4

A

■ *Graph each inequality. See Examples 1 and 2.*

1. $y > 2x + 4$
2. $y < 9 - 3x$
3. $3x - 2y \leq 12$
4. $2x + 5y \geq 10$
5. $x + 4y \geq -6$
6. $3x - y \leq -2$
7. $x < -3y + 1$
8. $x > 2y - 5$
9. $x \geq -3$
10. $y < 4$

11. $y > \dfrac{1}{2}$
12. $y > \dfrac{4}{3}x$
13. $0 \geq x - y$
14. $0 \geq x + 3y$
15. $-1 < y \leq 5$
16. $-2 \leq y < 0$

■ *Graph the solution to the system of inequalities. See Example 3.*

17. $y > 2, \quad x \geq -2$
18. $y \leq -1, \quad x > 2$
19. $y < x, \quad y \geq -3$
20. $y \geq -x, \quad y < 2$
21. $x + y \leq 6, \quad x + y \geq 4$
22. $x - y < 3, \quad x - y > -2$
23. $2x - y \leq 4, \quad x + 2y > 6$
24. $2y - x < 2, \quad x + y \leq 4$
25. $3y - 2x < 2, \quad y > x - 1$
26. $2x + y < 4, \quad y > 1 - x$

■ *Graph the solution set to the system of inequalities and find the coordinates of its vertices. See Example 4.*

27. $2x + 3y - 6 < 0$
    $x \geq 0, \quad y \geq 0$
28. $3x + 2y < 6$
    $x \geq 0, \quad y \geq 0$
29. $5y - 3x \leq 15$
    $x + y \leq 11$
    $x \geq 0, \quad y \geq 0$
30. $y - 2x \geq -4$
    $x + y \leq 5$
    $x \geq 0, \quad y \geq 0$
31. $2y \leq x$
    $2x \leq y + 12$
    $x \geq 0, \quad y \geq 0$
32. $y \geq 3x$
    $2y + x \leq 14$
    $x \geq 0, \quad y \geq 0$
33. $x + y \geq 3$
    $2y \leq x + 8$
    $2y + 3x \leq 24$
    $x \geq 0, \quad y \geq 0$
34. $2y + 3x \geq 6$
    $2y + x \leq 10$
    $y \geq 3x - 9$
    $x \geq 0, \quad y \geq 0$
35. $5y - x \geq 5$
    $y - 2x \geq -8$
    $3y - 9 \leq 4x$
    $x \geq 0, \quad y \geq 0$
36. $2y + x \leq 12$
    $4y \leq 2x + 8$
    $x \leq 4y + 4$
    $x \geq 0, \quad y \geq 0$

B

■ *Graph each inequality.*

37. $y \geq |x| + 2$
38. $y < |x| - 1$
39. $y < |x - 3|$
40. $y > |x + 4|$
41. $x < |y|$
42. $x \geq |x|$
43. $|x + y| > 1$
44. $|y - 2x| \leq 4$

■ *Graph each inequality. (Hint: Consider the graphs in each quadrant separately:* $x, y \geq 0$; $x \leq 0, y \geq 0$; $x, y \leq 0$; *and* $x \geq 0, y \leq 0$.)

45. $|x| + |y| \leq 1$
46. $|x| + |y| \geq 1$
47. $|x| - |y| \leq 1$
48. $|x| - |y| \geq 0$

CHAPTER REVIEW

A

[10.1]

■ *Solve each inequality and graph the solution set.*

1. $3x + 1 < 2 - 3(x - 1)$
2. $\dfrac{2x + 4}{3} \geq \dfrac{3x}{2}$
3. $\dfrac{1}{3}(x - 2) - \dfrac{1}{2}(x + 3) \leq -4$
4. $2(3x - 1) - \dfrac{1}{3}(x + 2) > \dfrac{5}{2}$
5. $-5 < 3x - 1 \leq 4$
6. $0.08 \leq \dfrac{x - 2.3}{2} < 1.2$

■ *Solve each compound inequality and graph the solution set.*

7. $x + 3 \leq 4$ and $x + 6 > 2$
8. $2x - 1 > 7$ and $-5 \geq -4x + 3$
9. $2x - 1 < 7$ or $3 - x < 10$
10. $4x + 3 \leq 11$ or $4 - 2x > 10$

[10.2]

■ *Solve.*

11. $|x - 2| = 5$
12. $|3x - 4| = 6$
13. $|3 - 5x| = 2$
14. $|x^2 - 3x + 1| = 1$

■ *Express using absolute value.*

15. x is four units from the origin.
16. The distance from y to -5 is three units.
17. p is within four units of 7.
18. q is at least three-tenths of a unit from -4.

■ *Solve each inequality and graph the solution set.*

19. $|3x - 2| < 4$
20. $|2x + 0.3| \leq 0.5$
21. $|3y + 1.2| \geq 1.5$
22. $|2 - 3y| > 0$
23. $\left|2z - \dfrac{1}{2}\right| < \dfrac{1}{4}$
24. $\left|3z + \dfrac{1}{2}\right| > \dfrac{1}{3}$

[10.3]
■ *Solve and graph the solution set.*

25. $(x - 3)(x + 2) > 0$
26. $y^2 - y - 12 \leq 0$
27. $2y^2 - y \leq 3$
28. $3z^2 - 5z > 2$
29. $s^2 \leq 4$
30. $4t^2 > 12$
31. $\dfrac{y - 4}{y + 3} < 0$
32. $\dfrac{3}{x - 4} > -2$
33. $(p - 1)(p + 2)(p - 4) < 0$
34. $q(q - 2)(q + 3) > 0$

[10.4]
■ *Graph each inequality.*

35. $3x - 4y < 12$
36. $x > 3y - 6$
37. $y < -\dfrac{1}{2}$
38. $-4 \leq x < 2$

■ *Graph the solution to the system of inequalities.*

39. $y > 3, \quad x \leq 2$
40. $y \geq x, \quad x > 2$
41. $3x - y < 6, \quad x + 2y > 6$
42. $x - 3y > 3, \quad y < x + 2$

■ *Graph the solution set to the system of inequalities and find the coordinates of its vertices.*

43. $3x - 4y \leq 12$
    $x \geq 0; \quad y \leq 0$
44. $x - 2y \leq 6$
    $y \geq x$
    $x \geq 0, \quad y \geq 0$
45. $x + y \leq 5$
    $y \geq x$
    $y \geq 2, \quad x \geq 0$
46. $x - y \leq -3$
    $x + y \leq 6$
    $x \leq 4$
    $x \geq 0; \quad y \geq 0$

[10.1–10.4]
■ *Solve.*

47. Amir has the following test grades in his algebra class: 72, 67, 84, 78, 82. What grade on the last test will give him a test average of B (less than 90 but greater than or equal to 80)?

48. Irene has a choice of two car rental plans: plan A costs $15 per day plus $0.30 per mile, and plan B costs $20 per day and $0.25 per mile. If she plans to keep the car for 5 days, how many miles must she drive in order for plan B to be the better bargain?

49. The scale at Paula's health club indicates that she weighs 54.5 kilograms, but the scale is accurate to within only 0.5 kilogram.
    a. Write an absolute-value inequality that describes Paula's weight.
    b. Find Paula's weight in pounds, including an error estimate.

50. Jeremy's older brother measures Jeremy's height at 51 inches, but the measurement is accurate only to the nearest inch.
    a. Write an absolute-value inequality that describes Jeremy's height.
    b. Find Jeremy's height in centimeters, including an error estimate.
51. The sum of two numbers is 15. When is their product less than 50?
52. When it charges p dollars for a certain item, Handy Hardware will sell $30 - \frac{1}{2}p$ items. How much should Handy charge per item if it wants the revenue from the item to be over $400?

B

■ *For Problems 53 and 54, express using absolute value.*

53. The difference of the distances from a to b and from a to c is equal to d.
54. If x is within d units of a, y is within t units of L.
55. Solve the following inequality by graphing a parabola: $2x^2 + 5x > 25$.
56. For what values of k does the equation $kx^2 - 2kx + 5 = 0$ have two real roots?

■ *Graph.*

57. $|x + y| \leq 2$
58. $|x - 2y| > 6$
59. $|x| - |y| \geq 1$
60. $|x| - |y| \leq 0$

Conic Sections

11

In Chapter 5 we found that the graph of any first-degree equation in two variables,

$$Ax + By = C,$$

is a line. We now turn our attention to second-degree equations in two variables. The most general form of such an equation is

$$Ax^2 + Bxy + Cy^2 + Dx + Ey + F = 0,$$

where A, B, and C cannot all be zero (since in that case the equation would not be of second degree). The graphs of such equations are curves called **conic sections** because they are formed by the intersection of a plane and a cone, as shown in Figure 11.1. Except for a few special cases called *degenerate* conics, which we shall describe later, the conic sections fall into four categories called **circles, ellipses, parabolas,** and **hyperbolas.**

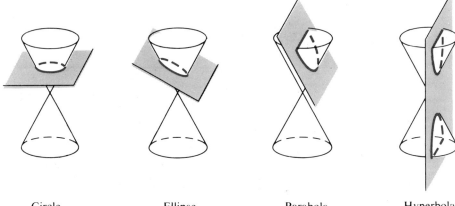

FIGURE 11.1 Circle   Ellipse   Parabola   Hyperbola

11.1 CIRCLES AND ELLIPSES

By considering their geometric properties we can find standard equations for each of the conic sections. To do this we first develop formulas for the distance between two points and for the midpoint of a line segment.

Distance and Midpoint Formulas

The distance between any two points P_1 and P_2 can be calculated in terms of their coordinates, (x_1, y_1) and (x_2, y_2), by using the Pythagorean theorem. We first create a right triangle by constructing a line parallel to the x-axis through P_1 and a line parallel to the y-axis through P_2. These lines meet at a point P_3, as shown in Figure 11.2. The x-coordinate of P_3 is the same as the x-coordinate of P_2, and the y-coordinate of P_3 is the same as the y-coordinate of P_1. Thus, the coordinates of P_3 are (x_2, y_1).

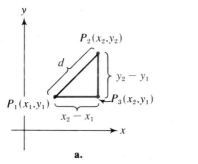
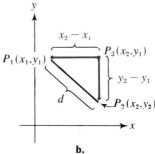

FIGURE 11.2

a. b.

Now, the distance between P_1 and P_3 is $|x_2 - x_1|$ and the distance between P_2 and P_3 is $|y_2 - y_1|$. These two numbers are the lengths of the legs of the right triangle; the length of the hypotenuse is the distance between P_1 and P_2, which we shall call d. So, by the Pythagorean theorem,

$$d^2 = (x_2 - x_1)^2 + (y_2 - y_1)^2.$$

Taking the (positive) square root of each side of the equation gives us the **distance formula**.

> The distance d between the points $P_1(x_1, y_1)$ and $P_2(x_2, y_2)$ is
>
> $$d = \sqrt{(x_2 - x_1)^2 + (y_2 - y_1)^2}.$$

EXAMPLE 1 Find the distance between $(2, -1)$ and $(4, 3)$.

Solution Substitute $(2, -1)$ for (x_1, y_1) and $(4, 3)$ for (x_2, y_2) in the distance formula to obtain

$$\begin{aligned} d &= \sqrt{(x_2 - x_1)^2 + (y_2 - y_1)^2} \\ &= \sqrt{(4 - 2)^2 + [3 - (-1)]^2} \\ &= \sqrt{4 + 16} \\ &= \sqrt{20} = 2\sqrt{5}. \end{aligned}$$

Notice that in the example above we would obtain the same answer if we used $(4, 3)$ for P_1 and $(2, -1)$ for P_2:

$$\begin{aligned} d &= \sqrt{(2 - 4)^2 + [(-1) - 3]^2} \\ &= \sqrt{4 + 16} = 2\sqrt{5}. \end{aligned}$$

Similar triangles can be used to derive the following formula for the midpoint of the line segment joining two points. The proof is left as an exercise.

The midpoint of the line segment joining the points $P_1(x_1, y_1)$ and $P_2(x_2, y_2)$ is the point $M(\bar{x}, \bar{y})$, where

$$\bar{x} = \frac{x_1 + x_2}{2} \quad \text{and} \quad \bar{y} = \frac{y_1 + y_2}{2}.$$

EXAMPLE 2 Find the midpoint of the line segment joining the points $(-2, 1)$ and $(4, 3)$.

Solution Substitute $(-2, 1)$ for (x_1, y_1) and $(4, 3)$ for (x_2, y_2) in the midpoint formula to obtain

$$\bar{x} = \frac{x_1 + x_2}{2} = \frac{-2 + 4}{2} = 1;$$

$$\bar{y} = \frac{y_1 + y_2}{2} = \frac{1 + 3}{2} = 2.$$

The midpoint of the segment is the point $(\bar{x}, \bar{y}) = (1, 2)$.

Circles

The circle is the most familiar of the conic sections. It is defined as follows.

> A **circle** is the set of points in a plane that lie at a given distance r called the **radius** from a fixed point called the **center**.

We can find an equation for a circle by using the distance formula and the definition above. First, consider the circle in Figure 11.3a. All points $P(x, y)$ on the circle lie at a distance r from the origin. Therefore,

$$(x - 0)^2 + (y - 0)^2 = r^2,$$

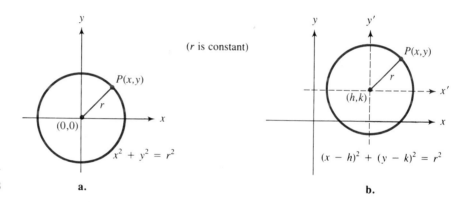

FIGURE 11.3   a.   b.

so the equation for a circle with radius r, centered at the origin, is

$$x^2 + y^2 = r^2.$$

Now consider the circle in Figure 11.3b, whose center is the point (h, k). Every point $P(x, y)$ on the circle lies at a distance r from the center, so

$$(x - h)^2 + (y - k)^2 = r^2. \tag{1}$$

Equation (1) is the **standard form** for a circle of radius r with center at (h, k).

EXAMPLE 3   a. The graph of $(x - 2)^2 + (y + 3)^2 = 16$ is a circle with radius 4 and center at $(2, -3)$.

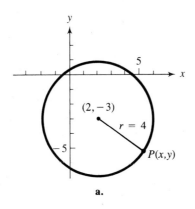
a.

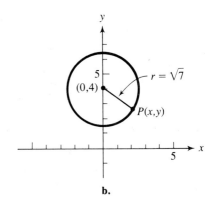
b.

b. The graph of $x^2 + (y - 4)^2 = 7$ is a circle with radius $\sqrt{7}$ and center at $(0, 4)$.

We can rewrite an equation of form (1) as an equivalent equation without parentheses. For example, the equation in Example 3a,

$$(x - 2)^2 + (y + 3)^2 = 16,$$

is equivalent to

$$x^2 - 4x + 4 + y^2 + 6y + 9 = 16,$$

or

$$x^2 + y^2 - 4x + 6y - 3 = 0.$$

Note that the coefficients of x^2 and y^2 are equal.

Conversely, an equation of the form $x^2 + y^2 + ax + by + c = 0$ can be written in standard form (1) by completing the square in both variables. Once this is done the center and radius of the circle can be determined directly from the equation.

EXAMPLE 4   a. Write the equation of the circle

$$x^2 + y^2 + 8x - 2y + 6 = 0$$

in standard form.

b. Graph the equation.

Solutions   a. Prepare to complete the square by writing

$$(x^2 + 8x \quad) + (y^2 - 2y \quad) = -6.$$

Complete the square in x by adding 16 to each side of the equation, and complete the square in y by adding 1 to each side, to get

$$(x^2 + 8x + 16) + (y^2 - 2y + 1) = -6 + 16 + 1,$$

from which we obtain the standard form

$$(x + 4)^2 + (y - 1)^2 = 11.$$

b. The center of the circle is at $(-4, 1)$ and the radius is $\sqrt{11}$.

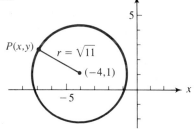

Using the standard form it is also possible to write an equation for a circle with known properties.

EXAMPLE 5   Find the equation of the circle whose diameter has endpoints $(7, 5)$ and $(1, -1)$.

Solution   Find the center and radius of the circle from the given information. The center of the circle is the midpoint of the diameter:

$$h = \bar{x} = \frac{7 + 1}{2} = 4;$$

$$k = \bar{y} = \frac{5 - 1}{2} = 2.$$

Thus, the center is the point $(h, k) = (4, 2)$. The radius is the distance from the center to either of the endpoints of the diameter, say, the point $(7, 5)$:

$$r = \sqrt{(7 - 4)^2 + (5 - 2)^2}$$
$$= \sqrt{3^2 + 3^2} = \sqrt{18}.$$

Substituting 4 for h and 2 for k (the coordinates of the center) and $\sqrt{18}$ for r (the radius) in

$$(x - h)^2 + (y - k)^2 = r^2$$

we obtain

$$(x - 4)^2 + (y - 2)^2 = 18.$$

Ellipses

An ellipse is defined as follows.

> An **ellipse** is the set of points in a plane the sum of whose distances from two fixed points (the **foci**) is a constant.

We can visualize the definition in the following way. Drive two nails into a board to represent the two foci. Attach the two ends of a piece of string to the two nails, and stretch the string taut with a pencil. Trace around the two nails, keeping the string taut, as illustrated in Figure 11.4. The figure described will be an ellipse because the sum of the distances from each point to the two foci is the length of the string, which is constant.

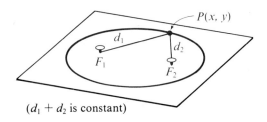

FIGURE 11.4   ($d_1 + d_2$ is constant)

11.1 ■ CIRCLES AND ELLIPSES

Ellipses appear in a variety of applications. The orbits of the planets about the sun and of satellites about the earth are ellipses. The arches in some bridges are elliptical in shape, as are certain bicycle gears and the styli in some stereo systems.

Using the distance formula and the definition above it can be shown that the equation of an ellipse centered at the origin has the form

$$\frac{x^2}{a^2} + \frac{y^2}{b^2} = 1 \quad (a > b) \tag{2}$$

when the foci lie on the x-axis (see Figure 11.5a). By setting y equal to zero in Equation (2) we find that the x-intercepts of this ellipse are a and $-a$; by setting x equal to zero we find that the y-intercepts are b and $-b$.

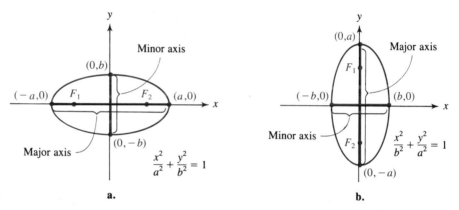

FIGURE 11.5

Figure 11.5b shows an ellipse whose foci lie on the y-axis. The equation of this ellipse is

$$\frac{x^2}{b^2} + \frac{y^2}{a^2} = 1 \quad (a > b). \tag{3}$$

In this case the x-intercepts are b and $-b$ and the y-intercepts are a and $-a$. In both cases the segment of length $2a$ is called the **major axis** and the segment of length $2b$ is called the **minor axis**. The endpoints of the major axis are called the **vertices** of the ellipse, and the endpoints of the minor axis are the **covertices**.

Equations (2) and (3) above are the standard forms for ellipses centered at the origin. As was the case with circles, the standard form of the equations gives us enough information to sketch the graph of the conic section.

EXAMPLE 6 Graph $\dfrac{x^2}{9} + \dfrac{y^2}{4} = 1$.

Solution The graph of the equation is an ellipse with its center at the origin. Since $9 > 4$, the equation is in standard form (2). Thus, $a^2 = 9$ and $b^2 = 4$, so the x-intercepts are 3 and -3 and the y-intercepts are 2 and -2. To sketch the graph we plot the x- and y-intercepts and draw a smooth closed curve through the points as shown in the figure.

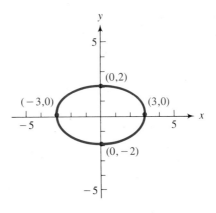

Translated Graphs

The standard forms for the equation of an ellipse centered at any point (h, k) can also be derived from the distance formula:

$$\frac{(x - h)^2}{a^2} + \frac{(y - k)^2}{b^2} = 1 \quad (a > b) ; \qquad (4)$$

$$\frac{(x - h)^2}{b^2} + \frac{(y - k)^2}{a^2} = 1 \quad (a > b) . \qquad (5)$$

Equation (4) describes an ellipse whose major axis is parallel to the x-axis, as shown in Figure 11.6a. Its vertices lie a units to the left and right of the center, and its covertices lie b units above and below the center. The ellipse described by Equation (5) and shown in Figure 11.6b has its major axis parallel to the y-axis; its vertices lie a units above and below the center, and its covertices lie b units to the left and right of the center.

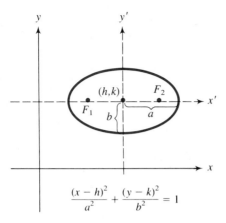
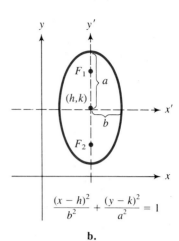

FIGURE 11.6   a.   b.

EXAMPLE 7   Graph $\dfrac{(x-3)^2}{8} + \dfrac{(y+2)^2}{25} = 1$.

Solution   The graph is an ellipse with center at $(3, -2)$. Since the equation is in standard form (5), $a^2 = 25$ and $b^2 = 8$, and the major axis is parallel to the y-axis. The vertices lie five units above and below the center, at $(3, 3)$ and $(3, -7)$. The covertices lie $\sqrt{8}$ units to the right and left of the center, at $(3 + \sqrt{8}, -2)$ and $(3 - \sqrt{8}, -2)$, or approximately $(5.8, -2)$ and $(0.2, -2)$.

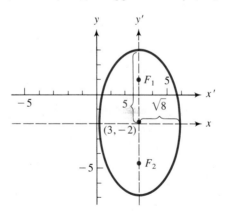

Second-degree equations in which the coefficients of x^2 and y^2 have the *same* sign can be written in one of the standard forms for an ellipse by completing the square. As with circles, the equation can be graphed easily from the standard form.

EXAMPLE 8 a. Write the equation

$$4x^2 + 9y^2 - 16x - 18y - 11 = 0$$

in standard form.

b. Graph the equation.

Solutions a. First, prepare to complete the square in both x and y by writing

$$4(x^2 - 4x \quad) + 9(y^2 - 2y \quad) = 11.$$

Note that the coefficients of x^2 and y^2 have been *factored* from their respective terms. Complete the square in x by adding $4 \cdot 4$, or 16, to each side of the equation, and complete the square in y by adding $9 \cdot 1$, or 9, to each side, to obtain

$$4(x^2 - 4x + 4) + 9(y^2 - 2y + 1) = 11 + 16 + 9.$$

Write each term on the left side as a perfect square to get

$$4(x - 2)^2 + 9(y - 1)^2 = 36.$$

To write this equation in standard form divide each side by 36 to get

$$\frac{(x - 2)^2}{9} + \frac{(y - 1)^2}{4} = 1.$$

b. The graph is an ellipse with center at $(2, 1)$, $a^2 = 9$, and $b^2 = 4$. The vertices lie three units to the right and left of the center at $(5, 1)$ and $(-1, 1)$, and the covertices lie two units above and below the center at $(2, 3)$ and $(2, -1)$.

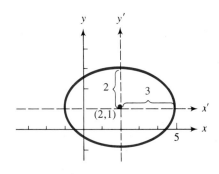

To write the equation of an ellipse from a description of its properties we must find the center of the ellipse and the lengths of its axes. We can then substitute this information into the appropriate equation in standard form.

EXAMPLE 9 Find the equation of the ellipse with vertices at $(3, 3)$ and $(3, -5)$ and covertices at $(1, -1)$ and $(5, -1)$.

Solution To find the center of the ellipse, find the midpoint of the major or minor axis:

$$h = \bar{x} = \frac{3 + 3}{2} = 3;$$

$$k = \bar{y} = \frac{3 - 5}{2} = -1.$$

The center of the ellipse is the point $(3, -1)$. The value of a is half the length of the major axis, or the distance from the center, $(3, -1)$, to one of the vertices, say, $(3, 3)$:

$$a = |3 - (-1)| = 4.$$

The value of b is half the length of the minor axis, or the distance from the center, $(3, -1)$, to one of the covertices, say, $(5, -1)$:

$$b = |5 - 3| = 2.$$

Since the major axis is a vertical line segment, the ellipse has an equation of standard form (5) on page 526, or

$$\frac{(x - h)^2}{b^2} + \frac{(y - k)^2}{a^2} = 1,$$

with $h = 3$, $k = -1$, $a = 4$, and $b = 2$. Thus, the equation is

$$\frac{(x - 3)^2}{2^2} + \frac{(y + 1)^2}{4^2} = 1,$$

or

$$\frac{(x - 3)^2}{4} + \frac{(y + 1)^2}{16} = 1$$

$$4(x - 3)^2 + (y + 1)^2 = 16$$

$$4(x^2 - 6x + 9) + (y^2 + 2y + 1) = 16$$

$$4x^2 - 24x + 36 + y^2 + 2y + 1 = 16$$

$$4x^2 + y^2 - 24x + 2y + 21 = 0.$$

EXERCISE 11.1

A

■ *Find the distance between each of the given pairs of points, and find the midpoint of the segment joining them. See Examples 1 and 2.*

1. $(1,1), (4,5)$
2. $(-1,1), (5,9)$
3. $(2,-3), (-2,-1)$
4. $(5,-4), (-1,1)$
5. $(3,5), (-2,5)$
6. $(-2,-5), (-2,3)$

■ *Graph each equation. See Examples 3 and 4.*

7. $x^2 + y^2 = 25$
8. $x^2 + y^2 = 16$
9. $4x^2 + 4y^2 = 16$
10. $2x^2 + 2y^2 = 18$
11. $(x-4)^2 + (y-2)^2 = 9$
12. $(x-1)^2 + (y-3)^2 = 16$
13. $(x+3)^2 + y^2 = 10$
14. $x^2 + (y+4)^2 = 12$
15. $x^2 + (y+3)^2 = 15$
16. $(x+2)^2 + y^2 = 18$
17. $(x-6)^2 + (y+2)^2 = 12$
18. $(x+6)^2 + (y-3)^2 = 8$
19. $x^2 + y^2 + 2x - 4y - 6 = 0$
20. $x^2 + y^2 - 6x + 2y - 4 = 0$
21. $x^2 + y^2 + 8x - 4 = 0$
22. $x^2 + y^2 - 10y - 2 = 0$
23. $x^2 + y^2 + 6y = 0$
24. $x^2 + y^2 - 6x = 0$

■ *Write an equation for the circle with the given properties. See Example 5.*

25. Center at $(-2,5)$, radius $2\sqrt{3}$
26. Center at $(4,-3)$, radius $2\sqrt{6}$
27. Center at $(-3,-1)$, tangent to the x-axis
28. Center at $(1,7)$, tangent to the y-axis
29. Endpoints of a diameter at $(1,5)$ and $(3,-1)$
30. Endpoints of a diameter at $(3,6)$ and $(-5,2)$
31. Endpoints of a diameter at $(-1,-5)$ and $(4,-3)$
32. Endpoints of a diameter at $(1,1)$ and $(-4,-2)$

■ *Graph each equation. See Examples 6 and 7.*

33. $\dfrac{x^2}{16} + \dfrac{y^2}{4} = 1$
34. $\dfrac{x^2}{9} + \dfrac{y^2}{16} = 1$
35. $\dfrac{x^2}{10} + \dfrac{y^2}{25} = 1$
36. $\dfrac{x^2}{16} + \dfrac{y^2}{12} = 1$
37. $x^2 + \dfrac{y^2}{14} = 1$
38. $\dfrac{x^2}{8} + y^2 = 1$
39. $\dfrac{(x-3)^2}{16} + \dfrac{(y-4)^2}{9} = 1$
40. $\dfrac{(x-2)^2}{4} + \dfrac{(y-5)^2}{25} = 1$
41. $\dfrac{(x+2)^2}{6} + \dfrac{(y-5)^2}{12} = 1$
42. $\dfrac{(x-5)^2}{15} + \dfrac{(y+3)^2}{8} = 1$
43. $\dfrac{x^2}{16} + \dfrac{(y+4)^2}{6} = 1$
44. $\dfrac{(x-5)^2}{15} + \dfrac{y^2}{25} = 1$

■ *Graph each equation. See Example 8.*

45. $9x^2 + 4y^2 - 16y = 20$
46. $x^2 + 16y^2 + 6x = 7$
47. $9x^2 + 16y^2 - 18x + 96y + 9 = 0$
48. $16x^2 + 9y^2 + 64x - 18y - 71 = 0$
49. $x^2 + 4y^2 + 4x - 16y + 4 = 0$
50. $2x^2 + y^2 - 16x + 6y + 11 = 0$
51. $6x^2 + 5y^2 - 12x + 20y - 4 = 0$
52. $5x^2 + 8y^2 - 20x + 16y - 12 = 0$
53. $8x^2 + y^2 - 48x + 4y + 68 = 0$
54. $x^2 + 10y^2 + 4x + 20y + 4 = 0$

■ *Write an equation for the ellipse with the given properties. See Example 9.*

55. Center at $(1, 6)$, $a = 3$, $b = 2$, major axis horizontal
56. Center at $(2, 3)$, $a = 4$, $b = 3$, major axis vertical
57. Vertices at $(3, 2)$ and $(-7, 2)$ and minor axis of length 6
58. Covertices at $(3, 7)$ and $(3, -1)$ and major axis of length 10
59. Vertices at $(-4, 9)$ and $(-4, -3)$ and covertices at $(-7, 3)$ and $(-1, 3)$
60. Vertices at $(-3, -5)$ and $(9, -5)$ and covertices at $(3, 0)$ and $(3, -10)$

B

■ *For Problems 61 and 62, use the distance formula to find the perimeter of the triangle with the given vertices.*

61. $(10, 1)$, $(3, 1)$, $(5, 9)$
62. $(-1, 5)$, $(8, -7)$, $(4, 1)$
63. Show that the rectangle with vertices $(-4, 1)$, $(2, 6)$, $(7, 0)$, and $(1, -5)$ is a square.
64. Show that the triangle with vertices $(0, 0)$, $(6, 0)$, and $(3, 3)$ is a right isosceles triangle—that is, a right triangle with two sides that have the same length.
65. Find the equation for the circle that passes through the points $(2, 3)$, $(3, 2)$, and $(-4, -5)$. (*Hint:* Find values for a, b, and c so that the three points lie on the graph of $x^2 + y^2 + ax + by + c = 0$.)
66. Find the equation for the circle that passes through the points $(0, 0)$, $(6, 0)$, and $(0, 8)$. (See the hint for Problem 65.)
67. In the adjoining figure $M(\bar{x}, \bar{y})$ is the midpoint of the line segment $\overline{AB}$. Using similar triangles show that

$$AC = 2AE \quad \text{and} \quad BC = 2BD.$$

(You may assume that $\angle ACB$, $\angle MDB$, and $\angle AEM$ are right angles.)

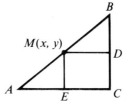

68. Use the result of Problem 67 and the distance formula to show that

$$\bar{x} = \frac{x_1 + x_2}{2} \quad \text{and} \quad \bar{y} = \frac{y_1 + y_2}{2}.$$

11.2

PARABOLAS

In Section 6.5 we considered quadratic equations of the form

$$y = ax^2 + bx + c,$$

whose graphs are called parabolas. Parabolas are conic sections defined as follows.

> A **parabola** is the set of points in a plane whose distances from a fixed line l and a fixed point F are equal.

The fixed line in the definition is called the **directrix**, and the fixed point is the **focus**. The **axis** of a parabola is the line running through the focus of the parabola and perpendicular to the directrix. (See Figure 11.7.)

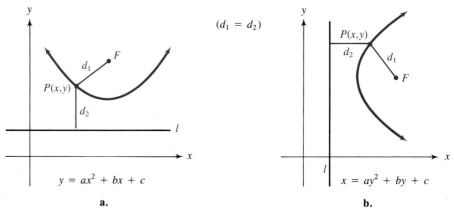

FIGURE 11.7

Parabolas have many applications in optics and in communications. Parabolic mirrors are used in telescopes because the light waves received reflect off the surface and form an image at the focus of the parabola. For similar reasons, radio antennae and television dish receivers are parabolic in shape. The parabolic mirrors in searchlights and automobile headlights reflect light from the focus into a beam of parallel rays.

We first consider parabolas with vertex at the origin. Using the distance formula and the definition above it can be shown that a parabola that "opens" upward or downward has an equation of the form

$$x^2 = 4py \quad \text{or} \quad x^2 = -4py,$$

respectively, where p is the distance between the vertex of the parabola and its focus or the distance between the vertex and the directrix. (See Figure 11.8.) Parabolas that open to the right or to the left have equations of the form

$$y^2 = 4px \quad \text{or} \quad y^2 = -4px,$$

respectively.

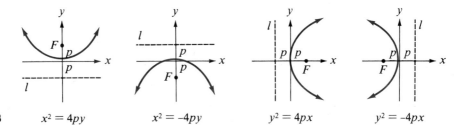

FIGURE 11.8 $x^2 = 4py$ $x^2 = -4py$ $y^2 = 4px$ $y^2 = -4px$

If we draw a line through the focus and parallel to the directrix, it will intersect the parabola at two points. These points are $2p$ units from the directrix, so they must also be $2p$ units from the focus. (Recall that by definition the points of a parabola are equidistant from the focus and the directrix. See Figure 11.9.) By locating these two "guide points" and the vertex we can make a reasonable sketch of the parabola, as illustrated in the next example.

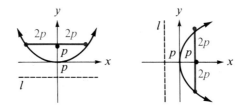

FIGURE 11.9

EXAMPLE 1 Graph $y^2 = -6x$.

Solution The parabola has its vertex at the origin and opens to the left, so its axis is the x-axis. Also, $-6 = -4p$, so $p = \frac{3}{2}$, and the focus is the point $(-\frac{3}{2}, 0)$. The directrix is the vertical line $x = \frac{3}{2}$.

To graph the parabola draw a line segment of length $4p = 6$ perpendicular to the axis and centered at the focus; since $2p = 3$, the endpoints of the segment are $(-\frac{3}{2}, 3)$ and $(-\frac{3}{2}, -3)$. Sketch the parabola through the vertex and the two guide points to obtain the graph below.

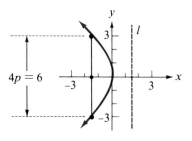

Translated Graphs

The equation of a parabola whose vertex is located at the point (h, k) has one of the following four standard forms:

$$(x - h)^2 = 4p(y - k) \quad \text{opens upward;} \qquad (1)$$
$$(x - h)^2 = -4p(y - k) \quad \text{opens downward;} \qquad (2)$$
$$(y - k)^2 = 4p(x - h) \quad \text{opens to the right;} \qquad (3)$$
$$(y - k)^2 = -4p(x - h) \quad \text{opens to the left.} \qquad (4)$$

EXAMPLE 2 Graph $(x + 2)^2 = -8(y - 3)$.

Solution The equation is in standard form (2), so the parabola opens downward. The vertex is the point $(-2, 3)$, and the axis of the parabola is the vertical line through $(-2, 3)$. Since $-4p = -8$, $p = 2$; and the focus, which lies two units below the vertex, is the point $(-2, 1)$. The directrix is the horizontal line $y = 5$.

To graph the parabola draw a line segment of length $4p = 8$ centered at the focus and perpendicular to the axis; since $2p = 4$, the endpoints of the segment are $(-6, 1)$ and $(2, 1)$. Sketch a parabola through the vertex, $(-2, 3)$, and the two guide points to obtain the graph shown on page 535.

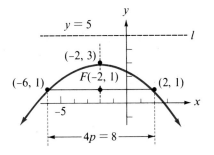

A quadratic equation that includes either an x^2 term or a y^2 term, but not both, can be put into one of the standard forms (1) through (4) by completing the square.

EXAMPLE 3 Write the equation

$$y^2 - 4y - 4x + 8 = 0$$

in standard form.

Solution First, prepare to complete the square in y by writing

$$(y^2 - 4y \quad) = 4x - 8.$$

Complete the square by adding 4 to each side of the equation to get

$$(y^2 - 4y + 4) = 4x - 8 + 4,$$

or

$$(y - 2)^2 = 4(x - 1).$$

To write the equation of a given parabola we must find the coordinates of the vertex and the value of the constant, p.

EXAMPLE 4   Find an equation for the parabola with vertex $(4, -1)$ and focus $(2, -1)$.

Solution   The axis of the parabola is the horizontal line $y = -1$, which passes through the vertex and the focus. The parabola opens to the left, and its equation has the standard form (4) on page 534,

$$(y - k)^2 = -4p(x - h).$$

The value of p is the distance from the vertex to the focus, so $p = 4 - 2 = 2$. Since the vertex is the point $(4, -1)$, the equation of the parabola is

$$(y + 1)^2 = -4(2)(x - 4),$$

or

$$(y + 1)^2 = -8(x - 4).$$

Or, alternatively, removing parentheses yields

$$y^2 + 2y + 8x - 31 = 0.$$

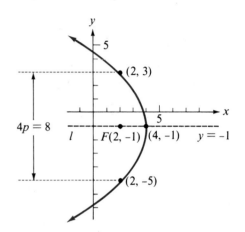

EXERCISE 11.2

A

■ *Find the focus and directrix for each parabola. Sketch the graph. See Example 1.*

1. $x^2 = 2y$
2. $y^2 = 4x$
3. $y^2 = -16x$
4. $x^2 = -18y$
5. $y^2 = 12x$
6. $4x^2 = 3y$
7. $x^2 + 8y = 0$
8. $4y^2 - 2x = 0$
9. $2y^2 - 3x = 0$
10. $3x^2 + 5y = 0$

■ *Graph each equation. See Example 2.*

11. $2x = (y + 3)^2$
12. $-3y = (x - 2)^2$
13. $-6(y + 4) = (x - 3)^2$
14. $4(x - 5) = (y + 1)^2$
15. $(y - 4)^2 + 3 = x$
16. $(x + 3)^2 - 2 = y$

■ *Write each equation in standard form. See Example 3.*

17. $x^2 + 6x - y + 5 = 0$
18. $x^2 - 8x - y + 2 = 0$
19. $2y^2 - 12y - x + 12 = 0$
20. $3y^2 + 6y - x + 4 = 0$
21. $3x^2 - 9x + y + 2 = 0$
22. $2x^2 - 5x + y + 1 = 0$

■ *Graph each equation. See Examples 2 and 3.*

23. $y^2 - 6y + 10x + 4 = 0$
24. $y^2 - 4y + 8x + 6 = 0$
25. $4x^2 - 4x = 23 - 12y$
26. $4x^2 + 4x = 8y - 5$
27. $9y^2 = 6y + 12x - 1$
28. $9y^2 + 12y - 12x = 0$

■ *Find an equation for the parabola with the given properties. See Example 4.*

29. Vertex at $(0, 0)$ and focus at $(0, 2)$
30. Vertex at $(0, 0)$ and focus at $(0, -2)$
31. Vertex at $(0, 0)$ and focus at $(-4, 0)$
32. Vertex at $(0, 0)$ and focus at $(4, 0)$
33. Vertex at $(1, 2)$ and focus at $(1, 1)$
34. Vertex at $(-1, 3)$ and focus at $(-1, 5)$
35. Vertex at $(-4, -2)$ and focus at $(-2, -2)$
36. Vertex at $(6, -3)$ and focus at $(3, -3)$
37. Vertex at $(2, 5)$ and directrix the line $x = 3$
38. Vertex at $(-4, 1)$ and directrix the line $y = 3$
39. Focus at $(3, -2)$ and directrix the line $y = 2$
40. Focus at $(-2, -2)$ and directrix the line $x = -6$

B

■ *Find an equation for the parabola with the given properties.*

41. The directrix is parallel to the x-axis, the vertex is at the origin, and the point $(2, 4)$ lies on the parabola.
42. The directrix is parallel to the y-axis, the vertex is at the origin, and the point $(1, 3)$ lies on the parabola.
43. The directrix is parallel to the y-axis, the vertex is at $(-3, 4)$, and the point $(-4, 5)$ lies on the parabola.
44. The directrix is parallel to the x-axis, the vertex is at $(1, -5)$, and the point $(3, 3)$ lies on the parabola.

11.3

HYPERBOLAS

If a cone is cut by a plane parallel to its axis the intersection is a hyperbola, the only conic section made of two separate pieces, or **branches**. As do the other conic sections, hyperbolas occur in a number of applied settings. The navigational system called loran (long-range navigation) uses radio signals to locate a ship or plane at the intersection of two hyperbolas. Satellites moving with sufficient speed will follow an orbit that is a branch of a hyperbola; for instance, a rocket sent to the moon must be fitted with retro-rockets to reduce its speed in order to achieve an elliptical, rather than a hyperbolic, orbit about the moon.

The hyperbola is defined as follows.

> A **hyperbola** is the set of points in the plane the difference of whose distances from two fixed points (the **foci**) is a constant.

If the origin is the center of a hyperbola and the foci lie on the x-axis, it can be shown that its equation may be written as

$$\frac{x^2}{a^2} - \frac{y^2}{b^2} = 1. \tag{1}$$

The graph has x-intercepts at a and $-a$ but no y-intercepts. (See Figure 11.10a.) A hyperbola centered at the origin with foci on the y-axis has the equation

$$\frac{y^2}{a^2} - \frac{x^2}{b^2} = 1. \tag{2}$$

In this case the graph has y-intercepts at a and $-a$ but no x-intercepts. (See Figure 11.10b.) In both orientations the line segment of length $2a$ is called

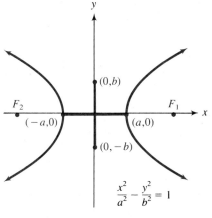
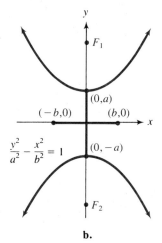

FIGURE 11.10   a.   b.

the **transverse axis** and the segment of length $2b$ is called the **conjugate axis**. The endpoints of the transverse axis, which lie on the hyperbola, are called the **vertices**. Note that, unlike the equations for an ellipse, a can be less than or equal to b for a hyperbola.

Asymptotes of Hyperbolas

The branches of the hyperbola approach two straight lines that intersect at its center. These lines are asymptotes of the graph, and they are useful as guidelines for sketching the hyperbola. We can obtain the asymptotes by first forming a rectangle (called the "cental rectangle") whose sides are parallel to and the same length as the transverse and conjugate axes. The asymptotes are determined by the diagonals of this rectangle.

EXAMPLE 1 Graph $\dfrac{x^2}{9} - \dfrac{y^2}{16} = 1$.

Solution The graph is a hyperbola with center at the origin. Since the equation is of form (1) on page 538, the branches of the hyperbola open to the left and right. Also, $a^2 = 9$ and $b^2 = 16$, so $a = 3$ and $b = 4$, and the vertices of the hyperbola are $(3, 0)$ and $(-3, 0)$.

Construct the central rectangle with dimensions $2a = 6$ and $2b = 8$, as shown in the figure. Draw the asymptotes through the opposite corners of the rectangle, and sketch the branches of the hyperbola through the vertices and approaching the asymptotes to obtain the graph shown.

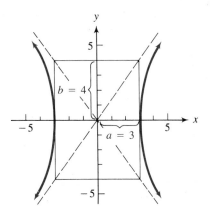

EXAMPLE 2 Graph $\dfrac{y^2}{9} - \dfrac{x^2}{4} = 1$.

Solution The equation is in form (2) on page 000, so the branches of the hyperbola open upward and downward. Since $a^2 = 9$ and $b^2 = 4$, $a = 3$ and $b = 2$, and the vertices are $(0, 3)$ and $(0, -3)$.

Construct the central rectangle with dimensions $2a = 6$ and $2b = 4$ and draw the asymptotes through its opposite corners. Sketch the branches of the hyperbola through the vertices and approaching the asymptotes as shown in the figure.

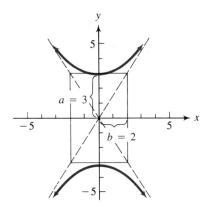

Translated Graphs

The following standard forms for equations of hyperbolas centered at the point (h, k) can be derived using the distance formula and the definition of hyperbola:

$$\frac{(x - h)^2}{a^2} - \frac{(y - k)^2}{b^2} = 1; \qquad (3)$$

$$\frac{(y - k)^2}{a^2} - \frac{(x - h)^2}{b^2} = 1. \qquad (4)$$

Equation (3) describes a hyperbola whose transverse axis is parallel to the x-axis, and Equation (4) describes a hyperbola whose transverse axis is parallel to the y-axis. (See Figure 11.11 on page 541.)

EXAMPLE 3 Graph $\dfrac{(x - 3)^2}{8} - \dfrac{(y + 2)^2}{10} = 1.$

11.3 ■ HYPERBOLAS   541

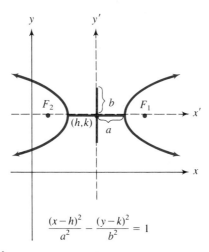
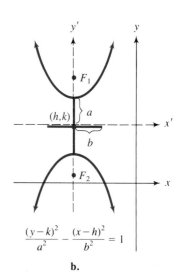

FIGURE 11.11   a.   b.

Solution   The graph is a hyperbola with center at $(3, -2)$. Since the equation is in standard form (3), the transverse axis is parallel to the x-axis, and

$$a = \sqrt{8}, \quad b = \sqrt{10}.$$

The coordinates of the vertices are thus $(3 + \sqrt{8}, -2)$ and $(3 - \sqrt{8}, -2)$, or approximately $(5.8, -2)$ and $(0.2, -2)$. The ends of the conjugate axes are $(3, -2 + \sqrt{10})$ and $(3, -2 - \sqrt{10})$, or approximately $(3, 1.2)$ and $(3, -5.2)$.

The central rectangle is centered at the point $(3, -2)$ and extends to the vertices in the horizontal direction and to the ends of the conjugate axis in the vertical direction, as shown in the figure. Draw the asymptotes through the opposite corners of the central rectangle, then sketch the hyperbola through the vertices and approaching the asymptotes to obtain the graph in the figure.

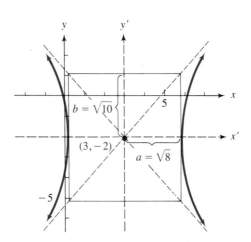

Quadratic equations in which the x^2 term and the y^2 term have *opposite* signs describe hyperbolas. We can write such equations in one of the standard forms (3) or (4) by completing the square in x and in y.

EXAMPLE 4   a. Write the equation

$$y^2 - 4x^2 + 4y - 8x - 9 = 0$$

in standard form.

b. Graph the equation.

Solutions   a. First, prepare to complete the square by writing

$$(y^2 + 4y \quad) - 4(x^2 + 2x \quad) = 9.$$

Complete the square in y by adding 4 to each side of the equation, and complete the square in x by adding $-4 \cdot 1$, or -4, to each side, to get

$$(y^2 + 4y + 4) - 4(x^2 + 2x + 1) = 9 + 4 - 4$$

and then

$$(y + 2)^2 - 4(x + 1)^2 = 9.$$

Divide each side by 9 to obtain the standard form

$$\frac{(y + 2)^2}{9} - \frac{(x + 1)^2}{\frac{9}{4}} = 1.$$

b. The graph is a hyperbola with center at $(-1, -2)$. Since the equation is in standard form (4), the transverse axis is parallel to the y-axis, and

$$a^2 = 9, \quad b^2 = \frac{9}{4}.$$

Thus, $a = 3$ and $b = \frac{3}{2}$, and the vertices are $(-1, 1)$ and $(-1, -5)$. The ends of the conjugate axis are $(-\frac{5}{2}, -2)$ and $(\frac{1}{2}, -2)$.

The central rectangle is centered at $(-1, -2)$ and extends to the vertices in the vertical direction and to the ends of the conjugate axis in the horizontal direction, as shown in the figure. Draw the asymptotes through the corners of the rectangle, and sketch the hyperbola through the vertices and approaching the asymptotes to obtain the graph in the figure.

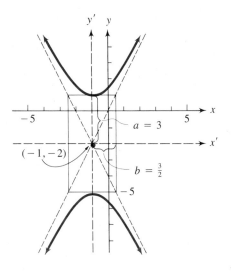

The General Quadratic Equation in Two Variables

We have considered graphs of second-degree equations in two variables,

$$Ax^2 + Bxy + Cy^2 + Dx + Ey + F = 0,$$

for which B, the coefficient of the xy term, is zero. Such graphs are conic sections with axes parallel to one or both of the coordinate axes. If B does not equal zero, the axes of the conic section are rotated with respect to the coordinate axes. The graphing of such equations is taken up in more advanced courses in analytic geometry.

The graph of a second-degree equation can also be a point, a line, a pair of lines, or no graph at all, depending on the values of the coefficients A through F. For example, the graph of the equation

$$x^2 + y^2 - 2x + 4y + 5 = 0,$$

or

$$(x - 1)^2 + (y + 2)^2 = 0,$$

is not a circle because $r^2 = 0$. The graph consists of the single point $(1, -2)$. Such graphs are called **degenerate conics**, and we will not include them in our discussion.

Given an equation of the form

$$Ax^2 + Cy^2 + Dx + Ey + F = 0$$

we can determine the nature of the graph from the coefficients of the quadratic terms. If the graph is not a degenerate conic, the following criteria apply.

The graph of $Ax^2 + Cy^2 + Dx + Ey + F = 0$ is

1. a circle if $A = C$;
2. a parabola if $A = 0$ or $C = 0$ (but not both);
3. an ellipse if $A \neq C$ and they have the same sign;
4. a hyperbola if A and C have opposite signs.

EXAMPLE 5 Name the graph of each equation, assuming that the graph is not degenerate.

a. $3x^2 + 3y^2 - 2x + 4y - 6 = 0$ b. $4y^2 + 8x^2 - 3y = 0$
c. $4x^2 - 6y^2 + x - 2y = 0$ d. $y + x^2 - 4x + 1 = 0$

Solutions
a. The graph is a circle because the coefficients of x^2 and y^2 are equal.
b. The graph is an ellipse because the coefficients of x^2 and y^2 are both positive.
c. The graph is a hyperbola because the coefficients of x^2 and y^2 are of opposite sign.
d. The graph is a parabola because y is of first degree and x is of second degree.

Note that the coefficients D, E, and F do not figure in determining the *type* of conic section the equation represents. They do, however, determine the *position* of the graph relative to the origin. Once we recognize the form of the graph it is helpful to write the equation in the appropriate standard form in order to discover more specific information about the graph. The standard forms for the conic sections are summarized in Table 11.1. For the parabola (h, k) is the vertex of the graph, while for the other conics (h, k) is the center.

11.3 ■ HYPERBOLAS

Name of Curve	Standard Form of Equation	Graph
Parabola: Opens upward Opens downward Opens to the right Opens to the left	$(x - h)^2 = 4p(y - k)$ $(x - h)^2 = -4p(y - k)$ $(y - k)^2 = 4p(x - h)$ $(y - k)^2 = -4p(x - h)$	
Circle	$(x - h)^2 + (y - k)^2 = r^2$	
Ellipse: Major axis parallel to *x*-axis Major axis parallel to *y*-axis	$\dfrac{(x - h)^2}{a^2} + \dfrac{(y - k)^2}{b^2} = 1$ $\dfrac{(x - h)^2}{b^2} + \dfrac{(y - k)^2}{a^2} = 1$	
Hyperbola: Transverse axis parallel to *x*-axis Transverse axis parallel to *y*-axis	$\dfrac{(x - h)^2}{a^2} - \dfrac{(y - k)^2}{b^2} = 1$ $\dfrac{(y - k)^2}{a^2} - \dfrac{(x - h)^2}{b^2} = 1$	

TABLE 11.1

EXAMPLE 6 Describe the graph of each equation without graphing.

a. $x^2 = 9y^2 - 9$
b. $x^2 - y = 2x + 4$
c. $x^2 + 9y^2 + 4x - 18y + 9 = 0$

Solutions a. $x^2 = 9y^2 - 9$ is equivalent to

$$9y^2 - x^2 = 9,$$

or

$$\frac{y^2}{1} - \frac{x^2}{9} = 1.$$

The graph is a hyperbola centered at the origin with vertices $(0, 1)$ and $(0, -1)$.

b. $x^2 - y = 2x + 4$ is equivalent to

$$(x^2 - 2x + 1) = y + 4 + 1,$$

or

$$(x - 1)^2 = (y + 5).$$

The graph is a parabola that opens upward from the vertex, $(1, -5)$.

c. $x^2 + 9y^2 + 4x - 18y + 9 = 0$ is equivalent to

$$(x^2 + 4x + 4) + 9(y^2 - 2y + 1) = -9 + 4 + 9,$$

or

$$(x + 2)^2 + 9(y - 1)^2 = 4,$$

or

$$\frac{(x + 2)^2}{4} + \frac{(y - 1)^2}{\frac{4}{9}} = 1.$$

The graph is an ellipse centered at $(-2, 1)$ with major axis of length $2a = 4$ parallel to the x-axis. It has vertices at $(0, 1)$ and $(-4, 1)$ and covertices at $(-2, 5/3)$ and $(-2, 1/3)$.

EXERCISE 11.3

A

■ *Graph each equation. See Examples 1, 2, and 3.*

1. $\dfrac{x^2}{25} - \dfrac{y^2}{9} = 1$

2. $\dfrac{y^2}{4} - \dfrac{x^2}{16} = 1$

3. $\dfrac{y^2}{12} - \dfrac{x^2}{8} = 1$

4. $\dfrac{x^2}{15} - \dfrac{y^2}{10} = 1$

5. $\dfrac{(x-4)^2}{9} - \dfrac{(y+2)^2}{16} = 1$

6. $\dfrac{(y+4)^2}{25} - \dfrac{(x-3)^2}{4} = 1$

7. $\dfrac{x^2}{4} - \dfrac{(y-3)^2}{8} = 1$

8. $\dfrac{y^2}{9} - \dfrac{(x+4)^2}{12} = 1$

9. $\dfrac{(y+2)^2}{6} - \dfrac{(x+2)^2}{10} = 1$

10. $\dfrac{(x-4)^2}{5} - \dfrac{(y-4)^2}{8} = 1$

11. $\dfrac{y^2}{6} - \dfrac{(x-3)^2}{15} = 1$

12. $\dfrac{x^2}{12} - \dfrac{(y+2)^2}{7} = 1$

■ Graph each equation. See Example 4.

13. $9x^2 - 4y^2 - 36x - 24y - 36 = 0$
14. $9y^2 - 4x^2 - 72y - 24x + 72 = 0$
15. $16y^2 - 4x^2 + 32x - 128 = 0$
16. $16x^2 - 9y^2 + 54y - 225 = 0$
17. $4x^2 - 6y^2 - 32x - 24y + 16 = 0$
18. $9y^2 - 8x^2 + 72y + 16x + 64 = 0$
19. $12x^2 - 3y^2 + 24y - 84 = 0$
20. $10y^2 - 5x^2 + 30x - 95 = 0$

■ Name the graph of each equation and describe its main features. See Examples 5 and 6.

21. $y^2 = 4 - x^2$
22. $y^2 = 6 - 4x^2$
23. $4y^2 = x^2 - 8$
24. $x^2 + 2y - 4 = 0$
25. $4x^2 = 12 - 2y^2$
26. $6x^2 = 8 - 6y^2$
27. $4x^2 = 6 + 4y$
28. $2x^2 = 5 + 4y^2$
29. $6 + \dfrac{x^2}{4} = y^2$
30. $y^2 = 6 - \dfrac{2x^2}{3}$
31. $\dfrac{1}{2}y^2 - x = 4$
32. $\dfrac{x^2}{4} = 4 + 6y^2$
33. $y = (x - 3)^2 + 2$
34. $\dfrac{(y-2)^2}{4} - \dfrac{(x+3)^2}{8} = 1$
35. $(x + 1)^2 + (y - 4)^2 = 16$
36. $\dfrac{(x+3)^2}{4} + \dfrac{y^2}{12} = 1$
37. $2x^2 + y^2 + 4x = 2$
38. $y^2 - 4x^2 + 2y - x = 0$
39. $x^2 + 6x = 4 - y^2$
40. $y - 2 = \dfrac{(x+4)^2}{4}$

B

41. Graph $x^2 - y^2 = 0$. (*Hint:* First, write as $(x - y)(x + y) = 0$, then graph $y = x$ and $y = -x$.)
42. Graph $4x^2 - y^2 = 0$. (See the hint for Problem 41.)
43. Graph $x^2 - y^2 = 4$, $x^2 - y^2 = 1$, and $x^2 - y^2 = 0$ on the same set of axes.
44. Graph $4x^2 - y^2 = 16$, $4x^2 - y^2 = 4$, and $4x^2 - y^2 = 0$ on the same set of axes.

■ Find an equation for the hyperbola with the given properties.

45. Center at $(-1, 5)$, $a = 8$, $b = 6$, transverse axis vertical
46. Center at $(6, -2)$, $a = 1$, $b = 4$, transverse axis horizontal
47. One vertex at $(-1, 3)$, one end of the conjugate axis at $(-5, 1)$
48. One vertex at $(1, -2)$, one end of the conjugate axis at $(-5, -4)$

11.4

SYSTEMS INVOLVING QUADRATIC EQUATIONS

In Chapter 9 we solved systems of *linear* equations in several variables. We now consider systems in which one or more of the equations is quadratic.

Recall that a solution of a system of two equations is an ordered pair that satisfies each equation in the system; it represents a point where the graphs of the equations intersect. For example, to find the solutions of the system

$$x^2 + y^2 = 26 \qquad (1)$$
$$x + y = 6 \qquad (2)$$

we graph the equations on the same set of axes, as shown in Figure 11.12. The graphs appear to intersect at the points $(1, 5)$ and $(5, 1)$. By substituting these ordered pairs into Equations (1) and (2) we can verify that the solutions of the system are in fact $(1, 5)$ and $(5, 1)$.

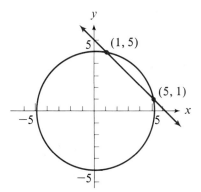

FIGURE 11.12

Solving second-degree systems by graphing may produce only approximations to real-valued solutions, and solutions in which one or both of the components is a complex number will not appear on the graph at all. It is therefore more practical to concentrate on algebraic methods of solution, since the results are exact and we can obtain complex-valued solutions. However, it is often helpful to sketch the graphs of the equations as a rough check on an algebraic solution.

Solution by Substitution

One of the most useful techniques for solving systems is **substitution**. This technique is particularly effective with systems containing one first-degree and one higher-degree equation.

EXAMPLE 1 Solve the system

$$x^2 + y^2 = 26 \tag{1}$$
$$x + y = 6 \tag{2}$$

by using substitution.

Solution First, solve the linear Equation (2) explicitly for y in terms of x:

$$y = 6 - x. \tag{2a}$$

Then replace y with $6 - x$ in Equation (1) to obtain

$$x^2 + (6 - x)^2 = 26, \tag{3}$$

an equation in one variable only. Solve Equation (3) for x to get

$$x^2 + 36 - 12x + x^2 = 26$$
$$2x^2 - 12x + 10 = 0$$
$$x^2 - 6x + 5 = 0$$
$$(x - 5)(x - 1) = 0,$$

from which $x = 1$ or $x = 5$. Each of these values is the first component of a solution of the system. To find the y-component for each solution, replace x in Equation (2a) with 1 and with 5:

if $x = 1$, $y = 6 - (1) = 5$, so $(1, 5)$ is a solution;
if $x = 5$, $y = 6 - (5) = 1$, so $(5, 1)$ is a solution.

Check the solutions by verifying that each ordered pair satisfies both Equations (1) and (2).

Note that in the example above if we had used Equation (1) rather than (2) or (2a) to find the y-components of the solutions, we would have found

$$(1)^2 + y^2 = 26 \qquad (5)^2 + y^2 = 26$$
$$y = \pm 5 \qquad\qquad y = \pm 1.$$

It appears that there are *four* solutions: $(1, 5)$, $(1, -5)$, $(5, 1)$, and $(5, -1)$. However, $(1, -5)$ and $(5, -1)$ are not solutions of Equation (2), and hence they are not solutions of the system. The only solutions to the system are $(1, 5)$ and $(5, 1)$.

To avoid encountering false solutions to a system it is best to *substitute one component of a solution into the equation of lower degree* to find only those ordered pairs that are solutions to both equations.

EXAMPLE 2 Solve the system

$$y = x^2 + 2x + 1 \qquad (1)$$
$$y - x = 3 \qquad (2)$$

by substitution.

Solution Solve (2), the linear equation, explicitly for y:

$$y = x + 3. \qquad (2a)$$

Substitute $x + 3$ for y in (1):

$$x + 3 = x^2 + 2x + 1.$$

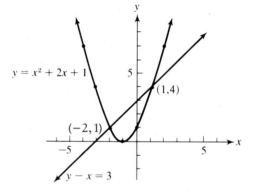

Solve for x:

$$x^2 + x - 2 = 0$$
$$(x + 2)(x - 1) = 0$$
$$x = -2 \quad \text{or} \quad x = 1.$$

Substitute each of these values in (2a) to determine the corresponding values for y:

if $x = -2$, then $y = (-2) + 3 = 1$;
if $x = 1$, then $y = (1) + 3 = 4$.

The solutions to the system are $(-2, 1)$ and $(1, 4)$.

Complex-Valued Solutions

In Examples 1 and 2 the components of each solution were real numbers, and they represented the intersection points of the graphs of the equations in the system. If one or both of the components of the solutions is a complex number, then the graphs of the equations do not have intersection points for these solutions.

11.4 ■ SYSTEMS INVOLVING QUADRATIC EQUATIONS

EXAMPLE 3 Solve the system

$$x^2 + y^2 = 26 \qquad (1)$$
$$x + y = 8. \qquad (2)$$

Solution Solve (2) explicitly for y to get

$$y = 8 - x. \qquad (2a)$$

Substitute $8 - x$ for y in (1) and simplify to get

$$x^2 + (8 - x)^2 = 26$$
$$x^2 + 64 - 16x + x^2 = 26$$
$$2x^2 - 16x + 38 = 0$$
$$x^2 - 8x + 19 = 0.$$

Since the left side does not factor, use the quadratic formula to solve for x:

$$x = \frac{8 \pm \sqrt{64 - 76}}{2(1)} = \frac{8 \pm \sqrt{-12}}{2}$$
$$= \frac{8 \pm 2i\sqrt{3}}{2} = \frac{2(4 \pm i\sqrt{3})}{2}$$
$$= 4 \pm i\sqrt{3}.$$

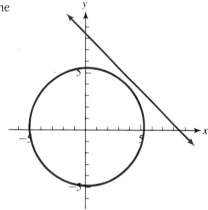

Substitute $4 + i\sqrt{3}$ for x in (2a) to find $y = 4 - i\sqrt{3}$, and then substitute $4 - i\sqrt{3}$ for x to find $y = 4 + i\sqrt{3}$. The solutions to the system are $(4 + i\sqrt{3}, 4 - i\sqrt{3})$ and $(4 - i\sqrt{3}, 4 + i\sqrt{3})$. The graphs of the equations, which *do not* intersect, are shown in the figure.

Solution by Other Methods

If both equations in a system are of second degree in both variables, we can use the method of linear combinations to solve the system.

EXAMPLE 4 Solve the system

$$4x^2 + y^2 = 25 \qquad (1)$$
$$x^2 - y^2 = -5. \qquad (2)$$

Solution Eliminate the y^2 terms by adding Equations (1) and (2) to obtain

$$\begin{array}{r}4x^2 + y^2 = 25 \\ x^2 - y^2 = -5 \\ \hline 5x^2 = 20,\end{array}$$

from which

$$x = 2 \quad \text{or} \quad x = -2.$$

Substitute these values into either (1) or (2)—we shall use (1)—to find the corresponding values for y:

$$4(2)^2 + y^2 = 25$$
$$y^2 = 25 - 16 = 9,$$

from which

$$y = 3 \quad \text{or} \quad y = -3.$$

Thus, the ordered pairs $(2, 3)$ and $(2, -3)$ are solutions of the system. Now substitute -2 for x in (1) or (2)—this time we shall use (2)—to find

$$(-2)^2 - y^2 = -5$$
$$-y^2 = -5 - 4$$
$$y^2 = 9,$$

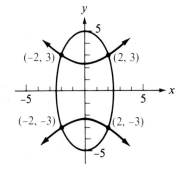

from which

$$y = 3 \quad \text{or} \quad y = -3.$$

The ordered pairs $(-2, 3)$ and $(-2, -3)$ are also solutions of the system. The graphs of the equations showing the four intersection points are given in the figure.

EXAMPLE 5 Solve the system

$$x^2 + y^2 = 5 \qquad (1)$$
$$x^2 - 2xy + y^2 = 1. \qquad (2)$$

Solution Eliminate both the x^2 and y^2 terms by subtracting Equation (2) from Equation (1):

$$\begin{array}{r} x^2 + y^2 = 5 \\ -x^2 + 2xy - y^2 = -1 \\ \hline 2xy = 4, \end{array}$$

or

$$xy = 2. \qquad (3)$$

Since all the solutions of (1) and (2) are also solutions of (3), we can form the new system

$$x^2 + y^2 = 5 \qquad (1)$$
$$xy = 2, \qquad (3)$$

which will have the same solutions as the original system.

We solve this new system by substitution. Solve (3) explicitly for y to get

$$y = \frac{2}{x}.$$

Replace y in (1) by $2/x$ to find

$$x^2 + \frac{4}{x^2} = 5. \qquad (4)$$

Solve Equation (4) for x. First, multiply each side by x^2 to obtain

$$x^4 + 4 = 5x^2$$
$$x^4 - 5x^2 + 4 = 0, \qquad (5)$$

which is quadratic in x^2. Factor the left side of (5) to get

$$(x^2 - 1)(x^2 - 4) = 0,$$

from which

$$x^2 - 1 = 0 \quad \text{or} \quad x^2 - 4 = 0.$$

Solve each of these equations to find

$$x = 1, \quad x = -1, \quad x = 2, \quad \text{or} \quad x = -2.$$

Since we multiplied Equation (4) by a variable, x^2, we must check for extraneous solutions. All of these values for x satisfy (4).

Finally, substitute 1, -1, 2, and -2 for x in Equation (3) or the equivalent equation $y = 2/x$ to find the y-components of each solution:

for $x = 1$, $y = 2$; for $x = -1$, $y = -2$;
for $x = 2$, $y = 1$; for $x = -2$, $y = -1$.

The solutions are the ordered pairs $(1, 2)$, $(-1, -2)$, $(2, 1)$, and $(-2, -1)$.

The methods of substitution and linear combinations can be used in conjunction to solve a variety of nonlinear systems. Each system should be scrutinized for an appropriate application of the techniques illustrated in the examples above.

EXERCISE 11.4

A

■ *Solve by the method of substitution. In Problems 1–10, sketch the graphs of the equations. See Examples 1, 2, and 3.*

1. $y = x^2 - 5$
   $y = 4x$

2. $y = x^2 - 2x + 1$
   $y + x = 3$

3. $x^2 + y^2 = 13$
   $x + y = 5$

4. $x^2 + 2y^2 = 12$
   $2x - y = 2$

5. $x + y = 1$
   $xy = -12$

6. $2x - y = 9$
   $xy = -4$

7. $xy = 4$
   $x^2 + y^2 = 8$

8. $x^2 - y^2 = 35$
   $xy = 6$

9. $x^2 + y^2 = 9$
   $y = 4$

10. $2x^2 - 4y^2 = 12$
    $x = 4$

11. $x^2 + y = 4$
    $x - y = -1$

12. $x^2 + 9y^2 = 36$
    $x - 2y = -8$

13. $x^2 - xy - 2y^2 = 4$
    $x - y = 2$

14. $x^2 - 2x + y^2 = 3$
    $2x + y = 4$

15. $2x^2 - 5xy + 2y^2 = 5$
    $2x - y = 1$

16. $2x^2 + xy + y^2 = 9$
    $-x + 3y = 9$

11.4 ■ SYSTEMS INVOLVING QUADRATIC EQUATIONS

■ Solve each system by linear combinations. See Example 4.

17. $x^2 + y^2 = 10$
    $9x^2 + y^2 = 18$

18. $x^2 + 4y^2 = 52$
    $x^2 + y^2 = 25$

19. $x^2 + 4y^2 = 17$
    $3x^2 - y^2 = -1$

20. $9x^2 + 16y^2 = 100$
    $x^2 + y^2 = 8$

21. $x^2 - y^2 = 7$
    $2x^2 + 3y^2 = 24$

22. $x^2 + 4y^2 = 25$
    $4x^2 + y^2 = 25$

23. $3x^2 + 4y^2 = 16$
    $x^2 - y^2 = 3$

24. $4x^2 + 3y^2 = 12$
    $x^2 + 3y^2 = 12$

25. $4x^2 - 9y^2 + 132 = 0$
    $x^2 + 4y^2 - 67 = 0$

26. $16y^2 + 5x^2 - 26 = 0$
    $25y^2 - 4x^2 - 17 = 0$

■ See Example 5.

27. $2x^2 + xy - 4y^2 = -12$
    $x^2 - 2y^2 = -4$

28. $x^2 + 2xy - y^2 = 14$
    $x^2 - y^2 = 8$

29. $x^2 + 3xy - y^2 = -3$
    $x^2 - xy - y^2 = 1$

30. $2x^2 + xy - 2y^2 = 16$
    $x^2 + 2xy - y^2 = 17$

31. $x^2 - xy + y^2 = 7$
    $x^2 + y^2 = 5$

32. $3x^2 - 2xy + 3y^2 = 34$
    $x^2 + y^2 = 17$

33. $3x^2 + 3xy - y^2 = 35$
    $x^2 - xy - 6y^2 = 0$

34. $x^2 - xy + y^2 = 21$
    $x^2 + 2xy - 8y^2 = 0$

35. $2x^2 - xy - 6y^2 = 0$
    $x^2 + 3xy + 2y^2 = 4$

36. $2x^2 + xy - y^2 = 0$
    $6x^2 + xy - y^2 = 1$

■ Solve each problem using a system of equations.

37. The sum of the squares of two positive numbers is 13. If 2 times the first number is added to the second, the sum is 7. Find the numbers.
38. The sum of two numbers is 6 and their product is $35/4$. Find the numbers.
39. The perimeter of a rectangle is 26 inches and the area is 12 square inches. Find the dimensions of the rectangle.
40. The area of a rectangle is 216 square feet. If the perimeter is 60 feet, find the dimensions of the rectangle.

B

41. The annual income from an investment is $32. If the amount invested were $200 more and the rate ½% less, the annual income would be $35. What are the amount and rate of the investment?
42. At a constant temperature the pressure P and volume V of a gas are related by the equation $PV = K$. The product of the pressure (in pounds per square inch) and the volume (in cubic inches) of a certain gas is 30 inch-pounds. If the temperature remains constant as the pressure is increased by 4 pounds per square inch, the

volume is decreased by 2 cubic inches. Find the original pressure and volume of the gas.

43. How many *real* solutions are possible for systems of *independent* equations that consist of the following?
   a. Two linear equations in two variables
   b. One linear equation and one quadratic equation in two variables
   c. Two quadratic equations in two variables

   (*Hint:* Sketch some typical systems.)

44. Consider the system

$$x^2 + y^2 = 8 \qquad (1)$$
$$xy = 4. \qquad (2)$$

We can solve this system by substituting $4/x$ for y in (1) to obtain

$$x^2 + \frac{16}{x^2} = 8,$$

from which we have $x = 2$ or $x = -2$. Now, if we obtain the y-components of the solution from (2), we find that

for $x = 2$, $y = 2$; for $x = -2$, $y = -2$.

But if we seek y-components from (1), we have

for $x = 2$, $y = \pm 2$; for $x = -2$, $y = \pm 2$.

Graph Equations (1) and (2) and discuss the fact that we seem to obtain more solutions from (1) than from (2). What is the solution set of the system?

CHAPTER REVIEW

A

[11.1]

■ *Find the distance between each of the given pairs of points, and find the midpoint of the segment joining them.*

1. $(-4, 6)$ and $(2, -8)$
2. $(3, -7)$ and $(-4, 0)$

■ *Graph each equation.*

3. $x^2 + y^2 = 9$
4. $(x - 2)^2 + (y + 3)^2 = 16$

5. $x^2 + y^2 - 4x + 2y - 4 = 0$
6. $x^2 + y^2 - 6y - 4 = 0$
7. $\dfrac{x^2}{9} + y^2 = 1$
8. $\dfrac{x^2}{4} + \dfrac{y^2}{16} = 1$
9. $\dfrac{(x-2)^2}{4} + \dfrac{(y+3)^2}{9} = 1$
10. $\dfrac{(x+4)^2}{12} + \dfrac{(y-2)^2}{6} = 1$
11. $4x^2 + y^2 - 16x + 4y + 4 = 0$
12. $8x^2 + 5y^2 + 16x - 20y - 12 = 0$

■ *Write an equation for the conic section with the given properties.*

13. Circle: center at $(-4, 3)$, radius $2\sqrt{5}$
14. Circle: endpoints of a diameter at $(-5, 2)$ and $(1, 6)$
15. Ellipse: center at $(-1, 4)$, $a = 4$, $b = 2$, major axis vertical
16. Ellipse: vertices at $(3, 6)$ and $(3, -4)$ and covertices at $(1, 1)$ and $(5, 1)$

[11.2]

■ *Find the focus and directrix for each parabola.*

17. $x^2 = 6y$
18. $y^2 = -12x$
19. $4y^2 + 2x = 0$
20. $4x^2 - 3y = 0$

■ *Graph each equation.*

21. $4(y - 2) = (x + 3)^2$
22. $(x - 2)^2 + 4y = 4$
23. $x^2 - 8x - y + 6 = 0$
24. $y^2 + 6y + 4x + 1 = 0$
25. $x^2 + y = 4x - 6$
26. $y^2 = 2y + 2x + 2$

■ *Find an equation for the parabola with the given properties.*

27. Vertex at $(0, 0)$ and focus at $(0, 4)$
28. Vertex at $(2, 4)$ and focus at $(-2, 4)$
29. Vertex at $(2, -3)$ and directrix the line $y = -5$
30. Focus at $(-4, 1)$ and directrix the line $x = 2$

[11.3]

■ *Graph each equation.*

31. $\dfrac{y^2}{6} - \dfrac{x^2}{8} = 1$
32. $\dfrac{(x-2)^2}{4} - \dfrac{(y+3)^2}{9} = 1$
33. $2y^2 - 3x^2 - 16y - 12x + 8 = 0$
34. $9x^2 - 4y^2 - 72x - 24y + 72 = 0$
35. $2x^2 - y^2 + 6y - 19 = 0$
36. $4y^2 - x^2 + 8x - 28 = 0$

■ *Name the graph of each equation and describe its main features.*

37. $x^2 + 4y^2 = 24$
38. $y^2 - 4x^2 = 12$
39. $\dfrac{(x-5)^2}{6} + \dfrac{(y+2)^2}{9} = 1$
40. $\dfrac{x^2}{4} - \dfrac{(y+2)^2}{12} = 1$
41. $x^2 + y^2 + 6x - 4y - 2 = 0$
42. $x^2 + 4y^2 - 4x + 8y = 0$

43. $x^2 + 48 = 8y$

44. $2x = y^2 + 2y + 3$

45. $y^2 - 4x^2 - 8x - 16 = 0$

46. $x^2 - 6y^2 - 12y - 18 = 0$

47. $x - 4 = \dfrac{(y + 2)^2}{2}$

48. $6 + \dfrac{y^2}{2} = x^2$

49. $(x - 3)^2 + (y + 3)^2 = 12$

50. $\dfrac{1}{4}y^2 - 2x = 2$

[11.4]

■ Solve each system by substitution or linear combination.

51. $x + 3y^2 = 4$
    $x = 3$

52. $x^2 + 2y^2 = -8$
    $y = -2$

53. $x^2 + y = 3$
    $5x + y = 7$

54. $x^2 + 3xy + x = -12$
    $2x - y = 7$

55. $6x^2 - y^2 = 1$
    $3x^2 + 2y^2 = 13$

56. $2x^2 + 5y^2 - 53 = 0$
    $4x^2 + 3y^2 - 43 = 0$

57. $x^2 - 2xy + 3y^2 = 17$
    $2x^2 + xy + 6y^2 = 24$

58. $x^2 - xy - y^2 = 1$
    $x^2 + 3xy - y^2 = 9$

59. The perimeter of a rectangle is 34 centimeters and the area is 70 square centimeters. Find the dimensions of the rectangle.

60. A rectangle has a perimeter of 18 feet. If the length is decreased by 5 feet and the width is increased by 12 feet, the area is doubled. Find the dimensions of the original rectangle.

B

61. Find the equation of the parabola $y = ax^2 + bx + c$ whose graph contains the points $(-1, -4)$, $(0, -6)$, and $(4, 6)$.

62. Find the equation of the circle $x^2 + y^2 + ax + by + c = 0$ whose graph contains the points $(-1, 2)$, $(1, 4)$, and $(3, -2)$.

63. Find the equation of a parabola if its directrix is parallel to the y-axis, its vertex is $(-2, 1)$, and one point on the parabola is $(2, 5)$.

64. Find the equation of a hyperbola for which one vertex is $(4, -1)$ and one end of the conjugate axis is $(3, 1)$.

65. a. What relationship must exist between the numbers a and b so that the solution set of the system

    $$y = x^2 - 4$$
    $$y = ax + b$$

    will have exactly one solution?

    b. Choose one set of values for a and b that satisfy the relationship found in (a). Sketch the graphs of the equations.

66. a. What relationship must exist between the numbers a and b so that the solution set of the system

$$x^2 + y^2 = 4$$
$$y = ax + b$$

will have exactly one solution?
b. Choose one set of values for a and b that satisfy the relationship found in (a). Sketch the graphs of the equations.

12 Sequences and Series

12.1

SEQUENCES

A **sequence** is a function whose domain is a set of successive positive integers. For example, the functions

$$s(n) = 2n - 1 \qquad (n \in \{3, 4, 5\})^* \tag{1}$$

and

$$s(n) = n + 3 \qquad (n \in \{1, 2, 3, \ldots\}) \tag{2}$$

are sequences. The elements in the range of a sequence are referred to as the **terms** of the sequence. The terms of Sequence (1) are

$$s(3) = 2(3) - 1 = 5,$$
$$s(4) = 2(4) - 1 = 7,$$
$$s(5) = 2(5) - 1 = 9.$$

The terms of Sequence (2) are

$$s(1) = 1 + 3 = 4,$$
$$s(2) = 2 + 3 = 5,$$
$$s(3) = 3 + 3 = 6,$$

and so on, for successive values of n. The expression $n + 3$ is called the **general term**, or **nth term**, of Sequence (2). Unless otherwise specified we will assume that the domain of a given sequence is the set of natural numbers, $\{1, 2, 3, \ldots\}$.

*The symbol $\in$ means "is an element of."

EXAMPLE 1 The first three terms of the sequence with general term $\dfrac{3}{2n-1}$ are

$$s(1) = \dfrac{3}{2(1)-1} = 3,$$
$$s(2) = \dfrac{3}{2(2)-1} = 1,$$
$$s(3) = \dfrac{3}{2(3)-1} = \dfrac{3}{5}.$$

The twenty-fifth term is

$$s(25) = \dfrac{3}{2(25)-1} = \dfrac{3}{49}.$$

We often describe a sequence by listing its terms in order. Thus, Sequence (1) above would appear as

$$5, 7, 9.$$

Sequence (1) is called a **finite sequence** because it has a finite number of terms. Sequence (2) is called an **infinite sequence**; it would appear as

$$4, 5, 6, 7, \ldots.$$

It is customary to denote the nth term of a sequence by s_n rather than $s(n)$, so that in general the notation

$$s_1, s_2, s_3, \ldots$$

denotes a sequence.

EXAMPLE 2 Find the first four terms in each sequence with the given general term.

a. $s_n = \dfrac{n(n+1)}{2}$
b. $s_n = (-1)^n 2^n$

Solutions a. $s_1 = \dfrac{1(1+1)}{2} = 1;$

$s_2 = \dfrac{2(2+1)}{2} = 3;$

$s_3 = \dfrac{3(3+1)}{2} = 6;$

$s_4 = \dfrac{4(4+1)}{2} = 10.$

The first four terms are $1, 3, 6, 10.$

b. $s_1 = (-1)^1 2^1 = -2;$
$s_2 = (-1)^2 2^2 = 4;$
$s_3 = (-1)^3 2^3 = -8;$
$s_4 = (-1)^4 2^4 = 16.$
The first four terms are $-2, 4, -8, 16.$

Recursively Defined Sequences

A sequence is said to be defined **recursively** if the general term of the sequence is given in terms of its predecessors. For example, the sequence defined by

$$s_1 = 2, \qquad s_{n+1} = 3s_n - 2$$

is a recursive sequence. Its first four terms are

$$s_1 = 2,$$
$$s_2 = 3s_1 - 2 = 3(2) - 2 = 4,$$
$$s_3 = 3s_2 - 2 = 3(4) - 2 = 10,$$
$$s_4 = 3s_3 - 2 = 3(10) - 2 = 28.$$

EXAMPLE 3   Find the first five terms of the recursive sequence

$$s_1 = -1, \qquad s_{n+1} = (s_n)^2 - 4.$$

Solution   The first term is given. Find each subsequent term by using the recursive formula:

$$s_2 = s_1^2 - 4 = (-1)^2 - 4 = -3,$$
$$s_3 = s_2^2 - 4 = (-3)^2 - 4 = 5,$$
$$s_4 = s_3^2 - 4 = 5^2 - 4 = 21,$$
$$s_5 = s_4^2 - 4 = (21)^2 - 4 = 437.$$

The first five terms are $-1, -3, 5, 21, 437.$

EXAMPLE 4 Karen joins a savings plan in which she deposits $100 per month and receives 12% annual interest compounded monthly.

   a. Find a recursively defined sequence that gives the amount of money in Karen's account after n months.
   b. Find the first four terms of the sequence found in (a).

Solutions a. In the first month Karen deposits $100, so $s_1 = 100$. Each month thereafter Karen receives 1% interest (one-twelfth of 12% annual interest) on the previous month's balance and then adds $100 to the total. First, compute the interest on the previous month's balance by using the formula $A = P(1 + r)$ with $P = s_n$ and $r = 0.01$. (See Section 8.1.) This gives

$$A = s_n(1 + 0.01) = 1.01s_n$$

for the amount. Now add Karen's $100 deposit to find

$$s_{n+1} = 100 + 1.01s_n.$$

   b. Evaluate the general formula found in (a) for $n = 1, 2, 3, 4$:

$$s_1 = 100,$$
$$s_2 = 100 + 1.01s_1$$
$$= 100 + 1.01(100) = 201,$$
$$s_3 = 100 + 1.01s_2$$
$$= 100 + 1.01(201) = 303.01,$$
$$s_4 = 100 + 1.01s_3$$
$$= 100 + 1.01(303.01) = 406.04.$$

EXERCISE 12.1

A

■ *Find the first four terms in a sequence with the general term as given. See Examples 1 and 2.*

1. $s_n = n - 5$
2. $s_n = 2n - 3$
3. $s_n = \dfrac{n^2 - 2}{2}$

4. $s_n = \dfrac{3}{n^2 + 1}$
5. $s_n = 1 + \dfrac{1}{n}$
6. $s_n = \dfrac{n}{2n - 1}$
7. $s_n = \dfrac{n(n - 1)}{2}$
8. $s_n = \dfrac{5}{n(n + 1)}$
9. $s_n = (-1)^n$
10. $s_n = (-1)^{n+1}$
11. $s_n = \dfrac{(-1)^n(n - 2)}{n}$
12. $s_n = (-1)^{n-1}3^{n+1}$

■ *Find the indicated term for each sequence.*

13. $s_n = 2^n - n$; s_6
14. $s_n = \sqrt{n + 1}$; s_{11}
15. $s_n = \log n$; s_{26}
16. $s_n = \log(n + 1)$; s_9
17. $s_n = 2\sqrt{n}$; s_{20}
18. $s_n = \dfrac{n + 1}{n - 1}$; s_{17}

■ *Find the first five terms of the recursive sequence. See Example 3.*

19. $s_1 = 3$, $s_{n+1} = s_n + 2$
20. $s_1 = 6$, $s_{n+1} = s_n - 4$
21. $s_1 = 24$, $s_{n+1} = \dfrac{-1}{2}s_n$
22. $s_1 = 27$, $s_{n+1} = \dfrac{2}{3}s_n$
23. $s_1 = 1$, $s_{n+1} = (n + 1)s_n$
24. $s_1 = 1$, $s_{n+1} = \left(\dfrac{n + 1}{n}\right)s_n$

■ *Write a recursive sequence to describe each situation. Find the first four terms of each sequence. See Example 4.*

25. A new car costs $14,000 and depreciates in value by 15% each year. How much is the car worth after n years?
26. Valerie takes a job as an executive secretary for $21,000 per year with a guaranteed 5% raise each year. What will her salary be after n years?
27. Geraldo inherits an annuity of $50,000 that draws 12% annual interest compounded monthly. If he withdraws $500 at the end of each month, what is the value of the annuity after n months?
28. Eve borrowed $18,000 for a new car at 6% annual interest compounded monthly. If she pays $400 per month toward the loan, how much does she owe after n months?

B

■ *Use a calculator to evaluate a large number of terms of the following recursive sequences. What happens to s_n as n gets large?*

29. $s_1 = 1$, $s_{n+1} = \dfrac{1}{1 + s_n} + 1$
30. $s_1 = 1$, $s_{n+1} = \dfrac{2}{1 + s_n} + 1$
31. $s_1 = 3$, $s_{n+1} = \dfrac{\sqrt{1 + s_n}}{2}$
32. $s_1 = 8$, $s_{n+1} = \dfrac{\sqrt{1 + s_n}}{2}$
33. $s_1 = 1$, $s_{n+1} = \dfrac{1}{2}\left(s_n + \dfrac{4}{s_n}\right)$
34. $s_1 = 1$, $s_{n+1} = \dfrac{1}{2}\left(s_n + \dfrac{9}{s_n}\right)$

35. The Fibonacci sequence is defined recursively by

$$s_1 = 1, \quad s_2 = 1, \quad s_{n+2} = s_n + s_{n+1}.$$

   a. Find the first 16 terms of the Fibonacci sequence.
   b. Calculate the quotients $\dfrac{s_{n+1}}{s_n}$ for $n = 1$ to $n = 15$. What do you observe? Find a decimal approximation for the "golden ratio," $\dfrac{1 + \sqrt{5}}{2}$.

36. The Lucas sequence is defined recursively by

$$s_1 = 2, \quad s_2 = 1, \quad s_{n+2} = s_n + s_{n+1}.$$

   a. Find the first 10 terms of the Lucas sequence.
   b. Calculate $(s_{n+1})^2 - s_n(s_{n+2})$ for $n = 1$ to $n = 8$. What do you observe?

12.2

SERIES

A **series** is the sum of the terms of a sequence. A series consisting of the first n terms of a sequence is usually denoted by S_n. Thus, the series associated with the sequence of five terms

$$2, 4, 8, 16, 32$$

is the *sum* of those five terms,

$$S_5 = 2 + 4 + 8 + 16 + 32 = 62.$$

Note that although the words "sequence" and "series" are often used interchangeably in everyday English, they have different and distinct meanings in mathematics. A sequence is a *list* of numbers, whereas a series is a *single* number, namely, a particular sum.

EXAMPLE 1 a. Find the series S_5 associated with the sequence with the general term $s_n = 3n + 1$.
   b. Find the series S_6 associated with the sequence with general term $s_n = x^n$.

Solutions a. $S_5 = 4 + 7 + 10 + 13 + 16 = 50$
b. $S_6 = x + x^2 + x^3 + x^4 + x^5 + x^6$

Summation Notation

There is a convenient notation for representing a series if a formula for the general term is known. For example, suppose we want to refer to the sum of the first six terms of the sequence

$$4, 7, 10, \ldots, 3n + 1, \ldots .$$

Since the terms of the series are obtained by replacing n in the general term $3n + 1$ with the numbers 1 through 6, we might express the sum as

"the sum, as n runs from 1 to 6, of $3n + 1$."

Thus, instead of writing out all the terms of the series we merely indicate which terms are to be included. The letter n is called the **index of summation**; it is like a variable because it represents numerical values. Any letter can be used for the index of summation; i, j, and k are other popular choices. The letter used for the index of summation does not affect the sum.

To abbreviate this expression further we use the Greek letter Σ (called "sigma") to stand for "sum" and indicate the first and last values of n, respectively, below and above the summation symbol Σ. Thus, the series above might appear as

$$S_6 = \sum_{n=1}^{6} (3n + 1).$$

EXAMPLE 2 Use summation notation to represent the sum of the first 20 terms of the sequence

$$-1, 2, 7, \ldots, j^2 - 2, \ldots .$$

Solution The general term of the sequence is $j^2 - 2$, and the first term, -1, is obtained by letting $j = 1$ in the formula for the general term. Thus,

$$S_{20} = \sum_{j=1}^{20} j^2 - 2.$$

The **expanded form** of a sum written with summation, or sigma, notation is obtained by writing out all the terms of the sum; thus,

$$\sum_{i=1}^{6} i(i-1) = 1(1-1) + 2(2-1) + 3(3-1) \\ + 4(4-1) + 5(5-1) + 6(6-1) \\ = 0 + 2 + 6 + 12 + 20 + 30 = 70.$$

The first value of the index of summation need not be 1. For example,

$$\sum_{j=3}^{5} \frac{j+1}{j} = \frac{3+1}{3} + \frac{4+1}{4} + \frac{5+1}{5} \\ = \frac{4}{3} + \frac{5}{4} + \frac{6}{5}.$$

Note that the series above has three terms, which is one more than the difference between the first and last replacements for j. In general, the series $\sum_{j=a}^{b} s_j$ includes $(b - a + 1)$ terms.

To show that a series has infinitely many terms—that is, that it has no last term—we use the symbol ∞ for the upper limit of summation. Thus,

$$S_\infty = \sum_{j=1}^{\infty} \frac{3}{j} = \frac{3}{1} + \frac{3}{2} + \frac{3}{3} + \frac{3}{4} + \cdots.$$

In later sections we will consider whether the sum of infinitely many terms can actually be evaluated.

EXAMPLE 3 Write each series in expanded form.

a. $\sum_{i=2}^{4} (i^2 + 1)$ b. $\sum_{k=1}^{\infty} (-1)^k 2^{k+1}$

Solutions a. Replace i with the values 2, 3, and 4 and form the sum of the terms:

$$\text{for } i = 2, \quad (2)^2 + 1 = 5; \\ \text{for } i = 3, \quad (3)^2 + 1 = 10; \\ \text{for } i = 4, \quad (4)^2 + 1 = 17.$$

The expanded form is $5 + 10 + 17$.

b. Replace k with the values $1, 2, 3, \ldots$ and form the sum of the terms:

$$\text{for } k = 1, \ (-1)^1 2^{1+1} = (-1)4 = -4;$$
$$\text{for } k = 2, \ (-1)^2 2^{2+1} = (1)(8) = 8;$$
$$\text{for } k = 3, \ (-1)^3 2^{3+1} = (-1)(16) = -16.$$

The expanded form is

$$-4 + 8 - 16 + \cdots.$$

Finding the General Term

As we have seen in the examples above, if the general term s_n of a sequence is known we can find the terms of the sequence by substituting the numbers $1, 2, 3, \ldots$ for n. However, if instead we know several terms of a sequence it is often difficult to find a formula for the general term. In Sections 12.3 and 12.4 we shall consider some special sequences for which the general term can always be found. For now, we shall attempt to find general terms by trial and error.

There may be more than one way to represent the terms of a sequence or series. For example, the series

$$1 + 3 + 5 + 7 + \cdots$$

can be denoted by either

$$\sum_{i=2}^{\infty} (2i - 3) \quad \text{or} \quad \sum_{i=0}^{\infty} (2i + 1).$$

This can be verified by writing the first few terms in each series.

EXAMPLE 4 Write each series using sigma notation.

a. $5 + 8 + 11 + 14$
b. $x^2 + x^4 + x^6 + \cdots + x^{2n}$

Solutions a. Note that each term is three greater than the previous term. One way to represent the series is

$$5 + 8 + 11 + 14 = \sum_{i=1}^{4} (3i + 2).$$

b. The exponents in each term are even integers, so we may represent the series by

$$x^2 + x^4 + x^6 + \cdots + x^{2n} = \sum_{i=1}^{n} x^{2i}.$$

EXERCISE 12.2

A

■ *Find the indicated series associated with each sequence. See Example 1.*

1. $s_n = 2n - 1$; S_6
2. $s_n = (-1)^n n$; S_8
3. $s_n = (-1)^{n+1} 2n$; S_7
4. $s_n = 2^{n-1}$; S_6
5. $s_n = \dfrac{1}{n}$; S_5
6. $s_n = \dfrac{1}{n^2}$; S_5

■ *Write each sum using sigma notation. See Example 2.*

7. The first 25 terms of the sequence $3, 9, 27, \ldots, 3^k, \ldots$
8. The first 100 terms of the sequence $2 \cdot 3, 3 \cdot 4, 4 \cdot 5, \ldots, (j+1)(j+2), \ldots$
9. The fifth through fifteenth terms of $x, \dfrac{x^2}{2}, \dfrac{x^3}{3}, \ldots, \dfrac{x^i}{i}, \ldots$
10. The second through ninth terms of $x^2, 8x^3, 27x^4, \ldots, k^3 x^{k+1}, \ldots$

■ *Write each sum in expanded form. See Example 3.*

11. $\sum_{i=1}^{4} i^2$
12. $\sum_{i=1}^{3} (3i - 2)$
13. $\sum_{j=5}^{7} (j - 2)$
14. $\sum_{j=2}^{6} (j^2 + 1)$
15. $\sum_{k=1}^{4} k(k+1)$
16. $\sum_{i=2}^{6} \dfrac{i}{2}(i+1)$
17. $\sum_{i=1}^{4} \dfrac{(-1)^i}{2^i}$
18. $\sum_{i=3}^{5} \dfrac{(-1)^{i+1}}{i-2}$
19. $\sum_{i=1}^{\infty} (2i - 1)$
20. $\sum_{j=1}^{\infty} \dfrac{1}{j}$
21. $\sum_{k=0}^{\infty} \dfrac{1}{2^k}$
22. $\sum_{k=0}^{\infty} \dfrac{k}{1+k}$

■ *Write each series in sigma notation. See Example 4. (There is more than one correct solution.)*

23. $1 + 2 + 3 + 4$
24. $2 + 4 + 6 + 8$
25. $x + x^3 + x^5 + x^7$
26. $x^3 + x^5 + x^7 + x^9 + x^{11}$
27. $1 + 4 + 9 + 16 + 25$
28. $1 + 8 + 27 + 64 + 125$
29. $\dfrac{1}{2} + \dfrac{2}{3} + \dfrac{3}{4} + \dfrac{4}{5} + \cdots$
30. $\dfrac{2}{1} + \dfrac{3}{2} + \dfrac{4}{3} + \dfrac{5}{4} + \cdots$
31. $\dfrac{1}{1} + \dfrac{2}{3} + \dfrac{3}{5} + \dfrac{4}{7} + \cdots$
32. $\dfrac{3}{1} + \dfrac{5}{3} + \dfrac{7}{5} + \dfrac{9}{7} + \cdots$
33. $\dfrac{1}{1} + \dfrac{2}{2} + \dfrac{4}{3} + \dfrac{8}{4} + \cdots$
34. $\dfrac{1}{2} + \dfrac{3}{4} + \dfrac{9}{6} + \dfrac{27}{8} + \cdots$

B

35. **a.** A car is accelerating so that it travels 10 feet in the first second, 30 feet in the second second, 50 feet in the third second, 70 feet in the fourth second, and so on. Find an expression s_k for the distance the car travels in the kth second.

    b. Find the total distance the car has traveled after 1 second, after 2 seconds, after 3 seconds, and after 4 seconds. Find a formula for the total distance S_n the car has traveled after n seconds.

36. **a.** A pebble falling from the top of Half Dome in Yosemite falls 16 feet in the first second, 48 feet in the second second, 80 feet in the third second, 112 feet in the fourth second, and so on. Find an expression for the distance the pebble falls during the kth second.

    b. Find the total distance the pebble has fallen after 1 second, after 2 seconds, after 3 seconds, and after 4 seconds. Find a formula for the total distance S_n the pebble has fallen after n seconds.

37. Calculate several sums $S_1, S_2, S_3, \ldots$ for the sequence with the general term
$$s_k = \frac{1}{k} - \frac{1}{k+1}.$$
Can you find a formula for S_n in terms of n?

38. Calculate several sums $S_1, S_2, S_3, \ldots$ for the sequence with the general term
$$s_k = \log\left(\frac{k}{k+1}\right).$$
Can you find a formula for S_n in terms of n?

39. Use a calculator to find several terms of the sequence with the general term $s_n = (1 + 1/n)^n$. As n increases, what value does s_n approach?

40. Calculate several values of sums S_n for the sequence with the general term $s_k = 1/k^2$. As n increases, what value does $\sqrt{6S_n}$ approach?

12.3

ARITHMETIC PROGRESSIONS

There are many situations in which the domain of a function consists of a set of whole numbers rather than an interval of real numbers. In such cases it may be more appropriate to describe the function as a sequence. For example, suppose an airport shuttle service charges $15 one way plus $5 for each passenger. The cost of a ride to the airport is a function of the number of passengers. Since the number of passengers n can be only a natural number, the domain of the function is the set of natural numbers $1, 2, 3, \ldots$, up to the capacity of the shuttle bus. If s_n represents the cost of a ride for n passengers, then

$$s_1 = 15 + 5(1) = 20,$$
$$s_2 = 15 + 5(2) = 25,$$
$$s_3 = 15 + 5(3) = 30,$$

and in general
$$s_n = 15 + 5n.$$

Note that each term of the sequence defined above can be obtained from the previous one by adding 5. Such a sequence, in which each term can be obtained from the previous one by adding a fixed quantity, is called an **arithmetic progression,** or **arithmetic sequence.**

The difference between two successive terms—in the example above the number 5—is called the **common difference.** If we denote the first term of the sequence by a and the common difference by d, then an arithmetic progression can be defined recursively by

$$s_1 = a,$$
$$s_{n+1} = s_n + d.$$

EXAMPLE 1 Find the first four terms of an arithmetic sequence with first term 6 and common difference 3.

Solution Since the first term is 6, we have $s_1 = 6$. To find each subsequent term add 3 to the previous term:

$$s_2 = s_1 + 3 = 6 + 3 = 9;$$
$$s_3 = s_2 + 3 = 9 + 3 = 12;$$
$$s_4 = s_3 + 3 = 12 + 3 = 15.$$

The first four terms are $6, 9, 12, 15$.

Observe that an arithmetic progression defines a linear function of n. In Figure 12.1a and b compare the graph of the linear function $f(x) = 2x + 3$, whose domain is the set of all real numbers, and the graph of the arithmetic progression $s_n = 2n + 3$, whose domain is the set of natural numbers. The common difference, 2, corresponds to the slope of the linear function.

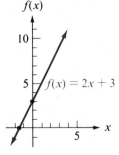
a.

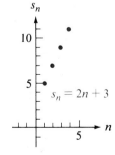
b.

FIGURE 12.1

General Term of an Arithmetic Progression

Consider an arithmetic progression with first term a and common difference d. The

$$\begin{aligned}
\text{first term is} \quad & a, \\
\text{second term is} \quad & a + d, \\
\text{third term is} \quad & a + d + d = a + 2d, \\
\text{fourth term is} \quad & a + d + d + d = a + 3d, \\
& \vdots \\
n\text{th term is} \quad & a + d + d + \cdots + d = a + (n - 1)d.
\end{aligned}$$

Thus, we have the following property.

The nth term of an arithmetic sequence is

$$s_n = a + (n - 1)d. \tag{1}$$

Equation (1) is a formula for the nth term of an arithmetic progression when the first term and common difference are known. For example, if the first term of an arithmetic progression is 7 and the common difference is 2, we have from Equation (1) that the nth term is

$$\begin{aligned}
s_n &= 7 + (n - 1)2 \\
&= 7 + 2n - 2 \\
&= 2n + 5.
\end{aligned}$$

EXAMPLE 2 Write the next three terms in each arithmetic progression. Find an expression for the general term.

a. $9, 5, \ldots$ \qquad b. $x, x + p, \ldots$

Solutions a. Find the common difference by subtracting any term from its successor:

$$d = 5 - 9 = -4.$$

The next three terms are $1, -3, -7$.

Use the formula $s_n = a + (n-1)d$ with $a = 9$ and $d = -4$ to find the general term:

$$s_n = 9 + (n-1)(-4)$$
$$= -4n + 13.$$

b. The common difference is

$$d = (x + p) - x = p.$$

The next three terms are $x + 2p$, $x + 3p$, $x + 4p$. Since $a = x$ and $d = p$, the general term is

$$s_n = x + (n-1)(p)$$
$$= x + p(n-1).$$

Equation (1) can also be used to find a particular term of an arithmetic progression if the first term and common difference are known.

EXAMPLE 3 Find the fourteenth term of the arithmetic progression $-6, -1, 4, \ldots$.

Solution First, find the common difference:

$$d = -1 - (-6) = 5.$$

Then use Equation (1) with $n = 14$:

$$s_{14} = -6 + (14-1)5 = 59.$$

Equation (1) can also be used to find a particular term if any two terms of an arithmetic progression are known.

EXAMPLE 4 Find the eighth term of an arithmetic progression in which the third term is 7 and the eleventh term is 55.

Solution Use Equation (1) with $n = 3$ and $n = 11$ to write a system of two equations in two unknowns, a and d:

$$s_3 = a + (3 - 1)d$$
$$7 = a + 2d \qquad (2)$$

and

$$s_{11} = a + (11 - 1)d$$
$$55 = a + 10d. \qquad (3)$$

Solve the system (2) and (3) to obtain $a = -5$ and $d = 6$. Then

$$s_8 = -5 + (8 - 1)6 = 37.$$

Arithmetic Series

It is usually very difficult to find a formula for the *sum* of n terms of a sequence, but for the special case of an arithmetic progression we can obtain such a formula. As an example, consider the sum of the first 12 terms of the sequence with general term $s_i = 4i + 1$. That is,

$$S_{12} = 5 + 9 + 13 + \cdots + 41 + 45 + 49. \qquad (1)$$

Now write the terms of the series in the opposite order:

$$S_{12} = 49 + 45 + 41 + \cdots + 13 + 9 + 5. \qquad (2)$$

By adding (1) and (2) term by term we discover an interesting phenomenon:

$$2S_{12} = 54 + 54 + 54 + \cdots + 54 + 54 + 54.$$

The term 54 occurs 12 times (since we are adding 12 terms), so

$$2S_{12} = 12(54),$$

or

$$S_{12} = \frac{12(54)}{2} = 324.$$

Note that the number 54 is the sum of the first term of the series, 5, and the last term, 49, or $54 = s_1 + s_{12}$. This observation is the key to the formula for a general arithmetic series. Consider the sum of the first n terms of the arithmetic progression

$$a, \quad a + d, \quad a + 2d, \quad a + 3d, \ldots, \quad a + (n - 1)d;$$

that is,

$$S_n = a + (a + d) + (a + 2d) + \cdots + [a + (n - 1)d]. \qquad (3)$$

We usually generate the terms of the series by successively *adding* the common difference to each term, starting with the first, as shown in (3). However, we can also start with the *last* term and *subtract* the common difference to get

$$S_n = s_n + (s_n - d) + (s_n - 2d) + \cdots + [s_n - (n - 1)d]. \qquad (4)$$

Now add (3) and (4) term by term as we did in the example above to find

$$2S_n = (a + s_n) + (a + s_n) + (a + s_n) + \cdots + (a + s_n),$$

where the term $(a + s_n)$ occurs n times. It follows that

$$2S_n = n(a + s_n),$$

from which we have the following property:

$$S_n = \frac{n}{2}(a + s_n). \qquad (5)$$

EXAMPLE 5 Find the sum of the first 12 terms of the arithmetic sequence $s_n = -2 + 3n$.

Solution The first term of the sequence is

$$a = s_1 = -2 + 3(1) = 1,$$

and the last term is

$$s_{12} = -2 + 3(12) = 34.$$

Use Equation (5) with $n = 12$, $a = 1$, and $s_{12} = 34$ to find

$$S_{12} = \frac{12}{2}(1 + 34) = 210.$$

We can also use Equation (5) to compute the sum of an arithmetic series expressed in summation notation.

EXAMPLE 6 Compute the sum of the series $\sum_{j=1}^{8} 3j - 10$.

Solution The general term for the series is $s_j = 3j - 10$, so

$$a = s_1 = 3(1) - 10 = -7$$

and

$$s_8 = 3(8) - 10 = 14.$$

Thus, by Equation (5),

$$S_8 = \frac{8}{2}(-7 + 14) = 28.$$

If (5) is rewritten as

$$S_n = n\left(\frac{a + s_n}{2}\right)$$

we can see that the sum is obtained by multiplying the number of terms times the average of the first and last terms.

Often we know the common difference for an arithmetic series but not the last term. If we substitute $a + (n - 1)d$ for s_n in Equation (5) we obtain

$$S_n = \frac{n}{2}(a + [a + (n - 1)d]),$$

which simplifies to the following property:

$$S_n = \frac{n}{2}[2a + (n - 1)d]. \qquad (6)$$

EXAMPLE 7 Find the sum of the first 15 odd integers.

Solution The odd integers form an arithmetic sequence with common difference $d = 2$. The first term of the sequence is $a = 1$. Thus, by Equation (6),

$$S_{15} = \frac{15}{2}[2(1) + (15 - 1)2]$$

$$= \frac{15}{2}[2 + (14)2] = 225.$$

EXAMPLE 8 Arlene starts a new job in a print shop at a salary of $800 per month. If she keeps up with the training program her salary will increase by $35 per month. How much will Arlene have earned at the end of the 18-month training program?

Solution Arlene's monthly salary is an arithmetic progression with $a = 800$ and $d = 35$. Find the sum of the first 18 terms of the progression, or S_{18}. Use Equation (6):

$$S_{18} = \frac{18}{2}[2(800) + (18 - 1)35]$$

$$= 9[1600 + 17(35)]$$

$$= 19{,}755.$$

Arlene's total earnings for the 18 months will be $19,755.

EXERCISE 12.3

A

■ *Find the first four terms of each arithmetic sequence. See Example 1.*

1. $a = 2, \ d = 4$
2. $a = 7, \ d = 3$
3. $a = 1, \ d = -3$
4. $a = 8, \ d = -5$
5. $a = \frac{1}{2}, \ d = \frac{1}{4}$
6. $a = \frac{2}{3}, \ d = \frac{1}{3}$
7. $a = 3, \ d = x + 2$
8. $a = 6, \ d = 1 - x$

■ *Write the next three terms in each arithmetic progression. Find an expression for the general term. See Example 2.*

9. $3, 7, \ldots$
10. $-6, -1, \ldots$
11. $-1, -5, \ldots$
12. $-10, -20, \ldots$
13. $x, x + 1, \ldots$
14. $h, h + 5, \ldots$
15. $x + p, x + 3p, \ldots$
16. $y - 2b, y, \ldots$
17. $2x + 1, 2x + 4, \ldots$
18. $q + 2b, q - 2b, \ldots$
19. $x, 2x, \ldots$
20. $3b, 5b, \ldots$

■ *Solve. See Example 3.*

21. Find the seventh term in the arithmetic progression $7, 11, 15, \ldots$.
22. Find the tenth term in the arithmetic progression $-3, -12, -21, \ldots$.
23. Find the twelfth term in the arithmetic progression $2, 5/2, 3, \ldots$.
24. Find the seventeenth term in the arithmetic progression $-5, -2, 1, \ldots$.
25. Find the twentieth term in the arithmetic progression $3, -2, -7, \ldots$.
26. Find the tenth term in the arithmetic progression $3/4, 2, 13/4, \ldots$.

■ *Solve. See Example 4.*

27. If the third term in an arithmetic progression is 7 and the eighth term is 17, find the common difference. What is the first term? What is the twentieth term?
28. If the fifth term of an arithmetic progression is -16 and the twentieth term is -46, what is the twelfth term?
29. What term in the arithmetic progression $4, 1, -2, \ldots$ is -77?
30. What term in the arithmetic progression $7, 3, -1, \ldots$ is -81?

■ *Find each sum. See Example 5.*

31. The first nine terms of the sequence $s_n = -4 + 3n$
32. The first 10 terms of the sequence $s_n = 5 - 2n$
33. The first 16 terms of the sequence $s_n = 18 - \dfrac{4}{3}n$
34. The first 13 terms of the sequence $s_n = -6 - \dfrac{1}{2}n$
35. The first 30 terms of the sequence $s_n = 1.6 + 0.2n$
36. The first 25 terms of the sequence $s_n = 2.5 + 0.3n$

■ *Find the sum of each series. See Example 6.*

37. $\sum_{i=1}^{7} (2i + 1)$
38. $\sum_{i=1}^{21} (3i - 2)$
39. $\sum_{j=1}^{15} (1 - 7j)$
40. $\sum_{j=1}^{20} (3 - 2j)$
41. $\sum_{k=1}^{8} \left(\dfrac{3}{2}k - 3\right)$
42. $\sum_{k=1}^{14} \left(\dfrac{4}{3}k - 5\right)$
43. $\sum_{n=1}^{100} n$
44. $\sum_{n=1}^{50} 3n$

■ *Solve. See Examples 7 and 8.*

45. a. Find the sum of all even integers n where $13 < n < 89$.
    b. Write the sum in (a) using sigma notation.
46. a. Find the sum of all integral multiples of 7 between 8 and 110.
    b. Write the sum in (a) using sigma notation.
47. Sales of Brussels Sprouts dolls peaked at $320,000 in 1990 and began to decline at a steady rate of $40,000 per year. What total revenue should the manufacturer expect to gain from sale of the dolls from 1991 to 1996?
48. Richard's water bill was $63.50 last month. If his bill increases by $2.30 per month, how much should he expect to pay for water during the next 10 months?
49. A computer takes 0.1 second to perform the first iteration of a certain loop, and each subsequent iteration takes 0.05 second longer than the previous one. How long will it take the computer to perform 50 iterations?
50. It took Alida 20 minutes to type the first page of her term paper, but each subsequent page takes her 40 seconds less than the previous one. How long will it take her to type a 30-page paper?
51. An auditorium seats 4340 people on the main floor. If the first row seats 50 people and each subsequent row has three more seats than the previous row, how many rows are there?
52. Clara started a jogging program by running ½ mile three times a week. Each week thereafter she increased her run by 0.2 mile. If Clara has now (at the end of a week) jogged a total of 136.8 miles, how long has she been jogging?

B

■ *Find the sum of each series.*

53. $\sum_{j=6}^{18}(4j-3)$
54. $\sum_{j=8}^{24}(5j+2)$
55. $\sum_{k=32}^{64}(28-0.4k)$
56. $\sum_{k=16}^{30}(15+0.3k)$

57. Find three numbers that form an arithmetic sequence such that their sum is 21 and their product is 168.
58. Find three numbers that form an arithmetic sequence such that their sum is 21 and their product is 231.
59. Find k if $\sum_{j=1}^{5} kj = 14$.
60. Find p and q if $\sum_{i=1}^{4}(pi+q) = 28$ and $\sum_{i=2}^{5}(pi+q) = 44$.
61. Show that the sum of the first n odd natural numbers is n^2.
62. Show that the sum of the first n even natural numbers is $n^2 + n$.

12.4 GEOMETRIC PROGRESSIONS

Consider a colony of 100 bacteria that doubles in size every hour. The sequence

$$100, 200, 400, 800, \ldots$$

gives the population of the colony at 1-hour intervals, where $s_1 = 100$ is the initial population and s_n is the population $n - 1$ hours later. Each term of the sequence can be obtained from the previous term by multiplying by 2. Such a sequence, in which each term is obtained from its predecessor by multiplying by a fixed quantity, is called a **geometric sequence,** or a **geometric progression.** The fixed multiplier (the number 2 in this example) is called the **common ratio.**

A geometric sequence is defined by the recursive equations

$$s_1 = a$$
$$s_{n+1} = rs_n,$$

where r is the common ratio. Thus, the sequence

$$2, 6, 18, 54, \ldots$$

is a geometric progression in which the common ratio is 3.

EXAMPLE 1 Identify the following sequences as arithmetic, geometric, or neither.

a. $3, 5, 7, \ldots$

b. $3, -6, 12, \ldots$

c. $3, 6, 10, \ldots$

d. $3, 1, \dfrac{1}{3}, \ldots$

Solutions

a. This sequence is arithmetic, since each term is obtained from the previous one by adding 2.

b. This sequence is geometric, since each term is obtained from the previous one by multiplying by -2.

c. This sequence is neither arithmetic nor geometric.

d. This sequence is geometric, since each term is obtained from the previous one by multiplying by ⅓.

General Term of a Geometric Progression

If we denote the first term of a geometric progression by a, then the

second term is ar,
third term is $ar \cdot r = ar^2$,
fourth term is $ar^2 \cdot r = ar^3$,

and in general:

> The nth term of a geometric sequence is
> $$s_n = ar^{n-1}. \tag{1}$$

EXAMPLE 2 Find the first four terms of the geometric sequence $s_n = 6(\frac{1}{3})^{n-1}$.

Solution Evaluate s_n for $n = 1, 2, 3,$ and 4:

$$s_1 = 6\left(\frac{1}{3}\right)^0 = 6; \quad s_2 = 6\left(\frac{1}{3}\right)^1 = 2;$$

$$s_3 = 6\left(\frac{1}{3}\right)^2 = \frac{2}{3}; \quad s_4 = 6\left(\frac{1}{3}\right)^3 = \frac{2}{9}.$$

The first four terms are $6, 2, \frac{2}{3},$ and $\frac{2}{9}$.

Note that a geometric sequence defines an exponential function of n. Compare the graphs of the exponential function $f(x) = 100(2)^{x-1}$, whose domain is the set of real numbers, and the geometric sequence $s_n = 100(2)^{n-1}$, whose domain is the set of natural numbers, shown in Figure 12.2a and b on page 582. The common ratio of the geometric sequence corresponds to the base of the exponential function.

Recall that an arithmetic sequence defines a linear function whose slope corresponds to the common difference. Just as an exponential function "grows" much faster in the long run than a linear function, so does a geometric progression grow much faster than an arithmetic progression.

We can find the common ratio for a given geometric progression by dividing any term by its predecessor. Note that in Example 2, s_2/s_1, s_3/s_2, and s_4/s_3

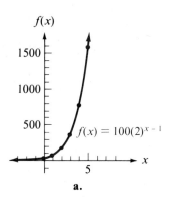
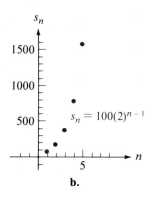

FIGURE 12.2 a.   b.

all equal the common ratio, $r = \frac{1}{3}$. Once the common ratio is known we can write the general term for the sequence and generate as many terms as we wish.

EXAMPLE 3 Write the next three terms in each geometric progression. Find the general term.

    a. $32, -8, 2, \ldots$ b. $x, 2, 4/x, \ldots$

Solutions   a.   First, find the common ratio:

$$r = \frac{-8}{32} = -\frac{1}{4}.$$

Multiply successively by r to find the next three terms:

$$-\frac{1}{2}, \frac{1}{8}, -\frac{1}{32}.$$

The general term is

$$s_n = ar^{n-1} = 32\left(-\frac{1}{4}\right)^{n-1}.$$

    b.   Find the common ratio:

$$r = \frac{2}{x}.$$

Multiply successively by r to find the next three terms:

$$\frac{8}{x^2}, \frac{16}{x^3}, \frac{32}{x^4}.$$

Use $s_n = ar^{n-1}$ to find the general term:

$$s_n = x\left(\frac{2}{x}\right)^{n-1}$$

$$= x\frac{2^{n-1}}{x^{n-1}} = \frac{2^{n-1}}{x^{n-2}}.$$

EXAMPLE 4 Find the ninth term of the geometric progression

$$-24, 12, -6, \ldots.$$

Solution Find the common ratio:

$$r = \frac{12}{-24} = -\frac{1}{2}.$$

Use $s_n = ar^{n-1}$ with $a = -24$, $r = -\frac{1}{2}$ and $n = 9$:

$$s_9 = -24\left(-\frac{1}{2}\right)^8 = -\frac{3}{32}.$$

Geometric Series

We can find a formula for the sum of the terms of a geometric progression as follows. Consider the geometric series (2) containing n terms and the series (3) obtained by multiplying both sides of (2) by r:

$$S_n = a + ar + ar^2 + ar^3 + \cdots + ar^{n-2} + ar^{n-1}, \qquad (2)$$

$$rS_n = ar + ar^2 + ar^3 + ar^4 + \cdots + ar^{n-1} + ar^n. \qquad (3)$$

If we subtract (3) from (2) we find that all terms on the right side vanish except for the first term in (2) and the last term in (3), yielding

$$S_n - rS_n = a - ar^n.$$

Factoring S_n from the left side and a from the right side gives

$$(1 - r)S_n = a(1 - r^n),$$

from which we have the following property:

$$S_n = \frac{a(1 - r^n)}{1 - r} \quad (r \neq 1). \qquad (4)$$

This is a general formula for the sum of the first n terms of a geometric progression. We must first identify the first term, a, and the common ratio, r, in order to use the formula.

EXAMPLE 5 Find the sum of the first four terms of a geometric progression with general term $s_n = 5(-3)^{n-1}$.

Solution From the general term note that $a = 5$ and $r = -3$. Substitute these values with $n = 4$ into Equation (4) to find

$$S_4 = \frac{5[1 - (-3)^4]}{1 - (-3)}$$
$$= \frac{5(1 - 81)}{4} = -100.$$

EXAMPLE 6 Find the sum of the first six terms of the sequence $3/16, 3/4, 3, \ldots$.

Solution The sequence is geometric, with $a = 3/16$. Find r by dividing 3 by the preceding term, $3/4$:

$$r = 3 \div \frac{3}{4} = 4.$$

From Equation (4),

$$S_6 = \frac{\frac{3}{16}(1 - 4^6)}{1 - 4}$$
$$= \frac{3(1 - 4096)}{16(-3)} = \frac{-4095}{-16} = \frac{4095}{16}.$$

If a geometric series is given in sigma notation, it helps to write the sum in expanded form in order to identify the first term a and the common ratio r.

12.4 ■ GEOMETRIC PROGRESSIONS

EXAMPLE 7 Compute $\sum_{i=2}^{7}\left(\frac{1}{3}\right)^{i}$.

Solution Write out the first two or three terms of the series:

$$\left(\frac{1}{3}\right)^{2} + \left(\frac{1}{3}\right)^{3} + \cdots.$$

The first term is $\frac{1}{9}$ and the common ratio is $\frac{1}{3}$. The given series contains $7 - 2 + 1 = 6$ terms (recall from page 567 that the series $\sum_{i=a}^{b} s_i$ contains $b - a + 1$ terms), so $n = 6$. Use Equation (4) to find

$$S_6 = \frac{\frac{1}{9}\left[1 - \left(\frac{1}{3}\right)^{6}\right]}{1 - \frac{1}{3}} = \frac{\frac{1}{9}\left(1 - \frac{1}{729}\right)}{\frac{2}{3}} = \frac{1}{9} \cdot \frac{728}{729} \cdot \frac{3}{2} = \frac{364}{2187}.$$

Geometric series can be useful in modeling a variety of applied problems.

EXAMPLE 8 Elmore's starting salary as an engineer is $20,000, with a 5% annual raise for each of his first 6 years, dependent on suitable progress. If Elmore receives each salary increase, how much will he make over the next 6 years?

Solution Elmore's salary is multiplied each year by a factor of 1.05, so its values form a geometric progression with common ratio $r = 1.05$. His total income over 6 years will be the sum of the first six terms of the sequence. Use Equation (4) to find

$$S_6 = \frac{20,000[1 - (1.05)^6]}{1 - (1.05)}$$

$$= \frac{20,000(-0.3401)}{-0.05} = 136,038.26.$$

Infinite Geometric Series

Is it possible to add infinitely many terms and arrive at a finite sum? In some cases, if the terms added become small enough, the answer is yes. Consider the infinite geometric series

$$\frac{1}{2} + \frac{1}{4} + \frac{1}{8} + \frac{1}{16} + \cdots.$$

The *n*th **partial sum** of the series is the sum of its first *n* terms; it is denoted by S_n. Thus,

$$S_1 = \frac{1}{2};$$

$$S_2 = \frac{1}{2} + \frac{1}{4} = \frac{3}{4};$$

$$S_3 = \frac{1}{2} + \frac{1}{4} + \frac{1}{8} = \frac{7}{8};$$

$$S_4 = \frac{1}{2} + \frac{1}{4} + \frac{1}{8} + \frac{1}{16} = \frac{15}{16};$$

$$\vdots$$

Note that as *n* increases—as we add more and more terms of the series—the partial sums appear to be "approaching" 1. That is, as *n* becomes very large S_n gets very close to 1. It seems reasonable that the sum of *all* the terms of the series is 1.

We can make this assertion more plausible by examining the formula for the *n*th partial sum. From Equation (4),

$$S_n = \frac{a(1 - r^n)}{1 - r} \quad (r \neq 1).$$

Notice that if $|r| < 1$, that is, if $-1 < r < 1$, then r^n gets closer to zero for increasingly large *n*. For example, if $r = \frac{1}{2}$, then

$$r^2 = \left(\frac{1}{2}\right)^2 = \frac{1}{4}, \quad r^3 = \left(\frac{1}{2}\right)^3 = \frac{1}{8}, \quad r^4 = \left(\frac{1}{2}\right)^4 = \frac{1}{16},$$

and so on, with $(\frac{1}{2})^n$ becoming smaller and smaller for larger values of *n*. If we write (4) in the form

$$S_n = \frac{a}{1 - r}(1 - r^n) \tag{5}$$

we see that the factor $(1 - r^n)$ will get closer and closer to 1 as n gets larger, provided that $|r| < 1$. Consequently, as we choose larger and larger values for n, S_n approaches the value

$$\frac{a}{1 - r}.$$

This motivates us to define the sum of an infinite geometric series as follows:

$$S_\infty = \frac{a}{1 - r} \quad (|r| < 1). \tag{6}$$

If $|r| \geq 1$, as in the infinite series

$$3 + 6 + 12 + 24 + \cdots$$

where $r = 2$, then the terms become larger as n increases and the sum of the series is not a finite number. In this case we say that the series does not have a sum.

EXAMPLE 9 Find the sum of each series, if the sum exists.

a. $3 + 2 + \dfrac{4}{3} + \cdots$ b. $\dfrac{1}{81} - \dfrac{1}{54} + \dfrac{1}{36} + \cdots$

Solutions a. $r = \frac{2}{3}$; the series has a sum because $|r| < 1$. By Equation (6) the sum is

$$S_\infty = \frac{a}{1 - r} = \frac{3}{1 - \dfrac{2}{3}} = 9.$$

b. $r = -\dfrac{1}{54} \div \dfrac{1}{81} = -\dfrac{3}{2}$; the series does not have a sum because $|r| > 1$.

Repeating Decimals

An interesting application of geometric series involves repeating decimals. Recall that the decimal representation of a rational number either terminates, as does 0.75, or repeats a pattern of digits. For example,

$$0.2121\overline{21}, \quad 0.333\overline{3}, \quad \text{and} \quad 0.138512512\overline{512}$$

are repeating decimals; in each case the bar indicates the repeating digits. Each of these decimal numbers represents a fraction of the form p/q, where p and q are integers.

In order to express a repeating decimal as a fraction we first write the number as an infinite geometric series. Consider the first repeating decimal above:

$$0.2121\overline{21}.$$

This decimal can be rewritten as

$$0.21 + 0.0021 + 0.000021 + \cdots$$

or as

$$\frac{21}{100} + \frac{21}{10{,}000} + \frac{21}{1{,}000{,}000} + \cdots. \tag{7}$$

This is a geometric series with common ratio 0.01, or $\tfrac{1}{100}$. Since $|r| < 1$, we can use (6) to find the sum of the infinite series (7). Thus,

$$S_\infty = \frac{a}{1-r} = \frac{\frac{21}{100}}{1 - \frac{1}{100}} = \frac{\frac{21}{100}}{\frac{99}{100}} = \frac{21}{99} = \frac{7}{33},$$

so the decimal number $0.2121\overline{21}$ is equivalent to $\tfrac{7}{33}$.

The calculations can also be performed in decimal form, using $a = 0.21$ and $r = 0.01$.

EXAMPLE 10 Find a fraction equivalent to $2.045045\overline{045}$.

Solution Rewrite the decimal as a series:

$$2 + \frac{45}{1000} + \frac{45}{1{,}000{,}000} + \cdots.$$

Consider only the repeating part of the series, that is,

$$\frac{45}{1000} + \frac{45}{1,000,000} + \cdots.$$

This is an infinite geometric series with $r = \frac{1}{1000}$, so its sum is given by $S_\infty = \dfrac{a}{1-r}$:

$$S_\infty = \frac{\frac{45}{1000}}{1 - \frac{1}{1000}} = \frac{\frac{45}{1000}}{\frac{999}{1000}} = \frac{45}{999} = \frac{5}{111}.$$

Hence,

$$2.045\overline{045} = 2 + \frac{5}{111} = \frac{227}{111}.$$

EXERCISE 12.4

A

■ *Identify the following sequences as arithmetic, geometric, or neither. See Example 1.*

1. $-2, -6, -18, -54, \ldots$
2. $-2, -6, -10, -14, \ldots$
3. $16, 8, 0, -8, \ldots$
4. $16, 8, 4, 2, \ldots$
5. $-1, 1, -1, 1, \ldots$
6. $1, 3, 6, 10, \ldots$
7. $1, 4, 9, 16, \ldots$
8. $5, -5, 5, -5, \ldots$
9. $27, 9, 3, 1, \ldots$
10. $-\dfrac{1}{3}, \dfrac{1}{3}, 1, \dfrac{5}{3}, \ldots$
11. $\dfrac{2}{3}, -1, \dfrac{3}{2}, -\dfrac{9}{4}, \ldots$
12. $2, -8, 32, -124, \ldots$

■ *Find the first four terms of each sequence. See Example 2.*

13. $s_n = 5(-2)^{n-1}$
14. $s_n = -4(3)^{n-1}$
15. $s_n = 9\left(\dfrac{2}{3}\right)^{n-1}$
16. $s_n = 25\left(\dfrac{4}{5}\right)^{n-1}$
17. $s_n = \left(\dfrac{3}{4}\right)^n$
18. $s_n = \left(\dfrac{5}{2}\right)^n$
19. $s_n = 60(0.4)^n$
20. $s_n = 10(0.3)^n$

■ *Write the next three terms in each geometric progression. Find the general term. See Example 3.*

21. $2, 8, 32, \ldots$
22. $4, 8, 16, \ldots$
23. $\dfrac{2}{3}, \dfrac{4}{3}, \dfrac{8}{3}, \ldots$
24. $6, 3, \dfrac{3}{2}, \ldots$
25. $4, -2, 1, \ldots$
26. $\dfrac{1}{2}, -\dfrac{3}{2}, \dfrac{9}{2}, \ldots$
27. $\dfrac{a}{x}, -1, \dfrac{x}{a}, \ldots$
28. $\dfrac{a}{b}, \dfrac{a}{bc}, \dfrac{a}{bc^2}, \ldots$

■ *Solve. See Example 4.*

29. Find the sixth term in the geometric progression $48, 96, 192, \ldots$.
30. Find the eighth term in the geometric progression $-3, \frac{3}{2}, -\frac{3}{4}, \ldots$.
31. Find the seventh term in the geometric progression $-\frac{1}{3}a^2, a^5, -3a^8, \ldots$.
32. Find the ninth term in the geometric progression $-81a, -27a^2, -9a^3, \ldots$.
33. Find the first term of a geometric progression with fifth term 48 and ratio 2.
34. Find the first term of a geometric progression with fifth term 1 and ratio $-\frac{1}{2}$.
35. Find the third term of a geometric progression with sixth term $-\frac{1}{128}$ and ratio $-\frac{1}{4}$.
36. Find the second term of a geometric progression with seventh term $\frac{64}{625}$ and ratio $\frac{2}{5}$.
37. How many terms are in the sequence $-\frac{1}{8}, \frac{1}{4}, -\frac{1}{2}, \ldots, -512$?
38. How many terms are in the sequence $\frac{27}{64}, \frac{9}{16}, \frac{3}{4}, \ldots, \frac{64}{27}$?

■ *Find each sum. See Examples 5 and 6.*

39. The first five terms of $s_n = 2(-4)^{n-1}$
40. The first eight terms of $s_n = 12(3)^{n-1}$
41. The first nine terms of $s_n = -48\left(\dfrac{1}{2}\right)^{n-1}$
42. The first six terms of $s_n = 81\left(-\dfrac{2}{3}\right)^{n-1}$
43. The first four terms of $s_n = 18\left(\dfrac{5}{3}\right)^{n-1}$
44. The first four terms of $s_n = -512\left(\dfrac{1}{6}\right)^{n-1}$

■ *Find each sum. See Example 7.*

45. $\sum_{i=1}^{6} 3^i$
46. $\sum_{j=1}^{4} (-2)^j$
47. $\sum_{k=3}^{7} \left(\dfrac{1}{2}\right)^{k-2}$
48. $\sum_{i=3}^{12} (2)^{i-5}$
49. $\sum_{j=1}^{6} \left(\dfrac{1}{3}\right)^j$
50. $\sum_{k=1}^{5} \left(\dfrac{1}{4}\right)^k$

■ *Solve. See Example 8.*

51. A rubber ball is dropped from a height of 24 feet and returns to three-fourths of its previous height on each bounce.
    a. How high does the ball bounce after hitting the floor for the third time?
    b. How far has the ball traveled when it hits the floor for the fourth time?
52. A Yorkshire terrier can jump 3 feet into the air on his first bounce and five-sixths the height of his previous jump on each successive bounce.
    a. How high can the terrier go on his fourth bounce?
    b. How far has the terrier traveled when he returns to the ground after his fourth bounce?
53. Richard's water bill was $63.50 last month. If his bill increases by 2% per month, how much should he expect to pay for water during the next 10 months?
54. Sales of Brussels Sprouts dolls peaked at $320,000 in 1990 and then began to decline by 8% per year. What total revenue should the manufacturer expect to gain from sale of the dolls from 1991 to 1996?
55. It took Alida 20 minutes to type the first page of her term paper, but each subsequent page takes her only 95% as long as the previous one. How long will it take her to type a 30-page paper?
56. A computer takes 0.1 second to perform the first iteration of a certain loop, and each subsequent iteration takes 20% longer than the previous one. How long will it take the computer to perform 50 iterations?

■ *Find the sum of each infinite geometric series. If the series has no sum, say so. See Example 9.*

57. $12 + 6 + 3 + \cdots$

58. $2 + 1 + \dfrac{1}{2} + \cdots$

59. $\dfrac{1}{36} + \dfrac{1}{30} + \dfrac{1}{25} + \cdots$

60. $1 - \dfrac{2}{3} + \dfrac{4}{9} - \cdots$

61. $\dfrac{3}{4} - \dfrac{1}{2} + \dfrac{1}{3} - \cdots$

62. $\dfrac{1}{49} + \dfrac{1}{56} + \dfrac{1}{64} + \cdots$

63. $\sum_{n=1}^{\infty} \dfrac{1}{16}(-2)^{n-1}$

64. $\sum_{n=1}^{\infty} 2\left(-\dfrac{3}{4}\right)^{n-1}$

65. $\sum_{i=1}^{\infty} \left(\dfrac{2}{3}\right)^{i}$

66. $\sum_{i=1}^{\infty} \left(-\dfrac{1}{4}\right)^{i}$

B

■ *Find a fraction equivalent to each of the given decimal numbers. See Example 10.*

67. $0.333\overline{3}$
68. $0.6666\overline{6}$
69. $0.3131\overline{31}$
70. $0.4545\overline{45}$
71. $2.410\overline{410}$
72. $3.027\overline{027}$
73. $0.1288\overline{8}$
74. $0.8333\overline{3}$

75. A force is applied to a particle moving in a straight line in such a fashion that each second it moves only one-half of the distance it moved the preceding second. If the particle moves 10 centimeters the first second, approximately how far will it move before coming to rest?

76. The arc length through which the bob on a pendulum moves is nine-tenths of its preceding arc length. Approximately how far will the bob move before coming to rest if the first arc length is 12 inches?

77. A ball returns two-thirds of its preceding height on each bounce. If the ball is dropped from a height of 6 feet, approximately what is the total distance the ball travels before coming to rest?

78. If a ball is dropped from a height of 10 feet and returns three-fifths of its preceding height on each bounce, approximately what is the total distance the ball travels before coming to rest?

12.5

THE BINOMIAL EXPANSION

In Chapter 1 we studied products of polynomials; in particular, we found expanded forms for powers of binomial expressions, including $(a + b)^2$ and $(a + b)^3$. The amount of work involved in expanding such powers increases as the exponent gets larger. In this section we develop a formula for the expanded form of a binomial raised to any positive integer power.

We begin by considering binomial expansions for several small powers. By direct multiplication we can show that

$$(a + b)^0 = 1$$
$$(a + b)^1 = a + b;$$
$$(a + b)^2 = a^2 + 2ab + b^2;$$
$$(a + b)^3 = a^3 + 3a^2b + 3ab^2 + b^3;$$
$$(a + b)^4 = a^4 + 4a^3b + 6a^2b^2 + 4ab^3 + b^4;$$
$$(a + b)^5 = a^5 + 5a^4b + 10a^3b^2 + 10a^2b^3 + 5ab^4 + b^5.$$

Studying these examples for a pattern, we observe the following.

1. The expansion of $(a + b)^n$ has $n + 1$ terms.
2. The first term in each expansion may be thought of as $a^n b^0$. In each of the following terms the exponent on *a decreases* by 1 and the exponent on *b increases* by 1. (The last term should be thought of as $a^0 b^n$.)
3. The sum of the exponents on each term is n, so the variable factors of each term may be expressed as $a^k b^{n-k}$, where k runs from 0 to n.

12.5 ■ THE BINOMIAL EXPANSION

These observations completely characterize the expansion of $(a + b)^n$ except for the numerical coefficients of each term. Thus, we know that the expansion of $(a + b)^7$, for example, will have the form

$$(a + b)^7 = a^7 + \_a^6b + \_a^5b^2 + \_a^4b^3 \\ + \_a^3b^4 + \_a^2b^5 + \_ab^6 + b^7, \quad (1)$$

where the blanks in front of each term stand for the coefficients we have yet to determine. If we let $\binom{n}{k}$ stand for the coefficient of the term $a^{n-k}b^k$ in the expansion of $(a + b)^n$, we can express the expanded form in summation notation as follows:

$$(a + b)^n = \sum_{k=0}^{n} \binom{n}{k} a^{n-k} b^k. \quad (2)$$

Equation (2) is called the **binomial formula**, and the numbers denoted by $\binom{n}{k}$ are called the **binomial coefficients**. As yet we do not know how to compute the values of the coefficients, but we can write binomial expansions using this notation.

EXAMPLE 1 Write $(a + b)^6$ with sigma notation and in expanded form.

Solution To write $(a + b)^6$ with sigma notation replace n with 6 in Equation (2):

$$(a + b)^6 = \sum_{k=0}^{6} \binom{6}{k} a^{6-k} b^k.$$

To expand the sum above replace the index k by the values 0 through 6:

$$(a + b)^6 = \binom{6}{0} a^6 b^0 + \binom{6}{1} a^5 b^1 + \binom{6}{2} a^4 b^2 + \binom{6}{3} a^3 b^3 \\ + \binom{6}{4} a^2 b^4 + \binom{6}{5} a^1 b^5 + \binom{6}{6} a^0 b^6.$$

Pascal's Triangle

To get a clearer picture of the binomial coefficients consider again the expansions of $(a + b)^n$ on page 592, but look only at the numerical coefficient of each term:

$$
\begin{array}{c}
n = 0 \qquad\qquad\qquad 1 \\
n = 1 \qquad\qquad\quad 1 \quad 1 \\
n = 2 \qquad\qquad 1 \quad 2 \quad 1 \\
n = 3 \qquad\quad 1 \quad 3 \quad 3 \quad 1 \\
n = 4 \quad\;\; 1 \quad 4 \quad 6 \quad 4 \quad 1 \\
n = 5 \;\; 1 \quad 5 \quad 10 \quad 10 \quad 5 \quad 1. \\
\vdots
\end{array}
$$

This triangular array of numbers is known as **Pascal's triangle.** It has many interesting and surprising properties that have been extensively studied. We might first notice the following.

1. Each row begins and ends with 1.
2. The second number and the next-to-last number in the nth row are n; for example, the second and fourth numbers in the row $n = 4$ are 4.
3. Each number (except the first and last 1s) in any row after the first two can be obtained by *adding* the two numbers diagonally adjacent in the row immediately above. For example, the number 10 in the row $n = 5$ is the sum of the adjacent numbers 4 and 6 in the row $n = 4$, as shown in Figure 12.3.

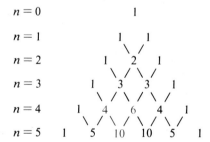

FIGURE 12.3

This last observation allows us to extend the triangle row by row as far as we like and hence to discover the numerical coefficients in the expansion of $(a + b)^n$ for higher values of n.

EXAMPLE 2 Expand $(a + b)^6$.

Solution Use Equation (1) to expand $(a + b)^6$ as in Example 1. To evaluate the binomial coefficients, write the row $n = 6$ of Pascal's triangle as shown.

$$\begin{array}{c} n = 4 \quad\quad 1 \quad 4 \quad 6 \quad 4 \quad 1 \\ n = 5 \quad 1 \quad 5 \quad 10 \quad 10 \quad 5 \quad 1 \\ n = 6 \quad 1 \quad 6 \quad 15 \quad 20 \quad 15 \quad 6 \quad 1 \end{array}$$

Finally, insert the coefficients in the appropriate positions in the expansion. Thus,

$$(a + b)^6 = a^6 + 6a^5b + 15a^4b^2 + 20a^3b^3 + 15a^2b^4 + 6ab^5 + b^6.$$

We can now use the binomial formula and Pascal's triangle to expand any positive integral power of a binomial.

EXAMPLE 3 Write $(x - 2)^4$ using summation notation and in expanded form.

Solution Substitute $a = x$, $b = -2$, and $n = 4$ in the binomial formula:

$$(x - 2)^4 = \sum_{k=0}^{4} \binom{4}{k} x^{4-k}(-2)^k$$

$$= \binom{4}{0} x^4(-2)^0 + \binom{4}{1} x^3(-2)^1 + \binom{4}{2} x^2(-2)^2$$

$$+ \binom{4}{3} x^1(-2)^3 + \binom{4}{4} x^0(-2)^4.$$

Use the row $n = 4$ of Pascal's triangle to evaluate the binomial coefficients and obtain

$$(x - 2)^4 = x^4 + 4x^3(-2) + 6x^2(4) + 4x(-8) + 16$$
$$= x^4 - 8x^3 + 24x^2 - 32x + 16.$$

Factorial Notation

Although we can find the binomial coefficients in the expansion of $(a + b)^n$ for any exponent n by extending Pascal's triangle, this process is tedious for large values of n. In such cases it would be more convenient to have a formula to calculate $\binom{n}{k}$ directly. To do this we use a special symbol, $n!$ (read "n factorial"), which is defined for nonnegative integers n as follows:

$$n! = n(n-1)(n-2) \cdots 3 \cdot 2 \cdot 1.$$

For example,

$$4! = 4 \cdot 3 \cdot 2 \cdot 1 = 24 \quad \text{and} \quad 6! = 6 \cdot 5 \cdot 4 \cdot 3 \cdot 2 \cdot 1 = 720.$$

The factorial symbol applies only to the variable or number it follows; for example, $3 \cdot 4!$ is not equal to $(3 \cdot 4)!$.

EXAMPLE 4 Write each expression in expanded form.

    a. $2n!$, for $n = 4$ **b.** $(2n-1)!$, for $n = 4$

Solutions   **a.** $2n! = 2 \cdot (4 \cdot 3 \cdot 2 \cdot 1) = 48$
           b. $(2n-1)! = 7! = 7 \cdot 6 \cdot 5 \cdot 4 \cdot 3 \cdot 2 \cdot 1 = 5040$

Note that

$$7! = 7 \cdot 6 \cdot 5 \cdot 4 \cdot 3 \cdot 2 \cdot 1$$

can be written as

$$7 \cdot 6! = 7 \cdot (6 \cdot 5 \cdot 4 \cdot 3 \cdot 2 \cdot 1)$$

and

$$5! = 5 \cdot 4 \cdot 3 \cdot 2 \cdot 1$$

can be written as

$$5 \cdot 4! = 5 \cdot (4 \cdot 3 \cdot 2 \cdot 1).$$

In general, we have the following relationship:

$$n! = n(n-1)!. \tag{3}$$

EXAMPLE 5 Write each expression in expanded form and simplify.

a. $\dfrac{6!}{3!}$
b. $\dfrac{4!\,6!}{8!}$

Solutions a. $\dfrac{6!}{3!} = \dfrac{6 \cdot 5 \cdot 4 \cdot 3!}{3!} = 120$

b. $\dfrac{4!\,6!}{8!} = \dfrac{4 \cdot 3 \cdot 2 \cdot 1 \cdot 6!}{8 \cdot 7 \cdot 6!} = \dfrac{3}{7}$

Common Error Note that in Example 5a $6!/3!$ is *not* equal to $2!$.

EXAMPLE 6 Write each product in factorial notation.

a. $1 \cdot 2 \cdot 3 \cdot 4 \cdot 5 \cdot 6$
b. $11 \cdot 12 \cdot 13 \cdot 14$
c. 150
d. $149 \cdot 150$

Solutions a. $1 \cdot 2 \cdot 3 \cdot 4 \cdot 5 \cdot 6 = 6!$
b. $11 \cdot 12 \cdot 13 \cdot 14 = \dfrac{14!}{10!}$

c. $150 = \dfrac{150!}{149!}$
d. $149 \cdot 150 = \dfrac{150!}{148!}$

EXAMPLE 7 Write $(2n + 1)!$ in factored form, showing the first three factors and the last three factors.

Solution $(2n + 1)! = (2n + 1)(2n)(2n - 1) \cdots\cdots 3 \cdot 2 \cdot 1$

If $n = 1$ in Equation (3) above we have

$$1! = 1 \cdot (1 - 1)!$$

or

$$1! = 1 \cdot 0!.$$

Therefore, we make the following definition for consistency:

$$0! = 1.$$

Note that both 1! and 0! are equal to 1.

The Binomial Coefficient

Earlier we introduced the notation $\binom{n}{k}$ for the numerical coefficients of the expansion of $(a + b)^n$. These binomial coefficients, which are given by Pascal's triangle, can also be expressed using factorial notation as follows:

$$\binom{n}{k} = \frac{n!}{(n-k)!\,k!}. \qquad (4)$$

EXAMPLE 8 Evaluate each binomial coefficient.

a. $\binom{6}{2}$ b. $\binom{9}{8}$

Solutions

a. $\binom{6}{2} = \dfrac{6!}{(6-2)!\,2!} = \dfrac{6!}{4!\,2!}$

$= \dfrac{6 \cdot 5 \cdot 4!}{4! \cdot 2 \cdot 1} = 15$

b. $\binom{9}{8} = \dfrac{9!}{(9-8)!\,8!} = \dfrac{9!}{1!\,8!}$

$= \dfrac{9 \cdot 8!}{1 \cdot 8!} = 9$

If we need to find all the terms in a binomial expansion, as in Example 8 above, it probably is still faster to use Pascal's triangle rather than the binomial formula to find the numerical coefficients. However, sometimes we need only one specific term of a binomial expansion. In this case Equation (4) is quite useful.

EXAMPLE 9 Find the numerical coefficient of x^6 in the expansion of $(x - 2)^{10}$.

Solution Apply Equation (2) on page 593 with $a = x$, $b = -2$, and $n = 10$:

$$(x - 2)^{10} = \sum_{k=0}^{10} \binom{n}{k} x^{10-k}(-2)^k.$$

The exponent on x in the formula is $10 - k$, so we are interested in the term for which $10 - k = 6$, or $k = 4$. This term is given by

$$\binom{10}{4} x^6 (-2)^4,$$

and by Equation (4) the coefficient of x^6 is

$$\frac{10!}{(10-4)!\, 4!}(-2)^4,$$

or

$$\frac{10 \cdot 9 \cdot 8 \cdot 7 \cdot 6!}{6!\, 4 \cdot 3 \cdot 2 \cdot 1}(16) = 3360.$$

Thus, the coefficient of x^6 is 3360.

Note in Equation (2) that the first term of the expansion corresponds to $k = 0$, the second term corresponds to $k = 1$, and in general the rth term of a binomial expansion corresponds to $k = r - 1$.

EXAMPLE 10 Find the seventh term in the expansion of $(2x + a^2)^9$.

Solution Use Equation (2), with $a = 2x$, $b = a^2$, $n = 9$, and $k = 7 - 1 = 6$. The seventh term is

$$\frac{9!}{3!\, 6!}(2x)^3 (a^2)^6 = \frac{9 \cdot 8 \cdot 7 \cdot 6!}{3 \cdot 2 \cdot 1 \cdot 6!}(8x^3)(a^{12})$$

$$= 672 x^3 a^{12}.$$

EXERCISE 12.5

A

■ Write each power with sigma notation and in expanded form, using $\binom{n}{k}$ for the binomial coefficients. See Example 1.

1. $(x + y)^5$
2. $(x + y)^4$
3. $(x - 2y)^4$
4. $(2x - y)^7$
5. $(x^2 - 3)^7$
6. $(1 - y^2)^5$

■ Write each power in expanded form, using Pascal's triangle to find the binomial coefficients. See Examples 2 and 3.

7. $(x + 3)^5$
8. $(2x + y)^4$
9. $(x - 3)^4$
10. $(2x - 1)^5$
11. $\left(2x - \dfrac{y}{2}\right)^3$
12. $\left(\dfrac{x}{3} + 3\right)^5$
13. $\left(\dfrac{x}{2} + 2\right)^6$
14. $\left(\dfrac{2}{3} - a^2\right)^4$

■ Write each expression in expanded form. See Example 4.

15. $(2n)!$ for $n = 4$
16. $(3n)!$ for $n = 4$
17. $2n!$ for $n = 4$
18. $3n!$ for $n = 4$
19. $n(n - 1)!$ for $n = 6$
20. $2n(2n - 1)!$ for $n = 2$

■ Write in expanded form and simplify. See Example 5.

21. $5!$
22. $7!$
23. $\dfrac{9!}{7!}$
24. $\dfrac{12!}{11!}$
25. $\dfrac{5!\,7!}{8!}$
26. $\dfrac{12!\,8!}{16!}$
27. $\dfrac{8!}{2!\,(8 - 2)!}$
28. $\dfrac{10!}{4!\,(10 - 4)!}$

■ Write each product in factorial notation. See Example 6.

29. $1 \cdot 2 \cdot 3$
30. $1 \cdot 2 \cdot 3 \cdot 4 \cdot 5$
31. $3 \cdot 4 \cdot 5 \cdot 6$
32. 7
33. $8 \cdot 7 \cdot 6$
34. $28 \cdot 27 \cdot 26 \cdot 25 \cdot 24$

■ Write each expression in factored form and show the first three factors and the last three factors. See Example 7.

35. $n!$
36. $(n + 4)!$
37. $(3n)!$
38. $3n!$
39. $(n - 2)!$
40. $(3n - 2)!$

■ Evaluate each binomial coefficient. See Example 8.

41. $\binom{9}{6}$
42. $\binom{8}{5}$
43. $\binom{12}{3}$
44. $\binom{13}{4}$
45. $\binom{20}{18}$
46. $\binom{18}{16}$

12.5 ■ THE BINOMIAL EXPANSION

■ *Find the coefficient of the indicated term. See Example 9.*

47. $(x + y)^{20}$; $x^{13}y^7$
48. $(x - y)^{15}$; $x^{12}y^3$
49. $(a - 2b)^{12}$; a^5b^7
50. $(2a - b)^{12}$; a^8b^4
51. $(x - \sqrt{2})^{10}$; x^4
52. $\left(\dfrac{x}{2} + 2\right)^8$; x^6

■ *Find each specified term. See Example 10.*

53. $(a - b)^{15}$; sixth term
54. $(x + 2)^{12}$; fifth term
55. $(x - 2y)^{10}$; fifth term
56. $(a^3 - b)^9$; seventh term
57. $(x^2 - y^2)^7$; third term
58. $\left(x - \dfrac{1}{2}\right)^8$; fourth term
59. $\left(\dfrac{a}{2} - 2b\right)^9$; fifth term
60. $\left(\dfrac{x}{2} + 4\right)^{10}$; eighth term

B

■ *Write the first four terms in each expansion.*

61. $\left(a - \dfrac{b}{2}\right)^8$
62. $\left(b - \dfrac{a}{3}\right)^7$
63. $(x^3 + 3y)^{10}$
64. $(x^2 + 2y)^9$
65. $(2\sqrt{x} - 1)^9$
66. $\left(\dfrac{1}{x} - 4\right)^6$
67. $\left(x + \dfrac{1}{x}\right)^{13}$
68. $(2 + \sqrt{y})^{11}$

■ *For Problems 69–72, find each power to the nearest hundredth. [Hint: Write the given power as $(1 + x)^n$ and use the first three terms of the binomial expansion.]*

69. $(1.02)^{10}$
70. $(1.01)^{15}$
71. $(0.99)^8$
72. $(0.95)^8$

73. Given that the binomial formula holds for $(1 + x)^n$, where n is a negative integer, do the following.

   a. Write the first four terms of $(1 + x)^{-1}$. *Hint:* Use the following alternate formula for the binomial coefficient:
   $$\binom{n}{k} = \dfrac{n(n - 1)(n - 2) \cdots (n - k + 1)}{k!}.$$

   b. Find the first four terms of the quotient $\dfrac{1}{1 + x}$ by dividing $(1 + x)$ into 1.

   c. Compare the results of (a) and (b).

74. Given that the binomial formula holds as an infinite "sum" for $(1 + x)^n$, where n is a noninteger rational number and $|x| < 1$, find the following to two decimal places.

   a. $\sqrt{1.02}$
   b. $\sqrt{0.99}$

   (See the hint for Problem 73.)

CHAPTER REVIEW

A

[12.1]
- Find the first four terms in a sequence with the general term as given.

1. $s_n = \dfrac{n}{n^2 - 1}$
2. $s_n = \dfrac{(-1)^{n-1}}{n}$

- Find the indicated term for each sequence.

3. $s_n = (-1)^n(n-2)^2$; s_7
4. $s_n = \sqrt{n^3 - 2}$; s_3

- Find the first five terms of the recursive sequence.

5. $s_1 = 5$; $s_{n+1} = s_{n-3}$
6. $s_1 = 1$; $s_{n+1} = -\dfrac{3}{4}s_n$

- Write a recursive sequence to describe each situation. Find the first four terms of each sequence.

7. Rick purchased a sailboat for $1800. How much is it worth after n years if it depreciates in value by 12% each year?
8. Sally earns $24,000 per year. If she receives a 6% raise each year, what will her salary be after n years?

[12.2]
- Find the indicated series associated with each sequence.

9. $s_n = (-1)^n(n+1)$; S_5
10. $s_n = \dfrac{n}{n+1}$; S_4

- Write each sum using sigma notation.

11. The first 12 terms of $1, 3, 7, \ldots, 2^k - 1, \ldots$
12. The fourth through fifteenth terms of $x, 4x^2, 9x^3, \ldots, k^2 x^k, \ldots$

- Write in expanded form.

13. $\displaystyle\sum_{k=2}^{5} k(k-1)$
14. $\displaystyle\sum_{j=2}^{\infty} \dfrac{j}{2j - 1}$

[12.3]
- Find the first four terms of each arithmetic sequence.

15. $a = 3$, $d = -4$
16. $a = \dfrac{1}{4}$, $d = \dfrac{1}{2}$

■ *Write the next three terms in each arithmetic progression. Find an expression for the general term.*

17. $-4, 0$
18. $x - a, x + a$

19. Find the twenty-third term of the arithmetic progression $-82, -74, -66, \ldots$.
20. Find the ninth term of the arithmetic progression $-\frac{1}{2}, 1, \frac{5}{2}, \ldots$.
21. The first term of an arithmetic progression is 8 and the twenty-eighth term is 89. Find the twenty-first term.
22. What term in the arithmetic progression $5, 2, -1, \ldots$ is -37?
23. Find the sum of the first 12 terms of the sequence $s_n = 3n - 2$.
24. Find the sum of the first 20 terms of the sequence $s_n = 1.4 + 0.1n$.

■ *Find the sum of each series.*

25. $\sum_{i=1}^{6} 3i - 1$
26. $\sum_{k=1}^{12} \frac{2}{3}k - 1$

27. a. Find the sum of all integral multiples of 6 between 10 and 100.
    b. Write the sum in (a) using sigma notation.
28. Kathy planted a 7-foot silver maple tree in 1988. If the tree grows 1.3 feet each year, in what year will it be 20 feet tall?

[12.4]

■ *Identify the following sequences as arithmetic, geometric, or neither.*

29. $-1, \frac{1}{2}, -\frac{1}{4}, \frac{1}{8}, \ldots$
30. $12, 9, 3, 1, \ldots$
31. $6, 1, -4, -9, \ldots$
32. $1, -4, 16, -64, \ldots$

■ *Find the first four terms of each sequence.*

33. $s_n = 3(-2)^{n+1}$
34. $s_n = 2\left(\frac{1}{3}\right)^{n-2}$

■ *Write the next three terms in each geometric progression. Find the general term.*

35. $-1, 2, -4, \ldots$
36. $\frac{2}{3}, \frac{1}{2}, \frac{3}{8}, \ldots$

37. Find the eighth term of the geometric progression $\frac{16}{27}, -\frac{8}{9}, \frac{4}{3}, \ldots$.
38. Find the fifth term of a geometric progression with third term $-\frac{2}{3}$ and sixth term $\frac{16}{81}$.
39. Find the sum of the first six terms of $s_n = 3(-2)^n$.
40. Find the sum of the first four terms of $3(\frac{2}{3})^{n-1}$.

■ *Find each sum.*

41. $\sum_{j=1}^{5} \left(\frac{1}{3}\right)^j$

42. $\sum_{k=1}^{6} (2)^{k-1}$

43. A rubber ball is dropped from a height of 12 feet and returns two-thirds of its previous height on each bounce. How high does the ball bounce after hitting the floor for the fourth time?

44. The property taxes on the Hardestys' family home were $840 in 1988. If the taxes increase by 2% each year, what will the taxes be in 1994?

■ *Find the sum of each infinite geometric series.*

45. $-3, 2, -4/3, \ldots$

46. $\sum_{i=1}^{\infty} 3\left(\frac{1}{3}\right)^{i-1}$

[12.5]

■ *Write each power in expanded form, using Pascal's triangle to find the binomial coefficients.*

47. $(x - 2)^5$

48. $\left(\frac{x}{2} - y\right)^4$

49. Write the first three factors and the last three factors of $(2n - 3)!$.

50. Write $\dfrac{6!}{3!(6-2)!}$ in expanded form and simplify.

■ *Evaluate each binomial coefficient.*

51. $\binom{7}{2}$

52. $\binom{16}{14}$

53. Find the fourth term of $(x - 2y)^9$.

54. Find the sixth term of $\left(\dfrac{x}{2} - 3\right)^8$.

B

55. Find the sum $\sum_{k=2}^{15} (6 - 0.2k)$.

56. Find three numbers that form an arithmetic sequence such that their sum is 21 and their product is 280.

■ *Find a fraction equivalent to each decimal.*

57. $3.2222\overline{2}$

58. $0.4181 \overline{81}$

59. If a ball is dropped from a height of 12 feet and returns three-fourths of its preceding height on each bounce, approximately what is the total distance the ball travels before coming to rest?

60. Find $(0.98)^6$ to the nearest hundredth.

Summary for Part III

The section in which the symbol or property is first used is shown in parentheses.

Symbols

(x, y, z): an ordered triple (9.2)

$$\begin{vmatrix} a_1 & b_1 \\ a_2 & b_2 \end{vmatrix} = a_1 b_2 - a_2 b_1:$$
second-order determinant (9.3)

$$\begin{vmatrix} a_1 & b_1 & c_1 \\ a_2 & b_2 & c_2 \\ a_3 & b_3 & c_3 \end{vmatrix} = a_1 \begin{vmatrix} b_2 & c_2 \\ b_3 & c_3 \end{vmatrix} - b_1 \begin{vmatrix} a_2 & c_2 \\ a_3 & c_3 \end{vmatrix} + c_1 \begin{vmatrix} a_2 & b_2 \\ a_3 & b_3 \end{vmatrix}:$$
third-order determinant (9.3)

For the system
$$a_1 x + b_1 y + c_1 z = d_1$$
$$a_2 x + b_2 y + c_2 z = d_2$$
$$a_3 x + b_3 y + c_3 z = d_3,$$

$$\begin{bmatrix} a_1 & b_1 & c_1 \\ a_2 & b_2 & c_2 \\ a_3 & b_3 & c_3 \end{bmatrix}$$ is the coefficient matrix (9.4)

$$\begin{bmatrix} a_1 & b_1 & c_1 & | & d_1 \\ a_2 & b_2 & c_2 & | & d_2 \\ a_3 & b_3 & c_3 & | & d_3 \end{bmatrix}$$ is the augmented matrix. (9.4)

∩: "the intersection of" (10.1)
∅: empty set, or null set (10.1)
∪: "the union of" (10.1)
$|a - b|$: distance between a and b (10.2)
s_n: nth term of a sequence (12.1)
S_n: sum of n terms of a sequence (12.2)

$$\sum_{k=1}^{n} a_k = a_1 + a_2 + \cdots + a_n: \text{ sum} \quad (12.2)$$

S_∞: infinite sum (12.2)
$n! = n(n-1)(n-2) \cdots \cdot 3 \cdot 2 \cdot 1:$
    n factorial (12.5)

$$\binom{n}{k} = \frac{n!}{k!(n-k)!}: \text{ binomial coefficient} \quad (12.5)$$

Properties

Systems of Linear Equations

A system of the form
$$a_1 x + b_1 y = c_1$$
$$a_2 x + b_2 y = c_2$$
has one and only one solution if
$$\frac{a_1}{a_2} \neq \frac{b_1}{b_2},$$
has no solution if
$$\frac{a_1}{a_2} = \frac{b_1}{b_2} \neq \frac{c_1}{c_2},$$
and has an infinite number of solutions if
$$\frac{a_1}{a_2} = \frac{b_1}{b_2} = \frac{c_1}{c_2}. \qquad (9.1)$$

Cramer's Rule

Cramer's rule for $a_1 x + b_1 y = c_1$
$\qquad\qquad\qquad\qquad a_2 x + b_2 y = c_2$:

$$x = \frac{D_x}{D} = \frac{\begin{vmatrix} c_1 & b_1 \\ c_2 & b_2 \end{vmatrix}}{\begin{vmatrix} a_1 & b_1 \\ a_2 & b_2 \end{vmatrix}};$$

$$y = \frac{D_y}{D} = \frac{\begin{vmatrix} a_1 & c_1 \\ a_2 & c_2 \end{vmatrix}}{\begin{vmatrix} a_1 & b_1 \\ a_2 & b_2 \end{vmatrix}}. \qquad (9.3)$$

Cramer's rule for a linear system of three equations in three variables:

$$x = \frac{D_x}{D}; \quad y = \frac{D_y}{D}; \quad z = \frac{D_z}{D}. \qquad (9.3)$$

Elementary Row Operations

A matrix can be written as a row-equivalent matrix by the following.

1. Multiplying the entries of any row by a nonzero real number.
2. Interchanging two rows.
3. Multiplying the entries of any row by a real number and adding the results to the corresponding elements of another row. (9.4)

Equations and Inequalities Involving Absolute Values

1. $|ax + b| = c \quad (c \geq 0)$
   is equivalent to
   $ax + b = c \quad \text{or} \quad -(ax + b) = c\,;$
2. $|ax + b| < c$
   is equivalent to
   $-c < ax + b < c\,;$
3. $|ax + b| > c$
   is equivalent to
   $ax + b > c \quad \text{or} \quad ax + b < -c\,.$ (10.2)

Analytic Geometry

Distance formula:
$d = \sqrt{(x_2 - x_1)^2 + (y_2 - y_1)^2}$ (11.1)

Midpoint formula:
$(\bar{x}, \bar{y}) = \left(\dfrac{x_1 + x_2}{2}, \dfrac{y_1 + y_2}{2}\right)$ (11.1)

Conic Sections

Name of Curve	Standard Form of Equation
Circle (11.1)	$(x - h)^2 + (y - k)^2 = r^2$
Ellipse: Major axis parallel to x-axis Major axis parallel to y-axis (11.1)	$\dfrac{(x - h)^2}{a^2} + \dfrac{(y - k)^2}{b^2} = 1$ $\dfrac{(x - h)^2}{b^2} + \dfrac{(y - k)^2}{a^2} = 1$
Parabola: Opens upward Opens downward Opens to the right Opens to the left (11.2)	$(x - h)^2 = 4p(y - k)$ $(x - h)^2 = -4p(y - k)$ $(y - k)^2 = 4p(x - h)$ $(y - k)^2 = -4p(x - h)$
Hyperbola: Transverse axis parallel to x-axis Transverse axis parallel to y-axis (11.3)	$\dfrac{(x - h)^2}{a^2} - \dfrac{(y - k)^2}{b^2} = 1$ $\dfrac{(y - k)^2}{a^2} - \dfrac{(x - h)^2}{b^2} = 1$

Sequences and Series

Arithmetic progression with common difference d:
$$s_n = a + (n - 1)d\,; \quad (12.3)$$
$$S_n = \dfrac{n}{2}(a + s_n)$$
$$= \dfrac{n}{2}[2a + (n - 1)d]\,. \quad (12.3)$$

Geometric progression with common ratio r ($r \neq 1$):
$$s_n = ar^{n-1}; \quad S_n = \dfrac{a(1 - r^n)}{1 - r}.$$

Infinite geometric progression with $|r| < 1$:
$$S_\infty = \dfrac{a}{1 - r}. \quad (12.4)$$

$$n! = n(n - 1)!; \quad 0! = 1 \quad (12.5)$$

Binomial expansion:
$$(a + b)^n = \sum_{k=0}^{n} \binom{n}{k} a^{n-k} b^k. \quad (12.5)$$

Cumulative Review Exercises for Part III

■ Solve the systems by linear combinations.

1. $\frac{1}{3}x + \frac{1}{2}y = 2$
   $\frac{1}{2}x - \frac{3}{4}y = 9$ [9.1]

2. $x + 3y - 2z = 2$
   $2x - 2y + z = 4$
   $x - 4y + 3z = -2$ [9.2]

■ Solve the inequalities and graph the solution.

3. $\frac{x-3}{x+2} \leq -4$ [10.3]

4. $x^2 + 12 > 7x$ [10.3]

■ Solve the systems by Cramer's rule.

5. $\frac{1}{4}x - \frac{2}{3}y = 6$
   $2x + 3y = -2$ [9.3]

6. $3x + 4y - z = 2$
   $\frac{1}{2}x + 2y + z = 5$
   $6x + y - \frac{3}{4}z = 9$ [9.3]

■ Solve the systems.

7. $2x^2 - xy + 2y^2 = 3$
   $2x - y = 1$ [11.4]

8. $x^2 - y^2 = 8$
   $xy = 3$ [11.4]

■ Solve, and graph the solution.

9. $|6 - 4x| \geq 8$ [10.2]

10. $2x^2 + 5x \geq 3$ [10.3]

■ Solve.

11. $0.75 \leq \frac{x - 1.3}{2} < 2.1$ [10.1]

12. $|2 - 3x| = 7$ [10.2]

13. Graph the inequality
$$3x - y < 2. \quad [10.4]$$

14. Graph the system
$$x - 3y \geq 3$$
$$2x + y < 3. \quad [10.4]$$

■ *Evaluate the following sums.*

15. $\sum_{k=10}^{100} (3k - 1) \quad [12.3]$

16. $\sum_{i=4}^{8} (-3)^{i-4} \quad [12.4]$

■ *Graph.*

17. $9x^2 + 8y^2 + 72x - 16y + 80 = 0 \quad [11.1]$
18. $x^2 - 2y^2 - 4x - 8y - 8 = 0 \quad [11.3]$
19. $x^2 + y^2 - 6x + 2y + 6 = 0 \quad [11.1]$
20. $x^2 - 4x - 2y - 2 = 0 \quad [11.2]$

21. Find the focus and directrix of $y = -2x^2 + x + 3$. $\quad [11.2]$
22. Write an equation for the ellipse with center at $(-2, 3)$, $a = 4$, $b = 2$, and major axis horizontal. $\quad [11.1]$

■ *Solve using matrices.*

23. $x + 3z = 2$
    $2y - 3z = 5$
    $2x - 7y = 3 \quad [9.4]$

24. $3x + 2y + z = -4$
    $x + y - z = 5$
    $2x - y - 3z = 5 \quad [9.4]$

25. Write in expanded form: $\sum_{k=2}^{8} (-1)^k \frac{2k - 1}{k - 1}$. $\quad [12.2]$
26. Write in sigma notation: $1 + 4x + 9x^2 + 16x^3 + \cdots$. $\quad [12.2]$
27. Find the sum of the infinite geometric series $-2, \frac{3}{2}, -\frac{9}{8}, \ldots$. $\quad [12.4]$
28. Find the first four terms of the sequence $s_n = \frac{(-1)^{n-1} 2^n}{n^2}$. $\quad [12.1]$
29. Find the equation of the circle whose graph contains the points $(4, 2)$, $(-2, -6)$, and $(-3, 1)$. $\quad [11.1]$
30. Find an equation for the circle whose diameter has endpoints $(2, -3)$ and $(-6, 1)$. $\quad [11.1]$
31. Find the sum of the first nine terms of the sequence $s_n = 4n - 1$. $\quad [12.3]$
32. Find the eighth term of $(2x - y^3)^{11}$. $\quad [12.5]$

■ *Solve the systems.*

33. $x + 2y = 3z$
    $x + \frac{1}{2}y + \frac{1}{2}z = \frac{1}{2}$
    $\frac{2}{3}z + x = \frac{1}{3}y - 1 \quad [9.2]$

34. $\frac{3}{x} + \frac{4}{y} = -1$
    $\frac{2}{3x} + \frac{5}{2y} = 3 \quad [9.1]$

■ Solve.

35. $3x^2 + 5y^2 = 63$
    $2x^2 - 3y^2 = 23$ [11.4]

36. $\dfrac{4}{2x+1} > \dfrac{3}{x-2}$ [10.3]

37. Evaluate $\begin{vmatrix} a & b & c \\ 0 & a & b \\ 0 & 0 & a \end{vmatrix}$. [9.3]

38. Solve for n: $\binom{n}{4} = \binom{n}{2}$. [12.5]

■ Write absolute-value inequalities to express the following.

39. x is within 0.5 unit of p. [10.2]
40. The length x of a steel cotter pin is 1.5 centimeters ± 0.01 centimeter. [10.2]
41. If the third term of an arithmetic progression is $-\frac{9}{2}$ and the thirteenth term is 3, what is the eighth term? [12.3]
42. Find the sum of all integral multiples of 14 between 25 and 500. [12.3]
43. Determine how many solutions the system has:

$$3x = 2y + 2$$
$$4y - 6x = 4. \quad [9.1]$$

44. Solve the compound inequality:

$$2x - 3 < 5 \quad \text{and} \quad 1 - 3x \leq 7. \quad [10.1]$$

45. Find the distance between $(-4, 2)$ and $(2, -10)$. [11.1]
46. Find the following sets. [10.1]
    a. $[-6, -1) \cap [-2, 2)$
    b. $(1, 5] \cup (4, \infty)$
47. Find a fraction equivalent to $1.6\overline{25}$. [12.4]
48. Use the binomial formula to estimate $\sqrt{4.1}$ to two decimal places. [12.5]
49. Give the first five terms of the recursive sequence

$$s_1 = -2, \quad s_{n+1} = s_n^2 - 4. \quad [12.1]$$

50. Find a formula for the general term of the sequence

$$\frac{9}{2}, -3, 2, \ldots. \quad [12.3]$$

51. Graph, and find the coordinates of the vertices:
    $y \leq x + 4$
    $3x + 2y \leq 18$
    $x \geq 0, \ y \geq 0.$ [10.4]

52. Graph:
    $|x| + 2|y| < 4.$ [10.4]

53. Student tickets to a concert sell for $5 and general admission tickets sell for $16. Three hundred and fifty tickets were sold, bringing in $4940. How many of each kind of ticket were sold?   [9.1]

54. One angle of a triangle measures twice the second angle, and the third angle measures 10° more than the larger of the other two. Find the measure of each angle.   [9.3]

55. The area of a rectangle is 208 square meters and its perimeter is 58 meters. Find the dimensions of the rectangle.   [11.4]

56. A company ships its product to three cities: Boston, Chicago, and Los Angeles. The cost of shipping is $10 per crate to Boston, $5 per crate to Chicago, and $12 per crate to Los Angeles. The company's shipping bill for April was $445. It shipped 55 crates in all, with twice as many crates going to Boston as to Los Angeles. How many crates were shipped to each destination?   [9.2]

57. Claudia made $16,000 this year, and she is scheduled to receive a 5% cost of living increase every year for the next 5 years. How much will Claudia make over the 6-year period?   [12.4]

58. Farmer Bob has 48 rows of corn planted in a trapezoidal field. There are 40 stalks of corn in the first row, 43 stalks in the second row, 46 stalks in the third row, and so on, for the 48 rows. How many stalks of corn are in the field?   [12.3]

59. Show that $\begin{vmatrix} x & y & 1 \\ x_1 & y_1 & 1 \\ x_2 & y_2 & 1 \end{vmatrix} = 0$ is the equation of the line passing through the points (x_1, y_1) and (x_2, y_2).   [9.3]

60. Show that $\begin{vmatrix} a_1 & b_1 \\ a_2 & b_2 \end{vmatrix} = \begin{vmatrix} a_1 & b_1 \\ a_1 + a_2 & b_1 + b_2 \end{vmatrix}$.   [9.3]

Appendix: Using Exponential and Logarithmic Tables

Before the advent of electronic calculators, exponential and logarithmic function values were usually obtained from tables. The following discussion on the use of tables is provided for students who are not using calculators. If tables are used for the exercises in Chapter 8, the answers obtained may vary slightly from those given in the answer section because of the greater accuracy available with calculators.

We shall first give our attention to logarithms and powers to base 10.

Common Logarithms

Recall from Section 8.3 that $\log_{10} x$ is the exponent that must be placed on 10 so that the resulting power is x. Values for $\log_{10} x$ are sometimes called **common logarithms**.

Values for $\log_{10} 10^k$, $k \in J$

Some values of $\log_{10} x$ can be obtained simply by considering the definition of a logarithm, while other values require tables. Let us first consider values of $\log_{10} x$ for all values of x that are integral powers of 10. These can be obtained by inspection.

$$
\begin{aligned}
\text{Since} \quad & 10^3 = 1000, & \log_{10} 1000 &= 3; \\
\text{since} \quad & 10^2 = 100, & \log_{10} 100 &= 2; \\
\text{since} \quad & 10^1 = 10, & \log_{10} 10 &= 1; \\
\text{since} \quad & 10^0 = 1, & \log_{10} 1 &= 0; \\
\text{since} \quad & 10^{-1} = 0.1, & \log_{10} 0.1 &= -1; \\
\text{since} \quad & 10^{-2} = 0.01, & \log_{10} 0.01 &= -2; \\
\text{since} \quad & 10^{-3} = 0.001, & \log_{10} 0.001 &= -3.
\end{aligned}
$$

Notice that the logarithm of a power of 10 is simply the exponent on the base 10. For example,

$$\log_{10} 100 = \log_{10} 10^2 = 2,$$
$$\log_{10} 0.01 = \log_{10} 10^{-2} = -2,$$

and so on.

EXAMPLE 1 Find each logarithm.

  a. $\log_{10} 10^6$ b. $\log_{10} 10^{-5}$

Solutions By definition we have the following.

  a. $\log_{10} 10^6$ is the exponent on 10 so that the power equals 10^6. Hence,

$$\log_{10} 10^6 = 6.$$

  b. $\log_{10} 10^{-5}$ is the exponent on 10 so that the power equals 10^{-5}. Hence,

$$\log_{10} 10^{-5} = -5.$$

Values for $\log_{10} x$, $1 < x < 10$

Table I on page 622 gives values for $\log_{10} x$ for $1 < x < 10$. Consider the following excerpt on page 614. Each number in the column headed x represents the first two digits of x, while each of the other column-head numbers represents the third significant digit of x. The number located at the intersection of a row and a column is the logarithm of x. For example, to find $\log_{10} 4.25$ we look at the intersection of the row starting with 4.2 and the column headed by 5. Thus,

$$\log_{10} 4.25 = 0.6284.$$

Similarly,

$$\log_{10} 4.02 = 0.6042,$$
$$\log_{10} 4.49 = 0.6522,$$

and so on.

x	0	1	2	3	4	5	6	7	8	9
3.8	.5798	.5809	.5821	.5832	.5843	.5855	.5866	.5877	.5888	.5899
3.9	.5911	.5922	.5933	.5944	.5955	.5966	.5977	.5988	.5999	.6010
4.0	.6021	.6031	.6042	.6053	.6064	.6075	.6085	.6096	.6107	.6117
4.1	.6128	.6138	.6149	.6160	.6170	.6180	.6191	.6201	.6212	.6222
4.2	.6232	.6243	.6253	.6263	.6274	.6284	.6294	.6304	.6314	.6325
4.3	.6335	.6345	.6355	.6365	.6375	.6385	.6395	.6405	.6415	.6425
4.4	.6435	.6444	.6454	.6464	.6474	.6484	.6493	.6503	.6513	.6522
4.5	.6532	.6542	.6551	.6561	.6571	.6580	.6590	.6599	.6609	.6618
4.6	.6628	.6637	.6646	.6656	.6665	.6675	.6684	.6693	.6702	.6712

The values in the tables are rational-number approximations of irrational numbers. We shall follow customary usage and write = instead of ≈.

EXAMPLE 2 Find an approximation for each logarithm using Table I.

a. $\log_{10} 1.68 = 0.2253$ b. $\log_{10} 4.3 = 0.6335$

Values for $\log_{10} x$, $x > 10$

Now, suppose we wish to find $\log_{10} x$ for values of x outside the range of the table—that is, for $x > 10$ or $0 < x < 1$. This can be done quite readily by first representing the number in scientific notation and then applying the first law of logarithms.

EXAMPLE 3
a. $\log_{10} = 42.5$
$= \log_{10}(4.25 \times 10^1)$
$= \log_{10} 4.25 + \log_{10} 10^1$
$= 0.6284 + 1$
$= 1.6284$

b. $\log_{10} 425$
$= \log_{10}(4.25 \times 10^2)$
$= \log_{10} 4.25 + \log_{10} 10^2$
$= 0.6284 + 2$
$= 2.6284$

c. $\log_{10} 4250$
$= \log_{10}(4.25 \times 10^3)$
$= \log_{10} 4.25 + \log_{10} 10^3$
$= 0.6284 + 3$
$= 3.6284$

d. $\log_{10} 42,500$
$= \log_{10}(4.25 \times 10^4)$
$= \log_{10} 4.25 + \log_{10} 10^4$
$= 0.6284 + 4$
$= 4.6284$

Observe that in Example 3 the integral portions of the logarithms are the exponents on 10 when the numbers are written in scientific notation. This integral part of $\log_{10} x$ is called the **characteristic**, and the (nonnegative) decimal fraction part is called the **mantissa**. Since the first factor of a number in scientific notation is always a number between 1 and 10, the mantissa (its logarithm) can always be found in Table I. Thus, we have the following steps for calculating $\log_{10} x$, where $x > 10$.

1. Write the number in scientific notation.
2. Use the first law of logarithms to write the logarithm as a sum of two terms: the mantissa and the characteristic.
3. Find the mantissa in Table I.
4. The characteristic is the exponent on 10.

Values for $\log_{10} x$, $0 < x < 1$

Now consider an example of the form $\log_{10} x$ for $0 < x < 1$. To find $\log_{10} 0.00425$ we write

$$\log_{10} 0.00425 = \log_{10}(4.25 \times 10^{-3})$$
$$= \log_{10} 4.25 + \log_{10} 10^{-3}.$$

We find from the table that $\log_{10} 4.25 = 0.6284$. On adding 0.6284 to the characteristic -3 we obtain

$$\log_{10} 0.00425 = 0.6284 + (-3)$$
$$= -2.3716,$$

where the decimal part of the logarithm is no longer 0.6284. The decimal part is -0.3716, a negative number.

However, if we want to use the table, which contains only positive entries, it is more convenient to write the logarithm in a form in which the decimal part is positive. In the example above we write

$$\log_{10} 0.00425 = 0.6284 - 3,$$

where the decimal part is positive. Because -3 can be written $1 - 4, 2 - 5, 3 - 6, 7 - 10$, and so on, the forms $1.6284 - 4, 2.6284 - 5, 3.6284 - 6, 7.6284 - 10$, and so on, are equally valid representations of the desired logarithm. It will sometimes be convenient to use these alternative forms.

EXAMPLE 4 a. $\log_{10} 0.294$
$= \log_{10}(2.94 \times 10^{-1})$
$= \log_{10} 2.94 + \log_{10} 10^{-1}$
$= 0.4683 - 1$

b. $\log_{10} 0.00294$
$= \log_{10}(2.94 \times 10^{-3})$
$= \log_{10} 2.94 + \log_{10} 10^{-3}$
$= 0.4683 - 3$

Antilog₁₀ N

Given a value for an exponent, $\log_{10} x$, we can use Table I to find the power x by reversing the process described to find the logarithm of a number. In this case the power x is called the **antilogarithm** of $\log_{10} x$. For example, if

$$\log_{10} x = 0.4409,$$

then

$$x = \text{antilog}_{10} 0.4409,$$

which can be obtained by locating 0.4409 in the body of Table I and observing that

$$\log_{10} 2.76 = 0.4409,$$

or

$$\text{antilog}_{10} 0.4409 = 2.76.$$

Note that taking the antilogarithm of a number is the same as exponentiating the number with base 10. That is,

$$\text{antilog}_{10} x = 10^x.$$

In particular,

$$\text{antilog}_{10} 0.4409 = 10^{0.4409} = 2.76.$$

If the $\log_{10} x$ is greater than 1, it can first be written as the *sum* of a positive decimal (the mantissa) and a positive integer (the characteristic). Antilog₁₀ x can then be written as the *product* of a number between 1 and 10 and a power of 10.

EXAMPLE 5   a. If $\log_{10} x = 2.7364,$ then

$$x = \text{antilog}_{10} 2.7364 = \text{antilog}_{10}(0.7364 + 2).$$

Locate the mantissa 0.7364 in the body of Table I and determine the associated antilog₁₀ (a number between 1 and 10). Write the characteristic 2 as an exponent on the factor with base 10:

$$x = \text{antilog}_{10}(0.7364 + 2)$$
$$= 5.45 \times 10^2$$
$$= 545.$$

b. If $\log_{10} x = 0.4409 - 3$, then

$$x = \text{antilog}_{10}(0.4409 - 3)$$
$$= 2.76 \times 10^{-3}$$
$$= 0.00276.$$

Common Error

Note that in Example 5a

$$\text{antilog}_{10}(0.7364 + 2) = 5.45 \times 10^2,$$

not $5.45 + 10^2$. This is because

$$\text{antilog}_{10}(0.7364 + 2) = 10^{(0.7364+2)}$$
$$= 10^{0.7364} \times 10^2$$

by the first law of exponents.

If the decimal part of $\log_{10} x$ is negative, we cannot use the table directly to obtain x. However, we can first write $\log_{10} x$ equivalently with a positive decimal part. For example, to find

$$\text{antilog}_{10}(-0.4522) \quad \text{or} \quad \text{antilog}_{10}(-2.4522)$$

we first add $(+1 - 1)$ to write -0.4522 as

$$-0.4522 + 1 - 1 = 0.5478 - 1$$

and then use the tables. Thus,

$$\text{antilog}_{10}(-0.4522) = \text{antilog}_{10}(0.5478 - 1)$$
$$= 3.53 \times 10^{-1} = 0.353.$$

To find $\text{antilog}_{10}(-2.4522)$ we first add $(+3 - 3)$ to write -2.4522 as

$$-2.4522 + 3 - 3 = 0.5478 - 3$$

and then use the tables to find

$$\text{antilog}_{10}(-2.4522) = \text{antilog}_{10}(0.5478 - 3)$$
$$= 3.53 \times 10^{-3} = 0.00353.$$

EXAMPLE 6 Use Table I to find the value of x.

a. $\log_{10} x = -0.7292$
b. $\log_{10} x = -1.4634$

Solutions

a. $x = \text{antilog}_{10}(-0.7292)$
$= \text{antilog}_{10}(\underbrace{-0.7292 + 1}_{} - 1)$
$= \text{antilog}_{10}(0.2708 - 1)$
$= 1.87 \times 10^{-1} = 0.187$

b. $x = \text{antilog}_{10}(-1.4634)$
$= \text{antilog}_{10}(\underbrace{-1.4634 + 2}_{} - 2)$
$= \text{antilog}_{10}(0.5366 - 2)$
$= 3.44 \times 10^{-2} = 0.0344$

In the examples above the mantissas, 0.2718 and 0.5366, were listed in Table I. If we seek the common logarithm of a number that is not an entry in the table (for example, $\log_{10} 23.42$), or if we seek x when $\log_{10} x$ is not an entry in the table, we shall simply use the entry in the table that is closest to the value we seek.

Powers to the Base 10

By the definition of a logarithm,

$$P = 10^E$$

can be written in logarithmic form as

$$\log_{10} P = E,$$

from which we see that the power P is the antilogarithm of the exponent E,

$$P = \text{antilog}_{10} E.$$

Since $P = 10^E$,

$$10^E = \text{antilog}_{10} E$$

and we can obtain a power 10^E simply by finding the antilogarithm of the exponent E.

EXAMPLE 7 Compute each power.

a. $10^{0.2148}$
b. $10^{-1.6345}$

Solutions   a. $10^{0.2148}$

$= \text{antilog}_{10}\, 0.2148$

$= 1.64$

b. $10^{-1.6345}$

$= \text{antilog}_{10}(-1.6345)$

$= \text{antilog}_{10}(\underline{-1.6345 + 2} - 2)$

$= \text{antilog}_{10}(0.3645 - 2)$

$= 2.32 \times 10^{-2}$

Powers to the Base e

Because of the importance of base $e \approx 2.7182818$, special tables are available for values of e^x and $\log_e x$.

Table II gives approximations for powers e^x and e^{-x} for $0 \leq x \leq 1.00$ in 0.01 intervals and for $1.00 < x \leq 10.00$ in 0.1 intervals.

EXAMPLE 8   Using Table II we have the following.

a. $e^{2.4} = 11.023$

b. $e^{-4.7} = 0.0091$

Although we can obtain values for e^x and e^{-x} outside the interval 0 to 10 by using the table along with the first law of exponents, at this time the function values in the table will be adequate for our work.

Natural Logarithms, ln x

The most-used values for $\log_e x$ $(x > 0)$ are printed in Table III. These values, like those for e^x, e^{-x}, and $\log_{10} x$, are *approximations* accurate to the number of decimals shown. The symbol $\log_e x$ is often written as **ln x** and read as "natural logarithm of x." Unlike Table I for common logarithms, which provides only the decimal part of $\log_{10} x$, the table for natural logarithms gives the entire value for ln x, both the integral and decimal portions.

EXAMPLE 9   Using Table III we have the following.

a. $\ln 6.6 = 1.8871$

b. $\ln 0.7 = -0.3567$

Powers to the Base e, Antilogarithms

If we seek a value for e^x or $\ln x$ for a value of x between two entries in the tables, we shall simply use the entry in the table that is closest to the value we seek.

Given an exponent $\ln x$ we can obtain the power x by finding the value of $\ln x$ in the body of Table II.

EXAMPLE 10 a. $\ln x = 1.3$ b. $\ln x = -0.47$

Solutions a. $\ln x = 1.3$ is equivalent to
$$x = e^{1.3}$$
$$= 3.6693.$$

b. $\ln x = -0.47$ is equivalent to
$$x = e^{-0.47}$$
$$= 0.6250.$$

As noted above, a power to the base b is called the antilog_b of the exponent. Thus, in Example 10a,

$$x = e^{1.3} = \text{antilog}_e\ 1.3$$
$$= 3.6693.$$

EXERCISE

A

- *Find each logarithm by inspection. See Example 1.*

1. $\log_{10} 10^2$
2. $\log_{10} 10^4$
3. $\log_{10} 10^{-4}$
4. $\log_{10} 10^{-6}$
5. $\log_{10} 10^0$
6. $\log_{10} 10^n$

- *Find an approximation for each logarithm using Table I. See Examples 2, 3, and 4.*

7. $\log_{10} 6.73$
8. $\log_{10} 891$
9. $\log_{10} 83.7$
10. $\log_{10} 21.4$
11. $\log_{10} 317$
12. $\log_{10} 219$
13. $\log_{10} 0.813$
14. $\log_{10} 0.00214$
15. $\log_{10} 0.08$
16. $\log_{10} 0.000413$
17. $\log_{10}(2.48 \times 10^2)$
18. $\log_{10}(5.39 \times 10^{-3})$

■ *Solve for x using Table I. See Example 5.*

19. $\log_{10} x = 0.6128$
20. $\log_{10} x = 0.2504$
21. $\log_{10} x = 1.5647$
22. $\log_{10} x = 3.9258$
23. $\log_{10} x = 0.8075 - 2$
24. $\log_{10} x = 0.9722 - 3$
25. $\log_{10} x = 7.8562 - 10$
26. $\log_{10} x = 1.8155 - 4$

■ *For Problems 27–32, see Example 6.*

27. $\log_{10} x = -0.5272$
28. $\log_{10} x = -0.4123$
29. $\log_{10} x = -1.2984$
30. $\log_{10} x = -1.0545$
31. $\log_{10} x = -2.6882$
32. $\log_{10} x = -2.0670$

■ *Compute each power. See Example 7.*

33. $10^{0.8762}$
34. $10^{1.6405}$
35. $10^{2.8943}$
36. $10^{4.3766}$
37. $10^{-1.4473}$
38. $10^{-2.0958}$

■ *Compute each power. See Example 8.*

39. $e^{0.43}$
40. $e^{0.62}$
41. $e^{-0.57}$
42. $e^{-0.08}$
43. $e^{1.5}$
44. $e^{2.6}$
45. $e^{-2.4}$
46. $e^{-1.2}$

■ *Find each logarithm. See Example 9.*

47. $\ln 3.9$
48. $\ln 6.3$
49. $\ln 16$
50. $\ln 55$
51. $\ln 0.4$
52. $\ln 0.7$

■ *Find each value of x. See Example 10.*

53. $\ln x = 0.16$
54. $\ln x = 0.25$
55. $\ln x = 1.8$
56. $\ln x = 2.4$
57. $\ln x = 4.5$
58. $\ln x = 6.0$

TABLE I

VALUES OF $\log_{10} x$ AND ANTILOG$_{10}$ x OR (10^x)

x	0	1	2	3	4	5	6	7	8	9
1.0	.0000	.0043	.0086	.0128	.0170	.0212	.0253	.0294	.0334	.0374
1.1	.0414	.0453	.0492	.0531	.0569	.0607	.0645	.0682	.0719	.0755
1.2	.0792	.0828	.0864	.0899	.0934	.0969	.1004	.1038	.1072	.1106
1.3	.1139	.1173	.1206	.1239	.1271	.1303	.1335	.1367	.1399	.1430
1.4	.1461	.1492	.1523	.1553	.1584	.1614	.1644	.1673	.1703	.1732
1.5	.1761	.1790	.1818	.1847	.1875	.1903	.1931	.1959	.1987	.2014
1.6	.2041	.2068	.2095	.2122	.2148	.2175	.2201	.2227	.2253	.2279
1.7	.2304	.2330	.2355	.2380	.2405	.2430	.2455	.2480	.2504	.2529
1.8	.2553	.2577	.2601	.2625	.2648	.2672	.2695	.2718	.2742	.2765
1.9	.2788	.2810	.2833	.2856	.2878	.2900	.2923	.2945	.2967	.2989
2.0	.3010	.3032	.3054	.3075	.3096	.3118	.3139	.3160	.3181	.3201
2.1	.3222	.3243	.3263	.3284	.3304	.3324	.3345	.3365	.3385	.3404
2.2	.3424	.3444	.3464	.3483	.3502	.3522	.3541	.3560	.3579	.3598
2.3	.3617	.3636	.3655	.3674	.3692	.3711	.3729	.3747	.3766	.3784
2.4	.3802	.3820	.3838	.3856	.3874	.3892	.3909	.3927	.3945	.3962
2.5	.3979	.3997	.4014	.4031	.4048	.4065	.4082	.4099	.4116	.4133
2.6	.4150	.4166	.4183	.4200	.4216	.4232	.4249	.4265	.4281	.4298
2.7	.4314	.4330	.4346	.4362	.4378	.4393	.4409	.4425	.4440	.4456
2.8	.4472	.4487	.4502	.4518	.4533	.4548	.4564	.4579	.4594	.4609
2.9	.4624	.4639	.4654	.4669	.4683	.4698	.4713	.4728	.4742	.4757
3.0	.4771	.4786	.4800	.4814	.4829	.4843	.4857	.4871	.4886	.4900
3.1	.4914	.4928	.4942	.4955	.4969	.4983	.4997	.5011	.5024	.5038
3.2	.5051	.5065	.5079	.5092	.5105	.5119	.5132	.5145	.5159	.5172
3.3	.5185	.5198	.5211	.5224	.5237	.5250	.5263	.5276	.5289	.5302
3.4	.5315	.5328	.5340	.5353	.5366	.5378	.5391	.5403	.5416	.5428
3.5	.5441	.5453	.5465	.5478	.5490	.5502	.5514	.5527	.5539	.5551
3.6	.5563	.5575	.5587	.5599	.5611	.5623	.5635	.5647	.5658	.5670
3.7	.5682	.5694	.5705	.5717	.5729	.5740	.5752	.5763	.5775	.5786
3.8	.5798	.5809	.5821	.5832	.5843	.5855	.5866	.5877	.5888	.5899
3.9	.5911	.5922	.5933	.5944	.5955	.5966	.5977	.5988	.5999	.6010
4.0	.6021	.6031	.6042	.6053	.6064	.6075	.6085	.6096	.6107	.6117
4.1	.6128	.6138	.6149	.6160	.6170	.6180	.6191	.6201	.6212	.6222
4.2	.6232	.6243	.6253	.6263	.6274	.6284	.6294	.6304	.6314	.6325
4.3	.6335	.6345	.6355	.6365	.6375	.6385	.6395	.6405	.6415	.6425
4.4	.6435	.6444	.6454	.6464	.6474	.6484	.6493	.6503	.6513	.6522
4.5	.6532	.6542	.6551	.6561	.6571	.6580	.6590	.6599	.6609	.6618
4.6	.6628	.6637	.6646	.6656	.6665	.6675	.6684	.6693	.6702	.6712
4.7	.6721	.6730	.6739	.6749	.6758	.6767	.6776	.6785	.6794	.6803
4.8	.6812	.6821	.6830	.6839	.6848	.6857	.6866	.6875	.6884	.6893
4.9	.6902	.6911	.6920	.6928	.6937	.6946	.6955	.6964	.6972	.6981
5.0	.6990	.6998	.7007	.7016	.7024	.7033	.7042	.7050	.7059	.7067
5.1	.7076	.7084	.7093	.7101	.7110	.7118	.7126	.7135	.7143	.7152
5.2	.7160	.7168	.7177	.7185	.7193	.7202	.7210	.7218	.7226	.7235
5.3	.7243	.7251	.7259	.7267	.7275	.7284	.7292	.7300	.7308	.7316
5.4	.7324	.7332	.7340	.7348	.7356	.7364	.7372	.7380	.7388	.7396
x	0	1	2	3	4	5	6	7	8	9

Table I *(continued)*

x	0	1	2	3	4	5	6	7	8	9
5.5	.7404	.7412	.7419	.7427	.7435	.7443	.7451	.7459	.7466	.7474
5.6	.7482	.7490	.7497	.7505	.7513	.7520	.7528	.7536	.7543	.7551
5.7	.7559	.7566	.7574	.7582	.7589	.7597	.7604	.7612	.7619	.7627
5.8	.7634	.7642	.7649	.7657	.7664	.7672	.7679	.7686	.7694	.7701
5.9	.7709	.7716	.7723	.7731	.7738	.7745	.7752	.7760	.7767	.7774
6.0	.7782	.7789	.7796	.7803	.7810	.7818	.7825	.7832	.7839	.7846
6.1	.7853	.7860	.7868	.7875	.7882	.7889	.7896	.7903	.7910	.7917
6.2	.7924	.7931	.7938	.7945	.7952	.7959	.7966	.7973	.7980	.7987
6.3	.7993	.8000	.8007	.8014	.8021	.8028	.8035	.8041	.8048	.8055
6.4	.8062	.8069	.8075	.8082	.8089	.8096	.8102	.8109	.8116	.8122
6.5	.8129	.8136	.8142	.8149	.8156	.8162	.8169	.8176	.8182	.8189
6.6	.8195	.8202	.8209	.8215	.8222	.8228	.8235	.8241	.8248	.8254
6.7	.8261	.8267	.8274	.8280	.8287	.8293	.8299	.8306	.8312	.8319
6.8	.8325	.8331	.8338	.8344	.8351	.8357	.8363	.8370	.8376	.8382
6.9	.8388	.8395	.8401	.8407	.8414	.8420	.8426	.8432	.8439	.8445
7.0	.8451	.8457	.8463	.8470	.8476	.8482	.8488	.8494	.8500	.8506
7.1	.8513	.8519	.8525	.8531	.8537	.8543	.8549	.8555	.8561	.8567
7.2	.8573	.8579	.8585	.8591	.8597	.8603	.8609	.8615	.8621	.8627
7.3	.8633	.8639	.8645	.8651	.8657	.8663	.8669	.8675	.8681	.8686
7.4	.8692	.8698	.8704	.8710	.8716	.8722	.8727	.8733	.8739	.8745
7.5	.8751	.8756	.8762	.8768	.8774	.8779	.8785	.8791	.8797	.8802
7.6	.8808	.8814	.8820	.8825	.8831	.8837	.8842	.8848	.8854	.8859
7.7	.8865	.8871	.8876	.8882	.8887	.8893	.8899	.8904	.8910	.8915
7.8	.8921	.8927	.8932	.8938	.8943	.8949	.8954	.8960	.8965	.8971
7.9	.8976	.8982	.8987	.8993	.8998	.9004	.9009	.9015	.9020	.9025
8.0	.9031	.9036	.9042	.9047	.9053	.9058	.9063	.9069	.9074	.9079
8.1	.9085	.9090	.9096	.9101	.9106	.9112	.9117	.9122	.9128	.9133
8.2	.9138	.9143	.9149	.9154	.9159	.9165	.9170	.9175	.9180	.9186
8.3	.9191	.9196	.9201	.9206	.9212	.9217	.9222	.9227	.9232	.9238
8.4	.9243	.9248	.9253	.9258	.9263	.9269	.9274	.9279	.9284	.9289
8.5	.9294	.9299	.9304	.9309	.9315	.9320	.9325	.9330	.9335	.9340
8.6	.9345	.9350	.9355	.9360	.9365	.9370	.9375	.9380	.9385	.9390
8.7	.9395	.9400	.9405	.9410	.9415	.9420	.9425	.9430	.9435	.9440
8.8	.9445	.9450	.9455	.9460	.9465	.9469	.9474	.9479	.9484	.9489
8.9	.9494	.9499	.9504	.9509	.9513	.9518	.9523	.9528	.9533	.9538
9.0	.9542	.9547	.9552	.9557	.9562	.9566	.9571	.9576	.9581	.9586
9.1	.9590	.9595	.9600	.9605	.9609	.9614	.9619	.9624	.9628	.9633
9.2	.9638	.9643	.9647	.9652	.9657	.9661	.9666	.9671	.9675	.9680
9.3	.9685	.9689	.9694	.9699	.9703	.9708	.9713	.9717	.9722	.9727
9.4	.9731	.9736	.9741	.9745	.9750	.9754	.9759	.9763	.9768	.9773
9.5	.9777	.9782	.9786	.9791	.9795	.9800	.9805	.9809	.9814	.9818
9.6	.9823	.9827	.9832	.9836	.9841	.9845	.9850	.9854	.9859	.9863
9.7	.9868	.9872	.9877	.9881	.9886	.9890	.9894	.9899	.9903	.9908
9.8	.9912	.9917	.9921	.9926	.9930	.9934	.9939	.9943	.9948	.9952
9.9	.9956	.9961	.9965	.9969	.9974	.9978	.9983	.9987	.9991	.9996
x	0	1	2	3	4	5	6	7	8	9

TABLE II

VALUES OF e^x

x	e^x	e^{-x}	x	e^x	e^{-x}
0.00	1.0000	1.0000	0.50	1.6487	0.6065
0.01	1.0101	0.9901	0.51	1.6653	0.6005
0.02	1.0202	0.9802	0.52	1.6820	0.5945
0.03	1.0305	0.9705	0.53	1.6990	0.5886
0.04	1.0408	0.9608	0.54	1.7160	0.5827
0.05	1.0513	0.9512	0.55	1.7333	0.5769
0.06	1.0618	0.9418	0.56	1.7507	0.5712
0.07	1.0725	0.9324	0.57	1.7683	0.5655
0.08	1.0833	0.9231	0.58	1.7860	0.5599
0.09	1.0942	0.9139	0.59	1.8040	0.5543
0.10	1.1052	0.9048	0.60	1.8221	0.5488
0.11	1.1163	0.8958	0.61	1.8404	0.5434
0.12	1.1275	0.8869	0.62	1.8590	0.5380
0.13	1.1388	0.8781	0.63	1.8776	0.5326
0.14	1.1503	0.8694	0.64	1.8965	0.5273
0.15	1.1618	0.8607	0.65	1.9155	0.5220
0.16	1.1735	0.8521	0.66	1.9348	0.5169
0.17	1.1853	0.8437	0.67	1.9542	0.5117
0.18	1.1972	0.8353	0.68	1.9739	0.5066
0.19	1.2092	0.8270	0.69	1.9937	0.5016
0.20	1.2214	0.8187	0.70	2.0138	0.4966
0.21	1.2337	0.8106	0.71	2.0340	0.4916
0.22	1.2461	0.8025	0.72	2.0544	0.4868
0.23	1.2586	0.7945	0.73	2.0751	0.4819
0.24	1.2712	0.7866	0.74	2.0959	0.4771
0.25	1.2840	0.7788	0.75	2.1170	0.4724
0.26	1.2969	0.7711	0.76	2.1383	0.4677
0.27	1.3100	0.7634	0.77	2.1598	0.4630
0.28	1.3231	0.7558	0.78	2.1815	0.4584
0.29	1.3364	0.7483	0.79	2.2034	0.4538
0.30	1.3499	0.7408	0.80	2.2255	0.4493
0.31	1.3634	0.7334	0.81	2.2479	0.4449
0.32	1.3771	0.7261	0.82	2.2705	0.4404
0.33	1.3910	0.7190	0.83	2.2933	0.4360
0.34	1.4050	0.7118	0.84	2.3164	0.4317
0.35	1.4191	0.7047	0.85	2.3396	0.4274
0.36	1.4333	0.6977	0.86	2.3632	0.4232
0.37	1.4477	0.6907	0.87	2.3869	0.4190
0.38	1.4623	0.6839	0.88	2.4109	0.4148
0.39	1.4770	0.6771	0.89	2.4351	0.4107
0.40	1.4918	0.6703	0.90	2.4596	0.4066
0.41	1.5068	0.6636	0.91	2.4843	0.4025
0.42	1.5220	0.6570	0.92	2.5093	0.3985
0.43	1.5373	0.6505	0.93	2.5345	0.3946
0.44	1.5527	0.6440	0.94	2.5600	0.3906
0.45	1.5683	0.6376	0.95	2.5857	0.3867
0.46	1.5841	0.6313	0.96	2.6117	0.3829
0.47	1.6000	0.6250	0.97	2.6379	0.3791
0.48	1.6160	0.6188	0.98	2.6645	0.3753
0.49	1.6323	0.6126	0.99	2.6912	0.3716

Table II *(continued)*

x	e^x	e^{-x}
1.0	2.7183	0.3679
1.1	3.0042	0.3329
1.2	3.3201	0.3012
1.3	3.6693	0.2725
1.4	4.0552	0.2466
1.5	4.4817	0.2231
1.6	4.9530	0.2019
1.7	5.4739	0.1827
1.8	6.0496	0.1653
1.9	6.6859	0.1496
2.0	7.3891	0.1353
2.1	8.1662	0.1225
2.2	9.0250	0.1108
2.3	9.9742	0.1003
2.4	11.023	0.0907
2.5	12.182	0.0821
2.6	13.464	0.0743
2.7	14.880	0.0672
2.8	16.445	0.0608
2.9	18.174	0.0550
3.0	20.086	0.0498
3.1	22.198	0.0450
3.2	24.533	0.0408
3.3	27.113	0.0369
3.4	29.964	0.0334
3.5	33.115	0.0302
3.6	36.598	0.0273
3.7	40.447	0.0247
3.8	44.701	0.0224
3.9	49.402	0.0202
4.0	54.598	0.0183
4.1	60.340	0.0166
4.2	66.686	0.0150
4.3	73.700	0.0136
4.4	81.451	0.0123
4.5	90.017	0.0111
4.6	99.484	0.0101
4.7	109.95	0.0091
4.8	121.51	0.0082
4.9	134.29	0.0074
5.0	148.41	0.0067
5.1	164.02	0.0061
5.2	181.27	0.0055
5.3	200.34	0.0050
5.4	221.41	0.0045

x	e^x	e^{-x}
5.5	244.69	0.0041
5.6	270.43	0.0037
5.7	298.87	0.0034
5.8	330.30	0.0030
5.9	365.04	0.0027
6.0	403.43	0.0025
6.1	445.86	0.0022
6.2	492.75	0.0020
6.3	544.57	0.0018
6.4	601.85	0.0017
6.5	665.14	0.0015
6.6	735.10	0.0014
6.7	812.41	0.0012
6.8	897.85	0.0011
6.9	992.27	0.0010
7.0	1096.6	0.0009
7.1	1212.0	0.0008
7.2	1339.5	0.0007
7.3	1480.3	0.0007
7.4	1636.0	0.0006
7.5	1808.0	0.0006
7.6	1998.2	0.0005
7.7	2208.4	0.0005
7.8	2440.6	0.0004
7.9	2697.3	0.0004
8.0	2981.0	0.0003
8.1	3294.5	0.0003
8.2	3641.0	0.0003
8.3	4023.9	0.0002
8.4	4447.1	0.0002
8.5	4914.8	0.0002
8.6	5431.7	0.0002
8.7	6002.9	0.0002
8.8	6634.2	0.0002
8.9	7332.0	0.0001
9.0	8103.1	0.0001
9.1	8955.3	0.0001
9.2	9897.1	0.0001
9.3	10938	0.0001
9.4	12088	0.0001
9.5	13360	0.0001
9.6	14765	0.0001
9.7	16318	0.0001
9.8	18034	0.0001
9.9	19930	0.0001

TABLE III

VALUES OF ln x

x	ln x	x	ln x	x	ln x
		4.5	1.5041	9.0	2.1972
0.1	−2.3026	4.6	1.5261	9.1	2.2083
0.2	−1.6094	4.7	1.5476	9.2	2.2192
0.3	−1.2040	4.8	1.5686	9.3	2.2300
0.4	−0.9163	4.9	1.5892	9.4	2.2407
0.5	−0.6931	5.0	1.6094	9.5	2.2513
0.6	−0.5108	5.1	1.6292	9.6	2.2618
0.7	−0.3567	5.2	1.6487	9.7	2.2721
0.8	−0.2231	5.3	1.6677	9.8	2.2824
0.9	−0.1054	5.4	1.6864	9.9	2.2925
1.0	0.0000	5.5	1.7047	10	2.3026
1.1	0.0953	5.6	1.7228	11	2.3979
1.2	0.1823	5.7	1.7405	12	2.4849
1.3	0.2624	5.8	1.7579	13	2.5649
1.4	0.3365	5.9	1.7750	14	2.6391
1.5	0.4055	6.0	1.7918	15	2.7081
1.6	0.4700	6.1	1.8083	16	2.7726
1.7	0.5306	6.2	1.8245	17	2.8332
1.8	0.5878	6.3	1.8405	18	2.8904
1.9	0.6419	6.4	1.8563	19	2.9444
2.0	0.6931	6.5	1.8718	20	2.9957
2.1	0.7419	6.6	1.8871	25	3.2189
2.2	0.7885	6.7	1.9021	30	3.4012
2.3	0.8329	6.8	1.9169	35	3.5553
2.4	0.8755	6.9	1.9315	40	3.6889
2.5	0.9163	7.0	1.9459	45	3.8067
2.6	0.9555	7.1	1.9601	50	3.9120
2.7	0.9933	7.2	1.9741	55	4.0073
2.8	1.0296	7.3	1.9879	60	4.0943
2.9	1.0647	7.4	2.0015	65	4.1744
3.0	1.0986	7.5	2.0149	70	4.2485
3.1	1.1314	7.6	2.0281	75	4.3175
3.2	1.1632	7.7	2.0412	80	4.3820
3.3	1.1939	7.8	2.0541	85	4.4427
3.4	1.2238	7.9	2.0669	90	4.4998
3.5	1.2528	8.0	2.0794	100	4.6052
3.6	1.2809	8.1	2.0919	110	4.7005
3.7	1.3083	8.2	2.1041	120	4.7875
3.8	1.3350	8.3	2.1163	130	4.8676
3.9	1.3610	8.4	2.1282	140	4.9416
4.0	1.3863	8.5	2.1401	150	5.0106
4.1	1.4110	8.6	2.1518	160	5.0752
4.2	1.4351	8.7	2.1633	170	5.1358
4.3	1.4586	8.8	2.1748	180	5.1930
4.4	1.4816	8.9	2.1861	190	5.2470

Answers to Odd-numbered Exercises in Each Section; Answers to All Exercises in Reviews

Exercise 1.1 [page 8] 1. Q, R 3. H, R 5. J, Q, R 7. W, J, Q, R 9. Q, R
11. H, R 13. -25 15. 81 17. -64 19. 6 21. -42 23. 5 25. -2
27. -2 29. 60 31. 10 33. 1 35. 50 37. -2 39. -5 41. $\dfrac{19}{30}$
43. 2 45. 100 47. 4 49. 1080 51. 64 53. 0 55. 72.5904
57. 108.39742 59. 57.060248 61. a. 7.23456 cubic meters   b. 6.1544 square centimeters
63. a. 2622.528 cubic meters   b. 1902.4632 square inches
65. a. $P = 2l + 2w$ b. 56 centimeters
67. a. $P = \dfrac{kt}{v}$
   b. 40 pounds per square inch
69. a. Interest: Prt
   amount: $P(1 + rt)$
   b. $\$1016.00$

Exercise 1.2 [page 17] 1. Binomial; third degree; $2, -1$ 3. Monomial; fourth degree; 5
5. Trinomial; second degree; $3, -1, 2$ 7. Trinomial; third degree; $1, -2, -1$ 9. (b), (c)
11. $7x^2$ 13. $-3y^3$ 15. z^2 17. $7x^2y - 2x$ 19. $6r^2 + 4r$ 21. $s^2 - 4s$
23. $t^2 + 3t - 2$ 25. $-2u^2 - u - 3$ 27. $-3x^3 + 2x^2 - 4x + 7$
29. $-2t^4 - 5t^3 - 4t^2 + 3t - 6$ 31. $3a^2 + a - 9$ 33. $8x^2y - xy + 3xy^2$
35. $-2x^2 + x - 3$ 37. $b^2 + 3b + 1$ 39. $-2y - 1$ 41. $2 - x$ 43. $x - 1$
45. $-x^2 - 3x - 1$ 47. $-2x^2 + x - 4$ 49. $2x - y$ 51. $-4x - 5$
53. a. -1 b. -21 55. a. $\dfrac{11}{4}$ b. $\dfrac{1}{9}$ 57. a. 14.016 b. -63.248
59. a. 2 b. 96 61. a. $-16t^2 + 16t + 8$ b. 8 feet; 12 feet
63. a. $2x^2 + 4xy$ b. 1224 square inches; 8.5 square feet
65. a. $0.12x^2 + 0.08xy$ b. $\$50.40$ 67. a. $-0.005x^2 + 8x - 400$ b. $\$2600$
69. a. Weekly profit: $-0.08x^2 + 200x - 200$ b. $\$11{,}512.00$
71. a. $6x^2 - \dfrac{5}{4}\pi x^2$ b. 132.8 square inches   73. a. $V = \dfrac{2}{3}\pi r^3 + \pi r^2 h$ b. $\dfrac{17}{3}\pi r^3$

A-1

Exercise 1.3 [page 26] 1. y^6z^3 3. $4x^2y^4z^2$ 5. $-8a^3b^9c^3$ 7. $-14t^3$
9. $-40a^3b^3c$ 11. $6x^5y^5$ 13. $-2r^6s^4t^2$ 15. $3x^2y^6z^4$ 17. $6r^3t^4$ 19. $x^4y^2 + x^3y^3$
21. $4r^2s^4t^2 - r^2s^2t^4$ 23. $m^2n^2 - m^3n^4$ 25. $4x^5y^3 + xy^2$ 27. $8x^2y^2 - 3x^5y^2$
29. $2u^4v^2$ 31. $4xy - 8y^2$ 33. $-12x^3 + 6x^2 - 6x$ 35. $3a^4b - 2a^3b^2 - a^2b^2$
37. $8x^3y^7 - 4x^4y^4 - 6x^5y^5$ 39. $y^3 - y + 6$ 41. $x^3 + 2x^2 - 21x + 18$ 43. $x^3 - 7x + 6$
45. $z^3 - 7z - 6$ 47. $6x^3 + x^2 - 8x + 6$ 49. $6a^4 - 5a^3 - 5a^2 + 5a - 1$
51. $n^2 + 10n + 16$ 53. $r^2 + 3r - 10$ 55. $2z^2 - 5z - 3$ 57. $8r^2 + 2rs - 3s^2$
59. $6x^2 - 13xy + 6y^2$ 61. $9t^2 - 16s^2$ 63. $2a^4 - 5a^2b^2 - 3b^4$ 65. 6   67. $-2a^2 - 2a$
69. $4x + 4$ 71. $-4x^2 - 11x$ 73. $8x - 16$ 75. $x^4 - 2x^3 + x^2 - 3x + 4$
77. a. $\frac{1}{2}n^2 - \frac{1}{2}n$ b. 45   79. a. $-10x^2 + 1200x$ b. \$20,000.00
81. a. $x + 60$; $-\frac{1}{2}x + 12$ b. $-\frac{1}{2}x^2 - 18x + 720$ c. 490 bushels;   850 bushels
83. a. $x + 20$; $-15x + 500$ b. $-15x^2 + 200x + 10,000$ c. \$10,625.00;   \$10,500.00
85. a^{3n-3} 87. y^{n+10} 89. $x^{6n}y^3$ 91. x^{6n+3} 93. $2x^{2n} - x^n$ 95. $a^{3n+1} + a^{2n+2}$
97. $-a^{2n} + a^n + 2$ 99. $2a^{2n} + 3a^nb^n - 2b^{2n}$

Exercise 1.4 [page 36] 1. $4xz(x + 2)$ 3. $3n^2(n^2 - 2n + 4)$ 5. $3r(5rs + 6s^2 - 1)$
7. $m^2n^2(3n^2 - 6mn + 14m)$ 9. $3a^2b^2c^4(5a^2b - 4c + 2b)$ 11. $(a + b)(a + 3)$
13. $(y - 3x)(y - 2)$ 15. $4(x - 2)^2(-2x^2 + 4x + 1)$ 17. $x(x - 5)^2(-x^2 + 5x + 1)$
19. $(x - 1)^2(-x - 2)$ 21. $2(x + 3)(3x^2 + 6x - 5)$ 23. $-(2n - 3m)$ 25. $-2(x - 1)$
27. $-a(b + c)$ 29. $-(-2x + y - 3z)$ 31. $(x + 2)(x + 3)$ 33. $(y - 3)(y - 4)$
35. $(x - 3)(x + 2)$ 37. $(2x - 1)(x + 2)$ 39. $(4x - 1)(x + 2)$ 41. $(3y + 1)(3y - 8)$
43. $(2u + 1)(5u - 3)$ 45. $(3x - 7)(7x + 2)$ 47. $(9a + 4)(8a - 3)$
49. $(2x - 3)(15x - 4)$ 51. $2(3t + 2)(9t - 11)$ 53. $(x - 2a)(3x - a)$
55. $(3x - 2y)(5x + 2y)$ 57. $(3u - 4v)(6u - 5v)$ 59. $(3a + 2b)(4a - 7b)$
61. $(5ab - 2)(2ab - 3)$ 63. $2(4xy + 1)(7xy - 2)$ 65. $(2az - 3)(11az + 7)$
67. $(a + b)(x + 1)$ 69. $(x + y)(x - a)$ 71. $(3x + y)(1 - 2x)$ 73. $(a^2 + 2b^2)(a - 2b)$
75. $(3x - 2y)(2xy + 1)$ 77. $(x^3 - 3)(y^2 + 1)$ 79. $(x + 2)(x^2 + 4)$ 81. $(2x - 3)(x^2 + 1)$
83. $3y(x + 2)^2$ 85. $a(2a + 1)(a + 7)$ 87. $40(a - b)^2$ 89. $xy(3x - y)(2x - 3y)$
91. $3u(2uv - 1)(uv - 2)$ 93. $2s^2t^2(2st - 1)(3st - 1)$ 95. $x^n(x^n - 1)$
97. $x^n(x^n + 1)(x^n - 2)$ 99. $2x^2(x^n + 2x^{n-2} - 1)$

Exercise 1.5 [page 43] 1. $x^2 + 6x + 9$ 3. $4y^2 - 20y + 25$ 5. $x^2 - 9$ 7. $9t^2 - 16s^2$
9. $25a^2 - 10ab + 4b^2$ 11. $64x^2z^2 + 48xz + 9$ 13. $2x^2 + 12x + 2$
15. $-2x^3 + 12x^2 - 11x$ 17. $-8x^3 + 6x^2 + 6x$ 19. $(x + 5)(x - 5)$ 21. $(x - 12)^2$
23. $(x + 2y)(x - 2y)$ 25. $(2x + 3)^2$ 27. $(3u - 5v)^2$ 29. $(2a + 5b)(2a - 5b)$
31. $(xy + 9)(xy - 9)$ 33. $(3xy + 1)^2$ 35. $(4xy - 1)(4xy + 1)$
37. $(x + 2 - y)(x + 2 + y)$ 39. $(x + 1 - y)(x + 1 + y)$ 41. $(y + x - 1)(y - x + 1)$

43. $(2x + 1 - 2y)(2x + 1 + 2y)$ 45. $x^3 - 1$ 47. $8x^3 + 1$ 49. $27a^3 - 8b^3$
51. $(x + 3)(x^2 - 3x + 9)$ 53. $(2x - y)(4x^2 + 2xy + y^2)$ 55. $(a - 2b)(a^2 + 2ab + 4b^2)$
57. $(xy - 1)(x^2y^2 + xy + 1)$ 59. $(3a + 4b)(9a^2 - 12ab + 16b^2)$
61. $(5ab - 1)(25a^2b^2 + 5ab + 1)$ 63. $(2x - y)(x^2 - xy + y^2)$
65. $(2y - x)(7x^2 + 8xy + 4y^2)$ 67. $2(3x^2 + 1)$ 69. $(y^2 + 3)(y^2 - 3)$
71. $(a^2 + 1)(a^2 + 2)$ 73. $(3x^2 + 1)(x^2 + 2)$ 75. $(x^2 + 4)(x + 2)(x - 2)$
77. $(x^3 + 4)(x - 1)(x^2 + x + 1)$ 79. $(u^2 - 2)(u^2 + 2)(u^2 + 1)(u + 1)(u - 1)$
81. $4y^3(x - 3)(x + 3)$ 83. $x^2y^2(x + 1)(x - 1)$ 85. $3b^2(2a^2 - 1)(4a^4 + 2a^2 + 1)$
87. $6a^2b^2(2a^2 + 1)(a + 1)(a - 1)$ 89. $9a^2x^4(x + a)(x - a)(x^2 + 2a^2)$
91. $2x^3(3x^3 + 1)(x^3 - 4)$ 93. $(x + y)(x^2 - xy + y^2)(x - y)(x^2 + xy + y^2)$
95. a. $500(1 + r)^2$; $500(1 + r)^3$; $500(1 + r)^4$ 97. a. Length, width: $9 - 2x$ feet; height: x feet
    b. $500r^2 + 1000r + 500$;   b. Volume: $4x^3 - 36x^2 + 81x$ cubic feet
       $500r^3 + 1500r^2 + 1500r + 500$;   c. Surface area: $81 - 4x^2$ square feet
       $500r^4 + 2000r^3 + 3000r^2 + 2000r + 500$
    c. $583.20; $629.86; $680.24
99. a. r centimeters, $r + 2$ centimeters, $r + 4$ centimeters
    b. $18\pi r^2$ cubic centimeters;
       $18\pi(r + 2)^2$ cubic centimeters;
       $18\pi(r + 4)^2$ cubic centimeters
    c. $54\pi r^2 + 216\pi r + 360\pi$ cubic centimeters
101. a. No   103. a. No
     b. $(x + y)(x - y)$ b. $(x - y)(x^2 + xy + y^2)$
     c. $x^2 - 2xy + y^2$ c. $x^3 - 3x^2y + 3xy^2 - y^3$
105. a. $(x + y)^2$ 107. a. $\frac{1}{2}x^2 - \frac{1}{2}y^2$
     b. $x^2 + 2xy + y^2$
     c. x^2, xy, xy, y^2 b. $\frac{1}{2}(x - y)(x + y)$
                                         c. 18 square feet

Chapter I Review [page 47] 1. -36 2. 36 3. -8 4. -8 5. 10 6. $\frac{-5}{2}$
7. 4 8. 15 9. -9 10. 13 11. 52 12. 3.675 13. $-x^2 - 3x + 7$
14. $y^3 - 3y^2 - 2y + 3$ 15. $4 - y$ 16. $12 - 3y$ 17. $y - 2x$ 18. $6y - 4$
19. -13 20. 1.416 21. 108 22. -36 23. $6a^6b^3$ 24. $-3a^5b^3$
25. $9x^3y^5 - x^3y$ 26. $4x^3y^3 + x^4y^2 - y^3$ 27. $2a^3b^2 - 3a^2b^3 + ab^3$
28. $-3a^4b + 2a^3b^3 + a^2b^3$ 29. $6x^2 + 8xy - 8y^2$ 30. $x^3 + x^2 - 10x + 8$
31. $8x^3 - 36x^2 + 54x - 27$ 32. $9x^6 - 1$ 33. $-2a^2 + a$ 34. $-3b^3$ 35. $-3a(a - 3)^2$
36. $2(3b - 1)(b + 1)^2$ 37. $(2x + 3)(7x - 1)$ 38. $(3xy + 1)(2xy - 5)$
39. $(3y + 2)(2x - 1)$ 40. $2(x^2 + 3)(x - 2)$ 41. $x^2(3x + 2)(x - 2)$

42. $xy(x + y)(x - 2y)$ 43. $y^n(y^n - 1)(y^{2n} + y^n + 1)$ 44. $x(x^n + x^{n-1} + 1)$
45. $2x^2 - 2x + 2$ 46. $y^3 - 5y^2 + y$ 47. $(2x + 7y)(2x - 7y)$ 48. $2y - 1$
49. $(a + b - 2)[(a + b)^2 + 2(a + b) + 4]$ 50. $[a^2 + a - b][a^2 - a(a^2 - b) + (a^2 - b)^2]$
51. $2{,}133{,}333\tfrac{1}{3}$ cubic yards 52. a. $\dfrac{2xy}{x + y}$ b. 8
53. a. $2w(w + 4) + 2hw + 2h(w + 4)$ square feet
    b. $2w(w + 4) + 4hw + 4h(w + 4) + 0.8w(w + 4)$
    c. $588.80
54. a. $-0.01x^2 + 6x - 700$ b. $-4700.00
55. a. $200x - 2x^2$ cents, or $2x - 0.02x^2$ dollars b.

Selling Price	Income
10¢	$18
20¢	$32
30¢	$42
40¢	$48
50¢	$50
60¢	$48
70¢	$42

56. $400 - 80x$ 57. a. $-5x^2 + 375x + 12{,}500$ dollars b. $16{,}280.00
58. a. $\dfrac{P}{4}r^2 + Pr + P$; $\dfrac{P}{8}r^3 + \dfrac{3P}{4}r^2 + \dfrac{3P}{2}r + P$ b. $530.45; $546.36
59. a. $4x^3 - 44x^2 + 120x$ cubic inches b. $-4x^2 + 120$ square inches
60. a. $16\pi r - 64\pi$ square feet; 1005.31 square feet
    b. $\dfrac{4}{3}\pi h^3 + 4\pi r^2 h + 4\pi r h^2$ cubic units; 5.347×10^{10} cubic miles
61. $2a^{2n} - a^{n+1}$ 62. $3a^{2n} - 7a^n - 6$ 63. $6b^{5n-5}$ 64. b^{6n-3} 67. $2x^n(1 - 2x^n)$
68. $x^n(2x + 1)(3x - 1)$
69. $A_1 = x^2 - 2\left(\dfrac{1}{2}y^2\right) = x^2 - y^2$;   70. $A_1 = \pi\left(\dfrac{2x + 2y}{2}\right)^2 - (\pi y^2 + \pi x^2)$
    $A_2 = (x - y)(x + y) = x^2 - y^2$; $= \pi[(x + y)^2 - (x^2 + y^2)]$
    therefore, $A_1 = A_2$. $= \pi(2xy) = 2\pi xy$;
    $A_2 = 2x(\pi y) = 2\pi xy$;
    therefore, $A_1 = A_2$.

Exercise 2.1 [page 58] 1. $\dfrac{3}{5}$ 3. $\dfrac{-3}{7}$ 5. $\dfrac{3x}{4y}$; undefined for $y = 0$

7. $\dfrac{-x - 1}{x}$; undefined for $x = 0$ 9. $\dfrac{y - 7}{3y + 2}$; undefined for $y = \dfrac{-2}{3}$

11. $\dfrac{3x - 4}{48 - 3x^2}$; undefined for $x = -4, 4$ 13. $\dfrac{4}{y - 3}$ 15. $\dfrac{-x - 1}{y - x}$ 17. $\dfrac{x - 2}{y - x}$

19. $\dfrac{a - 1}{3a + b}$ 21. $\dfrac{-2}{cd^2}$ 23. $\dfrac{2r}{t}$ 25. $\dfrac{x - 2}{3x^2}$ 27. $\dfrac{-5}{u^4}$ 29. $\dfrac{2x + 3}{3}$

31. $-3a^2 + 2a - 1$ 33. $\dfrac{2y^2 - 3x^2y}{3}$ 35. $\dfrac{6(1 + t)}{(1 - t)}$ 37. $y - 2$ 39. $\dfrac{-2}{y^2 + 3y + 9}$

41. $\dfrac{2(x^2 + 9)}{3(x + 3)}$ 43. $\dfrac{-3x - y}{3x - y}$ 45. $\dfrac{2x - 3}{x - 1}$ 47. $\dfrac{-2x - 3}{2x - 3}$ 49. $\dfrac{4y^2 + 6y + 9}{2y + 3}$

51. $\dfrac{3xy - 1}{2xy - 1}$ 53. $\dfrac{2a - b}{a - 2b}$ 55. $\dfrac{6(x^2 - 6)(x^2 + 1)}{(2x^2 + 3)(x^2 + 4)}$ 57. $\dfrac{q(2pq + 1)}{p(3pq + 1)}$ 59. $\dfrac{x + y}{2}$

61. $\dfrac{x + y}{a + 2b}$ 63. $\dfrac{x + 3}{2x + 1}$ 65. $\dfrac{8(3x - 4)}{x^3}$ 67. $\dfrac{x^2 - 1}{4}$ 69. (b)   71. None

73. a. $\dfrac{25}{r + 8}$ b. $\dfrac{25}{r - 8}$ c. $\dfrac{50r}{r^2 - 64}$

75. a. $\dfrac{900}{400 + w}$ b. $\dfrac{900}{400 - w}$ c. Orville, $\dfrac{1800w}{160{,}000 - w^2}$

77. a. $\dfrac{600}{w}$ yards   b. $2w + \dfrac{1200}{w}$; no   c. $20w + \dfrac{12{,}000}{w}$ dollars

79. a. $\dfrac{900}{w(w + 2)}$ b. $w^2 + 2w + \dfrac{3600(w + 1)}{w^2 + 2w}$

Exercise 2.2 [page 68]   1. $\dfrac{10}{3}$ 3. $\dfrac{25p}{n}$ 5. $\dfrac{-np^2}{2}$ 7. $\dfrac{2x^5}{7}$ 9. $\dfrac{x^3y^3}{2}$ 11. $\dfrac{-x^3y^2z^2}{3}$

13. $\dfrac{-b^2}{a}$ 15. $\dfrac{3c}{35ab}$ 17. $\dfrac{5}{ab}$ 19. 5 21. $\dfrac{a(2a - 1)}{a + 4}$ 23. $\dfrac{(3x + 1)(x - 2)}{(3x - 1)(x - 1)}$

25. $\dfrac{-2(a - 3)}{7(a - 1)}$ 27. $\dfrac{6(x - 2)(x - 1)^2}{(x^2 - 8)(x^2 - 2x + 4)}$ 29. $\dfrac{8v(u - 2v)(u + 2)}{(u + 2v)(u - 3v)}$ 31. $\dfrac{1}{5}x^2 - 3x$

33. $x^2 + \dfrac{2}{3}x + \dfrac{1}{9}$ 35. $y^2 - \dfrac{1}{2}y + \dfrac{1}{16}$ 37. $\dfrac{2}{9}$ 39. $\dfrac{a + 1}{a - 2}$ 41. $\dfrac{3x - 1}{x - 2}$

43. $3(x^2 - xy + y^2)$ 45. $\dfrac{(y - 3)(x + 2)}{(x + 1)(x - 3)}$ 47. $\dfrac{x + 2}{x^2 - 1}$ 49. $\dfrac{x^2(x - 4)}{(x + 1)}$ 51. $\dfrac{x + 3}{6y}$

53. $6rs - 5 + \dfrac{2}{rs}$ 55. $-5s^8 + 7s^3 - \dfrac{2}{s^2}$ 57. $9a + 3 + \dfrac{4a}{b}$ 59. $2y + 5 + \dfrac{2}{2y + 1}$

61. $x^2 + 4x + 9 + \dfrac{19}{x - 2}$ 63. $4z^3 - 2z^2 + 3z + 1 + \dfrac{2}{2z + 1}$

65. $x^3 + 2x^2 + 4x + 8 + \dfrac{15}{x - 2}$ 67. $x - 1 + \dfrac{-7x + 12}{x^2 - 2x + 7}$

69. $4a^2 - 9a + 31 + \dfrac{-104a + 32}{a^2 + 3a - 1}$ 71. $t - 1 + \dfrac{-t^2 - 3t + 3}{t^3 - 2t^2 + t + 2}$ 73. $k = -2$

A-6 ANSWERS

Exercise 2.3 [page 76] 1. $\dfrac{x-3}{2}$ 3. $\dfrac{a+b-5c}{6}$ 5. $\dfrac{2x-1}{2y}$ 7. $\dfrac{-2x+1}{x+2y}$
9. $\dfrac{4}{a-1}$ 11. $\dfrac{6}{18x}$ 13. $\dfrac{-a^2b^2}{b^3}$ 15. $\dfrac{xy^2}{xy}$ 17. $\dfrac{3y^2-9y}{y^2-y-6}$ 19. $\dfrac{-3b-3a}{b^2-a^2}$
21. $12xy^2(x+y)^2$ 23. $(a+1)^2(a+4)$ 25. $x(x-1)^3$ 27. $12x^3(x-1)^2$ 29. $\dfrac{7x}{6}$
31. $\dfrac{5x}{12}$ 33. $\dfrac{1}{12}y$ 35. $\dfrac{3}{4}y$ 37. $\dfrac{7xy-2x+3y}{6xy}$ 39. $\dfrac{8x-2}{(x-1)(x+1)}$
41. $\dfrac{-3y^2+3y}{(2y-1)(y+1)}$ 43. $\dfrac{y^2-4y+5}{(y+1)(2y-3)}$ 45. $\dfrac{-4}{15(x-2)}$ 47. $\dfrac{4x-2}{(x-2)(x+1)^2}$
49. $\dfrac{-6y-4}{(y+4)(y-4)(y-1)}$ 51. $\dfrac{3y+1}{y(y-3)(y+2)}$ 53. $\dfrac{-x^2-7x+2}{(x+2)(x-2)^2}$
55. $\dfrac{-1}{(z-4)(z-3)(z-2)}$ 57. $\dfrac{6}{a-4}$ 59. $\dfrac{x^2-1}{x}$ 61. $\dfrac{x^3-2x^2+2x-2}{(x-1)^2}$
63. $\dfrac{y^3-2y^2-y}{(y-1)(y+1)}$ 65. $\dfrac{x^2+x+1}{x+2}$ 67. $\dfrac{1}{x+5}$ 69. $\dfrac{x-2}{x+3}$

Exercise 2.4 [page 82] 1. $\dfrac{2}{21}$ 3. $\dfrac{4y}{3}$ 5. $\dfrac{1}{5}$ 7. $\dfrac{1}{10}$ 9. $\dfrac{7}{10a+2}$ 11. $\dfrac{a}{a-2}$
13. x 15. $\dfrac{x}{x-1}$ 17. $\dfrac{y}{y+2}$ 19. xy 21. $\dfrac{x^2(y-1)}{y^2(x+1)}$ 23. $\dfrac{-2(x+z)}{xz}$
25. $3a+b$ 27. $\dfrac{-3z}{2x}$ 29. $\dfrac{2x}{x-3}$ 31. $\dfrac{y^2}{(y-1)(y+1)^2}$ 33. $\dfrac{a+5}{a+4}$
35. $\dfrac{-(u+1)(u+5)}{(u+3)(u-1)}$ 37. $\dfrac{w^2-2w}{w^2-2}$ 39. 1 41. $\dfrac{2-y}{2y}$ 43. $\dfrac{3-a}{a}$ 45. $\dfrac{x+1}{x-2}$
47. a. $\dfrac{1}{q+60}+\dfrac{1}{q}$ b. $\dfrac{q^2+60q}{2q+60}$
49. a. $T_1 = \dfrac{d}{r_1}$, $T_2 = \dfrac{d}{r_2}$ b. Total distance: $D = 2d$;   c. Average speed: $\dfrac{2d}{\dfrac{d}{r_1}+\dfrac{d}{r_2}}$
total time: $\dfrac{d}{r_1}+\dfrac{d}{r_2}$
d. $\dfrac{2r_1r_2}{r_1+r_2}$ e. $58\tfrac{1}{3}$ miles per hour   51. $\dfrac{a-2b}{a+2b}$ 53. $\dfrac{x(x+2)}{(x-1)(x+1)}$ 55. 1

Exercise 2.5 [page 93] 1. x^2 3. $\dfrac{x}{y^4}$ 5. $\dfrac{x^3}{y^6}$ 7. $\dfrac{-8x^3}{27y^6}$ 9. $\dfrac{16}{x}$ 11. $-x^4y$
13. $\dfrac{-8x}{9y^2}$ 15. $\dfrac{9(x+y)}{x^6}$ 17. $\dfrac{1}{2}$ 19. 3 21. $\dfrac{-1}{8}$ 23. 9 25. $\dfrac{5}{3}$ 27. $\dfrac{9}{5}$
29. $\dfrac{82}{9}$ 31. $\dfrac{3}{16}$ 33. $\dfrac{-2}{x^3}$ 35. $\dfrac{-5x^5}{y^2}$ 37. $\dfrac{1}{(x-y)^2}$ 39. $\dfrac{b-c}{(b+c)^2}$ 41. $\dfrac{1}{x^4}$

43. x^7 45. $\dfrac{x^4}{9y^6}$ 47. $\dfrac{x^2}{4y^2}$ 49. a^6b^4 51. $\dfrac{1}{a+b}$ 53. $1 + x^{-1}y$ 55. $xy^{-1} + 1$

57. $(y^3 + 4x^4)$ 59. $(3x + 4y)$ 61. $\dfrac{x^2 + y^2}{x^2y^2}$ 63. $\dfrac{x^2y^2 + 1}{xy}$ 65. $\dfrac{y^2 - x^2}{xy}$

67. $1 - xy$ 69. $x + y$ 71. $\dfrac{xy}{x+y}$ 73. 2.85×10^2 75. 8.372×10^6

77. 2.4×10^{-2} 79. 5.23×10^{-4} 81. 240   83. 687,000   85. 0.005   87. 0.000202
89. a. 7.2×10^2 b. 8.4×10^2 91. a. 8.0×10^2 b. 5.6×10^2
93. a. 3.0×10^8 b. 1.18×10^{10} 95. a. 5.87×10^{12} b. More than 180 years

97. a. $4.003383263507 \times 10^{20}$ b. 4.2×10^{19} square feet   c. 7595 times   99. $\dfrac{s^{14}x^2}{t^2y^2}$

101. $\dfrac{-1}{a^3bx^3}$ 103. $\dfrac{4x^2}{9y^{16}}$ 105. $\dfrac{2^{10}y^7}{x^{12}}$ 107. $\dfrac{2x^4z}{27y^2}$ 109. $6 - x$ 111. $9x + 1$

Chapter 2 Review [page 96]   1. $\dfrac{a}{2(a-1)}$ 2. $\dfrac{-a}{4}$ 3. $\dfrac{2y-3}{3}$ 4. $\dfrac{y^2 - 2x}{2}$ 5. $\dfrac{x}{x+3}$

6. $\dfrac{-x + 2y}{x + 2y}$ 7. $\dfrac{a-3}{2(a+3)}$ 8. $\dfrac{2xy+1}{2xy-1}$ 9. $\dfrac{y+2}{x^2+1}$ 10. $a - 1$ 11. $\dfrac{5ab}{2}$

12. $\dfrac{-a^4b^3}{4}$ 13. $\dfrac{6x}{2x+3}$ 14. $\dfrac{(2x+3)(x+1)}{6}$ 15. $\dfrac{2(x-2)(x+1)}{(x-1)(x+3)}$

16. $\dfrac{(y-1)(x-1)}{4(x^2+2)(x-3)}$ 17. $\dfrac{a(a-2)}{(a+1)(a+2)}$ 18. $\dfrac{b(a^2+2ab+4b^2)}{a(a-2b)}$ 19. $\dfrac{1}{2x-1}$

20. $\dfrac{y+2}{12x}$ 21. $4yz + 2 - \dfrac{1}{yz}$ 22. $9x^2 - 7 + \dfrac{4}{x^2} - \dfrac{1}{x^4}$ 23. $y^2 + 2y - 4$

24. $x^2 - 2x - 2 - \dfrac{1}{x-2}$ 25. $x^2 + x + \dfrac{1}{2} - \dfrac{1}{2(2x-1)}$ 26. $y^2 - \dfrac{1}{3}y - \dfrac{2}{9} + \dfrac{38}{9(3y+1)}$

27. $2x + \dfrac{3x-5}{x^2-1}$ 28. $y^4 + y + \dfrac{y}{y^3-1}$ 29. $\dfrac{2}{x}$ 30. $\dfrac{y-2}{y+3}$ 31. $\dfrac{-1}{6}a$ 32. $\dfrac{5b}{4}$

33. $\dfrac{3x+1}{2(x+3)(x-3)}$ 34. $\dfrac{4(y+1)}{(y-2)(y+2)^2}$ 35. $\dfrac{2a^2 - 2a + 1}{(a-1)(a-3)}$ 36. $\dfrac{a^4 + a^3 - a^2 + a + 4}{(a-1)(a+1)^2}$

37. $\dfrac{1}{4}$ 38. $\dfrac{1}{5}$ 39. $\dfrac{y(x-2)}{(x+2)}$ 40. $\dfrac{x}{x+4}$ 41. $\dfrac{x^2}{(x+1)(x-1)^2}$ 42. $\dfrac{x+5}{x-7}$

43. $\dfrac{8}{9x}$ 44. $\dfrac{-(x-y)}{4}$ 45. $\dfrac{-2y^2}{3x^2}$ 46. $\dfrac{1}{(x-y)^4}$ 47. $\dfrac{a^6}{8b^9}$ 48. a^4b^6 49. $b + \dfrac{b^2}{a}$

50. $1 - \dfrac{1}{a^2}$ 51. $y^3 + 2x^3$ 52. $2x - y^3$ 53. $\dfrac{x^3 + y}{x^3y}$ 54. $\dfrac{1 - x^2y^2}{xy}$ 55. $\dfrac{y - xy^2}{x}$

56. $\dfrac{x+y}{y}$ 57. $\dfrac{(y-x)(x-y)}{xy}$ 58. $\dfrac{1}{y-x}$ 59. 7×10^{-1} 60. 4×10^1

61. $\dfrac{60x}{x^2 - 16}$ 62. $\dfrac{130x}{x^2 - 144}$ 63. $2L + \dfrac{60}{L}$ 64. $\dfrac{600}{L}$ 65. $\dfrac{2r - 15}{2}$ 66. 18 hours

67. a. 5.72674×10^7, 6.1×10^9 b. 1.06×10^2 persons per square mile

68. a. 9.2956×10^7, 1.86×10^5 b. 499.76 seconds or 8.33 minutes

69. $x - 1 + \dfrac{-2x + 3}{x^2 - x + 2}$ 70. $x^2 - 5 + \dfrac{18}{x^2 + 4}$ 71. $\dfrac{13}{10}$ 72. $\dfrac{3x - 7y}{2(x + y)}$ 73. 2

74. $2b^2$ 75. $\dfrac{1}{x^9 y^4}$ 76. $\dfrac{1}{x^5 y^{14}}$ 77. $\dfrac{-(3x + 2)(x + 2)}{(x + 1)^3}$ 78. $\dfrac{2x^2 - 9x + 24}{2(x + 2)^2(x - 3)}$

Exercise 3.1 [page 107] 1. 11 3. -3 5. Not real 7. $\dfrac{-2}{3}$ 9. $\dfrac{2}{3}$ 11. $\dfrac{2}{9}$

13. 3 15. Not real 17. -2 19. -2 21. $\dfrac{-1}{2}$ 23. $\dfrac{5}{8}$ 25. $\sqrt{3}$ 27. $4\sqrt[3]{x}$

29. $\sqrt[3]{4x}$ 31. $\dfrac{1}{\sqrt[4]{8}}$ 33. $\dfrac{3}{\sqrt[3]{xy}}$ 35. $\sqrt[4]{x - 2}$ 37. $7^{1/2}$ 39. $(2x)^{1/3}$ 41. $3(6)^{-1/4}$

43. $5x^{1/3}(yz)^{-1/4}$ 45. $x^{1/3} - 3y^{1/2}$ 47. $(x - 2y)^{-1/3}$ 49. 0.375; terminates

51. $0.8\overline{3}$; repeats a pattern 53. $0.\overline{285714}$; repeats a pattern 55. $3.\overline{09}$; repeats a pattern

57. 1.414 59. 4.217 61. -2.122 63. 1.125 65. 125 67. 2 69. -8

71. x 73. $12 < \sqrt{175} < 13$ 75. $7 < \sqrt[3]{423} < 8$ 77. $-5 < \sqrt[3]{-84.6} < -4$

79. $3 < \sqrt[4]{129} < 4$

81. a. 13 inches b. 3.873 feet c. 1.414 miles d. 0.48 centimeter

83. a. 12 millimeters b. 7.368063 yards c. 3.849 feet d. 2.5 inches

85. 2.22 seconds 87. 16 feet 89. a. 24,997.76 miles per hour b. 9,707.36 miles per hour

91. 0.283 centimeter 93. $16^{1/2}$; $a^{1/m} > a^{1/n}$

95. 1.41421, 1.25992, 1.18921, 1.07177, 1.00696, 1.00069; $2^{1/n}$ approaches 1 as n gets very large.

97. $x^{1/4} = (x^{1/2})^{1/2}$

Exercise 3.2 [page 115] 1. 27 3. 16 5. $\dfrac{1}{4}$ 7. $\dfrac{1}{625}$ 9. $\dfrac{-27}{125}$ 11. $\dfrac{16}{25}$

13. $\sqrt[5]{x^4}$ 15. $3\sqrt[5]{x^2}$ 17. $\dfrac{1}{\sqrt[6]{y^5}}$ 19. $\dfrac{1}{\sqrt[3]{(xy)^2}}$ 21. $\dfrac{3}{\sqrt[4]{y^2}}$ 23. $-2\sqrt[4]{xy^3}$

25. $6\sqrt{(x + 2y)^3}$ 27. $\dfrac{1}{\sqrt{(x^2 - 4)^5}}$ 29. $x^{2/3}$ 31. $(ab)^{2/3}$ 33. $2a^{1/5}b^{3/5}$

35. $-4x^{-3/4}y$ 37. $-(a^2 + b)^{3/4}$ 39. $5^{-1/5}ab(a - b^2)^{-2/5}$ 41. 8 43. $\dfrac{-81}{16}$ 45. $2y^3$

47. $-a^4 b^8$ 49. $\dfrac{-2}{5}x^3 y^9$ 51. $-3a^2 b^3$ 53. 7.931 55. -10.903 57. 0.089

59. 35.142 61. $2x^{7/4}$ 63. $\dfrac{x^{2/3}y}{3}$ 65. $2x$ 67. $\dfrac{9}{y^{1/3}}$ 69. $\dfrac{a^6 b^{1/2}}{8}$ 71. $\dfrac{1}{a^{0.3}b^{1.95}}$

ANSWERS A-9

73. $\dfrac{4y^5}{3x^4}$ 75. $a^{16/5}b^{1/15}$ 77. $2x^{3/2} - 2x$ 79. $\dfrac{3}{4}y^{-5/8} + 3y^{3/8}$ 81. $2x^{2/3} + x^{1/3} - 1$

83. $a^{3/2} - 4a^{3/4} + 4$ 85. $\dfrac{1}{2}b^{1/5} - \dfrac{1}{8}b^{-4/5}$ 87. $x^{2.3} - x^{3.4}$ 89. $x^{0.2} + 2x^{-1/2} - 3x^{1.4} - 6$

91. 161   93. 1.88 years   95. 3.16×10^{12} times as intense   97. $7114.32   99. 137

101. $28.24   103. $x(x^{1/2} + 1)$ 105. $\dfrac{x+1}{x^{3/2}}$ 107. $\dfrac{x}{(x+1)^{1/2}}$ 109. $\dfrac{y(y + y^{1/2} - 2)}{(y-2)^{1/2}}$

111. $\dfrac{8z + 3}{3z^{1/3}(2z + 1)^{2/3}}$

Exercise 3.3 [page 123]   1. $3\sqrt{2}$ 3. $2\sqrt[3]{3}$ 5. $-2\sqrt[4]{4}$ 7. $100\sqrt{6}$ 9. $10\sqrt[3]{900}$

11. $\dfrac{-2\sqrt[3]{5}}{3}$ 13. $x^3\sqrt[3]{x}$ 15. $3z\sqrt{3z}$ 17. $2a^2b^3\sqrt[4]{3a}$ 19. $2pqr^2\sqrt[5]{3p^2q^4r}$ 21. $-6s^2$

23. $-7b$ 25. $2\sqrt{4-x^2}$ 27. $a\sqrt[3]{8+a^3}$ 29. $\dfrac{-p^6v^4\sqrt{p}}{15a^2}$ 31. $\dfrac{-a^2\sqrt[5]{81a^2}}{2b^3}$ 33. $2\sqrt{3}$

35. $\dfrac{-\sqrt{21}}{7}$ 37. $\dfrac{\sqrt{14x}}{6}$ 39. $\dfrac{\sqrt{2ab}}{b}$ 41. $\dfrac{\sqrt{6k}}{k}$ 43. $\dfrac{-3x^3\sqrt{30xz}}{4z}$ 45. $\dfrac{\sqrt[3]{x}}{x}$

47. $\dfrac{\sqrt[3]{18y^2}}{3y}$ 49. $\dfrac{\sqrt[3]{2xy}}{2y}$ 51. $\dfrac{\sqrt[5]{48x^2}}{2x}$ 53. $3x^2\sqrt[4]{3x^3}$ 55. $\dfrac{xy\sqrt{x+y}}{x+y}$ 57. a^2b

59. $7y\sqrt{2x}$ 61. $\dfrac{2b\sqrt[3]{b}}{a^2}$ 63. $\sqrt[5]{b}$ 65. $\sqrt{3}$ 67. $\sqrt{3}$ 69. $\sqrt[3]{9}$ 71. $\sqrt{x}$

73. $\dfrac{2a^2\sqrt[3]{b^2}}{b}$ 75. $\dfrac{6\sqrt[4]{9xy^2}}{xy}$ 77. $x^2z\sqrt{2xz}$ 79. $6x^2(x^2 + 1)$ 81. $2a(a-b)\sqrt[3]{3b^2(a-b)}$

83. $2(x-1)\sqrt{2(x-1)}$ 85. $x(y-2)^2\sqrt{x(y-2)}$ 87. $\dfrac{(y-3)\sqrt{xy(y-3)}}{xy^2}$

89. $\dfrac{(x-1)\sqrt[3]{x^2y(x-1)}}{xy}$ 91. $x(x+1)\sqrt[3]{(4x^2-1)(x+1)}$ 93. $\dfrac{\sqrt{3x(x-1)}}{x^2}$ 95. $2|x|$

97. $|x+1|$ 99. $\dfrac{2}{|x+y|}$

Exercise 3.4 [page 130]   1. $5\sqrt{7}$ 3. $\sqrt{3}$ 5. $9\sqrt{2x}$ 7. $-\sqrt[3]{2}$ 9. $9\sqrt{5} + 5\sqrt[3]{5}$

11. $-6y\sqrt{x} + 4x\sqrt{y}$ 13. $-xy\sqrt[4]{2x^2} - 3xy\sqrt[4]{x^2}$ 15. $\dfrac{7}{2}\sqrt{3}$ 17. $-\dfrac{1}{2}\sqrt[3]{2x}$

19. $-\dfrac{1}{10}a\sqrt{5}$ 21. $6 - 2\sqrt{5}$ 23. $2\sqrt{3} + 2\sqrt{5}$ 25. $2\sqrt[3]{5} - 4\sqrt[3]{3}$

27. $4x\sqrt{6} + 4\sqrt{3x}$ 29. $\sqrt{x^2 - 3x} - \sqrt{x^2 - 9}$ 31. $x + 1 - \sqrt[3]{(x+1)^2}$ 33. $x - 9$

35. $-4 + \sqrt{6}$ 37. $7 - 2\sqrt{10}$ 39. $6x - 7\sqrt{2xy} - 6y$ 41. $a - 4\sqrt{ab} + 4b$ 43. $\sqrt[6]{75}$

45. $\sqrt[6]{500}$ 47. $\sqrt[12]{(4x)^7}$ 49. $y\sqrt[4]{x^3}$ 51. $2(1 + \sqrt{3})$ 53. $6(\sqrt{3} + 1)$ 55. $4(1 + \sqrt{y})$

57. $\sqrt{2}(1 - \sqrt{3})$ 59. $\sqrt{x}(2y + 3\sqrt{y})$ 61. $2\sqrt{x}(2\sqrt{x} - \sqrt{3})$ 63. $1 + \sqrt{3}$

65. $1 + \sqrt{2}$ 67. $1 - \sqrt{x}$ 69. $x - y$ 71. $\frac{1}{2}(4\sqrt{a} + \sqrt{2b})$ 73. $\sqrt[3]{x} - 3\sqrt[3]{x^2}$

75. $-2 + 2\sqrt{3}$ 77. $\frac{1}{3}(2\sqrt{7} + 4)$ 79. $\frac{x(\sqrt{x} + 3)}{x - 9}$ 81. $\frac{1}{2}\sqrt{6}$ 83. $\frac{1}{6}(\sqrt{15} - 3)$

85. $\frac{(2a + 2b)\sqrt{3} - 7\sqrt{ab}}{3a - 4b}$ 87. $\frac{x - 2}{(\sqrt{x} - 1)(\sqrt{x - 1} + 1)}$ 89. $\frac{1}{(x + \sqrt{x + 1})^2}$

91. $\frac{(x + 1)\sqrt{x}}{x}$ 93. $\frac{\sqrt{x + 1}}{x + 1}$ 95. $\frac{-\sqrt{x^2 + 1}}{x(x^2 + 1)}$ 97. $\frac{x - 2\sqrt{x} + 1}{x - 1}$

99. $\frac{(x + 2)\sqrt{x} - x\sqrt{x + 2}}{2x(x + 2)}$ 101. $x - 1$

Exercise 3.5 [page 136] 1. $2i$ 3. $4i\sqrt{2}$ 5. $6i\sqrt{2}$ 7. $6i\sqrt{6}$ 9. $40i$ 11. $-4i\sqrt{3}$
13. $4 + 2i$ 15. $2 + 15i\sqrt{2}$ 17. $2 + 2i$ 19. $5 + 5i$ 21. $-2 + i$ 23. $-1 - 2i$
25. $8 + i$ 27. $8 + 13i$ 29. $21 - 18i$ 31. $3 - 4i$ 33. 5 35. $-\frac{1}{3}i$
37. $-\frac{1}{5} - \frac{3}{5}i$ 39. $1 + i$ 41. $\frac{1}{2} - \frac{1}{2}i$ 43. $\frac{12}{13} - \frac{5}{13}i$ 45. $\frac{9}{34} + \frac{19}{34}i$ 47. $4 + 2i$
49. $15 + 3i$ 51. $-\frac{3}{2}i$ 53. $\frac{3}{5} - \frac{4}{5}i$ 55. $x \geq 5$; $x < 5$
57. a. -1 b. 1 c. $-i$ d. -1 59. $5 + 4i$

Chapter 3 Review [page 137] 1. -5 2. -2 3. Not real   4. $\frac{1}{2}$ 5. -3
6. -4 7. Not real   8. $\frac{2}{3}$ 9. $3\sqrt{x}$ 10. $\sqrt{3x}$ 11. $\frac{4}{\sqrt[3]{x}}$ 12. $\frac{1}{\sqrt[3]{4x}}$
13. $(5x)^{1/3}$ 14. $5x^{-1/3}$ 15. $3(xy)^{1/2}$ 16. $x^{1/2} - y^{1/3}$ 17. $\frac{4}{\sqrt[4]{x^3}}$ 18. $-2\sqrt[3]{xy^2}$
19. $4\sqrt{(x - 1)^3}$ 20. $\frac{2}{\sqrt{(x^2 - 1)^3}}$ 21. $4x^{3/2}$ 22. $2x^{2/3}y^{1/3}$ 23. $-2xy^{-2/3}$
24. $y(x - y)^{-2/3}$ 25. $3x^{5/4}$ 26. $\frac{x^{3/2}y}{2}$ 27. $2x^{1/4}$ 28. $\frac{4}{y^{1/3}}$ 29. $\frac{x^{3/2}y^{1/4}}{2}$
30. $\frac{1}{512xy^2}$ 31. $x^{13/3}y^{5/3}$ 32. $\frac{1}{y^4}$ 33. $y^{1/2} + 1$ 34. $x^{-1/3} - x^{2/3}$
35. $3x - 7x^{1/2} - 6$ 36. $x^{-2} + 4x^{-1} + 4$ 37. $2xy\sqrt[3]{2x}$ 38. $4x^3\sqrt{y}$ 39. $2a\sqrt{2a^2 + 1}$
40. $\frac{y^5\sqrt{y}}{2x^2}$ 41. $\frac{3\sqrt{x}}{x^2}$ 42. $\frac{\sqrt[3]{4xy^2}}{2y}$ 43. $y\sqrt{6x}$ 44. $x\sqrt{x}$ 45. $3x(2x - 1)^2\sqrt{2}$
46. $2y(y + 1)\sqrt[3]{6y^2(y + 1)}$ 47. $12\sqrt{2}$ 48. $5x\sqrt{xy}$ 49. $2\sqrt[3]{2x}$ 50. $\frac{-x\sqrt{3}}{6}$
51. $4y - 2y^2\sqrt{3}$ 52. $x - 2$ 53. $2a - \sqrt{2ab} - 6b$ 54. $2a - 2\sqrt{2ab} + b$

55. $x\sqrt[6]{xy^5}$ 56. $\sqrt[12]{2^{10}y^{11}}$ 57. $y(\sqrt{x}-1)$ 58. $2\sqrt[3]{y^2}-\sqrt[3]{y}$ 59. $\dfrac{x(\sqrt{x}-2)}{x-4}$

60. $\dfrac{(x-2\sqrt{y})^2}{x-4y}$ 61. $\dfrac{y}{y-1-\sqrt{y-1}}$ 62. $\dfrac{1}{y-\sqrt{y^2-y}}$ 63. $7-i$ 64. $-1+7i$

65. $15+5i$ 66. $15+8i$ 67. $\dfrac{-2}{5}i$ 68. $-\dfrac{12}{17}+\dfrac{3}{17}i$ 69. $\dfrac{1}{2}-\dfrac{1}{2}i$ 70. $\dfrac{7}{13}+\dfrac{17}{13}i$

71. $-6+i\sqrt{3}$ 72. $17-2\sqrt{2}i$ 73. $\dfrac{-5\sqrt{2}}{4}i$ 74. $\dfrac{3}{7}-\dfrac{5\sqrt{3}}{7}i$ 75. 252 pounds

76. 2.67 inches 77. 284 78. 12 79. 238 80. 1,122,369 discs 81. $94.48

82. $1364.23 83. $x^{-1/3}(1-x^{2/3})$ 84. $x^{1/4}(x+2)^{-3/4}[(x+2)^{3/2}+x^{1/4}]$

85. $x(x+1)\sqrt[3]{(x^3-1)(x+1)}$ 86. $|x-2|$ 87. $\dfrac{2x-1}{\sqrt{1-x}}$ 88. $\dfrac{\sqrt{x+1}+\sqrt{x}}{\sqrt{x(x+1)}}$

89. $\dfrac{(1+x\sqrt{x})^2}{1-x^3}$ 90. $\dfrac{\sqrt{x}+x}{x}$

Exercise 4.1 [page 149] 1. 7 3. $\dfrac{8}{3}$ 5. 4 7. $\dfrac{-13}{2}$ 9. $\dfrac{8}{3}$ 11. $\dfrac{1}{2}$ 13. 6

15. $-\dfrac{7}{9}$ 17. $\dfrac{1}{4}$ 19. 11 21. 24 23. -20 25. $-4.8\overline{3}$ 27. -0.1375

29. 2.325 31. $p=\dfrac{I}{rt}$ 33. $w=\dfrac{P-2l}{2}$ 35. $g=\dfrac{v-k}{t}$ 37. $r=\dfrac{A-P}{Pt}$

39. $b=\dfrac{R-2d-ha}{h}$ 41. $h=\dfrac{A-\pi r^2}{\pi r}$ 43. $h=\dfrac{S-2lw}{2(l+w)}$ 45. 26 47. 600,000

49. 38,611 51. a. 125,000 b. 115,741 53. a. 7.53% b. Yes; $0.93(1+x)=1$

55. 87% 57. 167 59. 22.5 pounds 61. $3181.82 at 8%; $1818.18 at 13.5%

63. a. 3 hours, 18.6 minutes b. 148.97 miles 65. 2 miles 67. 142.85 square feet

69. 37.5 hours

Exercise 4.2 [page 159] 1. 3 3. -30 5. 3 7. 4 9. $\dfrac{1}{2}$ 11. 5 13. 1

15. $-\dfrac{14}{5}$ 17. 4 19. 4 21. 40 23. $C=\dfrac{5}{9}(F-32)$ 25. $b=\dfrac{2A-hc}{h}$

27. $h=\dfrac{2E-mv^2}{2mg}$ 29. $R=\dfrac{E}{I}$ 31. $r=1.444h-10$ 33. $r=\dfrac{S-a}{S}$ 35. $w=\dfrac{Wd}{D}$

37. $h=\dfrac{Ct}{C-V}$ 39. $q=\dfrac{pf}{p-f}$ 41. $x=\dfrac{Hy}{2y-H}$ 43. $e=\dfrac{p-a}{p+a}$

45. 20 miles per hour 47. 144 miles 49. $6187.50 51. $37,500 53. 45 miles

55. 13.6 kilometers per liter 57. 889 59. 14 feet 61. 17.1 square feet 63. 89.23 feet

65. $e=\dfrac{v-c}{v-1}$ 67. $r=\dfrac{2aGm}{Gm+aV^2}$ 69. $v_2=\dfrac{v_1V}{2v_1-V}$

Exercise 4.3 [page 171] 1. $-\frac{5}{2}, 2$ 3. $-\frac{1}{2}, 0$ 5. $-\frac{3}{2}, 6$ 7. $0, 3$ 9. $-3, 3$
11. $-\frac{2}{3}, \frac{2}{3}$ 13. 1 15. $\frac{1}{2}, 1$ 17. $2, 3$ 19. $-1, 2$ 21. $-\frac{5}{2}, 0$ 23. $-3, 6$
25. $1, 4$ 27. $-3, 3$ 29. $-2, \frac{3}{2}$ 31. $-3, 1$ 33. $-\frac{10}{3}, 1$ 35. $-\frac{4}{3}, -\frac{1}{6}$
37. $-\frac{1}{2}$ 39. -2 41. $x^2 + x - 2 = 0$ 43. $x^2 + 5x = 0$ 45. $2x^2 + 5x - 3 = 0$
47. $8x^2 - 10x - 3 = 0$ 49. $-\frac{5}{3}, \frac{5}{3}$ 51. $-\sqrt{7}, \sqrt{7}$ 53. $-i\sqrt{6}, i\sqrt{6}$ 55. $-\sqrt{6}, \sqrt{6}$
57. $-1, 5$ 59. $-\frac{3}{2}, \frac{5}{2}$ 61. $-2 + i\sqrt{3}, -2 - i\sqrt{3}$ 63. $\frac{1 - \sqrt{3}}{2}, \frac{1 + \sqrt{3}}{2}$
65. $-\frac{1}{3} + \frac{1}{9}i, -\frac{1}{3} - \frac{1}{9}i$ 67. $\frac{7 + 2\sqrt{2}}{8}, \frac{7 - 2\sqrt{2}}{8}$ 69. a. $\frac{5}{4}$ seconds b. 5 seconds
71. 5.26 centimeters 73. 24 feet 75. $8\sqrt{2}$, or 11.3, inches 77. $\frac{5}{2}$ feet
79. 100 yards by 80 yards 81. 9 inches by 9 inches 83. $30 85. $2.50
87. 3 miles per hour 89. $a + b, -a - b$ 91. a, b 93. $\frac{b}{a}, \frac{c}{a}$ 95. $-\frac{3}{2}$
97. If $x^2 + bx + c = (x - r_1)(x - r_2) = x^2 - (r_1 + r_2)x + r_1 r_2$, then $b = -(r_1 + r_2)$ and $c = r_1 r_2$.

Exercise 4.4 [page 185] 1. $x^2 + 8x + 16 = (x + 4)^2$ 3. $x^2 - 7x + \frac{49}{4} = \left(x - \frac{7}{2}\right)^2$
5. $x^2 + \frac{3}{2}x + \frac{9}{16} = \left(x + \frac{3}{4}\right)^2$ 7. $x^2 - \frac{4}{5}x + \frac{4}{25} = \left(x - \frac{2}{5}\right)^2$ 9. 1 11. $-4, -5$
13. $\frac{-3 \pm \sqrt{21}}{2}$ 15. $-1 \pm \frac{\sqrt{10}}{2}$ 17. $-1, \frac{5}{2}$ 19. $\frac{-2 \pm i\sqrt{2}}{2}$ 21. $1, 4$ 23. $-4, 1$
25. $\frac{3 \pm \sqrt{5}}{2}$ 27. $\frac{5 \pm \sqrt{13}}{6}$ 29. $-\frac{2}{3}, \frac{3}{2}$ 31. $0, 5$ 33. $\pm i\sqrt{2}$ 35. $\frac{1 \pm i\sqrt{7}}{4}$
37. $\frac{3 \pm \sqrt{13}}{2}$ 39. $\frac{2 \pm \sqrt{7}}{3}$ 41. $\frac{1 \pm i\sqrt{23}}{6}$ 43. $\frac{1 \pm i\sqrt{3}}{4}$ 45. $-2, 1$ 47. $2 \pm \sqrt{5}$
49. $\pm 2i$ 51. $\pm \frac{1}{2}\sqrt{\frac{A}{\pi}}$ 53. $\pm \sqrt{\frac{Fr}{m}}$ 55. $-1 \pm \sqrt{\frac{A}{P}}$ 57. $\pm \sqrt{\frac{Gm_1 m_2}{F}}$
59. $\pm \sqrt{\frac{V - \pi r^2 h}{\pi h}}$ 61. $\pm \sqrt{c^2 - a^2}$ 63. $\pm \sqrt{\frac{48 - 6x^2}{2}}$ 65. $\pm \sqrt{\frac{2E - 2mgh}{m}}$
67. $\pm \sqrt{\frac{gr^2}{h + r}}$ 69. $\pm \sqrt{\frac{m(2a - r)}{ar}}$ 71. $\frac{-2l \pm \sqrt{4l^2 + 2A}}{2}$ 73. $\frac{1 \pm \sqrt{1 - 4h}}{8}$
75. $\frac{E \pm \sqrt{E^2 - 4RP}}{2R}$ 77. $\frac{3 \pm \sqrt{9 + 8D}}{2}$ 79. $\frac{-x \pm \sqrt{8 - 11x^2}}{2}$

81. 29.16 miles per hour   83. a. 24.5 seconds   b. 1.2 seconds   85. 11.8%
87. 18.09 feet by 4.61 feet, or 6.91 feet by 12.06 feet   89. a. 45 miles   b. 1.26 miles
91. $2 - \sqrt{5}$; $x^2 - 4x - 1 = 0$ 93. $4 + 3i$; $x^2 - 8x + 25 = 0$ 95. $\dfrac{-b \pm \sqrt{b^2 - 4c}}{2}$
97. Two complex solutions   99. One real solution   101. Two real solutions

Exercise 4.5 [page 196]   1. 64   3. -2 5. $-\dfrac{1}{3}$ 7. $-\dfrac{1}{2}, 2$ 9. 12   11. 5
13. 4   15. 5   17. 0   19. 4   21. 1, 3   23. -27 25. 17   27. ± 64
29. $\dfrac{13}{3}$ 31. $\dfrac{1}{243}$ 33. $\dfrac{19}{2}$ 35. $\pm\sqrt{30}$ 37. $-5, \dfrac{25}{2}$ 39. $\dfrac{gT^2}{4\pi^2}$ 41. $\dfrac{gr^2 - rS^2}{S^2}$
43. $\dfrac{4\pi r^3}{3}$ 45. $\dfrac{8Lvf}{\pi R^4}$ 47. $\pm\sqrt{t^2 - r^2}$ 49. $\dfrac{\pm\sqrt{(A - B)^2 - DC^2}}{C}$ 51. $\dfrac{\pm\sqrt{\pi a^2 - A^2p^2}}{\pi a^2}$
53. $\dfrac{\sqrt[3]{8mT^2\pi} - \pi R}{\pi}$ 55. $\pm\dfrac{c}{m}\sqrt{m^2 - M^2}$ 57. $-3, -1, 1, 3$ 59. $\dfrac{-\sqrt[3]{4}}{2}, 2$ 61. 25
63. 16, 256   65. $-2, 63$ 67. 1   69. $-1, 64$ 71. $\dfrac{1}{4}, 16$ 73. $-\dfrac{1}{3}, \dfrac{1}{4}$ 75. 626
77. 3.07 hours   79. 24 miles   81. 18 miles   83. $\dfrac{1}{2}, \dfrac{4}{5}$ 85. $-10, -4, 2$ 87. $-9, 0$
89. 2, 3   91. $-\dfrac{1}{4}$ 93. $\dfrac{\pm\sqrt{2}}{2}$

Chapter 4 Review [page 199]   1. 1   2. 600   3. $\dfrac{2}{3}$ 4. -5 5. 5   6. 8   7. $\dfrac{3}{2}$
8. $-\dfrac{1}{2}$ 9. 5   10. 200   11. $\dfrac{3(N + c)}{5}$ 12. $\dfrac{C + 2t - 10}{2}$ 13. $\dfrac{l - a + d}{d}$
14. $\dfrac{2s - k}{t}$ 15. $\dfrac{s + 2\pi r}{2\pi}$ 16. $\dfrac{9}{5}C + 32$ 17. $-\dfrac{27}{2}$ 18. -7 19. No solution
20. $\dfrac{3}{2}$ 21. -2 22. 2   23. $\dfrac{2S - ns}{n}$ 24. $n\left(1 - \dfrac{V}{C}\right)$ 25. $\dfrac{qf}{q - f}$ 26. $\dfrac{rp}{r - p}$
27. $-1, 2$ 28. 2   29. 1, 4   30. ± 1 31. $\dfrac{1}{8}, 4$ 32. $\dfrac{1}{2}, 6$ 33. $\pm\sqrt{6}$
34. $\pm i\sqrt{2}$ 35. 1, 4   36. $\dfrac{1 \pm i\sqrt{15}}{7}$ 37. $2 \pm \sqrt{10}$ 38. $\dfrac{-3 \pm \sqrt{21}}{2}$ 39. $\dfrac{1 \pm i\sqrt{7}}{2}$
40. $\dfrac{1 \pm i\sqrt{5}}{2}$ 41. 1, 2   42. $\dfrac{3 \pm \sqrt{5}}{2}$ 43. $\dfrac{1 \pm i\sqrt{7}}{2}$ 44. $\dfrac{-3 \pm i\sqrt{7}}{4}$ 45. $-1, 2$
46. $\dfrac{2 \pm \sqrt{22}}{3}$ 47. $\pm\dfrac{3}{2}\sqrt{x^2 - 144}$ 48. $\pm\sqrt{\dfrac{1 - sp^2}{2}}$ 49. $\dfrac{k \pm \sqrt{k^2 - 8}}{4}$
50. $-\dfrac{x}{2}(1 \pm i\sqrt{7})$ 51. 1, 4   52. 8   53. 9   54. 12   55. 7   56. 8   57. 5

A-14 ANSWERS

58. 4 59. $\dfrac{2v}{t^2}$ 60. $\dfrac{\pm\sqrt{4+3(q-1)^2}}{2}$ 61. $\pm 2\sqrt{R^2-R}$ 62. $\pm\sqrt{8q^3-1}$
63. $\pm 1, \pm 2$ 64. $\pm 1, \pm\sqrt{5}$ 65. ± 3 66. $\pm\sqrt{10}$ 67. $-8, 64$ 68. $-125, 343$
69. $-1, -1 \pm 2\sqrt{3}$ 70. $\pm 2, \pm\sqrt{5}$ 71. a. 2000 b. 3112 72. a. 5334 b. 2000
73. 22.08% 74. 24.2% 75. $7\dfrac{5}{7}$ minutes per mile 76. 84 miles 77. $\dfrac{5}{9}$ cup

78. $12\dfrac{4}{9}$ pounds 79. $19\dfrac{5}{7}$ feet 80. $\dfrac{24}{5}$ inches 81. 9 82. 13
83. 14 feet by 14 feet by 4 feet 84. 8 inches by 8 inches by 40 inches 85. 19 or 20
86. 120 87. Morning: 20 miles per hour 88. Light traffic: 40 miles per hour 89. 11%
    evening: 30 miles per hour   city traffic: 20 miles per hour

90. 8.5% 91. $r\left(\sqrt{\dfrac{s}{w}}-1\right)$ 92. $-r \pm \sqrt{\dfrac{E^2 r}{P}}$ 93. $\dfrac{gr \pm \sqrt{g^2 r^2 - 2grV^2}}{2g}$

94. $\dfrac{gr \pm \sqrt{g^2 r^2 - 2grV^2}}{2g}$ 95. $\dfrac{CD(D+2L)}{(D+L)^2 + L^2}$ 96. $\dfrac{D}{4\pi^2 f^2 DL - 1}$ 97. $-2, 6$

98. $-\dfrac{1}{2}, 2$ 99. $-1, 2$ 100. $\pm i\sqrt{2}$

Cumulative Review Exercises for Part I [page 206] 1. 4 2. $-3x - 15$ 3. $-2x - 1$
4. $\dfrac{y^6}{x^8}$ 5. $1 - \sqrt{x}$ 6. $\dfrac{1}{1-x}$ 7. $\dfrac{1}{x^2 y^3}$ 8. 80 9. $\dfrac{2x}{(2x+1)^2}$ 10. $19\sqrt{2x}$
11. $x^2 + 9x - 5$ 12. $-11a^4 b^5 - a^4 b^4$ 13. 8000 14. 0 15. $1, 3$ 16. 4
17. $\dfrac{3 \pm i\sqrt{3}}{6}$ 18. 16 19. $\dfrac{1 \pm i\sqrt{23}}{4}$ 20. -17 21. $\pm 2\sqrt{R^2 - R}$ 22. $\dfrac{S - 2\pi r^2}{2\pi r}$
23. $\dfrac{2y(y-1)}{y+1}$ 24. $-\dfrac{1}{2}$ 25. $\dfrac{-6y - 4}{(y-4)(y+4)(y-1)}$ 26. 1 27. $\dfrac{x^2 + xy + y^2}{x^2}$
28. $\dfrac{a^2 - 4a + 1}{a^2 + a - 5}$ 29. $x - 1 + \dfrac{-2x + 3}{x^2 - x + 2}$ 30. $(y - 2x + 1)(y + 2x - 1)$ 31. 56
32. $\dfrac{-x(x^2 + x + 2)}{(x-1)^2}$ 33. a. 2.3×10^{-8} b. 4.02×10^{11} 34. $\dfrac{y(2\sqrt{y} + 3)}{4y - 9}$ 35. 16
36. $x^4 + 2x^3 + x^2 + 2x + 4 + \dfrac{7}{x-2}$ 37. $3x(x+2)(x-3)^2(x-8)(x+1)$ 38. 7
39. $\dfrac{2}{x\sqrt{x+1}(1+\sqrt{x+1})}$ 40. 82 41. $2, -3$ 42. a. $x^{1/2} y^{-1/3}$ b. $2x^{2/3} y^{1/3}$
43. a. $-\dfrac{3}{5} - \dfrac{1}{5}i$ b. $9 + i\sqrt{3}$ 44. $-2x - 4y$ 45. -216 46. $\dfrac{x}{x+1}$

47. $\dfrac{y-3}{x^2+1}$ 48. $3a^2 - a + 1$ 49. $2x(x-1)\sqrt[3]{3x^2}$ 50. 7   51. $2000 + 60x - 2x^2$

52. 2.714×10^{12} gallons   53. 9 hours' driving time   54. $56\dfrac{2}{3}$ miles per hour   55. 64,740

56. 124,700,000   57. a. $P\left(1 + \dfrac{r}{2} + \dfrac{r^2}{16}\right)$; $P\left(1 + r + \dfrac{3r^2}{8} + \dfrac{r^3}{16} + \dfrac{r^4}{64}\right)$ b. $1261.13

58. 160 feet   59. $2\dfrac{2}{9}$ kilometers   60. 10 feet by 18 feet or 12 feet by 15 feet

Exercise 5.1 [page 220]

1. a. High: 7°F; low: −19°F
   b. Above 5°F from noon to 3 P.M.; below −5°F from midnight to 9 A.M. and from 7 P.M. to midnight
   c. 7 A.M.: −10°F; 2 P.M.: 6°F
      10 A.M. and 5 P.M.: 0°F;
      6 A.M. and 10 P.M.: −12°F.
   d. Between 3 A.M. and 6 A.M.: 8°F;
      between 9 A.M. and noon: 10°F;
      between 6 P.M. and 9 P.M.: 9°F
   e. Increased most rapidly: 9 A.M. to noon;
      decreased most rapidly: 6 P.M. to 9 P.M.

3. a. At 43 miles per hour: 28 miles per gallon
   b. 34 miles per gallon: at 47 miles per hour
   c. Best gas mileage: at 65 miles per hour. The graph seems to be "leveling off" for higher speeds; any improvement in mileage probably would not be appreciable and might in fact deteriorate.
   d. Road condition, weather conditions, traffic, weight in the car

5. a. 12 minutes
   b. First 38 minutes
   c. Approximately from 38 minutes to 55 minutes

7. a. (0,7)   b. (2,9)   c. (−2,5)   9. a. $\left(0, -\dfrac{3}{2}\right)$ b. (2,0)   c. $\left(-5, -\dfrac{21}{4}\right)$

11. (−3,−7), (0,−4), (3,−1)   13. (1,1), $\left(2, \dfrac{3}{4}\right)$, $\left(3, \dfrac{3}{5}\right)$ 15. (1,0), $(3, 2\sqrt{2})$, $(5, 2\sqrt{6})$

17. $y = 6 - 2x$; (2,2), (4,−2)   19. $y = \dfrac{x+2}{x}$; (−2,0), (2,2)

21. $y = \dfrac{4}{x-1}$; $\left(4, \dfrac{4}{3}\right)$, $\left(8, \dfrac{4}{7}\right)$ 23. $y = \dfrac{2}{x^2 - x - 4}$; (−1,−1), $\left(1, -\dfrac{1}{2}\right)$

25. $y = \pm\dfrac{1}{2}\sqrt{x+8}$; $\left(-1, \dfrac{1}{2}\sqrt{7}\right)$, $\left(-1, -\dfrac{1}{2}\sqrt{7}\right)$, $(4, \sqrt{3})$, $(4, -\sqrt{3})$

27. $y = \pm\dfrac{1}{2}\sqrt{3x^2 - 4}$; $(2, \sqrt{2})$, $(2, -\sqrt{2})$, $\left(3, \dfrac{1}{2}\sqrt{23}\right)$, $\left(3, -\dfrac{1}{2}\sqrt{23}\right)$

29.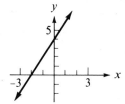

31.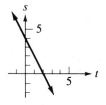

33.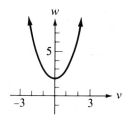

35.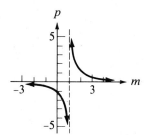

37.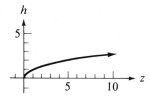

39.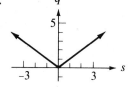

41.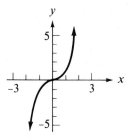

43. a. Graph (I) b. graph (II)

45.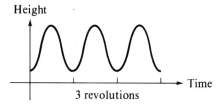

47.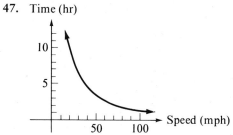

49. A solution of an equation in two variables is an ordered pair whose components satisfy the equation.

51. The graph of an equation in two variables is the graph of all the ordered pairs that are solutions of the equation.

Exercise 5.2 [page 231]

1. a. $n = 6 + 2t$
   b. h (in.)
   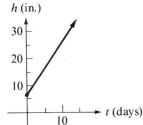
   c. 48 inches or 4 feet
   d. 33 days

3. a. $A = 250 - 15w$
   b. A (gal)
   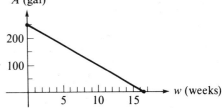
   c. 130 gallons
   d. 12 weeks

5. a. $t = 2 + \dfrac{n}{15}$
   b.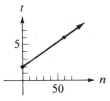
   c. 6 hours
   d. 90 houses

7. a. $C = 15w - 500$
   b.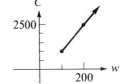
   c. 1375 calories
   d. 120 pounds

9. a. $0.60r + 0.80p = 4800$
   b. p (gal)
   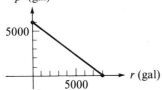
   c. 4000 gallons

11. a. $I = 10{,}000 + 0.03s$
    b.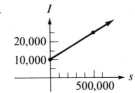
    c. $25,000
    d. $1,000,000

13.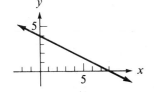

15.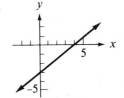

17.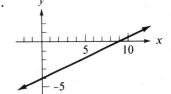

19. 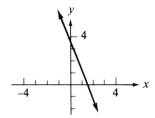 **21.** 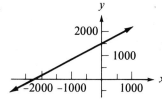 **23.**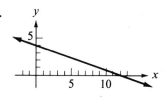

25. (Problem 1) t-int.: negative; not meaningful
                   h-int.: 6 (initial height)

27. (Problem 3) w-int.: $16.\overline{6}$ (weeks to empty tank)
                   A-int.: 250 (tank full)

29. (Problem 5) n-int.: -30 (not meaningful)
                   t-int.: 2 (length of shift if no deliveries)

31. (Problem 7) w-int.: $33.\overline{3}$ (no calorie intake; not meaningful)
                   C-int.: -500 (not meaningful)

33. (Problem 9) r-int.: 8000 (regular gas only)
                   p-int.: 6000 (premium gas only)

35. (Problem 11) s-int.: (negative; not meaningful)
                   A-int.: \$10,000 (basic salary only)

37. 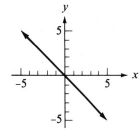 **39.** 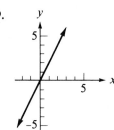 **41.**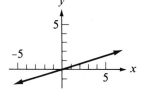

43. a. $d = 50t$
     b.

45. 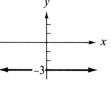 **47.** 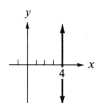 **49.**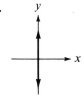

51. $2x + 3y = 12$

53. x-int.: $\dfrac{P}{M}$ y-int.: $\dfrac{P}{N}$

ANSWERS A-19

55.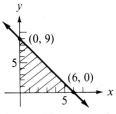

Area = 27 square units

57. The equation $ax + by = c$ is called linear because its graph is a straight line.

59. Horizontal and vertical lines cannot be graphed by the intercept method, nor can lines that pass through the origin.

Exercise 5.3 [page 243]

1. a. 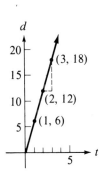 b. $\dfrac{\Delta d}{\Delta t} = \dfrac{(18 - 12) \text{ miles}}{(3 - 2) \text{ hours}} = 6$ c. Slope is the number of miles covered per hour.

3. $\dfrac{12}{5}$ 5. $-\dfrac{1}{2}$ 7. Undefined   9. -1.028 11. $-\dfrac{9}{2}$ 13. $-\dfrac{9}{19}$ 15. -1.05

17. Undefined

19. a.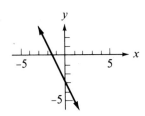

    b. Using $(0, -3)$ and $\left(-\dfrac{3}{2}, 0\right)$ the slope is -2.

21. a.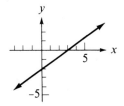

    b. Using $(0, -2)$ and $(3, 0)$ the slope is $\dfrac{2}{3}$.

A-20 ANSWERS

c. Using $(2, -7)$ and $(-2, 1)$ the slope is -2.

c. Using $\left(2, -\frac{2}{3}\right)$ and $\left(4, \frac{2}{3}\right)$ the slope is $\frac{2}{3}$.

23. (a) and (c) 25. a. l_1 negative; l_2 negative; l_3 positive; l_4 zero b. l_1, l_2, l_4, l_3
27. $3x + y = 1$ 29. $5x - 3y = 13$ 31. $0.27x + y = -5.228$ 33. $x - 7y = -18$
35. $x = -2$ 37. $1.15x - y = 8.04$
39. a. $C = 80x + 5000$
    b.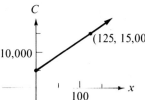

41. a. $k = 2.2p$
    b.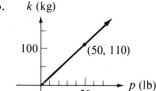

c. $m = 80$ dollars/bike is the cost of making each bike.

c. $m = 2.2$ kilograms/pound is the conversion factor from pounds to kilograms.

43. a. $d = 65t + 265$
    b.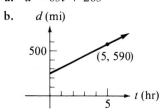

    c. $m = 65$ miles/hour is their average speed.

45. $-\dfrac{29}{4}$ 47. -7 49. a. 14 b. $23\dfrac{1}{3}$ c. -56 51. $14\dfrac{2}{7}$ feet

53.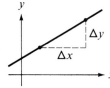

Slope is a number that measures the "steepness" of a line. It is defined as the ratio of change in vertical coordinate to change in horizontal coordinate, or $\Delta y/\Delta x$, between two points on the line.

55. For a fixed value of Δx, the larger the value of Δy, the steeper the line. Thus, the larger the slope, the more quickly the line rises from left to right. Lines that fall from upper left to lower right have negative slopes.

Exercise 5.4 [page 254]

1. a. $y = -\dfrac{3}{2}x + \dfrac{1}{2}$

3. a. $y = \dfrac{1}{3}x - \dfrac{2}{3}$

    b. Slope: $-\dfrac{3}{2}$

        y-int.: $\dfrac{1}{2}$

5. a. $y = -\dfrac{1}{6}x + \dfrac{1}{9}$

   b. Slope: $-\dfrac{1}{6}$

       y-int.: $\dfrac{1}{9}$

9. a. $y = -29$

   b. Slope: 0

       y-int.: -29

13. a.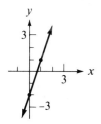

    b. $y = 3x - 2$

17. a.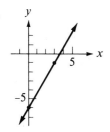

    b. $y = \dfrac{5}{3}x - 6$

21. a. Slope: 4

      y-int.: 38

    b. $y = 4x + 38$

    b. Slope: $\dfrac{1}{3}$

        y-int.: $-\dfrac{2}{3}$

7. a. $y = 14x - 22$

   b. Slope: 14

       y-int.: -22

11. a. $y = -\dfrac{5}{3}x + 16\dfrac{1}{3}$

    b. Slope: $-\dfrac{5}{3}$

        y-int.: $16\dfrac{1}{3}$

15. a.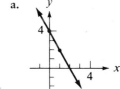

    b. $y = -2x + 4$

19. a.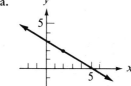

    b. $y = -\dfrac{1}{2}x + 3$

23. a. Slope: -87.5

      P-int.: -2000

    b. $P = -87.5t - 2000$

25. a. Slope: $\frac{1}{4}$

   V-int.: 0

   b. $V = \frac{1}{4}d$

27. a. Parallel   b. Neither   c. Neither   d. Perpendicular

29. Slope $\overline{AB}$: -1; slope $\overline{BC}$: 1; slope $\overline{AC}$: $\frac{1}{4}$. Hence, $\overline{AB} \perp \overline{BC}$, so the triangle is a right triangle.

31. Slope $\overline{PQ}$: 4; slope $\overline{QR}$: $-\frac{7}{2}$; slope $\overline{RS}$: 4; slope $\overline{SP}$: $-\frac{7}{2}$. Hence, $\overline{PQ} \parallel \overline{RS}$ and $\overline{QR} \parallel \overline{SP}$, so the points are the vertices of a parallelogram.

33. $x - 2y = 4$

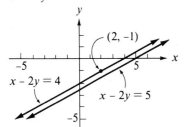

35. $2x + 3y = 14$

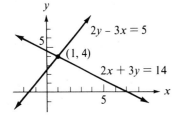

37. $k = -28$

39. a. Right angles are equal
    b. Alternate interior angles are equal
    c. Two angles of one triangle equal two angles of the other
    d. Definition of slope
    e. Corresponding sides of similar triangles are proportional

41. Find the slope of each line by transforming to slope-intercept form:

$$y = \frac{-a}{b}x + \frac{c}{b}, \text{ so } m_1 = \frac{-a}{b};$$

$$y = \frac{-p}{q}x + \frac{r}{q}, \text{ so } m_2 = \frac{-p}{q}.$$

Since the lines are parallel, $m_1 = m_2$ or $-a/b = -p/q$ or $-aq = -pb$ or $aq - pb = 0$.

43. First, locate the y-intercept, $(0, b)$. Write the slope as a ratio, $m = \Delta y/\Delta x$. From the y-intercept move Δy units in the vertical direction and from there move Δx units in the horizontal direction. Plot a second point at this location and draw the line.

Chapter 5 Review [page 258]
1. a. 9 feet
   b. 10 days
   c. 3 feet
   d. January 8 through January 9; January 13 through January 14
2. a. 32°F, 11°F
   b. 5 days
   c. January 8
   d. 14°F
3. a. $(0, -2)$ b. $\left(\dfrac{10}{3}, 0\right)$ c. $\left(2, -\dfrac{4}{5}\right)$
4. a. $\left(0, \dfrac{4}{3}\right)$ b. $(8, 0)$ c. $(20, -2)$
5. $y = \dfrac{4 - 2x}{x}$; $(2, 0), (4, -1)$
6. $y = \dfrac{2x - 6}{x}$; $(-2, 5), (2, -1)$
7. $y = \dfrac{6}{x - 3}$; $(-3, -1), (0, -2)$
8. $y = \dfrac{8}{x^2 + 2}$; $\left(-4, \dfrac{4}{9}\right), \left(-2, \dfrac{4}{3}\right)$
9. $y = \pm\dfrac{1}{2}\sqrt{x^2 - 16}$; $(4, 0), \left(5, \dfrac{3}{2}\right), \left(5, -\dfrac{3}{2}\right)$
10. $y = \pm\sqrt{\dfrac{8 - x^2}{2}}$; $\left(-1, \sqrt{\dfrac{7}{2}}\right), \left(-1, -\sqrt{\dfrac{7}{2}}\right), (0, 2), (0, -2)$

11.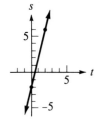

12.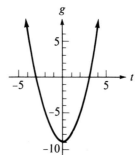

13.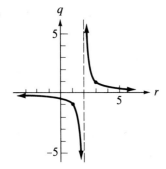

14.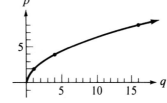

15.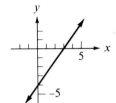
16.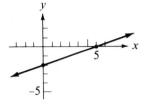
17.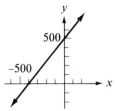

18.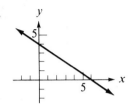
19.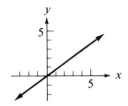
20.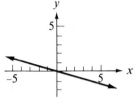

21.
22.

23. $-\dfrac{3}{2}$ **24.** 2 **25.** -0.4 **26.** -1.7 **27.** $2x + 3y = 10$ **28.** $3x - 2y = 16$

29. $9x + 5y = 2$ **30.** $5x - 2y = -16$

31. a. $y = \dfrac{1}{2}x - \dfrac{5}{4}$

b. Slope: $\dfrac{1}{2}$

y-int.: $-\dfrac{5}{4}$

32. a. $y = -\dfrac{3}{4}x + \dfrac{5}{4}$

b. Slope: $-\dfrac{3}{4}$

y-int.: $\dfrac{5}{4}$

33. a. $y = -4x + 3$
b. Slope: -4
y-int.: 3

34. a. $y = 3$
b. Slope: 0
y-int.: 3

35. a.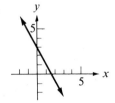

36. a.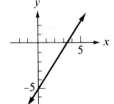

b. $2x + y = 3$ b. $3x - 2y = 10$
37. Parallel   38. Perpendicular   39. $2x + 3y = 14$ 40. $3x - 2y = -5$
41. a. $C = 20x + 2000$ 42. a. $W = 18x + 80$
    b.   b.

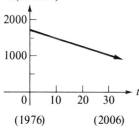

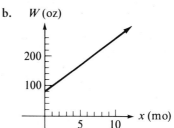

    c. $22,000   c. 242 ounces or 15 pounds, 2 ounces
    d. 400   d. $3\frac{5}{9}$ months or 3 months, 17 days

43. a. $R = 1660 - 20t$ 44. a. $R = 500 - 8t$
    b.   b.

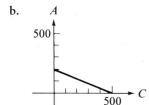

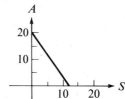

    c. 1180 billion barrels   c. 308 million tons
    d. Year 2059 (83 years from 1976)   d. Year 2038 $\left(62\frac{1}{2} \text{ years from } 1976\right)$

45. a. $5A + 2C = 1000$ 46. a. $60A + 100S = 1200$
    b.   b.

    c. 200 tickets   c. 6 days
    d. x-int.: (500) only children's tickets d. x-int.: all days in Saint-Tropez
       are sold   y-int.: all days in Atlantic City
       y-int.: (200) only adults' tickets are sold

A-26 ANSWERS

47. a. $F = 500 + 0.10R$
    b.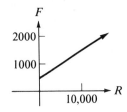
    c. $1700

48. a. $F = 35 + 0.02V$
    b.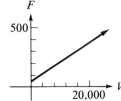
    c. $435

49. a. $P = 4800 + 132t$
    b.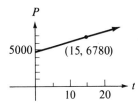
    c. Slope: 132 people/year is the rate of population growth.

50. a. $M = 112 + 28g$
    b.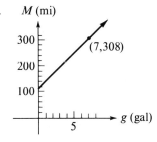
    c. Slope: 28 miles/gallon is the gas mileage for Cicely's car.

Exercise 6.1 [page 269]

1. Function; a specific sales tax for each price
3. Not a function; incomes may differ for the same number of years of education
5. Function; a specific weight for each volume
7. Independent: items purchased; dependent: price of item. Yes, a function.
9. Independent: topics; dependent: page or pages on which topic occurs. No, not a function.
11. Independent: students' names; dependent: students' scores on tests, quizzes, etc. No, not a function.
13. Independent: person stepping on scales; dependent: person's weight. Yes, a function.
15. No   17. Yes   19. Yes   21. Yes   23. No   25. Yes
27. a. t, independent; V, dependent

    b.
t	V
0	28,000
1	26,320
2	24,640
3	22,960
4	21,280

    c. A function

29. a. X, independent; R, dependent

    b.
X	R (thousands)
10	3,800
12	5,600
14	7,800
15	9,050
16	10,400

    c. A function

31. a. P, independent; N, dependent

b.
p	N
6,000	2000
8,000	1500
10,000	1200
12,000	1000
15,000	800

c. A function

33. a. d, independent; v, dependent

b.
d	v
10	10.95
20	15.49
50	24.49
100	34.64
150	42.43

c. A function

35. a. 60 b. 37.5 c. 30 37. a. 15% b. 14% c. 7010 − 9169
39. a. 0 b. 10 c. 31.4 d. $4\frac{2}{3}$ 41. a. 1 b. 6 c. $\frac{3}{8}$ d. 96.48
43. a. $\frac{5}{6}$ b. 9 c. $-\frac{1}{10}$ d. 0.923 45. a. $2\sqrt{3}$ b. 0 c. $\sqrt{3}$ d. 0.447
47. a. 4096 b. $\frac{1}{8}$ c. 28,988.79 d. 400,000
49. a. $27a^2 - 18a$ b. $3a^2 + 6a$ c. $3a^2 - 6a + 2$ d. $3a^2 + 6a$
51. a. 8 b. 8 c. 8 d. 8
53. a. $8x^3 - 1$ b. $2x^3 - 2$ c. $x^6 - 1$ d. $x^6 - 2x^3 + 1$ 55. $V(10) = \$11,200$
57. $R(40) = \$72,800,000$ 59. $D(6000) = 2000$ cars 61. $v(250) = 54.77$ miles per hour
63. a. 11 b. 13 c. $3a + 3b - 4$ d. $3a + 3b - 2$; $f(a) + f(b) \neq f(a + b)$
65. a. 19 b. 28 c. $a^2 + b^2 + 6$ d. $a^2 + 2ab + b^2 + 3$; $f(a) + f(b) \neq f(a + b)$
67. a. $\sqrt{3} + 2$ b. $\sqrt{6}$ c. $\sqrt{a+1} + \sqrt{b+1}$ d. $\sqrt{a+b+1}$; $f(a) + f(b) \neq f(a + b)$
69. a. $-\frac{5}{3}$ b. $-\frac{2}{5}$ c. $-\frac{2}{a} - \frac{2}{b}$ d. $\frac{-2}{a+b}$; $f(a) + f(b) \neq f(a + b)$
71. a. 0, 5 b. $0, -\frac{3}{2}$ c. $\frac{5}{6}$ d. $-5, \frac{1}{2}$
73. a. $\sqrt{2}, -4$ b. -2 c. $\frac{4}{3}$ d. $2, \frac{7}{9}$
75. a. $3x + 3h - 5$ b. $3h$ c. 3
77. a. $2x^2 + 4xh + 2h^2$ b. $4xh + 2h^2$ c. $4x + 2h$
79. a. $x^2 + 2xh + h^2 - 3x - 3h + 1$ b. $2xh + h^2 - 3h$ c. $2x + h - 3$
81. a. $\frac{3}{x + h + 2}$ b. $\frac{-3h}{(x + h + 2)(x + 2)}$ c. $\frac{-3}{(x + h + 2)(x + 2)}$
83. $f(x) = 3x - 2$ 85. $G(t) = t^2 + 1$

Exercise 6.2 [page 283]

1. a. $-2, 0, 5$ b. 2 c. x-int.: -2 and 1; y-int.: -2 d. 5 e. 3

3. **a.** $-1, 2$ **b.** 3 and $-\dfrac{3}{2}$ **c.** x-int.: $-2, 2,$ and 4; y-int.: 4 **d.** $4; -5$ **e.** $0; 5$

5. **a.** $0, \dfrac{1}{2}, 0$

   b. $\dfrac{3}{4}$

   c. $-\dfrac{5\pi}{6}, -\dfrac{\pi}{6}, \dfrac{7\pi}{6}, \dfrac{11\pi}{6}$

   d. $1, -1$

   e. Maximum values at $-\dfrac{3\pi}{2}, \dfrac{\pi}{2}$;

   minimum values at $-\dfrac{\pi}{2}, \dfrac{3\pi}{2}$

7. **a.** $2, 2, 1$
   b. $[-6, -4), [0, 2)$
   c. $2, -1$
   d. Maximum values at $[-3, -1), [3, 5)$;
   minimum values at $[-6, -4), [0, 2)$

9.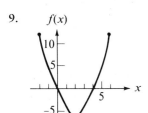

11.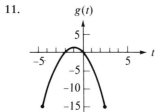

13.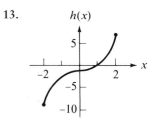

15.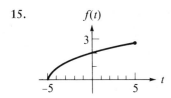

17.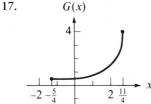

19.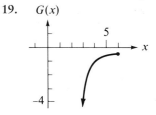

21. (Problem 1) domain: $[-4, 4]$
    range: $[-5, 5]$
23. (Problem 3) domain: $[-3, 5]$
    range: $[-5, 4]$
25. (Problem 5) domain: $[-2\pi, 2\pi]$
    range: $[-1, 1]$
27. (Problem 7) domain: $[-6, 5)$
    range: $-1, 1, 2$
29. (Problem 9) domain: $[-2, 6]$
    range: $[-6, 12]$
31. (Problem 11) domain: $[-5, 3]$
    range: $[-15, 1]$
33. (Problem 13) domain: $[-2, 2]$
    range: $[-9, 7]$
35. (Problem 15) domain: $[-4, 5]$
    range: $[0, 3]$
37. (Problem 17) domain: $\left[-\dfrac{5}{4}, \dfrac{11}{4}\right]$
    range: $\left[\dfrac{4}{17}, 4\right]$
39. (Problem 19) domain: $(3, 6]$
    range: $\left(\infty, -\dfrac{1}{3}\right]$

41. Function 43. Not a function 45. Not a function 47. Function 49. Not a function

Exercise 6.3 [page 291]

1.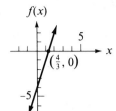

3.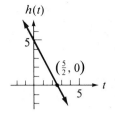

5.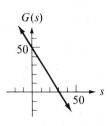

7.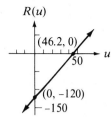

9.
x	y
−3	9
−2	4
−1	1
0	0
1	1
2	4
3	9

Domain: $(-\infty, \infty)$
range: $[0, \infty)$

11.
x	y
0	0
1	1
4	2
9	3
16	4
25	5

Domain: $[0, \infty)$
range: $[0, \infty)$

13.
x	y
−4	$-\frac{1}{4}$
−2	$-\frac{1}{2}$
−1	−1
$-\frac{1}{2}$	−2
$-\frac{1}{4}$	−4
⋮	⋮
0	Undef.
⋮	⋮
$\frac{1}{4}$	4
$\frac{1}{2}$	2
1	1
2	$\frac{1}{2}$
4	$\frac{1}{4}$

Domain: all real numbers except 0
range: all real numbers except 0

15.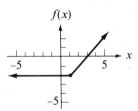

17.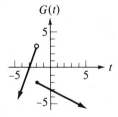

19.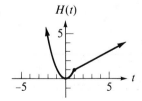

21.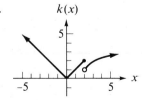

23.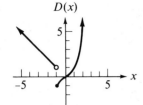

25.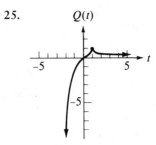

27. a. $f(x) > g(x)$ on $(0, 1)$ **b.** $g(x) > f(x)$ on $(1, \infty)$

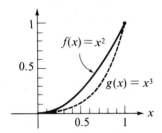 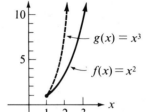

c.

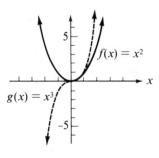

29. a. $g(x) > f(x)$ on $(0, 1]$ **b.** $f(x) > g(x)$ on $[1, \infty)$

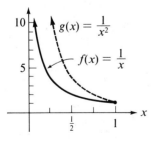 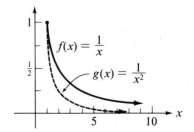

c.

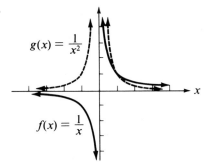

Exercise 6.4 [page 300]

1. $C = 10n + 5000$; $11,400
3. $C = 52.5s$; $3150
5. $P = 2w + \dfrac{6400}{w}$; 240 feet
7. $r = \dfrac{C}{2\pi}$; 7.48 meters (approx.)
9. $V = \dfrac{1}{4}\pi h^3$; 785.4 cubic centimeters
11. $A = 3000(1.08)^t$; $3779.14
13. $I = 0.125x + 0.06(20{,}000 - x)$; $1850
15. $L = \dfrac{1}{3}h\sqrt{5}$; 14.9 feet (approx.)
17. $D = \sqrt{625t^2 + (700 - 10t)^2}$; 1503 feet (approx.)
19. a. $y = \dfrac{2}{3}x^2$
    b. $\dfrac{8}{3}$
    c.
21. a. $y = \dfrac{11200}{x}$
    b. 1400
    c.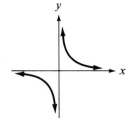
23. a. $d = 16t^2$
    b. 1600 feet
    c.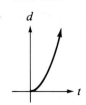
25. a. $P = \dfrac{7500}{V}$
    b. 120 kilograms per square centimeter
    c.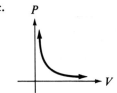

27. **a.** $p = 62.4d$
    b. 6240 pounds per square foot
    c.

29. **a.** $f = \dfrac{14300}{l}$
    b. 55.86 centimeters
    c.
    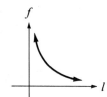

31. $C = 15w + \dfrac{1500}{w}$

33. $A = \dfrac{x^2}{16} + \dfrac{(200 - x)^2}{25}$

35. $C = 15\pi - 0.07\pi(10 - x)^2$

37. $t = \dfrac{4 - x}{4} + \dfrac{\sqrt{9 + x^2}}{3}$

39. One-fourth of original illumination

41. 81% of original resistance

43. **a.** Yes **b.** No

Exercise 6.5 [page 316]

1.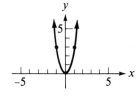

3.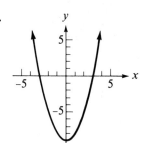

5.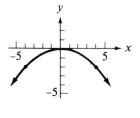

7.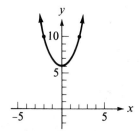

9.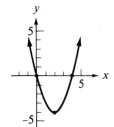

11.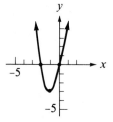

13.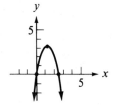

15. $(1, 1)$ **17.** $\left(\dfrac{5}{2}, -\dfrac{13}{4}\right)$ **19.** $\left(\dfrac{2}{3}, \dfrac{1}{9}\right)$ **21.** $(-4.5, 18.5)$

23. Vertex: $\left(\dfrac{5}{2}, -\dfrac{9}{4}\right)$

y-int.: 4

x-int.: 1, 4

25. Vertex: $\left(\dfrac{7}{4}, \dfrac{81}{8}\right)$

y-int.: 4

x-int.: $-\dfrac{1}{2}$, 4

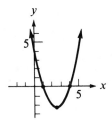
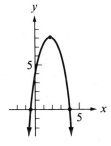

27. Vertex: $(-0.5, -1.35)$

y-int.: -1.2

x-int.: $-2, 1$

29. Vertex: $(-2, 3)$

y-int.: 7

x-int.: none

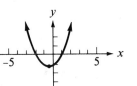

31. Vertex: $\left(\dfrac{1}{4}, -\dfrac{23}{8}\right)$

y-int.: -3

x-int.: none

33. Vertex: $(-1, -2)$

y-int.: -1

x-int.: $-2.4, 0.4$

35. Vertex: $\left(\dfrac{3}{2}, \dfrac{3}{2}\right)$

    y-int.: -3

    x-int.: $0.6, 2.4$

    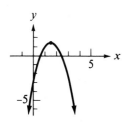

37. a. 2 seconds, 64 feet

    b. d (ft)

    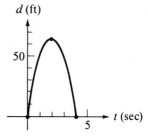

39. a. 625 square inches

    b.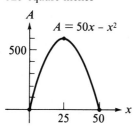
    $A = 50x - x^2$

41. a. 11,250 square yards

    b.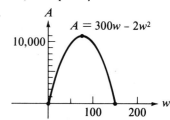
    $A = 300w - 2w^2$

43. a. $44

    b.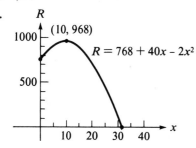
    $R = 768 + 40x - 2x^2$

45. a. 40¢ per pound

    b. 50¢ per pound

    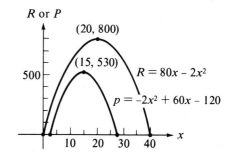
    $R = 80x - 2x^2$
    $p = -2x^2 + 60x - 120$

Many different answers besides those given here are possible for the following.

47. $y = x^2 + 2$ and $y = 2x^2 + 2$

49. $y = x^2 + 2x - 5$ and $y = -2x^2 - x - 5$

51. $y = x^2 + x - 6$ and $y = -x^2 - x + 6$
53. $y = x^2 - 4x + 1$ and $y = 2x^2 - 8x + 5$

Chapter 6 Review [page 318]
1. Function   2. Not a function   3. Not a function   4. Function
5. $f(-2) = 6$ and $f(0) = 1$ 6. $g(12) = 82$ and $g(15) = 89$
7. -1; 19 8. $-\dfrac{3}{7}$; $-\dfrac{11}{9}$ 9. 1; $\sqrt{37}$ 10. $4a^2 + 4a$; $a^2 + 4a + 3$
11. -11; -13 12. $2a^2 + 2b^2 - 8$; $2a^2 + 4ab + 2b^2 - 4$
13. 5; 1 and 5 14. 2; -2 and 2
15. a. $f(-2) = 3$; $f(2) = 5$
    b. 1 and 3
    c. x-int.: -3 and 4
       y-int.: 2
    d. 5; 2
16. a. $P(-3) = -2$; $P(3) = 3$
    b. -5, $-\dfrac{1}{2}$, and 4
    c. x-int.: -4, -1, 5
       y-int.: 3
    d. -3; -2

17.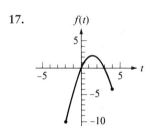

Range: $\left[-10, \dfrac{9}{4}\right]$

18.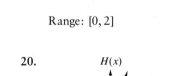

Range: $[0, 2]$

19.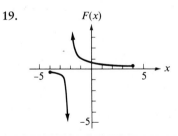

Range: $\left(-\infty, -\dfrac{1}{2}\right]$ and $\left[\dfrac{1}{6}, \infty\right)$

20.

Range: $\left(-\infty, -\dfrac{1}{2}\right]$ and $\left[\dfrac{1}{6}, \infty\right)$

21. Function   22. Not a function   23. Not a function   24. Function

25.
26.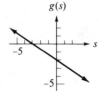
27.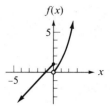
28.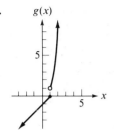
29.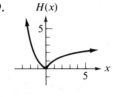
30.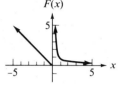
31.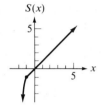
32.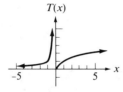

33. $V = 30{,}000 - 5000\left(\dfrac{t}{3}\right)$; $20,000
34. $C = 42l + 64$; $316
35. $R = 75{,}000 - 7x$; $73,250
36. $S = 0.25x + 0.40$; 0.90 ounce
37. $A = \dfrac{1}{4}s^2\sqrt{3}$; $4\sqrt{3}$ square centimeters
38. $A = \dfrac{1}{2}x\sqrt{144 - x^2}$; $16\sqrt{2}$ centimeters
39. $s = \dfrac{7}{4}t^2$; 63 centimeters
40. $V = \dfrac{4T}{P}$; 32 cubic units
41. 480 bottles
42. 14.0625 lumens

43.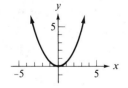

Vertex: (0, 0)
x- and y-int.: (0,0)

44.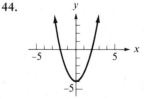

Vertex: (0, −4)
x-int.: −2, 2
y-int.: −4

45.

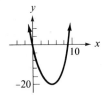

Vertex: $\left(\dfrac{9}{2}, -\dfrac{81}{4}\right)$

46.

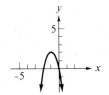

Vertex: $(-1, 2)$

x-int.: $-2, 0$

y-int.: 0

47.

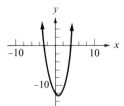

Vertex: $\left(\dfrac{1}{2}, -\dfrac{49}{4}\right)$

x-int.: $-3, 4$

y-int.: -12

48.

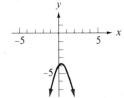

Vertex: $\left(\dfrac{1}{4}, -\dfrac{31}{8}\right)$

x-int.: none

y-int.: -4

49.

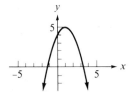

Vertex: $(1, 5)$

x-int.: $1 + \sqrt{5}, 1 - \sqrt{5}$

y-int.: 4

50.

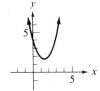

Vertex: $\left(\dfrac{3}{2}, \dfrac{7}{4}\right)$

x-int.: none

y-int.: 4

51. 42 trees per acre

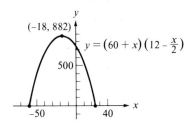

$y = (60 + x)\left(12 - \dfrac{x}{2}\right)$

52. $35

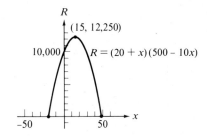

$R = (20 + x)(500 - 10x)$

A-38 ANSWERS

53. $2x + h + 2$ 54. $\dfrac{-1}{x(x+h)}$ 55. $g(x) = \dfrac{24}{x}$ 56. $F(x) = -x^3$

57. $\dfrac{2\sqrt{5}}{5}$ times the original period 58. $3963\sqrt{3}$, or 6864, miles

Exercise 7.1 [page 328]

1. 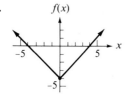 3. 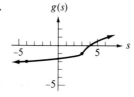 5.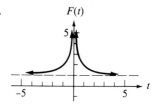

7. 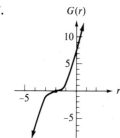 9. 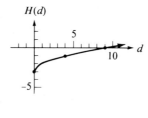 11.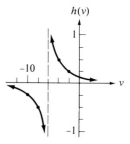

13. 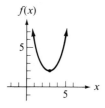 15. 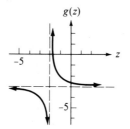 17.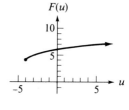

19. 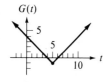 21. 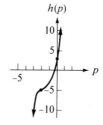 23.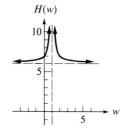

ANSWERS A-39

25.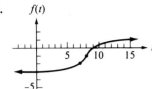

27.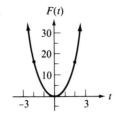

29.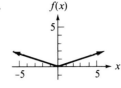

31.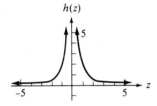

33.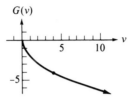

35.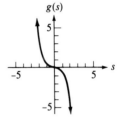

37.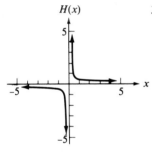

39. $y = |x + 1| - 2$ **41.** $y = -\sqrt{x} + 3$ **43.** $y = (x + 1)^2 + 1$

45. $y = (x - 2)^2 + 3$

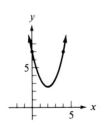

47. $y = (x + 1)^2 - 4$

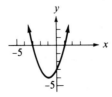

Exercise 7.2 [page 337]

1.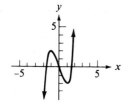
3.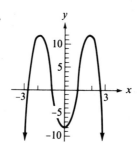

5. $74, -46$
7. $42, 5$

9.
x	-4	-3	-2	-1	0	1
$Q(x)$	-18	0	0	-6	-6	12

11.
x	-3	-2	-1	0	1	2	3	4	5
$D(x)$	72	-48	-30	0	12	0	-18	0	120

13. $10.99, 257.76$ 15. $-639, 355.7$ 17. 0 (multiplicity two)   19. $0, \pm 2\sqrt{2}$ 21. $-\dfrac{1}{2}$

23. $\dfrac{-3 \pm \sqrt{5}}{2}$, 0 (multiplicity two)   25. Not a factor   27. Factor   29. $\dfrac{1 \pm \sqrt{5}}{2}$

31. $0, 2, 4$

Exercise 7.3 [page 343]

1.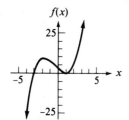
3.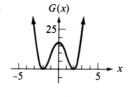
5.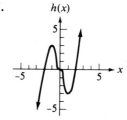

7.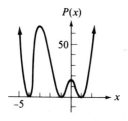
9.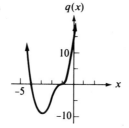
11.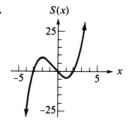

13.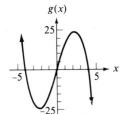
15.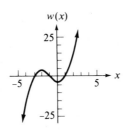
17.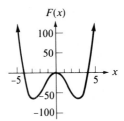
19.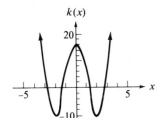
21.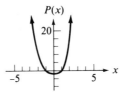
23.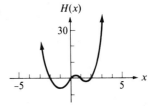
25.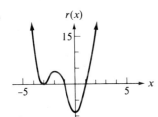
27.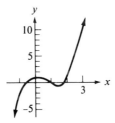
29.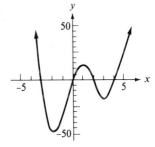

Exercise 7.4 [page 352] 1. $x = 3$; domain: all x, $x \neq 3$
3. $x = 2$, $x = -3$; domain: all x, $x \neq 2, -3$ 5. $x = -2$, $x = 3$; domain: all x, $x \neq -2, 3$

7. Vertical asymptotes: $x = 3$, $x = -3$
   horizontal asymptote: $y = 0$

9. Vertical asymptote: $x = \dfrac{1}{2}$
   horizontal asymptote: $y = \dfrac{1}{2}$

11. Vertical asymptotes: $x = -1$, $x = 4$
    horizontal asymptote: $y = 2$

13. Horizontal asymptote: $y = 0$
    vertical asymptote: $x = -3$
    y-intercept: $\dfrac{1}{3}$

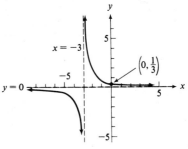

15. Horizontal asymptote: $y = 0$
    vertical asymptotes: $x = 4$, $x = -1$
    y-intercept: $-\dfrac{1}{2}$

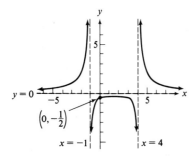

17. Horizontal asymptote: $y = 0$
    vertical asymptotes: $x = 4$, $x = 1$
    y-intercept: $\dfrac{1}{2}$

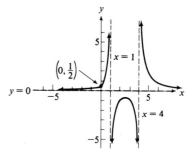

19. Horizontal asymptote: $y = 1$
    vertical asymptote: $x = -3$
    intercepts: $x = 0$, $y = 0$

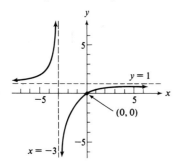

21. Horizontal asymptote: $y = 1$
    vertical asymptote: $x = -2$
    y-intercept: $\dfrac{1}{2}$
    x-intercept: -1

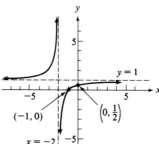

23. Horizontal asymptote: $y = 0$
    vertical asymptotes: $x = 2$, $x = -2$
    intercepts: $(0, 0)$

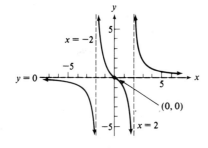

25. Horizontal asymptote: $y = 0$
    vertical asymptotes: $x = -1$, $x = -4$
    y-intercept: $-\dfrac{1}{2}$;
    x-intercept: 2

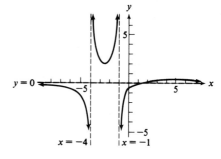

27. Horizontal asymptote: $y = 1$
vertical asymptotes: $x = 2$, $x = -2$
y-intercept: $\dfrac{1}{4}$
x-intercepts: -1 and 1

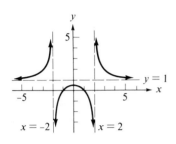

29. Horizontal asymptote: $y = 0$
vertical asymptote: none
intercepts: $(0, 0)$

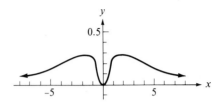

31.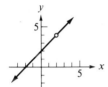

33.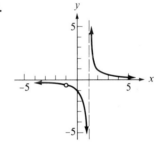

Exercise 7.5 [page 360] **1.** $g(x) = x - 2$ **3.** $g(x) = \dfrac{1}{2}x$ **5.** $g(x) = \dfrac{1}{2}x + 3$

7. $g(x) = 3 - 2x$ **9.** $g(x) = \sqrt[3]{x - 1}$ **11.** $g(x) = x^3$ **13.** $g(x) = \dfrac{x + 1}{x}$

15. $g(x) = (x + 4)^3$

17. a. $f(x) = (x - 2)^3$; $g(x) = \sqrt[3]{x} + 2$
  b. $f(4) = (4 - 2)^3 = 8$; $g(8) = \sqrt[3]{8} + 2 = 4$
  c. $g(-8) = \sqrt[3]{-8} + 2 = 0$; $f(0) = (0 - 2)^3 = -8$

19.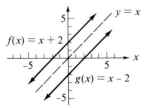

21.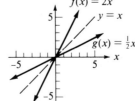

23.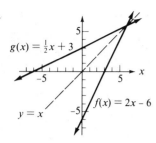

25.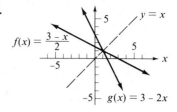

27.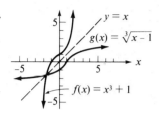

29.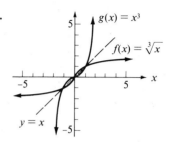

31.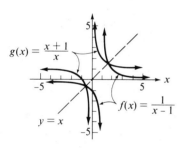

33.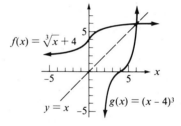

35. a. Yes b. No c. No d. Yes   37. a. Yes b. No   39. a. Yes b. No
41. 6   43. $\dfrac{2}{9}$ 45. $f^{-1}(x) = \dfrac{x+2}{x-1}$ 47. a. 0 b. 1   49. a. 0 b. 2

51. a.

x	f(x)
-2	$\dfrac{1}{4}$
-1	$\dfrac{1}{2}$
0	1
1	2
2	4

b, d.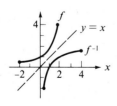

c.

x	$f^{-1}(x)$
$\dfrac{1}{4}$	-2
$\dfrac{1}{2}$	-1
1	0
2	1
4	2

Chapter 7 Review [page 362]

1.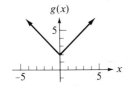

2.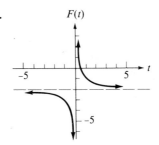

3.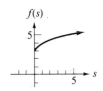

4.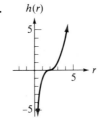

5.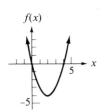

6.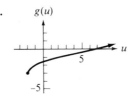

7.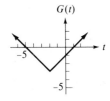

8.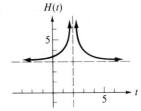

9.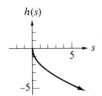

10.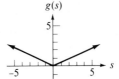

11. 25   12. −433   13. −4.656   14. 18.987   15. −2, 2, 0 (multiplicity three)

A-46 ANSWERS

16. $0, \dfrac{1}{2}, -2$

17.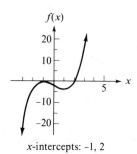
x-intercepts: −1, 2

18.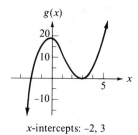
x-intercepts: −2, 3

19.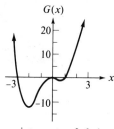
x-intercepts: −3, 0, 1

20.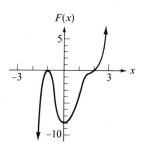

21.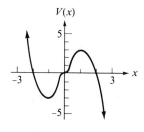

22.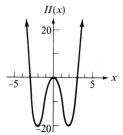

23.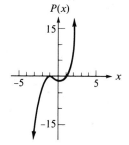

24.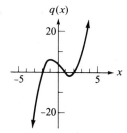

25.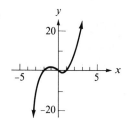

26.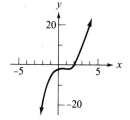

ANSWERS A-47

27.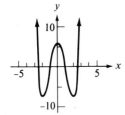

28.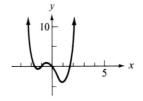

29. Asymptotes: $x = 4$, $y = 0$
y-intercept: $-\dfrac{1}{4}$

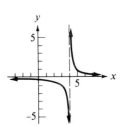

30. Asymptotes: $x = -2$, $x = 5$, $y = 0$
y-intercept: $-\dfrac{1}{5}$

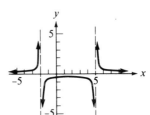

31. Asymptotes: $x = -3$, $y = 1$
x-intercept: 2
y-intercept: $-\dfrac{2}{3}$

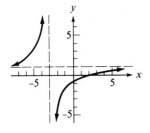

32. Asymptotes: $x = -1$, $x = 3$, $y = 0$
x-intercept: 1
y-intercept: $\dfrac{1}{3}$

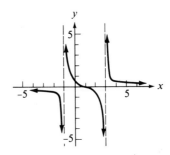

33. Asymptotes: $x = -2$, $x = 2$, $y = 3$
    x-intercept: 0
    y-intercept: 0

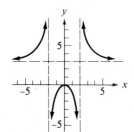

34. Asymptotes: $x = 3$, $x = -3$, $y = 2$
    x-intercepts: $-1, 1$
    y-intercept: $\dfrac{2}{9}$

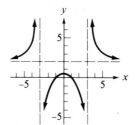

35. $g(x) = x - 4$ 36. $g(x) = 4x + 2$ 37. $g(x) = \sqrt[3]{x + 1}$ 38. $g(x) = \dfrac{1 - 2x}{x}$

39. $g(x) = \dfrac{1}{x - 2}$ 40. $g(x) = (x + 2)^3$

41. (Problem 35)

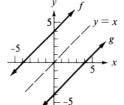

42. (Problem 37)

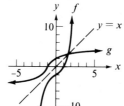

43. 0   44. $\dfrac{1}{7}$

45. $y = (x - 3)^2 - 3$

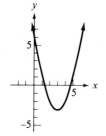

46. $y = (x + 1)^2 - 5$

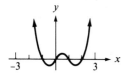

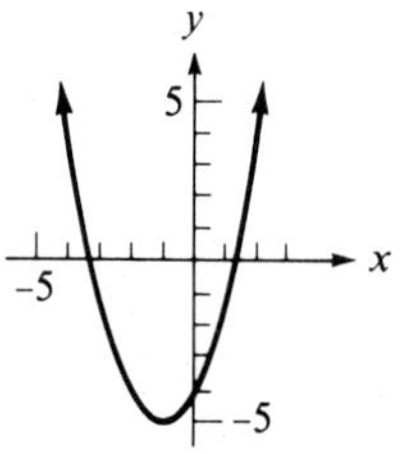

47. $-1, 0, 1, 2$

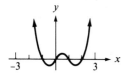

48. $-3, -1, 0, 3$

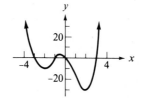

49. $f^{-1}(x) = \dfrac{3x + 1}{1 - x}$ 50. $f^{-1}(x) = \dfrac{2x + 4}{x - 2}$

Exercise 8.1 [page 375]

1. a. $P(t) = 300(2)^t$
   b. 76,800; 492
   c.

3. a. $P(t) = 20,000(2.5)^{t/6}$
   b. 36,840; 424,127
   c.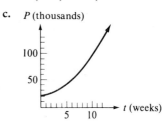

5. a. $A(t) = 4000(1.08)^t$
   b. $4,665.60; $8,635.70
   c.

7. a. $P(t) = 20,000(1.05)^t$
   b. $35,917.13; $74,669.13
   c.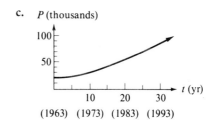

9. a. $S(t) = 1500(1.12)^t$
   b. 2960; 5844
   c.

11. a. $P(t) = 250,000(0.75)^{t/2}$
    b. 162,379; 79,101
    c.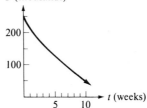

13. a. $L(d) = (0.85)^{d/4}$
    b. 44%; 16%
    c.

15. a. $P(t) = 50(0.992)^t$
    b. 46.7 pounds; 22.4 pounds
    c.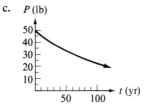

A-50 ANSWERS

17. Species B
19. a. $L(t) = 1.5t + 6$ b. $E(t) = 6(1.5)^{t/2}$

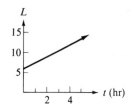
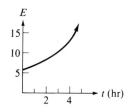

21. $163.84; $5,368,709.12
23. a. $466,560
    b. 279,936
    c. People lower on the list don't make money—there aren't enough victims to go around!
25. 1.57%   27. a. 3.53%   b. 3.53%   c. No   d. 3.53%

Exercise 8.2 [page 383]   1. 4.729   3. 19.470   5. −0.170   7. 0.018   9. 937.230
11. −140.169

13. 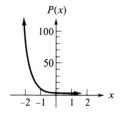   15. 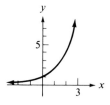   17.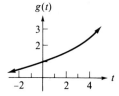

19. 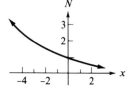   21. 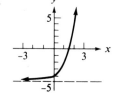   23.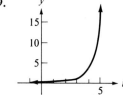

25.

27.

29.

31.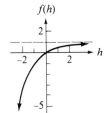

33.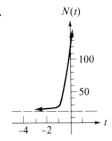

35. 5   **37.** $\dfrac{2}{3}$ **39.** $-\dfrac{1}{4}$ **41.** $\dfrac{1}{7}$ **43.** $-\dfrac{5}{4}$ **45.** ± 2 **47.** 12 days
49. 72 days later   **51.** 4 years   **53.** (a), (d)

55.

x	-2	-1	0	1	2	3	4	5	6
$f(x)$	4	1	0	1	4	9	16	25	36
$g(x)$	$\dfrac{1}{4}$	$\dfrac{1}{2}$	1	2	4	8	16	32	64

Exercise 8.3 [page 392]   **1.** 2   **3.** 3   **5.** $\dfrac{1}{2}$ **7.** -1 **9.** 1   **11.** 0   **13.** 5
15. -4 **17.** 4   **19.** -1 **21.** $16^w = 256$ **23.** $b^{-2} = 9$ **25.** $10^{-2.3} = A$
27. $4^{2q-1} = 36$ **29.** $u^w = v$ **31.** $\log_8 \dfrac{1}{2} = -\dfrac{1}{3}$ **33.** $\log_t 16 = \dfrac{3}{2}$ **35.** $\log_{0.8} M = 1.2$
37. $\log_x(w-3) = 5t$ **39.** $\log_3 2N_0 = -0.2t$ **41.** 2   **43.** 64   **45.** -1 **47.** 100
49. 11   **51.** $7^{2/3}$ **53.** $4 < \log_2 25 < 5$ **55.** $1 < \log_{10} 50 < 2$ **57.** $0 < \log_8 5 < 1$
59. $3 < \log_3 67.9 < 4$ **61.** 1.7348   **63.** 3.3700   **65.** -1.1367 **67.** -0.1585
69. 2.30   **71.** -0.23 **73.** 2.53   **75.** 0.77   **77.** -0.69 **79.** 3.65
81. 9.60 inches   **83.** 1.91 miles   **85.** 3.34 miles
87. a. 19,969,613   **b.** 25,372,873; 32,238,116; 40,960,915   **89. a.** 1970   **b.** 1987
91. 1   **93.** 0   **95.** 1   **97.** 0

Exercise 8.4 [page 400]

1.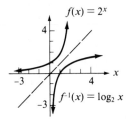

3.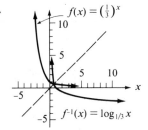

5. 2.688   7. 0.334   9. Undefined   11. 8.683   13. 15.614   15. 0.419   17. 12.04
19. -1.57×10^{-5} 21. 1.76   23. 25.70   25. 3.31   27. 0.50   29. 3.2
31. 1.3×10^{-2} 33. 100 decibels   35. 6.31×10^6 watts per square meter   37. 1000 times
39. 25,119 times   41. 4.7
43. a. 81   b. 4   c. $\log_3 x$ and 3^x are inverse functions.   d. 1.8; a
45. "Take the log base 6 of x."   47. 243   49. 100,000   51. $x > 9$

53. a.

x	x^2	$\log_{10} x$	$\log_{10} x^2$
1	1	0	0
2	4	0.3010	0.6020
3	9	0.4771	0.9542
4	16	0.6021	1.2041
5	25	0.6990	1.3979
6	36	0.7782	1.5563

b. $2 \log_{10} x = \log_{10} x^2$

55.

x	$y = \log_e x$
1	0
2	0.693
4	1.386
16	2.773
$\frac{1}{2}$	-0.693
$\frac{1}{4}$	-1.386
$\frac{1}{16}$	-2.773

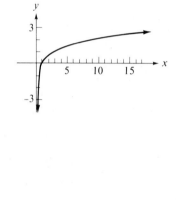

Exercise 8.5 [page 410]   1. $\log_b 2 + \log_b x$ 3. $\log_b x - \log_b y$
5. $\log_b x + \log_b y - \log_b z$ 7. $3 \log_b x$ 9. $\frac{1}{2} \log_b x$ 11. $\frac{2}{3} \log_b x$
13. $2 \log_b x + 3 \log_b y$ 15. $\frac{1}{2} \log_b x + \log_b y - 2 \log_b z$ 17. $\frac{1}{3}(\log_{10} x + 2 \log_{10} y - \log_{10} z)$
19. $\log_{10} 2 + \log_{10} \pi + \frac{1}{2} \log_{10} l - \frac{1}{2} \log_{10} g$ 21. $\frac{1}{2}[\log_{10}(s-a) + \log_{10}(s-b)]$ 23. 1.7917
25. -0.9163 27. 2.1972   29. 2.0149   31. -18.6432 33. 4.3174   35. $\log_b 4$
37. $\log_b x^2 y^3$ 39. $\log_b \frac{1}{x^2}$ 41. $\log_{10} \sqrt{\frac{xy}{z^3}}$ 43. $\log_b \frac{1}{4}$ 45. 500   47. 4   49. 11
51. 3   53. No solution   55. 2.8074   57. 0.8928   59. ± 1.3977 61. -1.6092

63. 0.2736   65. −12.4864   67. 56.97 hours   69. 19.08 years   71. 2.5 hours later
73. a. 12.49 hours   b. 24.97 hours; 37.46 hours   75. $t = \dfrac{1}{k} \log_{10}\left(\dfrac{A}{A_0} + 1\right)$
77. $q = \log_b\left(\dfrac{w}{p}\right)$ 79. $A = k(10^{t/T} - 1)$
81. $\log_b(4 \cdot 8) \stackrel{?}{=} \log_b \dfrac{64}{2}$ 83. $\log_b \dfrac{6^2}{9} \stackrel{?}{=} \log_b 2^2$
     $\log_b 32 = \log_b 32$ $\log_b 4 = \log_b 4$
85. $\log_b\left(\dfrac{12}{3}\right)^{1/2} \stackrel{?}{=} \log_b 8^{1/3}$ 87. $\log_{10}(10 + 100) \stackrel{?}{=} \log_{10} 10 + \log_{10} 100$
     $\log_b 2 = \log_b 2$ $\log_{10} 110 \neq 1 + 2$
89. $15,529.24; $16,310.19   91. 12 years; $11\dfrac{3}{4}$ years   93. 12.9%   95. 10 years, 361 days

Exercise 8.6 [page 418]   1. 1.3610   3. 2.7726   5. −1.2040   7. 0   9. 1.4918
11. 10.3812   13. 0.3012   15. 0.6703   17. 0.642   19. 3.807   21. −1.204
23. 4.137   25. 1.878   27. 0.0743   29. 0.8277   31. 7.5204   33. −2.9720
35. 1.6451   37. −3.0713   39. $t = \dfrac{1}{k} \ln y$ 41. $t = \ln\left(\dfrac{k}{k - y}\right)$ 43. $k = e^{T/T_0} - 10$
45. $N(t) = 100e^{0.6931t}$ 47. $N(t) = 1200e^{-0.5108t}$ 49. $N(t) = 10e^{0.1398t}$
51. a. 8950   b. 12.8 hours   53. a. 941.8 lumens   b. 2.2 centimeters
55. a. $3,888.98   b. 9.6 years   57. 7.5%   59. a. 1921 years   b. 5589.9 years
61. a. $P(t) = 20,000e^{0.05596t}$ b. 107,182   63. $N(t) = N_0 e^{-0.0866t}$
65. a. $8^x = 20$ b. $\log_{10} 8^x = \log_{10} 20$ c. $x = \dfrac{\log_{10} 20}{\log_{10} 8} = 1.4406$ 67. $\log_8 Q = \dfrac{\log_{10} Q}{\log_{10} 8}$
69. $\ln Q = \dfrac{\log_{10} Q}{\log_{10} e} = 2.3 \log_{10} Q$

Chapter 8 Review [page 421]   1. a. $D(t) = 8\left(\dfrac{3}{2}\right)^{t/5}$ b. 18; 43
2. a. $P(t) = 0.25(1.10)^t$ b. $0.65; $2.71
3. a. $M(t) = 100(0.85)^t$ b. 52.2 milligrams; 19.7 milligrams
4. a. $S(t) = 200,000(0.70)^t$ b. $48,020; $23,529.80   5. 15.6729   6. 94.6328
7. 3.2485   8. −0.2582

9.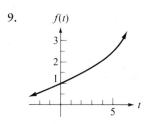

10.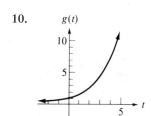

11.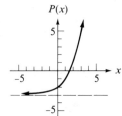

12.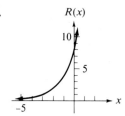

13. $-\dfrac{4}{3}$ 14. $\dfrac{1}{7}$ 15. -11 16. $\pm\sqrt{7}$ 17. 4 18. $\dfrac{1}{2}$ 19. -1 20. 1

21. -3 22. -4 23. $2^{x-2} = 3$ 24. $n^{p-1} = q$ 25. $\log_{0.3}(x+1) = -2$

26. $\log_4 3N_0 = 0.3t$ 27. -1 28. 81 29. 4 30. 3 31. 0.544 32. 3.716

33. 0.021 34. 0.502 35. 0.054 36. 2.959 37. 0.195 38. 2.823

39. $\log_b x + \dfrac{1}{3}\log_b y - 2\log_b z$ 40. $\log_b L - \dfrac{1}{2}[\log_b 2 + \log_b R]$ 41. $\dfrac{4}{3}\log_{10} x - \dfrac{1}{3}\log_{10} y$

42. $\dfrac{1}{2}\log_{10}(s-a) + \log_{10}(s-g)$ 43. $\log_{10}\sqrt[3]{\dfrac{x}{y^2}}$ 44. $\log_{10}\sqrt[6]{\dfrac{27x^3}{y^4}}$ 45. $\log_{10}\dfrac{1}{8}$

46. $\log_{10} 300$ 47. $\dfrac{9}{4}$ 48. 190 49. 2.001 50. $\dfrac{32}{9}$ 51. 3.771 52. 3.333

53. -7.278 54. 3.041 55. $t = \dfrac{1}{k}\log_{10}\dfrac{N}{N_0}$ 56. $t = \dfrac{1}{k}10^{Q-R_0/R}$ 57. 1.548

58. -0.693 59. 411.579 60. 0.247 61. 2.286 62. -1.470

63. $t = \dfrac{-1}{k}\ln\left(\dfrac{y-6}{12}\right)$ 64. $k = e^{N-N_0/4} - 10$ 65. $N(t) = 600e^{-0.9163t}$

66. $N(t) = 100e^{0.0583t}$

Cumulative Review Exercises for Part II [page 426] 1. 0 2. 1 3. $y = \pm\sqrt{x^2 - 1}$

4. 7.263 5. $2x + 3y = 11$ 6. $-\dfrac{1}{7}$ 7. 1 8. $\dfrac{4}{3}$ 9. 77 10. $2x + 3y = 1$

11. $9x - 2y = 3$ 12. Horizontal asymptote: $y = 3$
vertical asymptotes: $x = -1, 0, 1$

13. $0, \pm\sqrt{3}, \pm 2$ 14. Two sides have slope $4/5$; the two opposite sides have slope $-5/4$.

15. $9\dfrac{3}{8}$ 16. 0 17. -0.489 18. 5 19. 4 20. -0.791

21. 0.771   22. $y = \dfrac{-2}{2x^2 - x + 5}$

23. $k = \dfrac{-1}{t} \ln\left(\dfrac{P - C}{P_0}\right)$

24.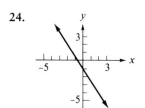

25.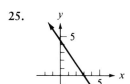

26.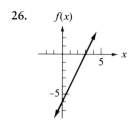

27.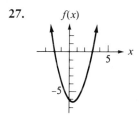

28.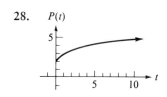

29.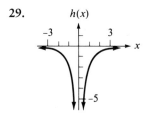

30.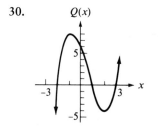

31.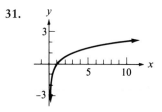

32.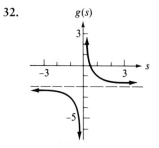

A-56   ANSWERS

33.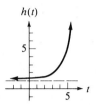

34.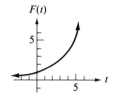

35.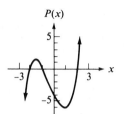

36.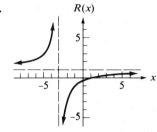

37.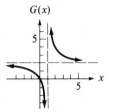

38.

39.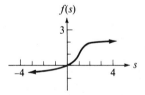

40.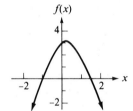

41.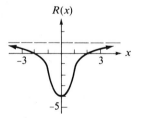

42.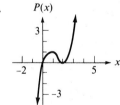

43.

44.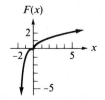

ANSWERS   A-57

45.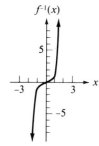

46.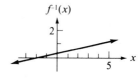

47. Domain: $[-4, 5]$
range: $[-3, 5]$

48. a. $-2, 4$
b. 6

49. 1.028

50. a. $p = 378 + \dfrac{2}{3}t$
b. $\dfrac{2}{3}$
c. Pages read per minute

51. 8.7%   **52.** 400 square inches   **53.** $253,377.52   **54.** $S(x) = 4x^2$; 1600 square inches

55. 4.096 times the cost of the smaller one   **56.** 6.93 seconds

57. A function is a relationship between two variables in which each value of the first variable is associated with only one value of the second variable. The domain is the set of permissible values for the first, or independent, variable, and the range is the set of associated values for the second, or dependent, variable.

58. A function grows linearly if it has a constant slope. For equal increments in the independent variable, equal amounts are *added* to the function value. A function that grows exponentially has a constant growth factor; the function values are *multiplied* by a constant amount for equal increments in the independent variable.

59. If the graph of a function passes the horizontal line test, that is, if no horizontal line intersects the graph more than once, the inverse of the function is also a function.

60. a.   **b.**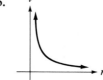

Exercise 9.1 [page 440]

1. $(3, 0)$

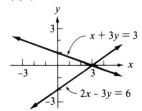

3. $(2, 1)$

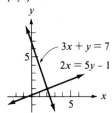

5. $(1, 2)$

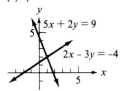

7. $(1, -2)$ 9. $(6, 0)$ 11. $(1, 2)$ 13. $(1, 2)$ 15. $(2.3, 1.6)$ 17. $(182, 134)$

19. $x = \dfrac{a+b}{2}$, $y = \dfrac{b-a}{2}$ 21. $x = \dfrac{ce - bf}{ae - bd}$, $y = \dfrac{af - cd}{ae - bd}$ 23. Dependent

25. Inconsistent   27. Consistent and independent   29. Dependent   31. 36 adults, 46 children

33. 3595 for winner, 3584 for loser   35. $800 at 8%, $1200 at 10%   37. 32 pounds

39. Airplane: 180 miles per hour   41. Airplane: 300 miles per hour
    automobile: 60 miles per hour   wind: 20 miles per hour

43. 0.6 cup oats, 0.4 cup wheat   45. 16 hardbacks, 48 paperbacks

47. a. $C = 0.4x + 20$
    b. $R = 1.2x$ c.
    d. 25

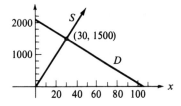

49. a, b.   c. $0.30; 1500 bushels

51. a. 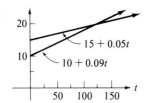   b. 125 minutes

53. $a = 2$, $b = 1$ 55. $y = x + 3$ 57. $\left(\dfrac{1}{5}, \dfrac{1}{2}\right)$ 59. $(-5, 3)$ 61. $\left(\dfrac{1}{6}, \dfrac{1}{2}\right)$

63. $y = \dfrac{-a_1}{b_1}x + \dfrac{c_1}{b_1}$; $y = \dfrac{-a_2}{b_2}x + \dfrac{c_2}{b_2}$.

$\dfrac{-a_1}{b_1} = \dfrac{-a_2}{b_2}$ if and only if $\dfrac{a_1}{a_2} = \dfrac{b_1}{b_2}$;

$\dfrac{c_1}{b_1} = \dfrac{c_2}{b_2}$ if and only if $\dfrac{c_1}{c_2} = \dfrac{b_1}{b_2}$.

Exercise 9.2 [page 455] 1. $(1, 2, -1)$ 3. $(2, -1, -1)$ 5. $(4, 4, -3)$ 7. $(2, -2, 0)$
9. $(2, 2, 1)$ 11. $(0, -2, 3)$ 13. $(-1, 1, 2)$ 15. $\left(\dfrac{1}{2}, \dfrac{2}{3}, -3\right)$ 17. $(4, -2, 2)$
19. $(1, 1, 0)$ 21. $\left(\dfrac{1}{2}, -\dfrac{1}{2}, \dfrac{1}{3}\right)$ 23. Inconsistent 25. $\left(\dfrac{1}{2}, 0, 3\right)$ 27. Dependent
29. $(-1, 3, 0)$ 31. Inconsistent 33. $\left(\dfrac{1}{2}, \dfrac{1}{2}, 3\right)$ 35. 60 nickels, 20 dimes, 5 quarters
37. $x = 40$ inches, $y = 60$ inches, $z = 55$ inches
39. 0.3 cup carrots, 0.4 cup green beans, 0.3 cup cauliflower
41. 40 score only, 20 evaluation, 80 narrative report   43. 20 tennis, 15 Ping-Pong, 10 squash
45. $(1, 1, 1)$ 47. $\left(\dfrac{7}{4}, \dfrac{14}{5}, -\dfrac{7}{6}\right)$ 49. $a = \dfrac{2}{9}$, $b = \dfrac{1}{3}$, $c = \dfrac{13}{36}$
51. $a = 3$, $b = 1$, $c = -2$

Exercise 9.3 [page 467] 1. 1   3. -12 5. 2   7. 5   9. $(1, 1)$ 11. $(2, 2)$
13. $(6, 4)$ 15. Inconsistent 17. $(4, 1)$ 19. $\left(\dfrac{1}{a+b}, \dfrac{1}{a+b}\right)$ 21. 3 23. 9 25. 0
27. -1 29. -5 31. 0 33. x^3 35. 0 37. $-2ab^2$ 39. $(1, 1, 1)$ 41. $(1, 1, 0)$
43. $(1, -2, 3)$ 45. $(3, -1, -2)$ 47. Dependent 49. $\left(1, -\dfrac{1}{3}, \dfrac{1}{2}\right)$ 51. $(-5, 3, 2)$

53. $\begin{vmatrix} a & a \\ b & b \end{vmatrix} = ab - ab = 0$

55. $\begin{vmatrix} a_1 & b_1 \\ a_2 & b_2 \end{vmatrix} = a_1 b_2 - a_2 b_1$; $-\begin{vmatrix} b_1 & a_1 \\ b_2 & a_2 \end{vmatrix} = -(b_1 a_2 - b_2 a_1) = -b_1 a_2 + b_2 a_1 = a_1 b_2 - a_2 b_1$

57. $\begin{vmatrix} ka & a \\ kb & b \end{vmatrix} = kab - kba = 0$

59. Since $D_x = \begin{vmatrix} c_1 & b_1 \\ c_2 & b_2 \end{vmatrix} = c_1 b_2 - c_2 b_1 = 0, \ b_1 c_2 = b_2 c_1.$

Since $D_y = \begin{vmatrix} a_1 & c_1 \\ a_2 & c_2 \end{vmatrix} = a_1 c_2 - a_2 c_1 = 0, \ a_1 c_2 = a_2 c_1.$

Forming the proportion $\dfrac{b_1 c_2}{a_1 c_2} = \dfrac{b_2 c_1}{a_2 c_1}$, if c_1 and c_2 are not both zero, then $\dfrac{b_1}{a_1} = \dfrac{b_2}{a_2}$, from which $a_1 b_2 = a_2 b_1$ and $a_1 b_2 - a_2 b_1 = 0$. Since $D = \begin{vmatrix} a_1 & b_1 \\ a_2 & b_2 \end{vmatrix} = a_1 b_2 - a_2 b_1$, it follows that $D = 0$.

61. Expanding about the elements of the third column of the given determinant produces

$$a \begin{vmatrix} y & y \\ z & z \end{vmatrix} - b \begin{vmatrix} x & x \\ z & z \end{vmatrix} + c \begin{vmatrix} x & x \\ y & y \end{vmatrix} = 0 + 0 + 0 = 0.$$

63. For the left side: $\begin{vmatrix} 1 & 2 & 3 \\ 4 & 5 & 6 \\ 0 & 0 & 1 \end{vmatrix} = 1 \begin{vmatrix} 1 & 2 \\ 4 & 5 \end{vmatrix} = 5 - 8 = -3.$

For the right side: $-\begin{vmatrix} 4 & 5 & 6 \\ 1 & 2 & 3 \\ 0 & 0 & 1 \end{vmatrix} = -1 \begin{vmatrix} 4 & 5 \\ 1 & 2 \end{vmatrix} = -1(8 - 5) = -1(3) = -3.$

65. Expanding about the elements of the first row produces

$$x \begin{vmatrix} -1 & 1 \\ 3 & 1 \end{vmatrix} - y \begin{vmatrix} 4 & 1 \\ 2 & 1 \end{vmatrix} + 1 \begin{vmatrix} 4 & -1 \\ 2 & 3 \end{vmatrix} = 0,$$

$$-4x - 2y + 14 = 0,$$

which is equivalent to

$$2x + y = 7$$

and is satisfied by $(4, -1)$ and $(2, 3)$.

Exercise 9.4 [page 476] **1.** $\begin{bmatrix} -2 & 1 & | & 0 \\ -9 & 3 & | & -6 \end{bmatrix}$ **3.** $\begin{bmatrix} 1 & -3 & | & 6 \\ 0 & -2 & | & 11 \end{bmatrix}$ **5.** $\begin{bmatrix} 1 & 0 & -2 & | & 5 \\ 2 & 6 & -1 & | & 4 \\ 0 & -3 & 2 & | & -3 \end{bmatrix}$

7. $\begin{bmatrix} 1 & 2 & 1 & | & -5 \\ 0 & 4 & -2 & | & 3 \\ 0 & -9 & 2 & | & 12 \end{bmatrix}$ **9.** $\begin{bmatrix} 1 & -3 & | & 2 \\ 0 & 7 & | & 0 \end{bmatrix}$ **11.** $\begin{bmatrix} 2 & 6 & | & -4 \\ -8 & 0 & | & -6 \end{bmatrix}$

13. $\begin{bmatrix} 1 & -2 & 2 & | & 1 \\ 0 & 7 & -5 & | & 4 \\ 0 & 9 & -11 & | & -1 \end{bmatrix}$ **15.** $\begin{bmatrix} -1 & 4 & 3 & | & 2 \\ 3 & 0 & -5 & | & 14 \\ -3 & 0 & -3 & | & 8 \end{bmatrix}$ **17.** $\begin{bmatrix} -2 & 1 & -3 & | & -2 \\ 0 & 4 & -6 & | & -3 \\ 0 & 0 & -4 & | & -5 \end{bmatrix}$

19. $(2, 3)$ **21.** $(-2, 1)$ **23.** $(3, -1)$ **25.** $\left(-\dfrac{7}{3}, \dfrac{17}{3}\right)$ **27.** $(1, 2, 2)$ **29.** $\left(2, -\dfrac{1}{2}, \dfrac{1}{2}\right)$

31. $(-3, 1, -3)$ 33. $\left(\dfrac{5}{4}, \dfrac{5}{2}, -\dfrac{1}{2}\right)$ 35. $\begin{bmatrix} 1 & 2 & 1 & 0 & | & -1 \\ 0 & -4 & -3 & 1 & | & 6 \\ 0 & 0 & -5 & 3 & | & 6 \\ 0 & 0 & 0 & 11 & | & 22 \end{bmatrix}$

37. $\begin{bmatrix} 2 & -1 & 1 & -2 & | & -6 \\ 0 & 1 & 3 & 1 & | & 2 \\ 0 & 0 & 1 & -3 & | & -4 \\ 0 & 0 & 0 & -5 & | & -5 \end{bmatrix}$ 39. $(-2, 3, 1, -3)$ 41. $(2, -5, 1, -1)$

Chapter 9 Review [page 418]   1. $\left(\dfrac{1}{2}, \dfrac{7}{2}\right)$ 2. $(1, 2)$ 3. $(12, 0)$ 4. $\left(\dfrac{1}{2}, \dfrac{3}{2}\right)$
5. Consistent and independent   6. Inconsistent   7. Dependent
8. Consistent and independent   9. $(2, 0, -1)$ 10. $(2, 1, -1)$ 11. $(2, -5, 3)$
12. $\left(\dfrac{4}{17}, -\dfrac{2}{17}, -\dfrac{84}{17}\right)$ 13. $(-2, 1, 3)$ 14. $(2, -1, 0)$ 15. -13 16. 24 17. $(8, 1)$
18. $(-7, 4)$ 19. $(1, 2)$ 20. $(6, 0)$ 21. -2 22. -19 23. $(1, 2, 3)$ 24. $(2, -1, 3)$
25. $(1, 2, -1)$ 26. $(1, 2, -1)$ 27. $(3, -1)$ 28. $(4, 0)$ 29. $(4, 1)$ 30. $(0, 3, -1)$
31. $(-1, 0, 2)$ 32. $(1, 0, 0)$
33. 32 dimes, 7 quarters   34. 16 first-class tickets, 48 tourist tickets
35. $5800 at 10%, $2200 at 12%   36. $1500 at 14%, $900 at 11%
37. Automobile: 60 miles per hour   38. First woman: 45 miles per hour
    airplane: 240 miles per hour   second woman: 50 miles per hour
39. 5 centimeters, 12 centimeters, 13 centimeters   40. $32°, 52°, 96°$
41. $\left(\dfrac{2}{5}, \dfrac{6}{5}\right)$ 42. $\left(\dfrac{22}{35}, \dfrac{11}{14}\right)$ 43. $a = \dfrac{1}{2},\ b = 3$ 44. $a = \dfrac{5}{2},\ b = \dfrac{1}{2}$
45. $a = \dfrac{1}{4},\ b = \dfrac{1}{2};\ c = 0$ 46. $a = \dfrac{1}{3},\ b = \dfrac{1}{3},\ c = \dfrac{1}{6}$

Exercise 10.1 [page 490]

1. $(3, \infty)$ 3. $\left(-\infty, -\dfrac{4}{9}\right]$
5. $\left(-\dfrac{17}{12}, \infty\right)$ 7. $(0, \infty)$
9. $\left(-\infty, \dfrac{6}{13}\right)$ 11. $\left(-\dfrac{3}{5}, \infty\right)$

13. $\left[\frac{17}{11}, \infty\right)$

15. $(-2, 3]$

17. $\left(-\frac{22}{3}, -4\right)$

19. $[7, 7.9)$

21. $[-6, -4]$

23. $\varnothing$

25. $\varnothing$

27. $(-3, 2)$

29. $(-2, 2)$

31. $[-4, 2]$

33. $(-7, -4]$

35. $[-5, 4]$

37. $(-4, 4)$

39. $(3, \infty)$

41. $(-3, -1) \cup (2, 3]$

43. $(-\infty, \infty)$

45. $(-2, \infty)$

47. $(-\infty, 1) \cup (4, \infty)$

49. Between $-9.4°F$ and $28.8°F$

51. At least 22 points

53. When first month's salary greater than $1000

55. $[79\%, 100\%]$

57. Interest rate less than 6.6%

Exercise 10.2 [page 498] 1. $1, 7$ 3. $-7, 3$ 5. $-\frac{3}{2}, \frac{5}{2}$ 7. $1, \frac{5}{3}$ 9. $-5, -3, -2, 0$

11. $|x| = 6$ 13. $|p + 3| = 5$ 15. $|t - 6| < 3$ 17. $|b + 1| > 0.5$

19. $(-2, 2)$

21. $[-7, 1]$

23. $\left(\frac{3}{2}, \frac{9}{2}\right)$

25. $[-4, 12]$

27. $(-0.3, -0.2)$

29. $(-\infty, -3) \cup (3, \infty)$

31. $(-\infty, -3) \cup (7, \infty)$

33. $(-\infty, -2] \cup [5, \infty)$

35. $\left(-\infty, -\frac{2}{3}\right) \cup \left(\frac{1}{3}, \infty\right)$

37. $\left(-\infty, -\frac{2}{3}\right) \cup \left(-\frac{2}{3}, \infty\right)$

39. $\left[-\dfrac{3}{2}, \dfrac{7}{6}\right]$ 41. $\left(-\dfrac{1}{12}, \dfrac{7}{12}\right)$

43. $(-\infty, -5.7] \cup [-2.7, \infty)$ 45. $(1.38, 1.42)$

47. $|x - 57| < 2$ 49. a. $|x - 24{,}900| < 2$ b. $r = 3962.96 \pm 0.32$ 51. $|x - 18| < \dfrac{1}{16}$

53. $|r - 30| < 0.12$; $r = 61.36$ miles per hour ± 0.24 mile per hour   55. $|a - b| = 2|b - c|$

57. $|y - x| < |y - z|$ 59. $|x| \leq 3$ 61. $|x - 4| < 4$ 63. $|x - 2| < 2$

65. $\left|x - \dfrac{9}{2}\right| \leq \dfrac{5}{2}$

Exercise 10.3 [page 507]

1. $(-\infty, -1) \cup (2, \infty)$ 3. $(-3, 4)$

5. $[0, 2]$ 7. $(-\infty, -1] \cup [4, \infty)$

9. $(-\infty, -1) \cup \left(\dfrac{2}{3}, \infty\right)$ 11. $[-5, 3]$

13. $\left(-\infty, -\dfrac{1}{2}\right) \cup (4, \infty)$ 15. $(-\infty, \infty)$

17. $(-\sqrt{5}, \sqrt{5})$ 19. $\varnothing$

21. $(-3, 2)$ 23. $\left(-\infty, -\dfrac{13}{2}\right) \cup (-5, \infty)$

25. $\left(-\infty, -\dfrac{8}{3}\right] \cup (-2, \infty)$ 27. $(-\infty, -1) \cup (1, 3)$

29. $\left[-\dfrac{3}{4}, -\dfrac{1}{2}\right] \cup \left[\dfrac{1}{2}, \infty\right)$ 31. $(-\infty, -1] \cup (1, 3)$

33. $\left(-\dfrac{2}{3}, \infty\right)$ 35. $(-\infty, -1] \cup (0, 6]$

37. $(-\infty, 0) \cup \left(\dfrac{2}{5}, \dfrac{3}{5}\right)$

39. $(4, 16)$ 41. $[0, 100) \cup (600, 700]$ 43. $(10, 30)$ 45. $(-\infty, -3) \cup (4, \infty)$ 47. $[-7, 4]$
49. -2 51. $(-\infty, -2) \cup (2, \infty)$ 53. $(-\infty, 0) \cup (1, \infty)$ 55. $(-\infty, -2) \cup (0, \infty)$

Exercise 10.4 [page 513]

1.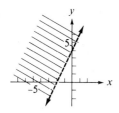

3.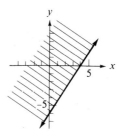

5.

7.

9.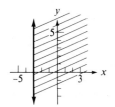

11.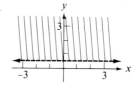

13.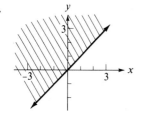

15.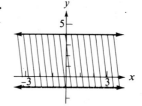

17.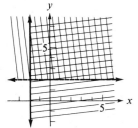

19.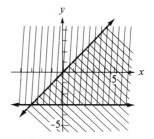

21.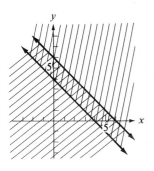

23.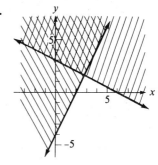

25.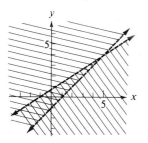

27.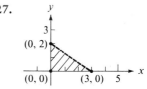

29.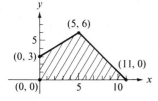

31.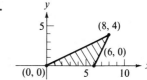

33.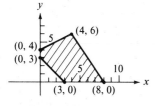

35.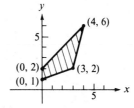

A-66 ANSWERS

37.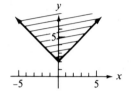

39.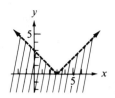

41.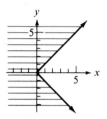

43.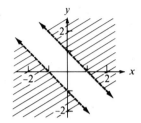

45.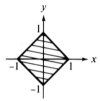

47.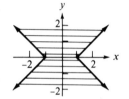

Chapter 10 Review [page 515]

1. $\left(-\infty, \dfrac{2}{3}\right)$

2. $\left(-\infty, \dfrac{8}{5}\right)$

3. $[9, \infty)$

4. $\left(\dfrac{31}{34}, \infty\right)$

5. $\left(-\dfrac{4}{3}, \dfrac{5}{3}\right]$

6. $[2.46, 4.7)$

7. $(-4, -1]$

8. $(4, \infty)$

9. $(-\infty, \infty)$

10. $(-\infty, 2]$

11. $-3, 7$ 12. $-\dfrac{2}{3}, \dfrac{10}{3}$ 13. $\dfrac{1}{5}, 1$ 14. $0, 1, 2, 3$ 15. $|x| = 4$ 16. $|y + 5| = 3$

17. $|p - 7| < 4$
18. $|q + 4| \geq 0.3$
19. $\left(-\dfrac{2}{3}, 2\right)$
20. $[-0.4, 0.1]$
21. $(-\infty, -0.9] \cup [0.1, \infty)$
22. $\left(-\infty, \dfrac{2}{3}\right) \cup \left(\dfrac{2}{3}, \infty\right)$
23. $\left(\dfrac{1}{8}, \dfrac{3}{8}\right)$
24. $\left(-\infty, -\dfrac{5}{18}\right] \cup \left[-\dfrac{1}{18}, \infty\right)$
25. $(-\infty, -2) \cup (3, \infty)$
26. $[-3, 4]$
27. $\left[-1, \dfrac{3}{2}\right]$
28. $\left(-\infty, -\dfrac{1}{3}\right) \cup (2, \infty)$
29. $[-2, 2]$
30. $(-\infty, -\sqrt{3}) \cup (\sqrt{3}, \infty)$
31. $(-3, 4)$
32. $\left(-\infty, \dfrac{5}{2}\right) \cup (4, \infty)$
33. $(-\infty, -2) \cup (1, 4)$
34. $(-3, 0) \cup (2, \infty)$
35.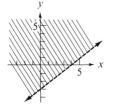
36.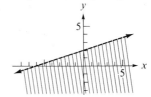
37.
38.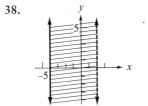

39.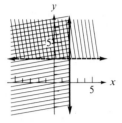
40.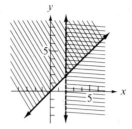
41.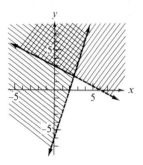
42.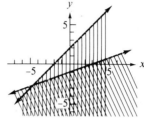
43.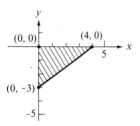
44.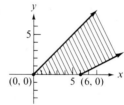
45.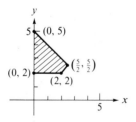
46.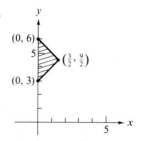

47. 97 or higher
48. More than 500 miles
49. a. $|x - 54.5| < 0.5$
    b. 119.9 pounds ± 1.1 pounds
50. a. $|x - 51| < 1$
    b. 129.54 centimeters ± 2.54 centimeters
51. When one number is less than 5
52. More than $20 but less than $40
53. $|a - b| - |a - c| = d$
54. If $|x - a| < d$, then $|y - L| < t$.

ANSWERS A-69

55. $(-\infty, -5) \cup \left(\dfrac{5}{2}, \infty\right)$

56. $(-\infty, 0) \cup (5, \infty)$

57.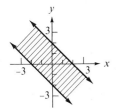

58.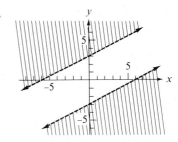

59.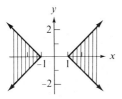

60.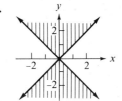

Exercise 11.1 [page 530]   1. 5; $\left(\dfrac{5}{2}, 3\right)$ 3. $2\sqrt{5}$; $(0, -2)$ 5. 5; $\left(\dfrac{1}{2}, 5\right)$

7.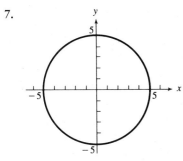

9.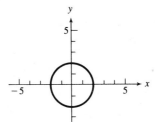

11.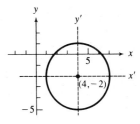

13.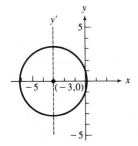

15.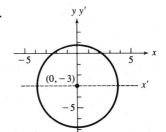

17.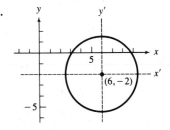

19. 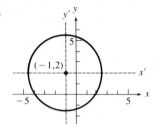 **21.** 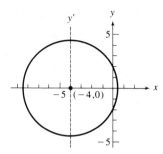 **23.**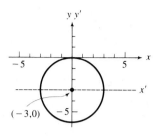

25. $x^2 + y^2 + 4x - 10y + 17 = 0$ **27.** $x^2 + y^2 + 6x + 2y + 9 = 0$
29. $x^2 + y^2 - 4x - 4y - 2 = 0$ **31.** $x^2 + y^2 - 3x + 8y + 11 = 0$

33. 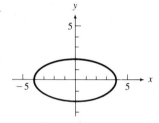 **35.** 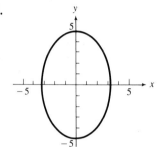 **37.**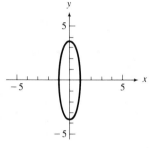

39. 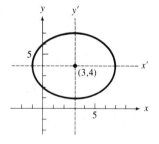 **41.** 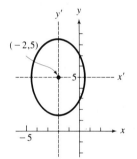 **43.**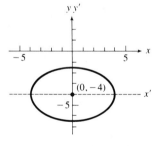

45. 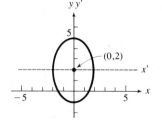 **47.** 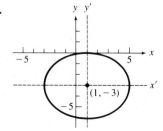 **49.**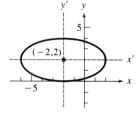

51.
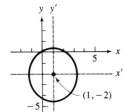

53.
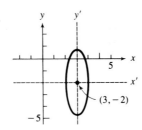

55. $4x^2 + 9y^2 - 8x - 108y + 292 = 0$ **57.** $9x^2 + 25y^2 + 36x - 100y - 89 = 0$
59. $4x^2 + y^2 + 32x - 6y + 37 = 0$ **61.** $7 + \sqrt{89} + 2\sqrt{17}$
63. The rectangle is a square because all sides are **65.** $x^2 + y^2 + 2x + 2y = 23$
equal:
$\sqrt{(-4-2)^2 + (1-6)^2} = \sqrt{61}$;
$\sqrt{(2-7)^2 + (6-0)^2} = \sqrt{61}$;
$\sqrt{(7-1)^2 + (0+5)^2} = \sqrt{61}$;
$\sqrt{(-4-1)^2 + (1+5)^2} = \sqrt{61}$.
67. $MD \parallel AC$; therefore, $\angle BAC = \angle BMD$ and $\triangle BAC \sim \triangle BMD$. Hence, $AB/AM = AC/AE$. Since M is the midpoint of AB, $AB = 2AM$. Therefore, $AC = 2AE$ as well. Similarly, $AB/MB = BC/BD$. Hence, $BC = 2BD$.

Exercise 11.2 [page 536]

1. Focus: $\left(0, \frac{1}{2}\right)$

directrix: $y = -\frac{1}{2}$

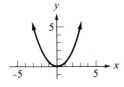

3. Focus: $(-4, 0)$

directrix: $x = 4$

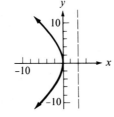

5. Focus: $(3, 0)$

directrix: $x = -3$

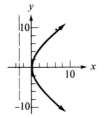

7. Focus: $(0, -2)$

directrix: $y = 2$

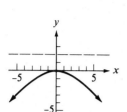

9. Focus: $\left(\dfrac{3}{8}, 0\right)$

directrix: $x = -\dfrac{3}{8}$

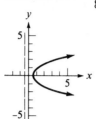

11.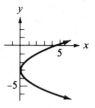

13.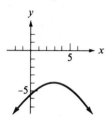

15.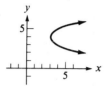

17. $(x + 3)^2 = y + 4$ **19.** $(y - 3)^2 = \dfrac{1}{2}(x + 6)$ **21.** $\left(x - \dfrac{3}{2}\right)^2 = -\dfrac{1}{3}\left(y - \dfrac{19}{4}\right)$

23. 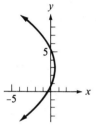   **25.** 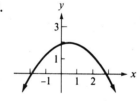   **27.**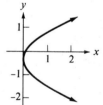

29. $x^2 = 8y$ **31.** $y^2 = -16x$ **33.** $x^2 - 2x - 4y + 9 = 0$ **35.** $y^2 + 4y - 8x - 28 = 0$
37. $y^2 - 10y + 4x + 17 = 0$ **39.** $x^2 - 6x + 16y + 41 = 0$ **41.** $x^2 = y$
43. $y^2 - 8y + x + 19 = 0$

Exercise 11.3 [page 546]

1. 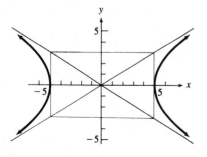   **3.**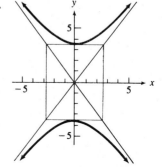

5. 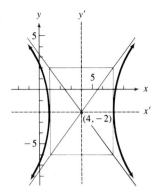   **7.** 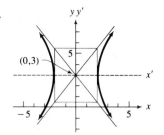   **9.**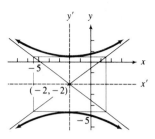

11. 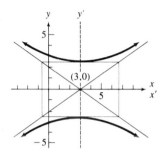   **13.** 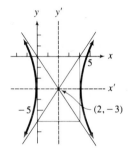   **15.**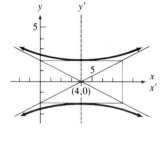

17. 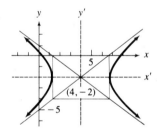   **19.**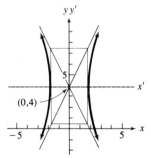

21. Circle; center $(0,0)$, radius 2

23. Hyperbola; center $(0,0)$, transverse axis on the x-axis, $a^2 = 8$, $b^2 = 2$

25. Ellipse; center $(0,0)$, major axis vertical, $a^2 = 6$, $b^2 = 3$

27. Parabola; vertex $\left(0, -\dfrac{3}{2}\right)$, opens upward, $p = \dfrac{1}{4}$

29. Hyperbola; center $(0,0)$, transverse axis on the y-axis, $a^2 = 6$, $b^2 = 24$

31. Parabola; vertex $(-4, 0)$, opens to the right, $p = \dfrac{1}{2}$

33. Parabola; vertex $(3, 2)$, opens downward, $p = -\dfrac{1}{4}$

35. Circle; center $(-1, 4)$, radius 4

A-74 ANSWERS

37. Ellipse; center $(-1, 0)$, major axis vertical, $a^2 = 4$, $b^2 = 2$

39. Circle; center $\left(-\frac{3}{2}, 0\right)$, radius $\frac{5}{2}$

41.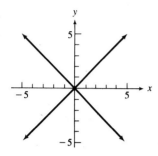

43.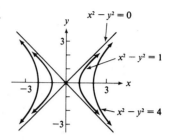

45. $9y^2 - 16x^2 - 90y + 32x - 367 = 0$ **47.** $4y^2 - x^2 - 8y + 2x - 13 = 0$

Exercise 11.4 [page 554]

1. $(-1, -4)$, $(5, 20)$ **3.** $(2, 3)$, $(3, 2)$

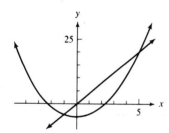   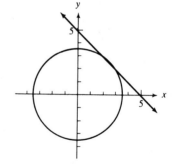

5. $(-3, 4)$, $(4, -3)$ **7.** $(2, 2)$, $(-2, -2)$ **9.** $(i\sqrt{7}, 4)$, $(-i\sqrt{7}, 4)$

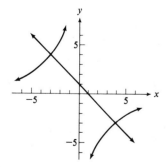   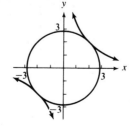   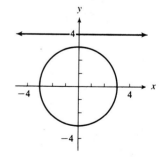

ANSWERS A-75

11. $\left(\dfrac{-1+\sqrt{13}}{2},\dfrac{1+\sqrt{13}}{2}\right), \left(\dfrac{-1-\sqrt{13}}{2},\dfrac{1-\sqrt{13}}{2}\right)$ 13. $(3,1), (2,0)$ 15. $(-1,-3)$

17. $(1,3), (-1,3), (1,-3), (-1,-3)$ 19. $(1,2), (-1,2), (1,-2), (-1,-2)$

21. $(3,\sqrt{2}), (-3,\sqrt{2}), (3,-\sqrt{2}), (-3,-\sqrt{2})$ 23. $(2,1), (-2,1), (2,-1), (-2,-1)$

25. $(\sqrt{3},4), (-\sqrt{3},4), (\sqrt{3},-4), (-\sqrt{3},-4)$ 27. $(2,-2), (-2,2), (2i\sqrt{2}, i\sqrt{2}), (-2i\sqrt{2}, -i\sqrt{2})$

29. $(1,-1), (-1,1), (i,i), (-i,-i)$ 31. $(1,-2), (-1,2), (2,-1), (-2,1)$

33. $(3,1), (-3,-1), (-2\sqrt{7},\sqrt{7}), (2\sqrt{7},-\sqrt{7})$

35. $\left(\dfrac{2\sqrt{3}}{3},\dfrac{\sqrt{3}}{3}\right), \left(\dfrac{-2\sqrt{3}}{3},-\dfrac{\sqrt{3}}{3}\right), (-6i,4i), (6i,-4i)$ 37. $2; 3$

39. Length 12 inches; width 1 inch   41. $800 invested at 4%

43. **a.** One or none   **b.** Two, one, or none   **c.** Four, three, two, one, or none

Chapter 11 Review [page 556]   1. $2\sqrt{58}; (-1,-1)$ 2. $7\sqrt{2}; \left(-\dfrac{1}{2},-\dfrac{7}{2}\right)$

3.

4.
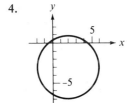

5.
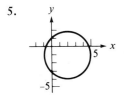

6.
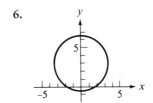

7.
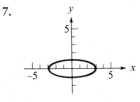

8.
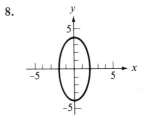

9.
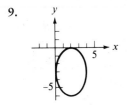

10.

A-76 ANSWERS

11.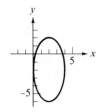
12.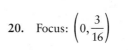

13. $x^2 + y^2 + 8x - 6y + 5 = 0$ 14. $x^2 + y^2 + 4x - 8y + 7 = 0$
15. $4x^2 + y^2 + 8x - 8y + 4 = 0$ 16. $25x^2 + 4y^2 - 150x - 8y + 129 = 0$

17. Focus: $\left(0, \dfrac{3}{2}\right)$

    directrix: $y = -\dfrac{3}{2}$

18. Focus: $(-3, 0)$

    directrix: $x = 3$

19. Focus: $\left(-\dfrac{1}{8}, 0\right)$

    directrix: $x = \dfrac{1}{8}$

20. Focus: $\left(0, \dfrac{3}{16}\right)$

    directrix: $y = -\dfrac{3}{16}$

21.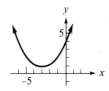
22.

23.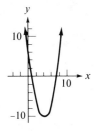
24.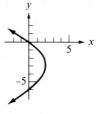

25.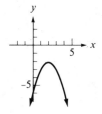
26.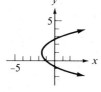

27. $x^2 = 16y$ 28. $y^2 - 8y + 16x - 16 = 0$ 29. $x^2 - 4x - 8y - 20 = 0$

30. $y^2 - 2y + 12x + 13 = 0$

31.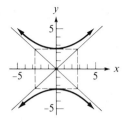

32.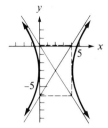

33.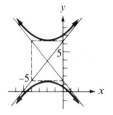

34.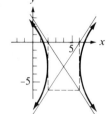

35.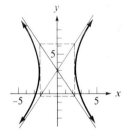

36.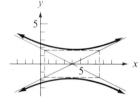

37. Ellipse; center (0,0), major axis horizontal, $a^2 = 24$, $b^2 = 6$

38. Hyperbola; center (0,0), transverse axis vertical, $a^2 = 12$, $b^2 = 3$

39. Ellipse; center (5, −2), major axis vertical, $a^2 = 9$, $b^2 = 6$

40. Hyperbola; center (0, −2), transverse axis horizontal, $a^2 = 4$, $b^2 = 12$

41. Circle; center (−3, 2), radius $\sqrt{15}$

42. Ellipse; center (2, −1), major axis horizontal, $a^2 = 8$, $b^2 = 2$

43. Parabola; opens upward, vertex (0, 6), focus (0, 8)

44. Parabola; opens to the right, vertex (1, −1), focus $\left(\dfrac{3}{2}, -1\right)$

45. Hyperbola; center (−1, 0), transverse axis vertical, $a^2 = 12$, $b^2 = 3$

46. Hyperbola; center (0, −1), transverse axis horizontal, $a^2 = 12$, $b^2 = 2$

47. Parabola; opens to the right, vertex (4, −2), focus $\left(\dfrac{9}{2}, -2\right)$

48. Hyperbola; center (0, 0), transverse axis horizontal, $a^2 = 6$, $b^2 = 12$

49. Circle; center $(3, -3)$, radius $2\sqrt{3}$

50. Parabola; opens to the right, vertex $(-1, 0)$, focus $(1, 0)$

51. $\left(3, \dfrac{\sqrt{3}}{3}\right), \left(3, -\dfrac{\sqrt{3}}{3}\right)$ **52.** $(4i, -2), (-4i, -2)$ **53.** $(4, -13), (1, 2)$

54. $\left(\dfrac{6}{7}, -\dfrac{37}{7}\right), (2, -3)$ **55.** $(1, \sqrt{5}), (1, -\sqrt{5}), (-1, \sqrt{5}), (-1, -\sqrt{5})$

56. $(2, 3), (2, -3), (-2, 3), (-2, -3)$ **57.** $(1, 2), (-1, 2), \left(2\sqrt{3}, -\dfrac{\sqrt{3}}{3}\right), \left(-2\sqrt{3}, \dfrac{\sqrt{3}}{3}\right)$

58. $(2, 1), (-2, -1), (i, -2i), (-i, 2i)$ **59.** Width: 7 centimeters
length: 10 centimeters

60. Width: 2 feet
length: 7 feet

61. $y = x^2 - x - 6$ **62.** $x^2 + y^2 - 4x - 2y - 5 = 0$ **63.** $y^2 - 2y - 4x - 7 = 0$

64. $y^2 - 4x^2 + 32x - 2y - 67 = 0$

65. a. $b = \dfrac{-a^2 - 16}{4}$

b. Let $a = 2$; hence, $b = -5$; the system is
$y = x^2 - 4$
$y = 2x - 5$.

66. a. $b = \pm 2\sqrt{a^2 + 1}$

b. Let $a = 2$; hence, $b = 2\sqrt{5}$; the system is
$x^2 + y^2 = 4$
$y = 2x + 2\sqrt{5}$.

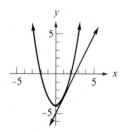

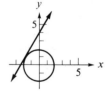

Exercise 12.1 [page 563] **1.** $-4, -3, -2, -1$ **3.** $-\dfrac{1}{2}, 1, \dfrac{7}{2}, 7$ **5.** $2, \dfrac{3}{2}, \dfrac{4}{3}, \dfrac{5}{4}$ **7.** $0, 1, 3, 6$

9. $-1, 1, -1, 1$ **11.** $1, 0, -\dfrac{1}{3}, \dfrac{1}{2}$ **13.** 58 **15.** $\log 26 \approx 1.415$ **17.** $4\sqrt{5}$

19. $3, 5, 7, 9, 11$ **21.** $24, -12, 6, -3, \dfrac{3}{2}$ **23.** $1, 2, 6, 24, 120$

25. a. $s_1 = 14{,}000;\ s_{n+1} = 0.85s_n$
b. $\$14{,}000, \$11{,}900, \$10{,}115, \$8{,}597.75$

27. a. $s_1 = 50{,}000;\ s_{n+1} = 1.01s_n - 500$
b. $s_n = \$50{,}000$ for all n

29. s_n approaches 1.41421. **31.** s_n approaches 0.64039. **33.** s_n approaches 2.

35. a. $1, 1, 2, 3, 5, 8, 13, 21, 34, 55, 89, 144, 233, 377, 610, 987$

b. s_n approaches 1.61803; $\dfrac{1 + \sqrt{5}}{2} \approx 1.61803$.

Exercise 12.2 [page 569] 1. 36   3. 8   5. $\frac{137}{60}$ 7. $S_{25} = \sum_{k=1}^{25} 3^k$ 9. $\sum_{i=5}^{15} \frac{x^i}{i}$

11. $1 + 4 + 9 + 16$ 13. $3 + 4 + 5$ 15. $2 + 6 + 12 + 20$ 17. $-\frac{1}{2} + \frac{1}{4} - \frac{1}{8} + \frac{1}{16}$

19. $1 + 3 + 5 + 7 + \cdots$ 21. $1 + \frac{1}{2} + \frac{1}{4} + \frac{1}{8} + \cdots$ 23. $S_4 = \sum_{i=1}^{4} i$ 25. $S_4 = \sum_{i=1}^{4} x^{2i-1}$

27. $S_5 = \sum_{n=1}^{5} n^2$ 29. $S_\infty = \sum_{n=1}^{\infty} \frac{n}{n+1}$ 31. $S_\infty = \sum_{k=1}^{\infty} \frac{k}{2k-1}$ 33. $S_\infty = \sum_{j=1}^{\infty} \frac{2^{j-1}}{j}$

35. a. $s_k = 20k - 10$ 37. $S_1 = \frac{1}{2}$, $S_2 = \frac{2}{3}$, $S_3 = \frac{3}{4}$, $S_4 = \frac{4}{5}$;
    b. 10 feet, 40 feet, 90 feet, 160 feet
    $S_n = \sum_{k=1}^{n} (20k - 10)$ $S_n = \frac{n}{n+1}$

39. e

Exercise 12.3 [page 577] 1. $2, 6, 10, 14$ 3. $1, -2, -5, -8$ 5. $\frac{1}{2}, \frac{3}{4}, 1, \frac{5}{4}$

7. $3, x + 5, 2x + 7, 3x + 9$ 9. $11, 15, 19$; $s_n = 4n - 1$ 11. $-9, -13, -17$; $s_n = -4n + 3$

13. $x + 2, x + 3, x + 4$; $s_n = x + n - 1$ 15. $x + 5p, x + 7p, x + 9p$; $s_n = x + (2n - 1)p$

17. $2x + 7, 2x + 10, 2x + 13$; $s_n = 2x + 3n - 2$ 19. $3x, 4x, 5x$; $s_n = nx$ 21. $s_7 = 31$

23. $s_{12} = \frac{15}{2}$ 25. $s_{20} = -92$ 27. $d = 2$, $a = 3$, $s_{20} = 41$ 29. Twenty-eighth

31. $S_9 = 99$ 33. $S_{16} = \frac{320}{3}$ 35. $S_{30} = 141$ 37. $S_7 = 63$ 39. $S_{15} = -825$

41. $S_8 = 30$ 43. $S_{100} = 5050$ 45. a. 1938   b. $\sum_{n=7}^{44} 2n$ 47. \$1,080,000.00

49. 66.25 seconds   51. 40 rows   53. 585   55. 290.4   57. $2, 7, 12$ 59. $k = \frac{14}{15}$

61. Find S_n for the arithmetic progression with
$a = 1$, $d = 2$:

$S_n = \frac{n}{2} [2(1) + (n - 1)(2)]$

$= \frac{n}{2}(2 + 2n - 2) = n^2$.

Exercise 12.4 [page 589] 1. Geometric   3. Arithmetic   5. Geometric   7. Neither

9. Geometric   11. Geometric   13. $5, -10, 20, -40$ 15. $9, 6, 4, \frac{8}{3}$ 17. $\frac{3}{4}, \frac{9}{16}, \frac{27}{64}, \frac{81}{256}$

19. $24, 9.6, 3.84, 1.536$ 21. $128, 512, 2{,}048$; $s_n = 2(4)^{n-1}$ 23. $\frac{16}{3}, \frac{32}{3}, \frac{64}{3}$; $s_n = \frac{2}{3}(2)^{n-1}$

25. $-\dfrac{1}{2}, \dfrac{1}{4}, -\dfrac{1}{8}$; $s_n = 4\left(-\dfrac{1}{2}\right)^{n-1}$ 27. $-\left(\dfrac{x}{a}\right)^2, \left(\dfrac{x}{a}\right)^3, -\left(\dfrac{x}{a}\right)^4$; $s_n = \left(\dfrac{-x}{a}\right)^{n-2}$ 29. 1536

31. $-243a^{20}$ 33. 3 35. $\dfrac{1}{2}$ 37. 13 39. 410 41. $-95\dfrac{13}{16}$ 43. $181\dfrac{1}{3}$

45. 1092 47. $\dfrac{31}{32}$ 49. $\dfrac{364}{729}$ 51. a. $10\dfrac{1}{8}$ feet b. $107\dfrac{1}{4}$ feet 53. \$709.21

55. 5 hours, 14.14 minutes 57. 24 59. No sum 61. $\dfrac{9}{4}$ 63. No sum 65. 2

67. $\dfrac{1}{3}$ 69. $\dfrac{31}{99}$ 71. $\dfrac{2408}{999}$ 73. $\dfrac{29}{225}$ 75. 20 centimeters 77. 30 feet

Exercise 12.5 [page 600]

1. $\displaystyle\sum_{k=0}^{5} \begin{bmatrix} 5 \\ k \end{bmatrix} x^{5-k} y^k$;

$\begin{bmatrix} 5 \\ 0 \end{bmatrix} x^5 + \begin{bmatrix} 5 \\ 1 \end{bmatrix} x^4 y + \begin{bmatrix} 5 \\ 2 \end{bmatrix} x^3 y^2 + \begin{bmatrix} 5 \\ 3 \end{bmatrix} x^2 y^3 + \begin{bmatrix} 5 \\ 4 \end{bmatrix} xy^4 + \begin{bmatrix} 5 \\ 5 \end{bmatrix} y^5$

3. $\displaystyle\sum_{k=0}^{4} \begin{bmatrix} 4 \\ k \end{bmatrix} x^{4-k}(-2y)^k$;

$\begin{bmatrix} 4 \\ 0 \end{bmatrix} x^4 + \begin{bmatrix} 4 \\ 1 \end{bmatrix} x^3(-2y) + \begin{bmatrix} 4 \\ 2 \end{bmatrix} x^2(-2y)^2 + \begin{bmatrix} 4 \\ 3 \end{bmatrix} x(-2y)^3 + \begin{bmatrix} 4 \\ 4 \end{bmatrix} (-2y)^4$

5. $\displaystyle\sum_{k=0}^{7} \begin{bmatrix} 7 \\ k \end{bmatrix} (x^2)^{7-k}(-3)^k$;

$\begin{bmatrix} 7 \\ 0 \end{bmatrix} (x^2)^7 + \begin{bmatrix} 7 \\ 1 \end{bmatrix} (x^2)^6(-3) + \begin{bmatrix} 7 \\ 2 \end{bmatrix} (x^2)^5(-3)^2 + \begin{bmatrix} 7 \\ 3 \end{bmatrix} (x^2)^4(-3)^3 + \begin{bmatrix} 7 \\ 4 \end{bmatrix} (x^2)^3(-3)^4 + \begin{bmatrix} 7 \\ 5 \end{bmatrix} (x^2)^2(-3)^5$
$+ \begin{bmatrix} 7 \\ 6 \end{bmatrix} (x^2)(-3)^6 + \begin{bmatrix} 7 \\ 7 \end{bmatrix} (-3)^7$

7. $x^5 + 15x^4 + 90x^3 + 270x^2 + 405x + 243$ 9. $x^4 - 12x^3 + 54x^2 - 108x + 81$

11. $8x^3 - 6x^2 y + \dfrac{3}{2} xy^2 - \dfrac{1}{8} y^3$ 13. $\dfrac{x^6}{64} + \dfrac{3x^5}{8} + \dfrac{15x^4}{4} + 20x^3 + 60x^2 + 96x + 64$

15. $8! = 8 \cdot 7 \cdot 6 \cdot 5 \cdot 4 \cdot 3 \cdot 2 \cdot 1$ 17. $2 \cdot 4! = 2 \cdot 4 \cdot 3 \cdot 2 \cdot 1$ 19. $6 \cdot 5! = 6 \cdot 5 \cdot 4 \cdot 3 \cdot 2 \cdot 1$

21. $5 \cdot 4 \cdot 3 \cdot 2 \cdot 1 = 120$ 23. $\dfrac{9 \cdot 8 \cdot 7!}{7!} = 9 \cdot 8 = 72$ 25. $\dfrac{5! \, 7!}{8 \cdot 7!} = \dfrac{5 \cdot 4 \cdot 3 \cdot 2 \cdot 1}{8} = 15$

27. $\dfrac{8!}{2! \, 6!} = \dfrac{8 \cdot 7 \cdot 6!}{2 \cdot 1 \cdot 6!} = \dfrac{8 \cdot 7}{2 \cdot 1} = 28$ 29. $3!$ 31. $\dfrac{6!}{2!}$ 33. $\dfrac{8!}{5!}$

35. $n(n-1)(n-2) \cdots 3 \cdot 2 \cdot 1$ 37. $3n(3n-1)(3n-2) \cdots 3 \cdot 2 \cdot 1$

39. $(n-2)(n-3)(n-4) \cdots 3 \cdot 2 \cdot 1$ 41. 84   43. 220   45. 190   47. 77,520

49. $-101,376$ 51. 1680   53. $-3003a^{10}b^5$ 55. $3360x^6 y^4$ 57. $21x^{10} y^4$ 59. $63a^5 b^4$

61. $a^8 - 4a^7 b + 7a^6 b^2 - 7a^5 b^3$ 63. $x^{30} + 30x^{27} y + 405x^{24} y^2 + 3240x^{21} y^3$

65. $512x^4 \sqrt{x} - 2304x^4 + 4608x^3 \sqrt{x} - 5376x^3$ 67. $x^{13} + 13x^{11} + 78x^9 + 286x^7$ 69. 1.22

71. 0.92

73. a. $1^{-1} + (-1)(1^{-2})x + 1(1^{-3})x^2 + (-1)(1^{-4})x^3$ or $1 - x + x^2 - x^3$
  b. $1 - x + x^2 - x^3$
  c. Results are equal.

Chapter 12 Review [page 602] **1.** Undefined, $\frac{2}{3}, \frac{3}{8}, \frac{4}{15}$ **2.** $1, -\frac{1}{2}, \frac{1}{3}, -\frac{1}{4}$ **3.** $s_7 = -25$

4. $s_3 = 5$ **5.** $5, 2, -1, -4, -7$ **6.** $1, -\frac{3}{4}, \frac{9}{16}, -\frac{27}{64}, \frac{81}{256}$

7. $s_1 = 1800$; $s_{n+1} = 0.88 s_n$; $\$1800, \$1584, \$1393.92, \$1226.65$
8. $s_1 = 24{,}000$; $s_{n+1} = 1.06 s_n$; $\$24{,}000, \$25{,}440, \$26{,}966.40, \$28{,}584.38$

9. $-2 + 3 - 4 + 5 - 6 = -4$ **10.** $\frac{1}{2} + \frac{2}{3} + \frac{3}{4} + \frac{4}{5} = \frac{163}{60}$ **11.** $\sum_{k=1}^{12}(2^k - 1)$

12. $\sum_{k=4}^{15} k^2 x^k$ **13.** $2 + 6 + 12 + 20$ **14.** $\frac{2}{3} + \frac{3}{5} + \frac{4}{7} + \frac{5}{9} + \cdots$ **15.** $3, -1, -5, -9$

16. $\frac{1}{4}, \frac{3}{4}, \frac{5}{4}, \frac{7}{4}$ **17.** $4, 8, 12$; $s_n = 4n - 8$

18. $x + 3a, x + 5a, x + 7a$; $s_n = x + (2n - 3)a$ **19.** 94   **20.** $\frac{23}{2}$ **21.** 68   **22.** s_{15}

23. 210   **24.** 49   **25.** 57   **26.** 40   **27.** a. 810   b. $\sum_{k=2}^{16} 6k$ **28.** 1998

29. Geometric   **30.** Neither   **31.** Arithmetic   **32.** Geometric   **33.** $12, -24, 48, -96$

34. $6, 2, \frac{2}{3}, \frac{2}{9}$ **35.** $8, -16, 32$; $s_n = -1(-2)^{n-1}$ **36.** $\frac{9}{32}, \frac{27}{128}, \frac{81}{512}$; $s_n = \frac{2}{3}\left(\frac{3}{4}\right)^{n-1}$

37. $-10\frac{1}{8}$ **38.** $-\frac{8}{27}$ **39.** 126   **40.** $\frac{65}{9}$ **41.** $\frac{121}{243}$ **42.** 63   **43.** $2\frac{10}{27}$ feet

44. $\$945.98$ **45.** $-\frac{9}{5}$ **46.** $\frac{9}{2}$ **47.** $x^5 - 10x^4 + 40x^3 - 80x^2 + 80x - 32$

48. $\frac{x^4}{16} - \frac{x^3 y}{2} + \frac{3x^2 y^2}{2} - 2xy^3 + y^4$ **49.** $(2n - 3)(2n - 4)(2n - 5) \cdots 3 \cdot 2 \cdot 1$

50. $\frac{6 \cdot 5 \cdot \cancel{4!}}{3 \cdot 2 \cdot 1 \cdot \cancel{4!}} = 5$ **51.** 21   **52.** 120   **53.** $-672 x^6 y^3$ **54.** $-1701 x^3$ **55.** 60.2

56. $4, 7, 10$ **57.** $\frac{29}{9}$ **58.** $\frac{23}{55}$

Cumulative Review Exercises for Part III [page 608]   **1.** $(12, -4)$ **2.** $(3, 1, 2)$
3. $[-2, -1]$ **4.** $(-\infty, 3) \cup (4, \infty)$

5. $(8, -6)$ **6.** $(2, 0, 4)$ **7.** $(1, 1), \left(-\frac{1}{8}, -\frac{5}{4}\right)$ **8.** $(3, 1), (-3, -1)$

A-82 ANSWERS

9. $\left(-\infty, -\dfrac{1}{2}\right] \cup \left[\dfrac{7}{2}, \infty\right)$

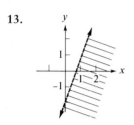

10. $(-\infty, -3] \cup \left[\dfrac{1}{2}, \infty\right)$

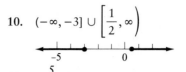

11. $(2.8, 5.5]$

12. $-\dfrac{5}{3}, 3$

13.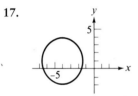

14.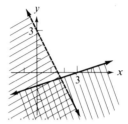

15. 14,924   16. 61

17.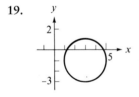

18.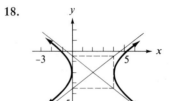

19.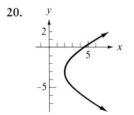

20. (graph of sideways parabola)

21. Focus: $\left(\dfrac{1}{4}, 3\right)$

    directrix: $y = \dfrac{13}{4}$

22. $x^2 + 4y^2 + 4x - 24y + 24 = 0$

23. $(5, 1, -1)$ 24. $(-2, 3, -4)$ 25. $3 - \dfrac{5}{2} + \dfrac{7}{3} - \dfrac{9}{4} + \dfrac{11}{5} - \dfrac{13}{6} + \dfrac{15}{7}$ 26. $\displaystyle\sum_{k=1}^{\infty} k^2 x^{k-1}$

27. $-\dfrac{8}{7}$ 28. $2, -1, \dfrac{8}{9}, -1$ 29. $x^2 + y^2 - 2x + 4y - 20 = 0$

30. $x^2 + y^2 + 4x + 2y - 15 = 0$ 31. 171   32. $-5280 x^4 y^{21}$ 33. $(-1, 2, 1)$

34. $\left(-\frac{1}{3}, \frac{1}{2}\right)$ **35.** $(4, \sqrt{3}), (4, -\sqrt{3}), (-4, \sqrt{3}), (-4, -\sqrt{3})$ **36.** $\left(-\infty, -\frac{11}{2}\right) \cup \left(-\frac{1}{2}, 2\right)$
37. a^3 **38.** 6 **39.** $|x - p| < 0.5$ **40.** $|x - 1.5| < 0.01$ **41.** $-\frac{3}{4}$ **42.** 8806
43. No solutions **44.** $[-2, 4)$ **45.** $6\sqrt{5}$ **46. a.** $[-2, -1)$ **b.** $(1, \infty)$ **47.** $\frac{1609}{990}$
48. 2.02 **49.** $-2, 0, -4, 12, 140$ **50.** $s_n = \frac{9}{2}\left(-\frac{2}{3}\right)^{n-1}$

51. 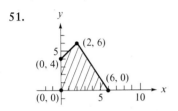   **52.**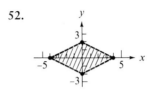

53. 60 student tickets, 290 general admission **54.** 68°, 34°, 78° **55.** 13 meters by 16 meters
56. 20 to Boston, 25 to Chicago, 10 to Los Angeles **57.** $108,830.61 **58.** 5304

59. $0 = \begin{vmatrix} x & y & 1 \\ x_1 & y_1 & 1 \\ x_2 & y_2 & 1 \end{vmatrix} = x(y_1 - y_2) - y(x_1 - x_2) + (x_1 y_2 - x_2 y_1),$

so $y(x_1 - x_2) + y_1 x_2 = x(y_1 - y_2) + x_1 y_2,$

or $y(x_1 - x_2) - y_1 x_1 + y_1 x_2 = x(y_1 - y_2) - x_1 y_1 + x_1 y_2,$

or $(y - y_1)(x_1 - x_2) = (x - x_1)(y_1 - y_2),$

or $y - y_1 = \frac{y_1 - y_2}{x_1 - x_2}(x - x_1),$ which is an equation for the line.

60. $\begin{vmatrix} a_1 & b_1 \\ a_1 + a_2 & b_1 + b_2 \end{vmatrix} = a_1(b_1 + b_2) - b_1(a_1 + a_2)$

$= a_1 b_1 + a_1 b_2 - a_1 b_1 - a_2 b_1$

$= a_1 b_2 - a_2 b_1 = \begin{vmatrix} a_1 & b_1 \\ a_2 & b_2 \end{vmatrix}$

Appendix Exercises [page 612] **1.** 2 **3.** -4 **5.** 0 **7.** 0.8280 **9.** 1.9227
11. 2.5011 **13.** $0.9101 - 1$ **15.** $0.9031 - 2$ **17.** 2.3945 **19.** 4.10 **21.** 36.7
23. 0.0642 **25.** 0.00718 **27.** 0.297 **29.** 0.0503 **31.** 0.00205 **33.** 7.52
35. 784 **37.** 0.0357 **39.** 1.5373 **41.** 0.5655 **43.** 4.4817 **45.** 0.0907
47. 1.3610 **49.** 2.7726 **51.** $0.0837 - 1$ or -0.9163 **53.** 1.1735 **55.** 6.0496
57. 90.017

Index to Review of Elementary Topics

Addition of signed numbers, xiv–xv
Additive inverse property, xiv
Associative property:
  of addition, xiv
  of multiplication, xiv

Commutative property:
  of addition, xiv
  of multiplication, xiv

Distributive property, xiv

Geometric formulas: xv
  area, xvi
  perimeter, xvi
  surface area, xvi
  volume, xvi

Identity properties:
  for addition, xiv
  for multiplication, xiv

Metric conversions, xvii
Multiplicative inverse property, xiv

Negative property, xiv
Notation:
  absolute value, xiii
  equality, xiii
  inequality, xiii

Properties of real numbers, xiv
Pythagorean theorem, xv

Reciprocal property, xiv

Substitution property, xiv

Transitive property, xiv

Index

Absolute value:
  and distance, 493
  equations involving, 492
  function, 291
  inequalities involving, 494
  and radicals, 123
Accuracy, 497
Addition:
  of complex numbers, 133
  of fractions, 70, 73
  of polynomials, 14
  of radicals, 125
Algebraic expression, 7
Antilogarithm, 616
Approximation(s):
  for irrational numbers, 105, 378
Arithmetic progression:
  definition, 571
  general term, 572
  sum of n terms, 575
Arithmetic series, 574
Asymptote(s):
  horizontal, 290, 347
  of a hyperbola, 539
  vertical, 289, 345
Augmented matrix, 470
Axis of a coordinate system, 217
Axis of symmetry, 305, 308

Back-substitution, 445
Base:
  of a logarithm, 386, 612
  of a power, 4, 618, 619
Binomial, 12
Binomial coefficients, 598

Binomial expansion, 592
Binomial formula, 593
Binomial products, 23
Break-even point, 294
Building factor, 71

Calculators:
  for base 10 logarithms, 390
  for exponential functions, 397
  for irrational powers, 378
  for polynomials, 335
  for radicals, 105, 111
  for scientific notation, 92
Cartesian coordinate system, 217
Circle, 521
Coefficient, 11
Coefficient matrix, 470
Common difference, 571
Common logarithms, 390
Common ratio, 581
Completing the square, 177
Complex conjugate, 134
Complex fraction, 78
Complex numbers, 132, 133
Component(s), 213
Compound interest, 42, 368
Conic sections, 518
Conjugate, 39, 128, 134, 183
Consistent system, 436
Coordinate(s), 213
Counting numbers, 3
Covertices, 525
Cramer's rule, 461, 465
Critical number, 501, 503
Cube root, 100

Decay factor, 370
Degree, 12
Demand function, 439
Denominator:
  least common, 72
  rationalizing, 128
Dependent equations, 436, 450
Determinant, 459, 462
Difference:
  of fractions, 70, 73
  of polynomials, 14
  of squares, 39
Dimension of a matrix, 470
Direct variation, 296
Directrix, 532
Discriminant, 188
Disjoint sets, 486
Distance formula, 519
Division of polynomials, 65
Domain, 278

Element, 4, 459
Elementary row operations, 471
Elementary transformation, 471
Ellipse, 524
Empty set, 487
Entry, 470
Equation(s):
  absolute-value, 492
  dependent, 436, 450
  equivalent, 141
  exponential, 381, 407
  first-degree, 141
  formulas, 143, 156, 183, 192, 409
  fractions in, 153, 166

Equation(s) (continued):
  linear, 141, 156, 225
  logarithmic, 406
  proportions, 155
  quadratic, 164
  quadratic in form, 192
  radicals in, 189
  solution of, 141, 215
  in two variables, 215
Equilibrium price, 439
Equivalent equations, 141
Evaluate, 7, 15, 267, 332
Explicitly related variables, 217
Exponent(s):
  decimal form, 112
  definition, 4
  fractional, 103
  integer, 84
  laws of, 21, 85
  negative, 87
  rational, 103, 110
  zero, 87
Exponential decay, 370
Exponential equation, 381, 407
Exponential function, 372, 378
Exponential growth, 367, 416
Extraction of roots, 168
Extraneous solution, 154, 189

Factor(s):
  building, 71
  common, 30, 55
  definition, 11
  of a polynomial, 11
  theorem, 335
Factor theorem, 336
Factorial notation, 596
Factoring:
  complete, 29, 30, 41
  differences of cubes, 40
  differences of squares, 39
  by grouping, 34, 40
  polynomials, 30 ff
  quadratic trinomials, 31
  solving equations by, 164, 194
  sums of cubes, 40
Focus, 532
Formula, quadratic, 180
Formulas, 143, 156, 183, 192, 409
Fraction(s):
  addition of, 70, 73
  algebraic, 51
  building, 71

  complex, 78
  fundamental principle of, 53
  least common denominator, 72
  products, 61
  quotients, 64
  reducing, 54
  subtraction of, 70, 73
Function(s):
  absolute-value, 291
  decreasing, 297
  defined piecewise, 290
  definition of, 264
  domain of, 278
  evaluation of, 267
  exponential, 372, 382, 413
  graph of, 275
  increasing, 297
  inverse, 353
  linear, 286, 293, 373
  logarithmic, 394, 413
  notation, 266
  polynomial, 330
  quadratic, 305
  range of, 278
  rational, 344
Fundamental principle of fractions, 53

Gaussian elimination, 473
General quadratic equation, 543
General term:
  arithmetic sequence, 572
  definition, 560
  geometric sequence, 581
Geometric progression:
  definition, 580
  general term, 581
  infinite, 586
  sum of n terms, 583
Graph(s):
  absolute value function, 291
  cubic function, 287
  of an equation, 217
  exponential functions, 379
  of a function, 275
  inverse functions, 356
  linear, 225
  linear functions, 286
  polynomial functions, 338
  quadratic function, 287, 305, 312
  radical functions, 288
  rational functions, 288, 344
Growth factor, 367

Horizontal line, 29, 240
Horizontal line test, 358
Hyperbola, 538

Imaginary numbers, 132
Implicitly related variables, 217
Inconsistent system, 436, 450
Independent equations, 436
Index:
  of a radical, 101
  of summation, 567
Inequalities:
  absolute-value, 494
  compound, 485
  graphs of, 483, 484
  linear, 482
  nonlinear, 503
  quadratic, 500
  solution set, 482
  systems, 512
  in two variables, 509
Infinite series, 586
Integers, 3
Intercepts, 227, 305, 338
Intersection, 486
Intervals, 279
Inverse function, 353
Inverse variation, 298
Irrational numbers, 3, 104

Laws of exponents, 21
Least common denominator, 72, 153
Like terms, 13
Linear combination, 433, 446
Linear equation(s):
  definition, 141
  graph, 225
  point-slope form, 241
  slope-intercept form, 248
Linear systems, solution by:
  determinants, 461, 465
  graphing, 432
  linear combinations, 433
  matrices, 470
  substitution, 432
Logarithm(s):
  base, 386
  characteristic, 615
  common, 390
  definition, 386
  equations, 406
  mantissa, 615

natural, 413
properties of, 395, 404
reading tables, 613
Logarithmic function, 394

Major axis, 525
Mathematical modeling, 7, 16, 25, 42, 56, 144, 146, 157, 169, 184, 242, 293, 364, 397, 438, 452, 489, 506
Matrix, 470
Maximum value, 315
Member, 4
Midpoint formula, 520
Minor(s), expansion of a determinant by, 462
Minor axis, 525
Monomial, 12
Multiplication:
 of complex numbers, 134
 of fractions, 61
 of polynomials, 22
 of radicals, 126
Multiplicity, 167, 337

nth root, 100
nth term, 560, 568, 581
Natural base, 413
Natural numbers, 3
Negative exponent, 87
Notation:
 exponential, 4, 103
 factorial, 595
 function, 266
 interval, 279
 inverse function, 359
 logarithmic, 386
 radical, 100
 scientific, 91
 sigma, 566
 summation, 566
Number(s):
 complex, 132
 counting, 3
 critical, 501
 imaginary, 132
 integers, 3
 irrational, 3, 104
 natural, 3
 ordered pairs of, 213
 rational, 3

real, 3
whole, 3
Numerical coefficient, 11

Operations, order of, 5
Order:
 determinant, 319
 matrix, 470
Ordered pair, 213
Ordered triple, 445
Origin, 217

Parabola, 287, 305, 312, 532
Parallel lines, 250
Partial sum, 586
Pascal's triangle, 594
Percent decrease, 371
Percent increase, 368
Perpendicular lines, 251
Point-slope form, 241
Polynomial(s):
 addition of, 14
 definition of, 12
 division of, 65
 evaluation of, 15, 332
 factoring, 41
 function, 330
 multiplication of, 22
 subtraction of, 14
Power, 4
Principal nth root, 102
Product(s):
 of binomials, 23
 of fractions, 61
 of monomials, 21
 of polynomials, 22
 of radicals, 118
Progression:
 arithmetic, 571
 geometric, 580
Proportion, 155

Quadrants, 217
Quadratic equation(s):
 definition, 164
 discriminant, 188
 in form, 192
 multiple solutions, 167
 solution by:
  completing the square, 177, 180
  extraction of roots, 168

 factoring, 164
 quadratic formula, 180
Quadratic formula, 180
Quadratic inequalities, 500, 502
Quotient(s):
 of fractions, 64
 of polynomials, 65
 of radicals, 119

Radical(s):
 addition, 125
 definition, 101
 division, 119, 128
 in equations, 189
 factors involving, 126
 index, 101
 multiplication, 118, 126
 simplest form, 121
Radicand, 101
Range, 278
Rational exponents, 103, 110
Rational expression, 51
Rational function, 344, 350
Rational numbers, 3
Rationalizing:
 denominators, 119, 128
 numerators, 129
Real numbers, 3
Rectangular coordinate system, 218
Recursive definition, 562
Reducing:
 fractions, 55, 56
 indices, 121
Repeating decimals, 588
Root(s):
 extraction of, 168
 cube, 100
 nth, 101
 square, 100

Scale factor, 327
Scientific notation, 91
Second-degree equations, see Quadratic equation(s)
Sequence:
 arithmetic, 571
 definition, 560
 geometric, 580
 recursive, 562
Series:
 arithmetic, 574, 575
 definition, 565

Series (continued):
   geometric, 583
   infinite geometric, 586
Set(s):
   definition, 3
   intersection, 486
   union, 487
Sigma notation, 566
Sign array of a determinant, 464
Slope-intercept form, 248
Slope of a line, 234, 236
Solution(s):
   complex, 550
   of equation in one variable, 141
   of equation in two variables, 215
   extraneous, 154, 189
   of systems, 431
Solution set, 482
Square root, 100
Subset, 4
Subtraction:
   of complex numbers, 133
   of fractions, 70, 73
   of polynomials, 14
   of radicals, 125

Sum(s):
   of polynomials, 14
   of terms, 12, 13
Summation notation, 566
Supply function, 439
Synthetic substitution, 332
System(s):
   of inequalities, 512
   of linear equations:
     in two variables, 431
     in three variables, 445
   quadratic, 548

Term(s):
   addition of like, 13
   arithmetic sequence, 571
   definition, 11
   geometric sequence, 580
   like, 13
   of a sequence, 560
   of a series, 565
Translation, 323, 526, 534, 540
   horizontal, 325
   vertical, 324

Trinomial(s):
   definition, 12
   perfect-square, 38
Turning points, 341

Union, 487
Upper triangular form, 471

Variable:
   definition of, 7
   dependent, 264
   independent, 264
Variation, 296, 298
Vertex, 305, 308
Vertical line, 229, 240
Vertical line test, 282
Vertices, 525

Weighted average, 146
Whole numbers, 3

Zero:
   exponents, 87
   of a polynomial, 336
Zero-factor principle, 164

STUDENT QUESTIONNAIRE

Your chance to rate **Modeling, Functions, and Graphs: Algebra for College Students** *(Franklin/Drooyan)*

In order to keep this text responsive to your needs, it would help us to know what you, the student, thought of this text. We would appreciate it if you would answer the following questions. Then cut out the page, fold, seal, and mail it; no postage is required. Thank you for your help.

Which chapters did you cover?   (circle)   1   2   3   4   5   6   7   8   9   10   11   12   All

Does the book have enough worked-out examples?   Yes _____   No _____

enough exercises?   Yes _____   No _____

Which helped most?

Explanations _____   Examples _____   Exercises _____   All three _____   Other _____
(fill in)

Were the answers at the back of the book helpful?   Yes _____   No _____

Did the answers have any typos or misprints? If so, where?

For you, was the course elective? _____   Required? _____

Do you plan to take more mathematics courses?   Yes _____   No _____

If yes, which ones?   Finite mathematics _____   Precalculus _____

Statistics _____   Calculus (for engineering and physics) _____

Trigonometry _____   Calculus (for business and social science) _____

Analytic geometry _____   Other _____

College algebra _____

How much algebra did you have before this course?   Terms in high school   (circle)   1   2   3   4

Courses in college   1   2   3

If you had algebra before, how long ago?

Last 2 years _____   3–5 years ago _____   5 years or longer _____

What is your major or your career goal? _____   Your age? _____

What did you like most about **Modeling, Functions, and Graphs?**

———————————————————— FOLD HERE ————————————————————

Can we quote you?   Yes _____   No _____

What did you like least about the book?

College _____   State _____

———————————————————— FOLD HERE ————————————————————

NO POSTAGE
NECESSARY
IF MAILED
IN THE
UNITED STATES

BUSINESS REPLY MAIL
First Class Permit No. 34 Belmont, CA

Postage will be paid by addressee

Mathematics Editor
WADSWORTH PUBLISHING COMPANY
10 Davis Drive
Belmont, CA   94002